NUTRITIC

AND DIET THERAPY REFERENCE DICTIONARY

FOURTH EDITION

NUTRITION AND DIET THERAPY REFERENCE DICTIONARY

FOURTH EDITION

ROSALINDA T. LAGUA, M.P.S., M.N.S., R.D.
NEW YORK STATE DEPARTMENT OF HEALTH

VIRGINIA S. CLAUDIO, Ph.D., M.N.S., R.D.
CONSULTANT, RIVERSIDE, CALIFORNIA

New York • Albany • Bonn • Boston • Cincinnati • Detroit • London • Madrid • Melbourne
Mexico City • Pacific Grove • Paris • San Francisco • Singapore • Tokyo • Toronto • Washington

Printed in the United States of America

For more information, contact:

Chapman & Hall
115 Fifth Avenue
New York, NY 10003

Thomas Nelson Australia
102 Dodds Street
South Melbourne, 3205
Victoria, Austrailia

Nelson Canada
1120 Birchmount Road
Scarborough, Ontario
Canada M1K 5G4

International Thomson Editores
Campos Eliseos 385, Piso 7
Col. Polanco
11560 Mexico D. F. Mexico

England

Chapman & Hall GmbH
Postfach 100 263
D-69442 Weinheim
Germany

International Thomson Publishing Asia
221 Henderson Road #05-10
Henderson Building
Singapore 0315

International Thomson Publishing - Japan
Hirakawacho-cho Kyowa Building, 3F
1-2-1 Hirakawacho-cho
Chiyoda-ku, 102 Tokyo
Japan

1 2 3 4 5 6 7 8 9 10 XXX 01 00 99 98 97 96 95

Library of Congress Cataloging-in-Publication Data

Lagua, Rosalinda T., 1937-
Nutrition and diet therapy reference dictionary / Rosalinda T.
Lagua, Virginia S. Claudio. --4th ed..
p. cm.
Rev. ed. of: Nutrition and diet therapy dictionary, 1991, with authors' names reversed.
Includes bibliographical references and index.
ISBN 0-412-07051-0 (hb : alk. paper). -- ISBN 0-412-07061-8 (pb : alk. paper)
1. Diet therapy--Dictionaries. 2. Nutrition--Dictionaries.
I. Claudio, Virginia Serrao, 1932- . II. Title
RM217.L34 1995 95-14630
613.2' 03--dc20 CIP

British Library Cataloguing in Publication Data available

To order for this or any other Chapman & Hall book, please contact **International Thomson Publishing, 7625 Empire Drive, Florence, KY 41042.** Phone: (606) 525-6600 or 1-800-842-3636.
Fax: (606) 525-7778. e-mail: order@chaphall.com.

For a complete listing of Chapman & Hall's titles, send your request to
Chapman & Hall, Dept. BC, 115 Fifth Avenue, New York, NY 10003.

Contents

Preface

This new fourth edition of the *Nutrition and Diet Therapy Reference Dictionary* covers all aspects of nutrition, including assessment of drug–nutrient interactions, laboratory interpretations, enteral and parenteral nutrition support, community and public health nutrition programs, nutrition throughout the life cycle, and such topics as nutrition and the immune system, nutrition labeling, chemical dependency, AIDS, and organ transplantation.

Special features of the Dictionary are entries on 130 different diets (listed under D); nutrition therapy for more than 350 disorders, including inborn errors of metabolism; 145 drugs and their effects on nutrition; and more than 150 nutritional products with their main uses and composition. Of particular importance are topics of public health concerns for the 1990s and the year 2000, and dietary recommendations for prevention of major degenerative diseases such as obesity, coronary heart disease, hypertension, diabetes mellitus, and cancer.

With more than 3000 carefully selected entries, the new Fourth Edition includes 380 new terms and more than 600 revised and expanded definitions. In choosing the words to be entered and defined, the authors used as their criterion the frequency of use or importance of a term in relation to nutrition. Definitions are cross-referenced to other word entries and the materials found in the Appendix to provide further details and information.

All practitioners in the fields of nutrition and dietetics, as well as educators, students, and others interested in nutrition will find this handy desk reference particularly useful. It is easy-to-use and provides instant access to nutrition information.

Rosalinda T. Lagua
Virginia S. Claudio

GUIDE TO THE USE OF THE DICTIONARY

Word entry. The word or term to be defined is set in boldface. All entries, including abbreviations and compounds of two or more words, have been entered in strict alphabetical order, regardless of any space or hyphens that may occur between them. If two or more variant spellings of a single word exist, the one most frequently used is entered in boldface and the variants are given in the definition. When usage is about evenly divided, both spellings are entered in boldface.

Subentries. Groups or classes of definitions related by a common root term appear under that term: *anemia, amino acid, dietitian,* etc. The series is slightly indented, and each subentry is set in the same boldface type as the main word entry.

Definitions. Innumerable definitions have been scrutinized, redefined, or expanded to conform to changing concepts of present-day knowledge. The definitions of a term are numbered when there is more than one distinct meaning or use. The most inclusive definition is presented first, followed by the more restricted meanings. Definitions restricted to specialized fields are preceded by field labels such as "In nutrition," "In medicine," and so on. Advertently omitted are definitions in certain specialized fields that do not have any application in the field of nutrition. A semicolon after a definition generally means that the material that follows is not part of the definition proper but is additional information enlarging on the factual content.

Abbreviations. Abbreviations with nutritional significance appear in their proper alphabetical sequence in the dictionary. They are defined in full in word entries for which such abbreviations stand.

Cross-references. The user is directed to additional or fuller information by such cross-reference terms as *see* or *see under*. Cross-references to related information are identified by the words *see also*. The word entry to which the user is directed is italicized; when a cross-reference appears under a group entry, the user is instructed to look for the subentry under the word entry for the definition of the specific term. Cross-references to the appendices are not italicized but presented in the same typeface as the definition; the user, however, is clearly directed to the appendices for additional information.

Italics. Some words are italicized to indicate to users that such words, in case they are not known to them, are defined elsewhere in the text. Cross-referenced words are also italicized.

Nutrient requirements. Unless specified otherwise, nutrient requirements stated are for adults.

A

AA. Abbreviation for 1. *Adenylic acid.* 2. *Amino acid.* 3. *Alcoholics Anonymous.*

AAA. Abbreviation for *aromatic amino acid.*

ABCD. Abbreviation for *Advocates for Better Children's Diets.*

Abetalipoproteinemia. A rare congenital disorder due to a lack of apoprotein B, which is needed to secrete chylomicrons or to export hepatic triglycerides. As a result, fat is not transported from the intestinal cells into the lacteals. There is fat malabsorption and steatorrhea, with progressive neuropathy and retinopathy in the first two decades of life. The blood lacks the chylomicrons and has very low levels of low-density and very-low-density lipoproteins, cholesterol, triglycerides, and free fatty acids, especially linoleic and arachidonic acids. *Nutrition therapy:* A low-fat diet is recommended; substitution of medium-chain triglycerides for long-chain fats in the diet may improve fat absorption, because the shorter-chain fats are absorbed by way of the portal vein rather than by the lymph. For infant feeding, use a formula high in medium-chain triglycerides such as Portagen®. Supplements of the fat-soluble vitamins A and K are necessary, and pharmacologic doses of vitamin E (200 to 300 mg/kg daily) may improve the retinal and neuromuscular abnormalities.

Absorption, nutrients. After food is digested, the end products (see Appendix 11) are absorbed mainly in the intestines through the *villi*. Each villus is connected to the circulatory and lymphatic systems. Almost all the water-soluble nutrients are absorbed by diffusion or osmosis (passive absorption). Glucose, galactose, and amino acids are absorbed with the help of energy supplied by the enzyme K^{+}/-ATPase and a cofactor, sodium ion. The process is carried out against a concentration gradient and it therefore needs ATP energy (active absorption). Fructose absorption, however, uses a carrier without expending energy. Fructose pulls water with it on entering the intestines. This type of absorption is called facilitative. The fourth absorption process is pinocytosis or phagocytosis, in which the absorptive cell engulfs the material. This process is used for fat absorption. Below is a summary of absorption in the gastrointestinal tract (GIT):

GIT ORGAN	NUTRIENTS AND OTHER SUBSTANCES
Stomach	Alcohol (20% of the total ingested) Some short-chain fatty acids
Intestines	
Duodenum	Vitamins A and B_1, iron, calcium, glycerol and fatty acids, monoglycerides, amino acids, monosaccharides, and disaccharides
Jejunum	Glucose; galactose; amino acids; glycerol and fatty acids; monoglycerides; diglycerides; dipeptides; copper; zinc; potassium; calcium; magnesium; phosphorus; iodine, iron; fat-soluble vitamins D, E, and K; most of the vitamin B complex; vitamin C; and the rest of the alcohol not absorbed by the stomach

Continued

GIT ORGAN	NUTRIENTS AND OTHER SUBSTANCES
Intestines *continued*	
Ileum	Disaccharides; sodium; potassium; chloride; calcium; magnesium; phosphorus; iodine; vitamins D, E, K, B_1, B_2, B_6, B_{12}, and ascorbic acid, and most of the water
Colon	Sodium, potassium, water, acids and gases, some short-chain fatty acids metabolized from plant fibers and undigested starch, and vitamins synthesized by bacteria (biotin and vitamin K)

Water-soluble nutrients pass directly into the circulatory system, whereas fat-soluble materials pass through the lymphatic system first before they are transported by the blood. Amino acids and peptides are actively absorbed by the absorptive cells of the villi; peptides are broken down into individual amino acids which then go, via the portal vein, to the liver, where they are metabolized. Water-soluble fatty acids (fewer than 12 carbons) and glycerol pass through the portal vein to the liver. Bile aids in the emulsification of fats to facilitate absorption. Fatty acids of longer carbon chain (more than 14 carbons) form triglycerides in the absorptive cells and combine with cholesterol, phospholipids, and similar substances with a protein coat (the compound formed is called a chylomicron). The chylomicrons pass through the lymphatic system before entering the bloodstream. Understanding the absorption process gives background knowledge and a rationale for nutritional therapy for disorders associated with *malabsorption syndrome*.

Accessory food factors. Earliest name given to vitamins by Hopkins, who demonstrated in 1906 that foods contain, in addition to the nutrients then recognized (i.e., carbohydrates, proteins, fats, minerals, and water), minute traces of unknown substances essential to health and life.

Accupep HPF®. Brand name of an elemental formula with short-chain peptides as the primary protein source. Used when intact nutrients are not indicated. See Appendix 39.

Acesulfame-K. A nonnutritive sweetener that is a potassium salt of the 6-methyl derivative of oxathiazinone dioxides. Its relative sweetness (based on that of sucrose, which is 1.0) is 130, and it is noncaloric. Sold under the brand name Sweet One® and Swiss Sweet®.

Acetoacetic acid. Monobasic ketone acid formed in the course of normal fatty acid catabolism and oxidized further to acetic acid, which is utilized in various metabolic reactions. It accumulates in the blood when fatty acids are incompletely oxidized. The reduction of acetoacetic acid yields beta-hydroxybutyric acid, and its decarboxylation yields acetone. See also *Ketone bodies*.

Acetohexamide. A sulfonylurea oral hypoglycemic agent used for the management of non-insulin-dependent diabetes mellitus; sold under the brand name Dymelor®.

Acetone. Dimethylketone. A colorless liquid with a sweetish, ethereal odor; formed by the decarboxylation of acetoacetic acid. It is normally present in minute quantities in blood and urine but may accumulate when fatty acid degradation is excessive or incomplete. See also *Ketone bodies*.

Acetonemia. Presence of large amounts of acetone (ketone bodies) in the blood. See also *Ketosis*.

Acetonuria. Excretion of large amounts of acetone bodies in the urine. Although normally present in trace amounts in the urine, the levels excreted may increase from 0.02 g/day to as much as 6 g/day in certain pathologic conditions.

Acetylcholine (ACh). Acetic acid ester of choline. It is released from nerve endings to initiate a series of reactions leading to the transmission of a nerve impulse. Its actions correspond to those of the cholinergic fibers, including a depressant effect on the blood pressure and stimulation of intestinal peristalsis.

Acetylcoenzyme A (acetyl-CoA). Acetyl derivative of coenzyme A. An important component of the *Krebs cycle* as a precursor for the biosynthesis of fatty acids and sterols, it gives rise to acetoacetic acid and is the biologic acetylating agent in the synthesis of acetylcholine.

Achalasia. Neuromuscular disorder of the esophagus that causes dyspepsia, esophageal regurgitation, esophageal pain, vomiting, and eventually weight loss. It is also called esophageal dysnergia or cardiospasm. The main problem is failure of the lower esophageal sphincter to relax and open after swallowing to permit food to enter the stomach. The lowest part of the esophagus therefore becomes narrowed and blocked with food. The main treatment is the dilatation of the sphincter or a surgical procedure to cut the circular muscle of the cardiac sphincter. *Nutrition therapy:* depending on the degree of dysphagia, foods and beverages per os have to be carefully selected for each individual; some may require tube feeding. Avoid foods that stimulate secretion of gastrin, the hormone that controls the lower esophageal sphincter. Proteins and carbohydrates promote gastrin production. Fats, chocolate, and coffee lower esophageal pressure. Give liquids and semisolid foods in small, frequent feedings, avoiding extremes of temperature and acidic or highly spiced foods. For other dietary guidelines, see *Dysphagia*.

Achlorhydria. Absence or lack of hydrochloric acid in the gastric juice. Primarily due to a decline in the number of acid-producing cells in the stomach, as in the elderly. It also occurs in pernicious anemia and severe iron deficiency anemia. Digestion of protein that needs the enzyme *pepsin* is lessened in achlorhydria. In *achylia gastrica,* hydrochloric acid and pepsin are both absent. Overall digestion of protein may still be adequate because of the action of trypsin and other enzymes of the pancreas and small intestines.

Acid-ash residue. Inorganic radicals (chiefly chloride, sulfate, and phosphate) that form acid ions in the body. When excreted in the urine, they lower the pH, making the urine more acid.

Acid–base balance. Equalization of total acid and total base in body fluids at levels compatible with life. Normally, the blood is kept within a narrow pH range of 7.35 to 7.45. Adjusting mechanisms come into play to neutralize or remove excess acid or base to maintain the balance. These include the buffer systems of the blood, the excretion of carbon dioxide by the lungs, and the excretion of fixed acid or base by the kidneys.

Acid-forming foods. Foods in which the acidic residue exceeds the alkaline residue. These include meats, fish, poultry, eggs, and cereals. See also *Alkaline-forming foods* and *Diet, ash*.

Acidosis. An abnormal condition characterized by a fall in the pH of the blood or a decrease in the *alkali reserve* of the body. A reduction in blood bicarbonate (alkali reserve) indicates that an excess of fixed acids is being produced or retained in the body at a rate exceeding that of neutralization or elimination. Various acids retained in different conditions are the acidic ketone bodies (as in diabetes mellitus); phosphoric, sulfuric, and hydrochloric acids (as in renal insufficiency); lactic acid (as in anoxia, ether anesthesia, and prolonged strenuous exercise); and carbonic acid (as in respiratory disease).

Acne vulgaris. Skin condition characterized by pimples or eruptions occurring most frequently on the face, back, and chest. An acne pimple is an obstructed and infected oil gland, and the pimples are more numerous where the oil glands are most abundant. For many years, people had associated diets high in fat or carbohydrate (particularly chocolate, nuts, candies, carbonated drinks, and fried foods) with acne. There is no basis for such beliefs. Studies have shown that foods do not produce major flareups of acne. However, some cases respond to high doses of vitamin A (100,000 IU) and retinoic acid applied topically, both of which should be used only under a physician's order. The use of 13-*cis*-retinoic acid has been effective for some cases of severe acne but has adverse side effects. See *Isotretinoin*.

Acquired immunodeficiency syndrome (AIDS). A life-threatening infectious disease that attacks the immune system of the body. The causative agent is the human immunodeficiency virus (HIV), which destroys the T cells of the defense system of the body. Incubation period for the HIV infection averages 10 years, which can be delayed by good nutrition. The progression of HIV infection to AIDS is monitored mainly by measuring the T4 lymphocytes. The normal level is $1000/mm^3$; if values fall below $500/mm^3$, the HIV-infected person is diagnosed as having AIDS. To date there is no known cure for AIDS, and due to its mode of transmission and the increasing mortality rate, it has become a major health problem. *Clinical signs and symptoms:* night sweating, easy fatigability, poor digestion, anorexia, fever, cachexia, and eventually protein-energy malnutrition (PEM). Lymph nodes are swollen, and there is dysphagia, diarrhea, and malabsorption. The two most common problems are *Kaposi's sarcoma,* which is a form of cancer, and *Pneumonitis carinii* pneumonia, which develops in the latter stage of AIDS. It is the opportunistic infections and severe cachexia that most patients with AIDS eventually die of, rather than the HIV infection itself. Death occurs when the body weight falls to about 56% of desirable weight. The clinical picture is similar to prolonged starvation. Oral and esophageal lesions cause pain during eating and alter food intake; fever increases energy requirements; and various drugs interfere with nutrient absorption or metabolism. The breakdown of the immune system enables a variety of parasitic microorganisms to grow in the intestines, causing diarrhea, impaired absorption of nutrients, and cachexia. *Nutrition therapy:* the hallmark of delaying the progression of AIDS is a high-calorie, high-protein intake ($\geq$ 150% of recommended dietary allowances). Supplement with vitamins and minerals up to 200% of the Recommended Dietary Allowances, especially for selenium and vitamins B_6, B_{12}, and folate. Nutritional care should be highly individualized. Determine the most appropriate nutritional support option (enteral or parenteral) for patients experiencing oral lesions, dysphagia, anorexia, and protein-energy malnutrition (PEM). If tolerated, however, encourage feeding by mouth, avoiding foods the patient finds irritating. Serve foods at moderate temperatures. Experiment with different flavors if the patient has lost taste acuity. Perk up the appetite with mild spices and favorite foods. In cases of diarrhea, modify the diet accordingly to avoid dehydration and electrolyte imbalance. There are many choices of supplementary feedings to increase nutrient density. Serve foods in small, frequent feedings. Monitor any malabsorption, and restrict raw fruits and vegetables to protect against contamination with microorganisms. Educate the patient about food safety and give tips on dining out, such as ordering only well-cooked foods that are served hot. Watch out for side effects of drug therapy. The multifunctional role of nutrition in the care of AIDS patients cannot be overemphasized. See *Diet, neutropenic* and *Human immunodeficiency virus (HIV)*. See also *Nutrition, immune system*.

Acrodermatitis enteropathica. An autosomal recessive disease characterized by zinc malabsorption, resulting in alopecia, eczematous skin lesions, diarrhea, failure to thrive, and death if not treated. Symptoms are generally first observed at weaning from human milk to cow's milk. *Nutrition therapy:* maintain normal protein and energy intakes, based on age and activity level. Give zinc supplement (1 mg/kg body weight per day).

Acromegaly. Chronic condition resulting from the hypersecretion of growth hormone during adulthood. The characteristic features are overgrowth of the bones of the face and extremities; protrusion of the chin; enlargement of the hands, feet, and fingers; thickening of the scalp; bowing of the spine; glycosuria; and suppression of sexual function. Treatment is by radiation or surgery, usually by partial resection of the pituitary gland. *Nutrition therapy* is directed to surgical care. Maintain an adequate diet and a desirable body weight. If the patient develops diabetes, which has been observed in 20% to 25% of cases, use nutritional therapy to control diabetes mellitus.

ACS. Abbreviation for *American Cancer Society*.

ACTH. Abbreviation for *adrenocorticotropic hormone*.

Activator. A substance that renders another substance active, either by being part of the reaction system or by combining with the inactive substance. An enzyme activator is called a *cofactor*.

Active transport. Also called biologic "pump." The process by which a substance is moved across a cell membrane from a lower to a higher electrochemical potential. It involves an expenditure of metabolic energy derived from the breakdown of adenosine triphosphate (ATP). Active transport is mediated by a carrier molecule that combines with the substance to be transported. See *Absorption, nutrients,* for examples.

Acute-phase proteins. These are the proteins needed during significant metabolic stress, such as immunoglobulins, leukocytes, albumin, hemoglobin, and enzymes necessary for protein synthesis.

Acyclovir. An antiviral used in treating herpes simplex. It interferes with DNA synthesis. Adverse side effects include diarrhea, nausea, vomiting, anorexia, glossitis, anemia, abdominal pain, and colitis. Brand name is Zovirax®.

ADA. Abbreviation for 1. American Dietetic Association. 2. American Diabetes Association. 3. American Dental Association.

Addison's disease. Metabolic disorder due to adrenal insufficiency. It is characterized by rapid loss of weight, emaciation, weakness, poor appetite, anemia, deep bronzing of the skin, low blood pressure, hypoglycemia, electrolyte imbalance with excessive loss of NaCl in the urine, and retention of potassium. It is a life-threatening disease because of the loss of adrenal cortex functions, which are vital life processes (see *Adrenal glands*). There is failure of the regulation of the autoimmune processes and hemorrhaging of the gland. The main treatment is continued administration of corticoid drugs. *Nutrition therapy:* maintain adequate fluids and a regular diet to meet the nutritional needs of the patient. Monitor blood glucose, sodium, and potassium levels. If the patient has infection or trauma and is underweight, increase the protein and calorie intake to about 1.3 times the recommended daily allowances.

Adenine. One of the major purine bases of nucleic acids. As *adenosine phosphate,* it provides energy for muscular movement.

Adenohypophysis. Anterior lobe of the *hypophysis* or pituitary gland. It secretes vital hormones that regulate other endocrine glands. These are the growth, thyrotropic, adrenocorticotropic, lactogenic, and gonadotropic hormones.

Adenosine phosphates. Adenosine monophosphate (AMP), adenosine diphosphate (ADP), and adenosine triphosphate (ATP) are present in practically all tissues, especially the muscles and liver. Cyclic adenosine monophosphate, also called adenylic acid (AA), is important in the activation of phosphorylase, whereas the di- and triphosphates are important sources of high-energy phosphate for cellular activity.

Adenyl cobamide. One of the coenzyme forms of vitamin B_{12}. See *Cobamide*.

ADH. Abbreviation for *antidiuretic hormone*.

Adipose tissue. Fatty tissue that acts as depot fat for storage of energy and serves as insulation against heat loss and as padding for protection and support of organs. It is found largely in subcutaneous tissues and around visceral organs. Like other body constituents, it is not inert but is in a *dynamic state*. Fat is constantly being formed and hydrolyzed in adipose tissue. Adiposity is another term for *obesity*.

ADP. Abbreviation for adenosine diphosphate. See *Adenosine phosphates*.

Adrenal glands. Also called suprarenal glands; these two small endocrine glands are located at the upper end of each kidney. Each gland consists of two parts: the *cortex,* which secretes estrogen, androgen, progesterone, and the adrenocortical hormones; and the *medulla,* which elaborates epinephrine and norepinephrine. The two groups of adrenocortical hormones that maintain nutrient homeostasis in the body are glucocorticoids and mineralocorticoids. Glucocorticoids promote the release of amino acids from muscles and lipolysis as needed and increase glyconeogenesis. Mineralocorticoids maintain electrolyte balance by increasing urinary potassium excretion and retaining sodium by promoting kidney reabsorption of sodium back to the plasma.

Adrenocortical insufficiency. See *Addison's disease*.

Adrenocorticotropic hormone (ACTH). Or corticotropin. A hormone secreted by the anterior pituitary gland that stimulates the adrenal cortex to produce *corticosteroid*. The ACTH level increases when a person is under stress, during surgery, and in acute hypoglycemia.

Adrenoleukodystrophy (ALD). An inherited metabolic defect characterized by an unusual accumulation of very-long-chain fatty acids (VLFA), especially C26:0, a hexacosanate. These are normally present in small amounts in the diet and are also synthesized within the body. Accumulation of the VLFA results in demyelination of the central nervous system (CNS), which leads to loss of voluntary movements and death. Although this disorder usually occurs in young children, there is an adult form designated *adrenomyeloneuropathy* (AMN), indicating adrenal insufficiency, one of the clinical signs of ALD. *Nutrition therapy:* restrict intake of very-long-chain fatty acids to less than 3 mg/day and increase intake of monounsaturated fatty acids, especially oleic acid (C18:1). A commercial preparation of oleic acid is available as glyceryl trioleate (GTO). Other nutrient sources are provided by nonfat milk, simple sugars or syrups, and vitamin–mineral supplements. Depending on the individual's progress, give a regular diet adequate for his or her nutrient needs but continue to alter dietary fat, restricting VLFA.

Advance®. Registered name of a milk formula used as a transition between infant formula and cow's milk; combines soy protein isolates and cow's milk to help reduce the risk of cow milk-induced enteric blood loss. Available in ready-to-use concentrated liquid. See Appendix 37.

Advera®. Brand name of a specialized liquid nutritional product for individuals with HIV infection or AIDS. It is high in calories and protein, low in fat, and contains 8.9 g fiber/L. For oral or tube feeding. See Appendix 39.

Advocates for Better Children's Diets (ABCD). A coalition of 35 organizations with the American Dietetic Association as a charter member. Newly formed in 1994, this group aims to improve the nutritional quality and administrative efficiency of child nutrition programs by easy access to all children who can benefit from them. Dietary Guidelines for Americans are implemented with school meals and nutrition education and training is a component of all the programs. The group recommends that the current *child nutrition programs* be administratively managed into one comprehensive school nutrition program.

A/E ratio. Number of milligrams of each essential amino acid per gram of total essential amino acids. This ratio is a method of evaluating protein quality.

African-American food practices. See Appendix 47-A.

Aging. Theoretically, aging is a continuous process from conception until death. But in the young and growing organism the building-up processes exceed the breaking-down processes, so that the net result is a picture of growth and development. Once the body reaches adulthood, the process is reversed. Aging proceeds at different rates in different individuals. Environmental factors—chemical, physical, and biologic—influence the aging process. Certain physiologic functions show a gradual decrement with age. These include basal metabolic rate, cardiac output, renal blood flow, lung capacity and immune function. However, other physiologic functions remain quite stable over the entire life span unless the individual is subjected to stress factors. For example, fasting blood glucose levels do not change significantly with age, and blood volume and red cell content remain relatively constant. See also *Nutrition, aging process*.

A/G ratio. See *Albumin/globulin ratio*.

AHA. Abbreviation for the *American Heart Association*.

AIDS. Abbreviation for *acquired immunodeficiency syndrome*.

AIN. Abbreviation for *American Institute of Nutrition*.

Al. Chemical symbol for aluminum.

Alanine (Ala). Alpha-aminopropionic acid, a nonessential amino acid readily formed from carbohydrates by its reversible conversion to pyruvic acid. Beta-alanine is the only naturally occurring beta-amino acid. It is found in pantothenic acid, carnosine, and anserine.

Albumin. A simple protein soluble in water and dilute salt solutions and coagulable by heat. Examples are lactalbumin in milk and ovalbumin in egg. Albumin is found in practically all animal tissues and in many plant tissues. The most abundant protein in human plasma, albumin plays an important role in acid–base balance and maintenance of colloid oncotic pressure; it also acts as a carrier of metals, ions, fatty acids, amino acids, metabolites, bilirubin, enzymes, hormones, and drugs. Serum albumin is commonly used as an indicator of visceral protein nutriture. The normal range in serum is 3.5 to 5.0 g/dL (35 to 50 g/L). Values of 2.8 to 3.4 g/dL (28 to 34 g/L) indicate mild depletion; 2.1 to 2.7 g/dL (21 to 27 g/L) moderate depletion; and less than 2.1 g/dL (21 g/L) severe depletion. However, serum albumin is a poor indicator of short-term changes in nutritional status due to its long half-life (18 to 21 days) and large body pool size (about 5 g/kg body weight). Serum albumin is also not a good indicator of protein status in patients with chronic liver disease. Albumin is synthesized by the liver and constitutes as much as 50% of the liver's total protein production. Serum albumin is decreased in malabsorption, renal disease, pancreatic insufficiency, congestive heart failure, cirrhosis, acute stress, edema, ascites, burns, hemorrhage, nephrotic syndrome, protein-losing enteropathy, neoplasms, trauma, and infection. See also *Albuminuria* and *Hypoalbuminemia*.

Albumin/globulin ratio (A/G). Ratio of albumin to globulin concentration in the serum. The normal value ranges from 1.8 to 2.5. The ratio is increased in chronic infection, acute liver disease, lupus erythematosus, rheumatoid arthritis, multiple myeloma, hyperlipidemia, leukemia, collagen disease, lymphomas, and advanced carcinoma; it is decreased in malabsorption, severe liver disease, nephrotic syndrome, diarrhea, severe burns, malnutrition, and exfoliative dermatitis.

Albuminoid. Also called scleroprotein; simple protein characteristic of skeletal structures and protective tissues such as skin and hair. It is of three distinct types: *elastin* in tendons and ligaments; *collagen* in tendons and bones; and *keratin* in hair, nails, and hooves.

Albuminuria. Presence of albumin in the urine. It occurs in kidney disease, toxemia of pregnancy, and certain conditions in which circulation to the kidney is inadequate. Normally, the kidneys reabsorb plasma albumin, which filters out from the glomerulus into the Bowman's capsule. See also *Proteinuria*.

Albuterol. A synthetic sympathomimetic amine; used as a bronchodilator in the symptomatic relief of bronchospasm in obstructive airway diseases. It has an unusual taste and may cause nausea and vomiting; it may also cause hyperglycemia and in large doses may aggravate preexisting diabetes mellitus. The trade names are Proventil® and Ventolin.®

Alcohol. 1. Aliphatic hydrocarbon derivative containing a hydroxyl (—OH) group. 2. Groups of organic compounds derived from carbohydrate fermentation. 3. Unqualified, it refers to ethyl alcohol (ethanol) in wines and liquors. Alcoholic beverages supply calories but few or no nutrients. One gram (or milliliter) of alcohol yields 7 calories. When using the food exchange list, an alcoholic beverage is considered more as fat rather than a fruit or bread exchange. Alcohol is not a true food and does not require insulin for its utilization. Because it is rapidly absorbed in the stomach and small intestines, it gets metabolized in the liver in preference to fats. Excess alcohol consumption can cause fatty liver, decreased lipid oxidation, hyperuricemia, and hypoglycemia. Alcohol is a depressant, affecting the central nervous system and other vital organs. Excessive intake is a prominent contributor to three of the ten leading causes of death in the United States, namely, cirrhosis of the liver, motor vehicle and other accidents, and suicide. An average adult man can tolerate 14 g of alcohol per hour. Women are generally less able to tolerate alcohol than men and have more serious medical consequences from heavy drinking. Women have a higher percentage of body fat and less body water than men, leading to higher blood-alcohol concentrations for a given amount of alcohol. Individuals who consume alcoholic beverages are advised to drink in moderation. "Moderate " drinking is defined for normal healthy women as no more than one drink per day and for normal healthy men as no more than two drinks per day. Count as a drink 12 oz of regular beer, 5 oz of wine, or 1½ oz of distilled spirit (80% proof). Individuals who are cautioned not to drink alcoholic beverages include women who are pregnant or trying to conceive, individuals using certain medications that should not be taken with alcohol, and those who plan to drive or engage in other activities that require a high level of attention or skill. See also Appendix 33.

Alcoholics Anonymous (AA). An international nonprofit organization that regularly meets for group support of alcoholics. The main purpose is to help its volunteer members cope with abstinence from drinking.

Alcohol dependence. Chronic excessive use of alcohol, which eventually leads to irreversible disorders affecting the liver, digestive system, pancreas, and nervous system. Thiamin, folacin, pyridoxine, and vitamin B_{12} absorption are impaired, causing multiple vitamin deficiencies. Clinical signs and symptoms include Wernicke–Korsakoff syndrome, peripheral neuropathy, pedal edema, ascites, anemia, glossitis, cardiac arrhythmias, and electrolyte imbalance. Liver cells are damaged, resulting in reduced enzyme production and poor nutrient metabolism. Steatorrhea is common due to pancreatic dysfunction. Most alcoholics suffer from malnutrition because of poor eating habits and diets that are usually deficient in calories and essential nutrients. As a consequence, anemias and liver *cirrhosis* are serious problems that need nutritional intervention. Alcohol may induce hypoglycemia in some insulin-dependent diabetics, leading to complications. Generally, alcohol causes dehydration due to increased diuresis. *Nutrition therapy:* withdrawal from alcohol use is often associated with anorexia, nausea, and gastrointestinal problems, all of which need individualized assessment. Electrolyte imbalance and severe hypoglycemia are corrected intravenously. When the patient's condition improves, the diet progresses gradually from full liquid to regular meals with vitamin B supplements, particularly thiamin, folic acid, niacin, pyridoxine, pantothenic acid, and vitamins B_2 and B_{12}. Protein and calories are increased according to the patient's needs and state of recovery. With alcoholic hepatitis, a minimum of 30 kcal/kg body weight and 1.5 g protein/kg is ample for liver regeneration and protein anabolism. For other dietary guidelines, see also *Ascites, Cirrhosis,* and *Hepatitis*.

Alcoholic liver disease (ALD). Liver disorder caused by the hepatotoxic effects of alcohol. Whether the alcoholic develops fatty liver, alcoholic hepatitis, or cirrhosis depends on the amount and duration of alcohol intake. Fewer than 30% of heavy drinkers develop ALD. An adequate diet does not protect against liver degeneration, although nutritional factors may influence the toxic effect of alcohol. *Nutrition therapy:* the aim is to help liver tissue regenerate, replenish nutritional deficits, and correct fluid and electrolyte imbalances. For specific guidelines, see *Ascites, Cirrhosis,* and *Hepatitis*. See also *Alcohol dependence* and *Fatty liver*.

ALD. Abbreviation for *adrenoleukodystrophy* and *alcoholic liver disease*.

Aldosterone. An adrenocortical hormone. It plays an important role in the regulation of electrolyte balance. Aldosterone acts on the distal convoluted tubules of the kidneys to reabsorb sodium and water and excrete potassium.

Alimentary tract. The digestive tract or gastrointestinal tract (GIT) extending from the mouth to the anus. Site of *digestion* and *absorption* of nutrients. See Appendix 11.

Alimentum®. Brand name of a hypoallergenic formula for infants and children with severe food allergies, fat malabsorption, and sensitivity to intact protein.

Alitame. A compound of alanine and aspartic acid; a nonnutritive sweetener that is 2000 times sweeter than sugar. It is heat stable and has no after-taste, except in acid foods at high temperature. Has 1.4 kcal/g. Pending approval by the Food and Drug Administration.

AlitraQ®. Brand name of a specialized elemental formula for metabolically stressed patients with impaired gastrointestinal function. It is low in fat and fortified with glutamine and arginine. For oral or tube feeding. See Appendix 39.

Alkaline-ash residue. Inorganic elements, chiefly sodium, potassium, calcium, and magnesium, that form basic ions in the body. When excreted in the urine, they increase the pH or make the urine more alkaline or basic.

Alkaline-forming foods. Foods in which the alkaline residue exceeds the acid residue. These include milk, vegetables, and fruits, except for cranberries, plums, and prunes. Most fruits, despite their acidity, exert a basic effect on the body since the organic acids in them, such as citric acid and malic acid, can be completely oxidized to carbon dioxide and water, leaving the salts to contribute to the supply of basic elements. Cranberries, plums, and prunes contain benzoic and quinic acids, which are not oxidized in the body but are converted in the liver into hippuric acid and excreted as such in the urine. These fruits are therefore acid-forming. See also *Diet, ash*.

Alkaline phosphatase. An enzyme that is present in the bone, liver, kidneys, and the intestines. It functions best at a pH of 9. Normal values in serum are 30 to 120 U/L (SI 0.5 to 2.0 μkat/L); it is higher in infants and children, especially during growth spurts. Serum alkaline phosphatase is increased in liver disease, bone disease, bone cancer, hyperparathyroidism, infectious mononucleosis, and leukemia; it is decreased in hypophosphatemia, malnutrition, hypothyroidism, vitamin D excess, scurvy, pernicious anemia, and celiac disease.

Alkali reserve. Buffer compounds in the blood, e.g., sodium bicarbonate, dipotassium phosphate, and proteins, that are capable of neutralizing acids. Sometimes called blood bicarbonate because bicarbonate is the chief alkali reserve of the body.

Alkalosis. An abnormal condition characterized by a rise in the pH or a fall in the hydrogen ion concentration of the blood. *Metabolic alkalosis* results from excessive loss of acids from the body without comparable loss of base or the formation or excessive intake of bicarbonate at a rate faster than its neutralization or elimination. *Respiratory alkalosis* may be caused by hyperventilation, resulting in excess loss of carbon dioxide.

Alkaptonuria. Inborn error of metabolism characterized by excretion of urine that darkens on contact with air due to the presence of abnormal amounts of homogentisic acid. Phenylalanine and tyrosine are not completely oxidized because of a lack of hepatic homogentisic acid oxidase. *Nutrition therapy:* the precise dietary treatment is not known, although ascorbic acid supplementation and low intake of phenylalanine and tyrosine may be of some value.

Allergy. Unusual or exaggerated susceptibility to a substance (allergen) that is harmless in similar amounts to most people. See *Food allergy*.

Allopurinol. A structural isomer of hypoxanthine; used in the treatment of gout and other hyperuricemias. Allopurinol has a metallic taste, may decrease iron absorption and cause anemia, prolongs the action of anticoagulants, and may cause anorexia, nausea and vomiting, abdominal pain, and diarrhea. Brand names are Lopurin® and Zyloprim.®

Alloxan. A red crystalline substance produced by oxidation of uric acid. It can cause necrosis of the islets of Langerhans in the pancreas. Alloxan has been used to treat cases of hyperinsulinism due to pancreatic tumors, but it is toxic in high amounts.

Alopecia. Baldness or loss of hair. This is seen in experimentally induced *biotin* deficiency in rats, which begins with dermatitis around the eyes and progresses to general loss of hair. Alopecia in humans is not corrected by administration of biotin. It is due to various causes, including seborrheic dermatitis; dandruff; effect of certain drugs or chemicals; syphilis; and other bacterial, fungal, or viral infections. Baldness is also seen in *myxedema* and other cases of pituitary insufficiency.

Alpha-fetoprotein (AFP). The fetal counterpart of adult albumin; the two share similar properties. AFP measurements in amniotic fluid are used for early diagnosis of fetal neural tube defects, such as spina bifida and anencephaly. Elevated serum levels may be present in hereditary tyrosinemia, cirrhosis, alcoholic hepatitis and viral hepatitis, and certain malignancies.

Alterna®. Brand name of a low-protein milk powder substitute for patients with renal disease; tastes like low-fat milk. Can be used as beverage and in place of milk in cooking.

Alternative sweeteners. Sweetening agents other than table sugar or sucrose, syrups, honey, and molasses. Grouped as nutritive or nonnutritive sweeteners. The latter are sometimes called "low-calorie sugar substitutes," although most of them have almost no calories and are useful for persons who have to avoid nutritive sugars, such as diabetics and obese persons. The other advantage of nonnutritive sweeteners relate to reduced tooth decay. For nutritive sweeteners, see *Fructose, Glucose, Glycine, Mannitol, Sorbitol,* and *Xylitol*. For nonnutritive sweeteners, see *Acesulfame-K. Alitame, Aspartame, Cyclamate, Dulcin,* and *Saccharin*.

Aluminum (Al). The third most abundant element in the earth's crust; its compounds are widespread in soils. Although present in trace amounts in biologic material, aluminum does not appear to be an essential element. It is very poorly absorbed. Large intake, however, is known to produce gastrointestinal irritation. Aluminum combines with the phosphates present in food to form insoluble aluminum phosphate, which is excreted in the feces. The ability of the kidneys to excrete aluminum is decreased in renal failure. Aluminum poisoning is an important potential hazard to patients with chronic renal failure on hemodialysis; aluminum present in the dialysis fluid may be transferred to the patient. Some aluminum is also absorbed when aluminum hydroxide is used as an antacid or oral phosphate binder to control plasma phosphorus. Accumulation of aluminum may also result from the parenteral administration of nutritional solution contaminated with aluminum. Toxic effects are manifested in neurologic, bone, and hematopoietic functions. Aluminum intoxication intensifies anemia in uremic patients and can cause uremic osteomalacia and encephalopathy. Excess aluminum accumulation in the brain has been associated with Alzheimer's disease and other dementias. Contrary to earlier beliefs, trace amounts of aluminum from cooking utensils are harmless and do not cause chronic poisoning. There is no evidence that aluminum cookware, foil wrapping, pie plates, or cookie sheets contribute to increased aluminum levels in food.

Alzheimer's disease (AD). A form of dementia characterized by a group of symptoms that include loss of memory, thinking, and reasoning power; disorientation, confusion, and sometimes speech disturbances. The exact causes of Alzheimer's disease are not yet known, although structural changes in the brain have been observed, consisting of neurofibrillary tangles and neuritic plaques in the cortex. Recent studies have shown an abnormal accumulation of aluminum in the brain cells of persons with Alzheimer's disease. There is currently no scientific evidence in support of decreasing aluminum intake to prevent the disease, and dietary supplements of choline and lecithin have not been shown to improve cognitive function. While AD patients are generally over 65 years old, a few are in their forties. Food intake is affected and weight loss is common. The following are typical feeding problems seen during the three stages of the disease: Stage 1—difficulty in shopping and cooking, unusual food choices, decreased appetite, changes in taste and smell, and forgetting to eat. Stage 2—increased energy requirement from agitation, losing ability to use utensils, eating with hands, holding food in the mouth, forgetting to swallow, and failing to chew food before swallowing. Stage 3—no recognition of food, refusing to eat or open mouth, and dysphagia. *Nutrition therapy:* feeding must be individualized. Constant pacing and agitation may require up to 1600 kcal/day on the average. Recognize and assess feeding problems associated at each stage of the disease. Encourage and promote as much independence in eating by selecting appropriate food consistency, providing adequate time to eat, using the appropriate feeding equipment and technique, and by providing the proper dining environment. Give small, frequent meals of nutrient-dense foods. Supplemental vitamins and minerals may be necessary. Monitor body weight and guard against dehydration, constipation, and aspiration of food. Delay tube feeding unless absolutely necessary. See also *Dysphagia*.

Amantadine. A synthetic compound used in the treatment of parkinsonian syndrome and drug-related extrapyramidal reactions and in the symptomatic treatment of influenza caused by influenza A virus strains. Constipation and dry mouth may become problems because of its anticholinergic effect. Brand names are Symmetrel® and Symadine®.

Amblyopia. Dimness of vision. Nutritional amblyopia appears after a period of many months when diets are grossly deficient in nutrients, particularly the B complex vitamins. Common presenting complaints are blurred or dim vision, difficulty in reading, photophobia, and discomfort in the retrobulbar region on moving the eyes. When the symptoms are not severe, they are relieved rapidly by improved nutrition and B vitamin supplementation; little or no improvement results from any form of treatment when vision is already markedly affected.

American Cancer Society (ACS). In 1990, the American Cancer Society and eight other organizations recommended the following dietary guidelines for reducing the risk for certain types of cancer: maintain a desirable body weight; eat a varied diet; include a variety of both vegetables and fruits in the daily diet; eat more high-fiber foods, such as whole-grain cereals, legumes, vegetables, and fruits; cut down on total fat intake (to 30% or less of total calorie intake); limit consumption of alcoholic beverages; and limit consumption of salt-cured and nitrite-preserved foods.

American Dietetic Association (ADA). Professional organization whose objectives are to improve the nutrition of humans, to advance the science of dietetics and nutrition, and to promote education in these and allied areas. It publishes monthly the *Journal of the American Dietetic Association* and the *ADA Courier*. The ADA is an authority on national food and nutrition issues. It is the largest and oldest professional organization concerned with the practice of dietetics.

American Heart Association (AHA) dietary guidelines. The AHA dietary guidelines for healthy Americans are the following: Total fat intake should be less than 30% of calories; saturated fat intake should be less than 10% of calories; polyunsaturated fat intake should not exceed 10% of calories; cholesterol intake should not exceed 300 mg/day; carbohydrate intake should constitute 50% or more of total calories per day, with emphasis on complex carbohydrates; protein intake should provide the remainder of the calories; sodium intake should not exceed 3 g/day; alcoholic consumption should not exceed 1 to 2 fluid oz of ethanol per day. Two oz of 100 proof whisky, 8 oz of wine, or 24 oz of beer each contains about 1 oz of ethanol. Total calories should be sufficient to maintain the individual's recommended body weight; and a wide variety of foods should be consumed.

American Institute of Nutrition (AIN). Professional organization founded in 1928 to develop and extend nutrition knowledge and to promote personal contact between researchers in nutrition and related fields. It publishes the *Journal of Nutrition* and the *American Journal of Clinical Nutrition*.

American Society for Clinical Nutrition (ASCN). Division of the *American Institute of Nutrition (AIN)* that aims to promote education about human nutrition in health and disease and to promote the presentation and discussion of research in human nutrition.

American Society for Parenteral and Enteral Nutrition (ASPEN). Group of physicians, dietitians, nurses, pharmacists, nutritionists, and others dedicated to fostering good nutritional support of patients during hospitalization, rehabilitation, and home care. By promoting the team approach and by educating health care professionals at all levels, ASPEN encourages the development of improved nutritional support procedures. It publishes the *Journal of Parenteral and Enteral Nutrition (JPEN)* and *Nutrition in Clinical Practice (NCP)*.

Amethopterin. A potent metabolic antagonist of *folic acid;* it inhibits the enzyme dihydrofolate reductase and interferes with the conversion of folic acid to its active form, *folinic acid*. Amethopterin is used in the treatment of leukemia. Adverse gastrointestinal side effects include oral lesions, glossitis, stomatitis, abdominal distress, anorexia, nausea, vomiting, and diarrhea. Trade names are Folex® and Mexate®.

Amiloride. A potassium-sparing diuretic; used concomitantly with other diuretics in the treatment of edema associated with congestive heart failure and hepatic cirrhosis. The potassium-sparing effect of amiloride may cause hyperkalemia, especially in geriatric patients and in those with renal insufficiency and diabetes mellitus; it may also cause hyponatremia, appetite changes, gas pain and abdominal bloating, and heartburn. Brand names are Midamor® and Moduretic®.

Amin-Aid®. Brand name of an instant drink that is high in calories but low in protein, sodium, and potassium for use in acute or chronic renal failure; contains essential amino acids and diglycerides. It requires vitamin and mineral supplementation if used as the principal source of nutrition. See Appendix 39.

Amine. Compound that has the characteristic amino (NH_2) group. It is formed by replacing one or more of the hydrogen atoms of ammonia with one or more organic radicals. Amines are classified as *primary, secondary,* or *tertiary,* depending on whether one, two, or three hydrogens are replaced.

Aminess®. Brand name of a 5.2% amino acid solution for total parenteral nutrition. Contains essential amino acids plus histidine, based on the minimal requirement for each. It is low in aromatic amino acids (tryptophan, phenylalanine, and tyrosine) and is used for nutritional support of uremic patients, particularly when oral feeding is not possible. See Appendix 40.

Aminess® Tablet. Brand name of a dietary supplement in tablet form for patients with renal disease. Contains all the essential amino acids plus histidine. Used in conjunction with a low-protein diet.

Amino acid (AA). Fundamental structural unit of protein with the general formula

$$\begin{array}{c} NH_2 \\ | \\ R\text{—}C\text{—}COOH \\ | \\ H \end{array}$$

(R is a radical, the basis for classifying amino acids)

Amino acids may be acidic, basic, or neutral, depending on the number of acidic or basic groups in the molecule. According to their structure, amino acids may be aliphatic, aromatic, or heterocyclic. Of the 20 amino acids considered to be physiologically important, 9 are known to be essential for the human adult. The others are dietary nonessential amino acids. For a list of the amino acids and human requirements, see Appendix 9. Amino nitrogen comprises about 16% of the weight of protein. The biological value of protein is based on its amino acids. See *Protein*.

Acidic AA. The two acidic amino acids, aspartic acid and glutamic acid, are important brain neurotransmitters.

Aliphatic AA. Amino acid with a carbon-to-carbon chain. Examples are threonine, glycine, serine, and alanine.

Antiketogenic AA. See *Glucogenic AA*.

Aromatic AA (AAA). Amino acid in which the carbon atoms are arranged in a ring. Examples are tryptophan, phenylalanine, and tyrosine, which are biosynthetic precursors for the neurotransmitters dopamine, norepinephrine, and serotonin.

Branched-chain AA (BCAA). Amino acid that is aliphatic but whose side chain is branched. Examples are leucine, isoleucine, and valine. The three branched-chain amino acids (leucine, isoleucine, and valine) are not catabolized by the liver but are taken up by extrahepatic tissues.

Conditionally essential AA. Also called conditionally indispensable amino acid. An amino acid that becomes indispensable under certain conditions, such as severe stress, sepsis, and thermal injury. Arginine, cysteine, glutamine, taurine and tyrosine, are conditionally essential AAs.

Dibasic AA. Amino acid that has a second nitrogen atom. Examples are lysine, arginine, and histidine.

Dispensable AA. See *Nonessential AA*.

Essential AA (EAA). Also called indispensable amino acid; an amino acid that cannot be synthesized by the body from materials readily available at a speed commensurate with the demands for normal growth and other physiologic needs, as in hypermetabolism. It must therefore be supplied preformed in the diet. If the essential amino acids supplied by the diet are not adequate, the body will break down its own proteins. Nonessential amino acids may become essential if the diet fails to supply the precursors required for protein synthesis. The nine essential amino acids for humans are *histidine, isoleucine, leucine, lysine, methionine, phenylalanine, threonine, tryptophan,* and *valine*.

Glucogenic AA. An amino acid that can be converted to an alpha-keto acid, a carbohydrate former. Examples are glycine, alanine, serine, threonine, aspartic acid, and glutamic acid.

Indispensable AA. See *Essential AA*.

Ketogenic AA. An amino acid that can be converted to acetate or acetoacetate, a ketone body. Examples are leucine, isoleucine, and lysine.

Limiting AA. The essential amino acid that is most deficient in a protein, in comparison with the amino acids of a standard protein. Lysine is the limiting amino acid in rice and other cereals; tryptophan is limiting in corn; and methionine and cystine are limiting in beans. The limiting AAs in many nuts and seeds are lysine and isoleucine; peanuts are low in methionine and threonine. Most vegetable proteins are low in cystine, methionine, and isoleucine.

Neutral AA. The aromatic AAs and the branched-chain AAs are considered neutral in their pH reactions.

Nonessential AA. Also called dispensable amino acid; an amino acid that can be synthesized in the body, provided that there is an adequate source of nitrogen. It need not be supplied preformed in the diet. Examples are alanine, asparagine, aspartic acid, cystine, glycine, glutamic acid, hydroxyproline, proline, and serine.

Semidispensable AA. See *Semiessential AA*.

Semiessential AA. Also called semidispensable amino acid; an amino acid that, when present in the diet, reduces the need for an essential amino acid. For example, cystine reduces the need for methionine, and tyrosine reduces the need for phenylalanine.

Sulfur-containing AA. Amino acid that contains sulfur in its molecule. Examples are methionine, cystine, and cysteine.

Amino acid, functions. The physiologic roles of amino acids as integral components (building blocks) of protein are summarized under *Protein, functions*. However, some amino acids have specialized functions, and these are discussed under the specific amino acid. For example, tryptophan is a precursor of niacin; methionine is a methylating agent (donates the CH_3—radical); cystine is a component of bile acid; and tyrosine is needed for the synthesis of thyroxine.

Amino acid pool. Reservoir or metabolic pool of amino acids that come from the diet (exogenous source), that are synthesized in cells, and that are derived from the breakdown of tissue proteins (endogenous sources). The size of the pool is determined by the quantity of the constituents instantaneously present and available for all of the reactions leading into and from the pool (i.e., anabolic and catabolic reactions).

Amino acid reference pattern. The ideal combination of amino acids in total quantity and proportion to meet all physiologic requirements. The Food and Agriculture Organization reference pattern was derived from the minimal daily requirements for the essential amino acids for infants and adults. Other reference patterns have been based on the amino acids present in eggs and human milk.

Amino acid requirements. Established values for the nine essential amino acids are given in Appendix 9, using nitrogen balance as the criterion for adults. See also *Protein requirement*.

Aminoaciduria. Increase in the urinary excretion of amino acids due to elevated concentrations of amino acids in the plasma. The condition is caused by a defect in the renal tubular reabsorption of amino acids. The renal defect may be congenital, or it may be acquired as a result of toxic agents, metabolic disorders such as acidosis and hypercalcemia, and deficiencies of vitamins B, C, and D.

Aminoglycoside. An antibiotic derived from cultures of *Streptomyces* or *Micromonospora*. The aminoglycosides include amikacin, gentamicin, kanamycin, neomycin, streptomycin, and tobramycin. Adverse effects of orally administered aminoglycosides include nausea, vomiting, and diarrhea; malabsorption of lipids, protein, sodium, calcium, iron, and vitamin B_{12}; increased fecal excretion of bile acids, potassium, and calcium; and decreased serum concentrations of carotene and vitamin K. A sprue-like syndrome with steatorrhea, malabsorption, and electrolyte imbalances has occurred following oral administration of kanamycin or neomycin. The effects of malabsorption may be pronounced if the drug is taken for prolonged periods.

Aminogram. Amino acid pattern showing the quantitative relationship between the essential amino acids in a dietary protein and those found in egg protein. Because egg is an unreasonably high protein standard for the world food supply, the Food and Agriculture Organization uses a theoretical ideal aminogram as the protein standard.

Aminopterin. Folic acid antagonist used clinically in the treatment of leukemia and other neoplastic diseases. See also *Methotrexate*.

Aminosyn®. Brand name of a standard parenteral amino acid solution in five different concentrations (3.5%, 5%, 7%, 8.5%, and 10%); includes all the essential amino acids and several nonessential amino acids. Also available with high levels of branched-chain amino acids for hepatic diseases, stress, and trauma (Aminosyn®-HBC) and with low levels of aromatic amino acids for renal disease (Aminosyn®-RF). See Appendix 40.

Amitriptyline. A tricyclic antidepressant. The drug interferes with riboflavin metabolism; it may also cause altered taste acuity, anorexia, stomatitis, dry mouth, constipation, and changes in blood glucose. Brand names include Amitril®, Elavil®, Endep®, Enovil®, Etrafon®, and Triavil®.

Ammonia. 1. Volatile alkaline gas that is soluble in water. 2. By-product of protein metabolism by deamination of amino acids. In the body, ammonia may be used in the reductive amination of alpha-keto acids to form new amino acids, or it may be used in the synthesis of purines and pyrimidines. Ammonia is toxic in large concentrations and normally is not allowed to accumulate in the cells. It is either excreted directly in the urine or eliminated via glutamine or urea formation. Blood ammonia is elevated in cirrhosis and liver failure. Normal range is 40 to 70 mcg/dL (24 to 41 μmol/liter).

Amphotericin B. Antifungal antibiotic used to treat severe systemic infections and meningitis caused by susceptible fungi. Weight loss is common with this drug; it may also cause proteinuria, decreased serum potassium and magnesium, increased blood urea nitrogen and creatinine, anorexia, dyspepsia, cramping, and epigastric pain. Brand name is Fungizone®.

Amputation. The surgical removal of a part of the body or a limb for medical treatment, as in the removal of malignant tumors, gangrene of the toe, a portion of a limb due to a severe injury, as from an auto accident, and so on. The percentage of body weight loss in amputees is considered in determining desirable body weight. The following data are close estimates:

BODY PART AMPUTATED	(PERCENT OF BODY WEIGHT LOSS)
Whole arm (5.0)	Total leg and foot (16.0)
Upper arm (2.7)	Upper leg (10.1)
Lower arm (1.6)	Lower leg and foot (5.9)
Hand (0.7)	Foot (1.5)

Amylase. Enzyme that catalyzes the hydrolysis of starch to sugar. There are two types: *alpha-amylase* found in saliva and pancreatic juice and *beta-amylase* found in grains and vegetables. An increased level of amylase in the blood is diagnostic of acute pancreatitis. Normal range in the serum is 25 to 125 mU/dL or less (25 to 125 U/liter).

Amylophagia. A form of *pica* characterized by consumption of an excessive amount of starch, such as laundry starch. See *Pica*.

Amyotrophic lateral sclerosis (ALS). Also known as Lou Gehrig's disease or motor neuron disease; characterized by progressive spinal muscular atrophy, which may be fatal due to respiratory failure. Early symptoms are loss of reflexes and gait, difficulty in chewing and swallowing, and negative nitrogen balance, inadequate food intake, weight loss, and nutritional deficiencies. *Nutrition therapy:* correct protein losses and other nutritional deficiencies. Vitamin and mineral supplementation may be necessary. Small frequent meals consisting of foods that are easy to chew may increase intake and help conserve energy. Assess swallowing ability and modify food consistencies as tolerated. Emphasize proper posture for swallowing, with the head in a chin-down position. Dry crumbly foods tend to break apart and cause choking. As dysphagia progresses, thin liquids become difficult to swallow. Use thickened liquids and soft foods high in water content to maintain fluid balance. Gastrostomy feeding may be necessary if swallowing reflex is severely affected. Provide liberal fluid intake (about 2 liters daily). When the swallowing reflex is severely affected, tube feeding is recommended. A well-balanced, regular diet in six to eight small feedings, consisting of foods that are easy to chew and swallow, should be provided as the patient's condition improves. Supplement with zinc, phosphorus, potassium, magnesium, and vitamins, particularly vitamin E.

Anabolic agent. Compound that promotes synthesis, growth, and weight gain. The endogenous anabolic hormones, produced throughout every human's lifetime, are the steroid hormones—estradiol, progesterone, and testosterone. There are synthetic steroid and nonsteroid hormones which are also anabolic. Those used in the food industry are approved by the Food and Drug Administration. The levels used and the residues left on food (e.g., meats) are closely monitored by the Food Safety and Inspection Service, U.S. Department of Agriculture.

Anabolism. Synthesis; process by which simple substances are converted by living cells into more complex substances. Also referred to as constructive metabolism. Compare with *Catabolism*.

Androgen. Generic name for the hormones secreted by the testes that are responsible for the development of male accessory sex organs and secondary sex characteristics. Androgenic hormones also have anabolic influence on nitrogen and calcium metabolism. The two major naturally occurring androgens are androsterone and testosterone.

Anemia. Reduction in the size or number of red blood cells, the quantity of hemoglobin, or both, resulting in decreased capacity of the blood to carry oxygen. The symptoms vary according to the etiologic factors, but the common clinical signs and symptoms include pallor, breathlessness on exertion, easy fatigue, dizziness, insomnia, and lack of appetite. Anemias may be classified according to cell size, which may be *large* (macrocytic), *small* (microcytic), or *normal* (normocytic). Another classification is based on the color index of the blood, which may be *high* (hyperchromic), *low* (hypochromic), or *normal* (normochromic). Anemia may be due to excessive loss of blood; to excessive blood cell destruction as a result of chemical poisons, such as lead or specific infections such as malaria; or to congenital abnormalities of the red cells, as in sickle cell anemia. Anemias may also be due to a defect in blood formation. The defect may be nutritional in origin or may be due to aplasia of the bone marrow, toxic inhibition, or diseases that affect the bone marrow, spleen, liver, or lymph nodes.

Anemia, nonnutritional. The anemia seen in infection is not caused by deficient intake of iron but by increased hepatic uptake. Supplemental iron is contraindicated; increased serum levels may worsen the infection. *Anemia of chronic disease* is due to abnormalities in the recycling of iron from hemoglobin of old blood cells that frequently accompany infections, inflammatory disorders, or connective tissue disease such as arthritis. "*Sports anemia*" is a transient anemia seen in strenuous exercise or heavy physical training, characterized by a significant decrease in RBC count, hemoglobin concentration, and packed cell volume; it does not respond to iron supplementation. *Sickle cell anemia* is an inherited anemia in which defective hemoglobin causes the erythrocytes to be sickle-shaped. Iron stores are frequently in excess. The diet should therefore restrict iron-rich foods but should be high in folic acid, vitamin E, and zinc. *Sideroblastic anemia* is a microcytic, hypochromic anemia similar to iron-deficiency anemia, except that serum iron is normal or elevated. It may be caused by cancer, chronic alcoholism, and collagen disorders. Anemia of *end-stage renal disease* (ESRD) is caused primarily by decreased production of the hormone erythropoietin, which is normally produced by the kidneys. This hormone stimulates the bone marrow to produce red blood cells. ESRD patients receiving erythropoietin therapy require an adequate supply of available iron.

Anemia, nutritional. Anemia due to a deficiency of nutrients necessary in the formation of blood. Iron, protein, folic acid, vitamin B_{12}, and vitamin C are the major nutrients essential in blood formation. Copper and cobalt are also essential, but the amounts needed are so small that they are more than amply supplied by the normal adequate diet. The deficiency in these nutrients may be caused by inadequate intake, defective absorption, imperfect utilization, increased requirement, or increased excretion. See also *Hemopoiesis* and *Nutrition, anemias*.

Copper-deficiency a. Copper is essential for the formation of hemoglobin and mobilization of iron from its storage sites to the plasma. Iron cannot be released if copper is deficient, leading to low serum iron and hemoglobin levels even in the presence of normal iron stores. Copper deficiency anemia may occur in infants who are fed cow's milk or a copper-deficient infant formula. It is also seen in children or adults who have a malabsorption syndrome or who are receiving long-term total parenteral nutrition that does not contain copper. *Nutrition therapy:* foods rich in copper are liver, shellfish, whole grain cereals, nuts, and legumes. Ensure an adequate intake of calories, protein, vitamins, and other minerals.

Folic acid-deficiency a. Normal folate stores are depleted within 2 to 4 months on a folate-deficient diet. Folate deficiency may also be due to faulty absorption, as in intestinal resection and congenital folate malabsorption; increased requirements due to growth and pregnancy; and drug-induced, as by anticonvulsants, methotrexate, and oral contraceptives. Results in a megaloblastic anemia because the erythrocyte protein cannot be synthesized properly in the deficient state, resulting in large (macrocytic) and immature (megaloblastic) blood cells. *Nutrition therapy:* to replenish folate stores, 1 mg of folate every day for 2 to 3 weeks is recommended. A safe maintenance dose is 50 to 100 μg/day, either in food or in a supplement. Foods rich in folate are dark green leafy vegetables, whole grains, legumes, nuts, and organ meats.

Iron-deficiency a. Form of anemia characterized by small (microcytic), pale (hypochromic) erythrocytes. It is generally due to chronic blood loss, as in excessive or prolonged menstruation, repeated pregnancies, and parasitic infestation; faulty iron intake; impaired iron absorption, as in achlorhydria and chronic diarrhea; and increased blood volume, which occurs during infancy and pregnancy. *Nutrition therapy:* a daily dose of 50 to 200 mg of ferrous iron is the chief treatment. Iron is best absorbed when the stomach is empty, but large doses can cause gastric irritation and other gastrointestinal side effects. Start with a small dose and gradually increase the dose. Good food sources of iron are: liver, beef, egg, dried fruits, dried peas and beans, nuts, green leafy vegetables, whole grain breads and cereals, and fortified cereals. Include a good source of vitamin C at every meal to enhance iron absorption. Avoid large amounts of coffee or tea with meals if iron is taken after meals.

Protein-deficiency a. Macrocytic type of anemia seen in association with protein malnutrition. Patients with this type of anemia also show signs of multiple nutritional deficiencies, especially of folic acid, vitamin B_{12}, and iron. Protein is essential for the proper production of hemoglobin and red blood cells.

Vitamin B_6-responsive a. A microcytic, hypochromic anemia that responds to vitamin B_6 therapy. Characterized by high serum iron, increased saturation of iron-binding protein, and hemosiderin deposits in bone marrow and liver. *Nutrition therapy:* a therapeutic trial dose of pyridoxine (50 to 200 mg/day) is given. If the anemia responds, pyridoxine therapy is continued for life. Good sources of pyridoxine include liver, pork, poultry, whole grains, fortified cereals, and legumes.

Vitamin B_{12}-deficiency a. A megaloblastic, macrocytic type of anemia that is most commonly due to a lack of the *intrinsic factor* which is necessary for the absorption of vitamin B_{12}. Other causes include: decreased absorption, as in partial or total gastrectomy, small bowel bacterial overgrowth, achlorhydria, fish tapeworm disease; diseases affecting the ileum, as in regional enteritis and ileal resection and bypass; and drug-induced. Very rarely, vitamin B_{12} deficiency anemia occurs in strict vegetarians. Vitamin B_{12} deficiency is slow to develop; the vitamin stores are depleted only after several years. Signs and symptoms are similar to that of folic acid-deficiency anemia, except that in vitamin B_{12} deficiency there is also inadequate myelinization of the nerves which causes paresthesia, poor muscle coordination, poor memory, and hallucination. *Nutrition therapy:* treatment usually consists of intramuscular or subcutaneous injections of 50 to 100 μg/day of vitamin B_{12} for 1 to 2 weeks, followed by a monthly injection of 100 μg. Large oral doses of vitamin B_{12} (1000 μg) are also effective, even in the absence of intrinsic factor, because about 1% of the vitamin is absorbed by diffusion. Liver, meat, fish, poultry, milk, and eggs are good sources; the vitamin is not present in plant foods. A high-protein diet (1.5 g/kg body weight) is desirable for blood regeneration. Most cases of vitamin B_{12} deficiency that are not of dietary origin require treatment for life.

Vitamin C-deficiency a. Macrocytic type of anemia seen in severe cases of vitamin C deficiency (scurvy). Vitamin C is necessary for the absorption of iron and the conversion of folic acid to its biological active form, folinic acid. Good food sources of vitamin C include citrus fruits, melon, papaya, cabbage, green pepper, tomatoes, and vitamin C-fortified juices and beverages.

Vitamin E-responsive a. Also called hemolytic anemia. It is characterized by oxidative destruction of red blood cells; the red blood cells have an abnormal fragile membrane that results in hemolysis. Infants who receive polyunsaturated fatty acids (PUFAs) but lack vitamin E risk having hemolytic anemia. *Nutrition therapy:* it is recommended that preterm infants receive an additional 5 to 25 IU/day of vitamin E to prevent hemolytic anemia.

Anergy. 1. Lack of energy or activity. 2. Diminished reactivity to specific antigen(s); the total absence of an immune response under conditions that would otherwise be expected to lead to such a response. Anergy is seen in protein-energy malnutrition, stress, cancer, and sepsis. Anergy is also seen in deficiencies of vitamin A, zinc, iron, and vitamin B_6. See *Hypersensitivity skin test*.

Angina pectoris. A sudden, severe pain radiating from the heart region to the left shoulder and down the arm into the fingers. It tends to occur suddenly following emotional stress, physical exertion, and other conditions subjecting the heart to heavy strain. *Nutrition therapy:* maintain desirable body weight. Give small feedings with rest periods when eating. Avoid caffeine and increase dietary fiber. If the angina is associated with coronary heart disease, restrict intake of sodium, cholesterol, and saturated fats.

Angular stomatitis. Inflammation of the oral mucosa at the angles of the mouth, giving the appearance of fissures radiating outward from the mouth. It may extend into the buccal mucosa as whitish patches on the mucous membrane lining the cheeks. The tongue is often red and smooth or has patchy areas of white coating. Angular stomatitis often responds to large doses of riboflavin and sometimes to pyridoxine. It is often seen in patients receiving long-term antibiotic therapy, especially chloramphenicol, and also occurs in association with iron-deficiency anemia. Nonnutritional factors, such as ill-fitting dentures, may also cause angular stomatitis.

Anorexia. Lack or loss of appetite.

Anorexia nervosa. A disorder characterized by an aversion to food and a self-imposed restriction of food intake. There is a preoccupation with food as a fear of getting fat, and the individual denies being excessively thin despite extreme emaciation, denies hunger despite malnutrition, and denies fatigue despite frantic exercise. A spectrum of the disorder occurs, ranging from mild conditions that require little or no intervention to very severe forms that can require hospitalization. Anorexia nervosa is a life-threatening disorder due to the effects of starvation and extreme inanition, which lead to secondary endocrine disorders resulting in amenorrhea, lowered basal metabolic rate, slow pulse, electrolyte imbalance, and other metabolic consequences; immune functions can be compromised. Some individuals with anorexia nervosa go on periodic binge eating, which is followed by self-induced purging by vomiting or the use of enemas or cathartics. Many treatments of anorexia nervosa have evolved, focusing on biologic interventions to bring about weight gain, psychological techniques to deal with personal and family conflicts, or a combination of these methods. *Nutrition therapy:* the main goal is to change the patient's dietary habits and to normalize his/her dramatic weight loss. This is a lengthy process that involves a multidisciplinary approach, focused on emotional and psychological aspects of nutrition, called "psychonutritional." Start with small feedings and gradually increase the calories. Establish a long-term weight goal. Consider the previous weight history and the weight the patient will accept. Parenteral or nasogastric feeding should be reserved for life-threatening states and usually are not necessary. See also *Eating disorders*.

Antacid. Inorganic salt that dissolves in acid gastric secretions, releasing anions that partially neutralize gastric hydrochloric acid. Antacids inactivate thiamin, decrease iron and vitamin A absorption, and cause phosphate depletion. Undesirable gastrointestinal effects are bloating, constipation and fecal impaction, and stomach cramps. Brand names include Amphojel®, Alu-Cap®, Basaljel®, Di-Gel®, Gelusil®, Maalox®, and Mylanta®.

Anthranilic acid. Product resulting from the hydrolysis of *kynurenine* by the enzyme kynureninase with pyridoxal phosphate as a cofactor. In pyridoxine deficiency, hydrolysis of kynurenine results instead in the production of *xanthurenic acid,* which is excreted in the urine.

Anthropometry. Scientific measurement of the various parts of the body. This includes the measurement of body weight and height, as well as the chest, arms, head, and other body parts. Anthropometry is a useful aid in assessing the nutritional status of individuals and groups. See also *Nutritional assessment*. See Appendices 18 to 23.

Antibiotic. An antimicrobial agent produced by certain microorganisms or semisynthetically. Many antibiotics have direct effects on the gastrointestinal tract (e.g., nausea, anorexia, glossitis, stomatitis, and diarrhea); some antibiotics bind nutrients (e.g., tetracycline binds protein, and penicillin and sulfonamides bind serum albumin); and other antibiotics increase the volume and/or frequency of stools. In general, antibiotics alter microflora and inhibit bacterial synthesis of certain vitamins. See *Cephalosporin, Chloramphenicol, Erythromycin, Penicillin,* and *Tetracycline*.

Antibody. Specific substance produced in the body in response to invasion by a foreign or antagonistic substance known as an antigen. Antibodies are serum proteins elicited by the lymphoid cell system. These proteins, called immunoglobulins (Ig), protect the body by reacting as agglutinins, lysins, precipitins, or antitoxins.

Anticoagulant. A substance that inhibits or prevents blood coagulation by interfering with the clotting mechanism. Examples are Dicumarol and heparin, which inhibit prothrombin formation, and oxalate and citrate, which combine with calcium. See *Blood clotting*.

Antidiuretic hormone (ADH). A hormone produced by the posterior portion of the pituitary gland (neurohypophysis). It has a marked antidiuretic action by increasing the rate of reabsorption of water from the kidney tubules, thus decreasing water excretion. A deficiency of this hormone results in a condition known as *diabetes insipidus*.

Antiketogenic factor. See *Ketogenic/antiketogenic ratio*.

Antimetabolite. Structurally related compound that interferes with the metabolism or function of a chemical compound (metabolite) in the body. Also called metabolic antagonist.

Antioxidant. A substance that delays or prevents oxidation. Nutrients with antioxidant properties include beta-carotene, vitamin C, vitamin E, and the trace element selenium. These antioxidants are believed to have protective effects against cancer and may also prevent or retard coronary artery disease by reducing serum low-density lipoprotein-cholesterol and increasing high-density lipoprotein-cholesterol. A limited number of studies in humans suggest potential health benefits of increased antioxidant status in several other age-related conditions, including cataracts, macular degeneration, inflammatory disorders such as arthritis, immunologic deficits linked to impaired wound healing, and Parkinsonism. A scientific advisory panel convened by the Alliance for Aging Research recently advised healthy adults to increase their intake of selected antioxidants and recommended daily consumption of 250 to 1000 mg vitamin C, 100 to 400 IU vitamin E, and 10 to 30 mg beta-carotene to prevent chronic, age-related diseases. Medical groups and federal agencies, including the Food and Drug Administration and the National Academy of Sciences, maintain that antioxidant supplement recommendations are premature.

Antipyrine. A chemical used as an antipyretic and analgesic; also widely used in total body water determinations. See *Water determination, body*.

Antithyroid agents. A large number of substances that inhibit normal thyroid function either by inhibiting the synthesis of thyroid hormones or by preventing their release from the thyroid gland. Examples are thiourea, thiouracil, and *goitrogens* in foods.

Antivitamin. Natural or synthetic substance similar in structure to a vitamin that interferes with its normal functioning by competitive inhibition, inactivation, or chemical destruction. Some of the most common antivitamins are Dicumarol® (vitamin K), thiaminase (vitamin B_1), Atabrine® (vitamin B_2), aminopterin (folic acid), and avidin (biotin).

Anuria. Suppression of renal secretion; absence of urinary excretion. It may occur in the final stages of glomerulonephritis or after severe trauma, surgery, or transfusion of incompatible blood. Sometimes anuria is nervous in origin.

AOAC. Abbreviation for *Association of Official Agricultural Chemists*.

Aphagia. Loss of ability or power to swallow. See also *Dysphagia*.

Apparent digestibility. Difference between the measured intake of food and the portion recovered in the feces. Expressed as a percentage, apparent digestibility is called *coefficient of digestibility*.

Appetite. Natural desire or craving for food. Loss of appetite (anorexia) accompanies many disorders; certain drugs decrease appetite, e.g., amphetamines, alcohol, and insulin. Some drugs increase appetite, e.g., antihistamines, steroids, and psychotropic drugs.

Applied Nutrition Programs. Practical nutrition programs aimed at strengthening the national nutrition services and group feeding practices of a country; sponsored by the specialized agencies of the United Nations—the World Health Organization, the Food and Agriculture Organization, and the United Nations Children's Fund.

Arabinose. A pentose sugar widely distributed in root vegetables and plants, usually as a component of a complex polysaccharide. It has no known physiologic function in humans, although it is used in studies of bacterial metabolism.

Arachidonic acid. Unsaturated fatty acid containing 20 carbon atoms and four double bonds. It is an important constituent of lecithin and cephalin and occurs in the lipids of the brain, liver, and other organs. This fatty acid is considered one of the essential fatty acids. See *Essential FA* under *Fatty acid*.

Arginine (Arg). Aminoguanidovaleric acid, a dibasic amino acid. Its major metabolic roles include the synthesis of urea and creatine. It is not considered an essential amino acid for humans because the urea cycle provides sufficient arginine for maintenance in adults. However, endogenous synthesis of arginine is insufficient to meet the increased demands in immaturity and severe stress, such as thermal injury, wound healing, sepsis, and trauma. Dietary arginine supplementation may be beneficial under these conditions.

Argininosuccinic acid. Intermediate product in the conversion of citrulline to arginine. It is formed by the condensation of citrulline and aspartic acid in the presence of adenosine triphosphate and magnesium.

Argininosuccinic aciduria (ASA). Inborn metabolic defect due to a lack of the enzyme *argininosuccinase*. The condition is characterized by increased excretion of arginosuccinic acid in the urine, hair abnormalities, intermittent ataxia, seizures, coma, mental retardation, and ammonia intoxication. *Nutrition therapy:* a diet moderately low in protein (1 to 2 g/kg/day depending on individual tolerance) with arginine supplementation (1 g/day for infants and 2 g/day for older children) to prevent arginine deficiency. Sodium benzoate at 0.25 g/kg/day may aid in ammonia excretion.

Ariboflavinosis. Term denoting *riboflavin* deficiency. It is characterized by inflammation of the lips with cracking at the angles (cheilosis), sore mouth with purplish red or magenta tongue, and seborrheic dermatitis around the nasolabial folds. Visual symptoms include photophobia, lacrimation, burning and itching of the eyes, dimness of vision, and a normochromic, normocytic anemia. *Nutrition therapy:* give daily oral riboflavin until clinical symptoms resolve (5 to 25 mg for adults, 3 mg for children, and 1 mg for infants). When malabsorption is present, a prophylactic dose of 3 mg/day is recommended. Supplemental B-complex vitamins is recommended because other vitamin B deficiencies may also be present. Maintain a normal intake of protein and kilocalories. Emphasize good sources of riboflavin, such as milk and dairy products, dark green vegetables, lean meats, and liver.

Arm muscle circumference. See *Midarm muscle circumference*.

Armspan measurement. A practical way of approximating height in nonambulatory and elderly persons. Armspan and stature are nearly equal at maturity and do not change significantly with age. In a standing or recumbent position and the arms extended at right angles to the body with the palms facing upward, measure the full span (in centimeters) between the tips of the middle fingers, excluding the fingernails.

Arsenic (As). A trace element that is better known as a poison, although minute amounts are essential in maintaining the life span of red blood cells. In animals, arsenic deficiency results in stunted growth, rough hair, anemia, and increased erythrocyte fragility. It remains unclear whether arsenic is also an essential element for humans. Various forms of arsenic are readily absorbed from the diet; seafoods, especially shellfish and shrimp, are rich in arsenic. Intake is quite variable and depends on the amount of seafood consumed, the natural arsenic content of water, and inhalation from environmental exposure. Chronic toxicity is characterized by weakness, prostration, muscle aches, gastrointestinal symptoms, peripheral neuropathy, and changes in the pigmentation of fingernails and skin. Arsenic is fatal in large amounts: The estimated fatal acute dose of arsenic trioxide is about 0.76 to 1.95 mg/kg body weight. Although it is naturally present in some foods, the amount usually consumed is too minute to cause toxicity.

Arteriosclerosis. Cardiovascular disease characterized by thickening, loss of elasticity, and hardening (calcification) of the walls of the arteries. It is generally a part of the aging process, although factors other than advancing age are believed to hasten the condition. Among these are high blood pressure, diabetes mellitus, nephrosclerosis, hyperlipidemia, excessive nerve strain, certain infectious diseases, and several other factors not clearly understood. There is no specific treatment except for relief of symptoms. Dietary management depends on the attending disorders and causative factors. See also *Atherosclerosis*.

Arteritis. Also referred to as "cranial or giant cell arteritis," it consists of chronic thickening of the temporal arteries causing constricted blood flow. Slight but persistent temple throbs with swollen, painful arteries are common symptoms, accompanied by fever and anorexia. *Nutrition therapy:* cut down on sodium and fat intake. Maintain nutritional adequacy, increasing or decreasing the calorie level for weight control. In some cases, high carnitine intake has been effective. Monitor protein (serum albumin) and potassium levels when steroids are used.

Arthritis. Acute or chronic inflammation of a joint. It occurs in varying forms according to the severity, location, deformity, and cause. The most common are rheumatoid arthritis (also called arthritis deformans or atrophic arthritis), osteoarthritis (also called degenerative or hypertrophic arthritis), and gouty arthritis. Arthritis is the principal crippler in the United States, occurring mostly after 45 years of age, with a higher incidence in women. The inflammatory process is initiated by the production of prostaglandins which are produced from arachidonic acid, an omega-6 fatty acid. A diet rich in omega-3 fatty acids has been shown to decrease the severity of inflammation. For specific nutrition therapy, see *Gout, Osteoarthritis,* and *Rheumatoid arthritis*. See also omega fatty acid under *Fatty acid*.

Artificial feeding. Introduction of food by a nonnatural method, such as by *tube feeding* or *parenteral feeding*. In infant feeding, it refers to the nourishment of the baby other than by breast feeding. See *Infant feeding*.

Artificial sweeteners. Synthetic sweetening agents used as sugar substitutes. See *Alternative sweeteners*.

As. Chemical symbol for arsenic.

ASA. Abbreviation for *argininosuccinic aciduria*.

Ascites. Accumulation of fluid in the peritoneal cavity due to portal hypertension, low blood protein levels, or sodium retention. The condition is often associated with cirrhosis of the liver, cardiac failure, and renal insufficiency. *Nutrition therapy:* institute rigid sodium restriction of 500 mg/day (22 mEq/day) initially until fluid retention subsides. Gradually increase the sodium level to 1000 mg/day and eventually to 2000 mg/day, depending on the rate of diuresis. Dietary sodium level should be kept as high as possible to maintain palatability and improve the appetite. Give small, frequent feedings. Potassium-losing diuretic may require potassium supplementation. Adjust fluids based on output. Fluid intake is restricted to 1500 ml/day if edema is present. See *Diet, sodium-restricted*.

ASCN. Abbreviation for *American Society for Clinical Nutrition*.

Ascorbic acid. Also known as *vitamin C;* a white crystalline compound closely related to the monosaccharide sugars. It exists in two forms: as L-ascorbic acid (reduced form) and as dehydroascorbic acid (oxidized form). Both forms are biologically active, although dehydroascorbic acid is somewhat less stable. They are readily reversibly reduced and oxidized, but further oxidation of dehydroascorbic acid results in complete loss of activity. See *Vitamin C* and Appendix 16.

Ash. Incombustible *mineral* residue remaining after all the organic matter has been burned or oxidized. See also *Diet, ash*.

Asian-American food practices. See Appendix 47A.

Asparaginase. An antineoplastic agent; used as a component of various chemotherapeutic regimens for treatment of lymphosarcoma and leukemia. Its side effects include impaired pancreatic function and decreased insulin synthesis, hyperglycemia and glucosuria, anorexia, weight loss, and malabsorption syndrome. The brand name is Elspar®.

Asparagine. The beta amide of aspartic acid present in most tissues and occurring abundantly in higher plants. It participates in transamination reactions. The enzyme *asparaginase* has anticancer activity in guinea pigs but produces side effects in humans. It depresses protein synthesis and causes nausea, anorexia, and weight loss.

Aspartame. A low-calorie nutritive sweetener composed of the two naturally occurring amino acids aspartic acid and phenylalanine. Its relative sweetness is 200 times greater than that of sucrose. Its sweetness is decreased by heat and acid. Has 4 kcal/packet and has the sweetening equivalent of 2 teaspoons of sugar. Contraindicated in persons with phenylketonuria (PKU). Sold under the brand names Equal®, Nutrasweet®, Natra-Taste®, and SweetMate®.

Aspartic acid. Aminosuccinic acid; a nonessential glucogenic amino acid involved in transamination reactions and the formation of urea, purines, and pyrimidines. It is hydrolyzed by the enzyme *aspartase* to fumaric acid and ammonia.

Aspirin. Acetylsalicylic acid; an analgesic, antipyretic, and anti-inflammatory drug. It may cause iron-deficiency anemia due to gastrointestinal blood loss, decreased absorption of glucose and amino acids, and depletion of folate, vitamin C, vitamin K, thiamin, and potassium. Supplementation with these nutrients may be necessary when aspirin is taken in large doses or for long-term use. Brand names include Bayer®, Ecotrin®, and Empirin®; aspirin with buffers includes Ascriptin® and Bufferin®; and combination drugs with aspirin include Anacin®, Excedrin®, Percodan®, Talwin®, and Vanquish®.

Association of Official Agricultural Chemists (AOAC). Voluntary organization of chemists that sponsors the development and testing of methods for analyzing nutrients, foods, food additives, animal feeds, liquors, beverages, drugs, cosmetics, pesticides, and many other commodities.

Atabrine. Antimalarial drug; a riboflavin antagonist.

Ataxia. Inability to coordinate bodily or muscular movements. It is generally due to a disorder in the brain or spinal cord, or it may be due to nutritional deficiencies, especially of the B complex vitamins.

Atenolol. A beta-adrenergic agent used in the management of hypertension or chronic angina pectoris. Drug uptake is enhanced by ingestion of food; a decrease in dosage is needed if the patient is hypoalbuminemic to avoid central nervous system side effects; the drug may also increase serum triglycerides. Brand names are Tenormin® and Tenoretic®.

Atherosclerosis. Term denoting a number of different processes resulting in atheroma or patchy deposits of various materials in the intima of the arteries. These deposits are produced by an accumulation of fatty substances (cholesterol, phospholipids, and triglycerides), complex carbohydrates, lipoproteins, calcium and calcified plaques, fibrin, and the formed elements of the blood. Areas of thickening in the intima of affected arteries look like patchy tiny lumps, and lead to narrowing of the lumen and diminution of blood-carrying capacity. Atherosclerosis is the result of an interplay of several factors, including elevated blood lipids, high blood pressure, cigarette smoking, sedentary living, obesity, psychologic tension, and endocrine disorders. Foods influence many of these risk factors. A diet high in saturated fat and cholesterol increases blood cholesterol and lipid levels; habitual overeating coupled with inactivity leads to obesity. *Nutrition therapy:* the recommended dietary modifications for the general public in the prevention of atherosclerotic diseases include adjustment of caloric intake to achieve and maintain healthy weight, reduction of dietary cholesterol intake to about 300 mg/day or less, and control of saturated fat intake to less than 10% of total calories, with calories from fat accounting for no more than 30% of total caloric intake. See *Diet, cholesterol-restricted, fat-controlled;* see also *Hyperlipoproteinemia*.

ATP. Abbreviation for adenosine triphosphate. See under *Adenosine phosphates*.

Atresia. An abnormal condition due to the absence of a body opening or duct, such as the anus, bile duct, external ear canal, etc. In *biliary atresia,* there is total degeneration or incomplete development of one or more bile duct components due to stagnated fetal growth. Clinical signs are jaundice, swollen abdomen, and dark-colored urine. There is hypoproteinemia and malabsorption of fat-soluble vitamins and calcium, resulting in malnutrition and failure to thrive. *Nutrition therapy:* in infants, human milk or infant formula (20 kcal/oz) with a blend of medium-chain triglycerides and long-chain triglycerides. Use a formula low in sodium if there is edema or ascites. (Alimentum®, Pregestimil®). Increase formula concentration to 24 to 27 kcal/oz to promote weight gain. Continuous enteral or parenteral feeding may be required if the individual is unable to take sufficient amounts orally. Supplement with vitamins A, D, E, and K and other vitamins and minerals as needed.

Attain®. Brand name of an isotonic, lactose-free, low-residue liquid formula for oral or tube feeding; made of calcium and sodium caseinates, maltodextrin, and corn oil. See Appendix 39.

Attention deficit disorder. Also called "hyperactivity" and "hyperkinesis." Characterized by restlessness, short attention span, and impulsivity. The exact etiology is not known, although dietary factors such as food additives, food allergies, and sugar have been implicated. *Nutrition therapy:* avoidance of the offending food or substances in food, such as artificial color and flavor, salicylates, and some preservatives such as BHA and BHT. See *Diet, Feingold.*

Atwater respiration calorimeter. Apparatus for measuring the total energy expenditure of the body by confining the subject inside the chamber. The original Atwater apparatus was later modified by Rosa and Benedict. See *Calorimetry*.

Atwater values. Average physiologic fuel values of carbohydrate, protein, and fat based on experiments conducted by Atwater. He found that on a typical American diet, each gram of carbohydrate, fat, and protein yields 4, 9, and 4 calories, respectively. The Atwater values are used extensively in dietary calculations and food analysis. See also *Food, energy value*.

Avidin. A glycoprotein in raw egg white that binds firmly with the vitamin *biotin* and prevents its absorption from the digestive tract. Excessive intake of raw egg whites can result in a biotin deficiency syndrome known as *egg white injury*, which is characterized by loss of hair, scaly dermatitis, and cirrhosis of the liver. Avidin is inactivated by heat and other agents that denature proteins.

Avitaminosis. Literally, it means "without vitamin," but the term is more commonly used in reference to a vitamin lack or deficiency which is more appropriately called hypovitaminosis. The condition may be due to inadequate intake, deficient absorption, increased body requirement or excretion, or ingestion of vitamin antagonists.

Azathioprine. A purine antagonist metabolite used mainly for its immune suppressive activity and as an adjunct for prevention of the rejection of kidney allografts. It is a folacin antagonist and inhibits RNA and DNA synthesis. Large doses may cause macrocytic anemia, anorexia, oral lesions, steatorrhea, diarrhea, nausea, and vomiting. The brand name is Imuran®.

Azidothymidine (AZT). Also called Zidovudine. An antiviral agent used for treating AIDS. May cause anorexia, nausea, vomiting, and megaloblastic anemia. The brand name is Retrovir®.

Azotemia. Also called *uremia*. Retention of urea and other nitrogenous substances in the blood. It is a manifestation of kidney failure. *Prerenal* azotemia is caused by volume depletion due to excessive diuresis, burns, hemorrhage, and gastrointestinal loss, or cardiac and vascular disorders such as congestive heart failure and acute myocardial infarction.

B

B. Chemical symbol for boron.

Baby bottle tooth decay (BBTD). Dental caries of the maxillary incisors and molars. The main causes are prolonged, inappropriate bottle feeding beyond the normal time babies are weaned from the milk bottle and putting the baby to sleep with a bottle containing sweetened beverage. The national incidence of BBTD has been reported as ranging from 4% to 20%. It occurs more frequently in the American Indian and Native Alaskan communities. Nutrition education of parents is crucial in preventing this disorder.

Bacterial translocation. The passage or migration of indigenous bacteria in the gastrointestinal tract across the mucosal barrier to invade distant organs. Translocation can spread systematically and lead to septicemia. The three basic mechanisms for translocation are intestinal bacterial overgrowth, impairment of host immune defenses, and physical disruption of the gut mucosal barrier. Nutrition depletion may predispose to bacterial translocation in several ways: malnutrition is associated with impaired immune response and a loss of gut mass and function; starvation and protein depletion disrupt the ecology of the gut microflora by allowing intestinal overgrowth or colonization with pathogens; and the composition of the diet alters intestinal morphology and function. The addition of insoluble dietary fiber in formulas for long-term tube feeding has been shown to reduce diet-induced bacterial translocation by providing short-chain fatty acids necessary to maintain gut microflora. Groups at high risk for translocation include immunocompromised and malnourished patients, burn and trauma victims, and critically ill patients with multisystem organ failure.

Balance study. Quantitative method of measuring the amount of a nutrient ingested and the amount of the same nutrient or its metabolic end product(s) excreted in order to determine whether there has been a gain (positive balance) or loss (negative balance) in the body. At equilibrium, nutrient intake equals output. Balance studies are generally classified into two types: balance of matter (those dealing with nutrients that can be weighed) and balance of energy (those dealing with heat and energy). See *Energy balance, Nitrogen balance,* and *Water balance.*

Bariatrics. The study of *weight control.* Bariatric surgery is the surgical treatment of obesity. Surgical procedures vary and may induce partial malabsorption of ingested nutrients or limit the anatomical ability to overeat. See *Gastric bypass.* See also *Nutrition, weight control.*

Baryophobia. An eating disorder resulting in the reduced growth rate of a child caused by underfeeding because the parents are afraid he or she may develop obesity and/or cardiovascular diseases later in life. Nutrition education of the parents is needed.

Basal energy expenditure (BEE). The amount of energy used in 24 hours by a person at complete rest, lying down in a comfortable environment and temperature, and 12 hours after the last meal. BEE correlates directly with measures of lean body mass and is the largest component of *total energy expenditure*. There are many formulas for estimating BEE but the most common and widely used is the Harris–Benedict equation. See *basal metabolic rate* and *resting energy expenditure*. See also *Harris–Benedict formula*.

Basal metabolic rate (BMR). The amount of energy expended per unit of time under basal conditions. The adult basal metabolic rate is approximately 1 kcal/kg body weight per hour. The rate is affected by the size, shape, and weight of the individual; body composition (amount of active protoplasmic tissue); age (highest during infancy, with a gradual decline with advancing age); activity of the endocrine glands; state of nutrition; rate of growth; and pregnancy. Clinically, the BMR is reported as a percentage above or below normal. See also *Resting metabolic rate (RMR)*.

Basal metabolism. Energy expended in the maintenance of "basal metabolic" processes, or involuntary activities in the body (respiration, circulation, gastrointestinal contractions, and maintenance of muscle tonus and body temperature) and the functional activities of various organs (kidneys, liver, endocrine glands, etc.). It is taken under "basal" conditions, i.e., at complete physical and mental rest, in the postabsorptive state (12 to 16 hours after taking food), and at a temperature within the zone of thermal neutrality (20° to 25°C). Measurement is performed when the person is awake but lying down and relaxed in a quiet atmosphere.

Basal metabolism determination. The amount of heat produced by the body may be measured in two ways: directly, by measuring the amount of heat given off with the use of an apparatus called a *calorimeter,* or indirectly, by measuring the amount of oxygen consumed over a given period of time with the use of a *respirometer*. Basal metabolism may also be determined by using various prediction formulas developed by Boothby, DuBois, Berkson, and Dunn (based on *body surface area*); by Harris and Benedict (based on body weight and standing height); and by Kleiber (based on *metabolic body size*). Another indirect clinical method is to analyze T_3 and T_4 blood levels.

Basic food groups. Classes of foods listed together under one heading because of their similarities as good sources of certain nutrients. The number of groupings and the foods in the groups may vary in different nations, depending on the food habits, food economics, and dietary needs of a country. The basic food groups are used in planning and evaluating diets for nutritional adequacy. See Appendix 4.

Basic-forming foods. See *Alkaline-forming foods*.

BCAA. Abbreviation for *branched-chain amino acid*.

BEE. Abbreviation for *basal energy expenditure*.

Behavioral Risk Factor Surveillance System (BRFSS). Coordinated by the Centers for Disease Control (CDC) and carried out in 35 states and the District of Columbia from 1983 to 1993. Data collected reflected the changing life styles of the American population and revealed health problems related to undernutrition and nutritional excesses such as obesity, hypertension, and diabetes.

Benedict–Roth spirometer. Closed-circuit apparatus for measuring oxygen consumption over a period of time to determine basal metabolism. The energy equivalent is calculated by multiplying the volume of oxygen consumed by 4.825, the caloric equivalent of 1 liter of oxygen. See also *Indirect calorimetry* under *Calorimetry*.

Benztropine. An antiparkinsonian agent; has an anticholinergic effect and may cause constipation and dry mouth. The brand name is Cogentin®.

Beriberi. Nutritional deficiency disease due to lack of thiamin, or vitamin B_1. It is characterized by loss of appetite and general malaise associated with heaviness and weakness of the legs, which may be followed by cramping of the calf muscles and burning and numbness of the feet. There may also be some edema of the legs, heart palpitations, and precordial pains. There are three forms of beriberi: dry beriberi, a form in which polyneuropathy and progressive paralysis are the essential features; wet beriberi, a form characterized by pitting edema, enlarged heart, rapid pulse, and circulatory failure (beriberi heart disease); and infantile beriberi, seen in infants breast-fed by mothers suffering from beriberi. The onset is often acute and is characterized by pallor, facial edema, irritability, and abdominal pain. The infant may suddenly become cyanotic, with dyspnea and tachycardia, and die within a few hours. Recovery with thiamin therapy is dramatic. *Nutrition therapy:* A daily thiamin dose of 50 to 100 mg/day is the usual range, although a B-complex concentrate is frequently prescribed because most patients suffer from multiple deficiencies. Neurologic symptoms usually resolve within 24 hours, cardiac symptoms in 24 to 48 hours, and edema in 48 to 72 hours. Motor weakness may require 1 to 3 months before it is completely resolved. Adults with mild beriberi respond to 10 mg orally three times a day. In fulminant heart disease, 100 mg of thiamin hydrochloride is necessary. Good food sources of thiamin are lean pork, legumes, wheat germ, whole grain, and enriched breads and cereals. See *Thiamin*.

Beta-carotene. A precursor of vitamin A; yields two molecules of vitamin A per molecule. See *Carotene*.

Bezoar. Hard ball of vegetable or hair fiber that may develop in the intestines or stomach. Decreased gastric acidity and loss of normal pyloric function can cause the formation of bezoar. Common foods leading to bezoar formation are apples, berries, coconuts, figs, oranges, persimmons, brussels sprouts, green beans, potato peels, and sauerkraut.

BIA. Abbreviation for *bioelectrical impedance analysis*.

Bicarbonate. Salt of carbonic acid, characterized by the radical—HCO_3. Blood bicarbonate is the chief *alkali reserve* of the body. It plays a key role in the maintenance of a constant hydrogen ion concentration in body fluids.

Bifidus factor. Collective term for growth factors needed by *Lactobacillus bifidus* var. *pennsylvanicus,* found in human milk and growing in the intestines of breast-fed infants. It is believed to be beneficial to young infants in preventing the growth of less desirable bacteria that cause intestinal putrefaction. See *Breast feeding*.

Bile. Fluid produced and secreted by the liver, stored and concentrated in the gallbladder, and poured into the duodenum at intervals, particularly during fat digestion. It aids in the emulsification and absorption of fat, activates the pancreatic lipase, and prevents putrefaction. Among its constituents are *bile acids, bile salts, bile pigments, cholesterol,* and *lecithin*.

Bile acids. Glycocholic and taurocholic acids formed by the conjugation of glycine or taurine with cholic acid and produced during the metabolism of cholesterol. Bile acids are responsible for the emulsification of fatty acids into micelles for absorption.

Binge eating. Previously called "compulsive eating." Recurrent episodes of eating an amount of food that is larger than most people would eat in a similar period of time under similar circumstances, and a sense of lack of control over eating during the episode. The binge eating episodes are associated with at least three of the following: eating much more rapidly than normal; eating until feeling uncomfortably full; eating large amounts of food when not feeling physically hungry; eating alone because of being embarrassed by how much one is eating; and feeling disgusted with oneself, depressed, or feeling guilty after overeating. The binge eating occurs, on the average, 2 days a week for 6 months. It is not associated with the regular use of inappropriate compensatory behaviors (such as purging, fasting, and excessive exercise).

Bioassay. Also called biologic assay; measurement of the activity of a drug, substance, or nutrient by noting its effect on test animals or microorganisms.

Bioavailability. Descriptive term for the extent of digestion and absorption of a nutrient; therefore the amount actually available for cell utilization.

Biocytin. Biotinyl lysine; a naturally occurring complex of biotin and lysine. It is resistant to hydrolysis by proteolytic enzymes in the intestinal tract, as is the complex of biotin and avidin. An enzyme in the plasma and erythrocytes, called biocytinase, catalyzes the hydrolysis of biocytin to yield free biotin.

Bioelectrical impedance analysis (BIA). Total body fat measurement using low-energy electric current. Higher impedance to the current indicates more fat tissues stored because fat has resistance to electric flow. See also *TOBEC*.

Bioflavonoids. Group of naturally occurring substances belonging to the flavin and flavonoid groups of compounds; originally designated as vitamin P (permeability vitamin) or vitamin C_2 (synergist of vitamin C). Some of the flavonoids exhibit biologic activities, including reduction of capillary fragility and protection of biologically important compounds through antioxidant activity, but none has been shown to be essential for humans or to cause deficiency syndromes when they are lacking. Beneficial effects from therapeutic doses include prevention of neonatal deaths due to erythroblastosis, reduced bleeding associated with capillary fragility, and lowering of cholesterol and high blood pressure. Hence bioflavonoids are now considered to be pharmacologic rather than nutritional agents.

Biologic oxidation. Also called physiologic oxidation; the cellular reactions liberating energy by the transfer of electrons via the *redox systems*.

Biologic value (BV). Relative nutritional value of individual proteins compared to that of a standard protein. It is the amount of protein digested and absorbed from food, i.e., utilized by the body and not excreted in the urine. BV is therefore a measure of how efficiently food proteins are retained to become part of body tissues. BV is calculated as follows:

$$BV = \frac{\text{N retained}}{\text{N absorbed}} = \frac{\text{dietary N} - (\text{UN} + \text{FN})}{\text{dietary N} - \text{FN}}$$

where

N = nitrogen
FN = fecal nitrogen
UN = urinary nitrogen

Biotin. Sulfur-containing, water-soluble vitamin; one of the B complex vitamins.

Biotin is an integral part of enzymes that transport carboxyl units and fix carbon dioxide in animal tissues. Biotin enzymes are involved in gluconeogenesis, fatty acid synthesis, and catabolism of branched-chain amino acids. Biotin is widely distributed in low concentration in both animal and plant foods; the best sources are liver, egg yolk, whole-grain cereals, soy flour, and yeast. It is synthesized in the lower gastrointestinal tract by microorganisms; hence, treatment with large doses of antibiotics, notably streptomycin, can decrease biotin levels. Dietary biotin deficiency is rare in humans, with the exception of seborrheic dermatitis seen in infants under 6 months of age and during hemodialysis and long-term total parenteral nutrition without added biotin and following extensive bowel resection. Deficiency has also been reported after long-term anticonvulsant therapy. Ingestion of large amounts of raw egg white can induce biotin deficiency, producing a syndrome characterized by scaly dermatitis, muscle pains, alopecia, glossitis, mental depression, and general malaise. There is one rare disorder with biotin dependency in which an abnormality in the carboxylation of propionic acid leads to a ketotic hyperglycemia. There are no reports of biotin toxicity with excessive intake. See Appendix 2 for the estimated safe and adequate daily dietary intake of biotin.

Biotinidase. An enzyme that releases protein-bound biotin and cleaves *biocytin* so that biotin can be recycled. Infants with biotinidase deficiency show features of biotin deficiency, including vomiting, hypotonia, metabolic acidosis, and neurologic changes; the condition responds to treatment with oral D-biotin.

Bisacodyl. A stimulant laxative. Abuse of this drug can cause fluid and electrolyte loss (especially potassium) and malabsorption with weight loss. Undesirable side effects are abdominal cramps, nausea, and belching. Oral bisacodyl should not be taken with milk because milk dissolves the enteric coating, causing gastric irritation. Brand names are Bisco-Lax®, Bisac-Evac®, and Dulcolax®.

Bitot's spots. Small, triangular, grayish or glistening white plaques, sometimes with a foamy surface, on the conjunctiva. The spots are generally bilateral on the temporal sides of the cornea and are associated with vitamin A deficiency. Bitot's spots may be treated with therapeutic vitamin A (about 20,000 to 50,000 IU/day).

Bleomycin. An antineoplastic antibiotic active against gram-positive and gram-negative bacteria and fungi; used in the treatment of lymphomas and squamous cell carcinomas. Its side effects include nausea, vomiting, oral lesions, anorexia, and weight loss which may persist after therapy is stopped. The brand name is Blenoxane®.

Blind loop syndrome. Bacterial overgrowth in a redundant segment of the intestine caused by an obstructive disease, fistula formation, radiation, or surgery. Injury to the brush border by the toxic effects of bacterial catabolism may cause enzyme loss and carbohydrate malabsorption. There may also be poor fat absorption and steatorrhea. Overgrowth of bacteria may decrease available vitamin B_{12}. *Nutrition therapy:* a lactose-restricted diet and medium-chain triglycerides may be beneficial. Monitor vitamin B_{12} status and supplement if necessary.

Blood. Fluid medium that carries oxygen and nutritive materials to the tissues, removes carbon dioxide and waste products for elimination by the excretory organs, and distributes other substances (such as clotting factors, regulatory agents, and body defense mechanisms) throughout the body for utilization or action. It consists of *formed elements* (erythrocytes, or red blood cells; leukocytes, or white blood cells; and thrombocytes, or blood platelets) and a pale yellow portion, *plasma,* which contains a large number of organic and inorganic substances in solution. A normal adult has a total blood volume of about 8% of body weight. For normal constituents of the blood, see Appendix 30.

Blood glucose (BG). See *Blood sugar level.*

Blood lipids. Principally cholesterol, phospholipid, and triglyceride. These lipids circulate in the plasma bound to proteins. As lipoprotein complexes, the otherwise insoluble lipids are solubilized, thus enabling their transport into and out of the plasma. Five major types of lipoproteins in the blood have been identified. Each type contains phospholipid, triglyceride, cholesterol, and protein in varying proportions. For details, see *Lipoprotein.* See also *Hyperlipoproteinemia.*

Blood platelet. Also called thrombocyte; one of the three formed elements of the blood that is necessary in the clotting of blood. Blood platelets liberate small amounts of thromboplastin, which activates the proenzyme prothrombin to its active form, thrombin, the enzyme that catalyzes clot formation. See *Blood clotting.*

Blood pressure (BP). The force that blood exerts against the walls of the blood vessels during each heartbeat. As measured by a sphygmomanometer, it is the pressure of the blood on the walls of the arteries of the upper arm. The human heart pumps intermittently by means of a sudden contraction of the entire ventricular musculature, followed by a period of relaxation. The pressure during the contraction phase is called systolic pressure and the pressure during the resting phase is called diastolic pressure. In recording blood pressure, the systolic pressure is written first, followed by the diastolic pressure. A normal blood pressure reading for adults is 120/80 mm Hg (systolic over diastolic blood pressure). The values are lower for children (110/65) and infants (90/60). Blood pressure is maintained by homeostatic mechanisms in the body, mainly by the renin–angiotensin–aldosterone system, the kinin–prostaglandin system, and the neuroendocrine adrenergic system. It is affected by blood volume, the lumen of the arteries and arterioles, the force of heart pumping, and any illness or disorder that alters these factors. See also *Hypertension.*

Blood sugar level (BSL). The normal fasting level of sugar (glucose) is 70 to 110 mg/dL (3.9 to 6.1 mmol/L) for adults and 80 to 115 mg/dL (4.4 to 6.4 mmol/L) for those over 60 years; it is slightly lower at 60 to 90 mg/dL (3.3 to 5.0 mmol/L) during pregnancy. Postprandial levels are 90 to 150 mg/dL (5.0 to 8.3 mmol/L) 1 hour after meals and 80 to 140 mg/dL (4.4 to 7.8 mmol/L) 2 hours after meals. Among several factors that *maintain* the blood sugar level are glycogen–glucose interconversion in the liver, conversion of carbohydrate to fat, formation of muscle glycogen and its utilization, and glucose excretion in the urine (renal threshold). Conditions that *increase* the blood sugar level include diabetes mellitus, hyperfunction of the anterior pituitary gland, hyperfunction of the adrenal cortex, insufficient insulin production, hyperfunction of the thyroid gland, head injury, multiple trauma, sepsis, and hypermetabolic states. Conditions that *decrease* the blood sugar level include hyperinsulinism, anterior pituitary deficiency, adrenal insufficiency, hypothyroidism, prolonged undernutrition, tumor of the pancreas, and abnormal kidney function (renal glycosuria). Clinically, the plasma sugar level would be expected to increase moderately (in individuals without a history of diabetes) following general surgery or multiple trauma (BS of 150 ± 25 mg/dL). Sepsis and hypermetabolic states are characterized by markedly elevated plasma glucose (about 250 ± 50 mg/dL). See also *Glucose tolerance test, Hyperglycemia,* and *Hypoglycemia.*

Blood urea nitrogen (BUN). The amount of nitrogenous substance present in the blood as urea; measured to indicate kidney function. It is increased in kidney failure, diabetes mellitus, shock, gastrointestinal bleeding, dehydration, and increased protein catabolism. It is decreased during pregnancy, malnutrition, overhydration, and severe liver damage. Many drugs may cause increased or decreased BUN levels. The normal value of BUN is 10 to 20 mg/dL (3.7 to 7.4 mmol/L). Values may be slightly higher in the elderly and slightly lower in children. See also *Azotemia*.See Appendix 30.

BMI. Abbreviation for *body mass index.*

BMR. Abbreviation for *basal metabolic rate.*

Body composition. The representative percentage composition of a human adult is about 16% protein, 15 to 20% fat, 0.5% carbohydrate, 4.5% ash, and 60% water. These body components are distributed through four separate compartments: lean body mass, extracellular fluid, mineral of the skeleton, and adipose tissue. Body composition varies with age, sex, and nutriture. Females generally have a higher fat content than males; infants and young children have a relatively higher water content than adults, which decreases as they grow older. The degree of leanness and fatness (e.g., mild, moderate, and morbid obesity) varies widely among individuals. Also, pathologic conditions alter body composition, as in osteopenia, marasmus, edema, and cancer.

Body composition determination. The living body can be partitioned into essentially two compartments: the fat-free portion and the fat portion, or adipose tissue. The fat-free weight of an individual remains relatively constant; variability in total weight is attributed to varying degrees of fatness. Measurements of body composition include *physical methods:* underwater weighing for body density, whole body counter, total body electric conductivity (TOBEC), computed tomography, magnetic resonance imaging, and bioelectrical impedance. *Anthropometric methods* of measuring body composition are more economical and speedier. These include skinfold thickness, midarm muscle circumference, height and weight measurements, ponderal index, and body mass index.

Body mass index (BMI). Or the Quetelet index. Ratio of weight (in kilograms) over height squared (in meters). For those who do not wish to make the conversion to metric equivalents, calculate the BMI as shown in the equation below:

$$\text{BMI} = \frac{\text{Weight in pounds} \times 704}{\text{Height in inches} \times 2}$$

Multiply the weight in pounds by 704; divide the result by the height in inches, and divide that result by the height in inches again. The BMI increases with age. For adults, normal BMI values are between 19 and 25 for ages 19 to 34, 21 to 27 for ages 35 to 65 years, and 24 to 29 for age 65+ years. The body mass index is a good estimate of the degree of obesity or amount of total body fat. BMI values of 27 to 30 indicate overweight; values of 30 to 40, moderate obesity; and values above 40, gross or morbid obesity. BMI values below 18 are life-threatening.

Body surface area (BSA). Area covered by the exterior of the body. The surface area of the body was first determined by wrapping the body with a gauze tape, removing the tape, and measuring the area covered. BSA is estimated by plotting a person's height and weight on a standard chart originally developed by DuBois. An improvement was made by Boothby and Sandiford, who devised a nomogram for determining BSA.

Bomb calorimeter. Apparatus that measures directly the energy value of foods. It consists of an inner chamber that holds the food sample and a double-walled, insulated jacket that holds a can containing water. An electric connection ignites the weighed sample of food. A differential thermometer records the rise in the temperature of the water surrounding the chamber.

Bone. Also called osseous tissue; a mineralized connective tissue consisting of an organic matrix in which inorganic elements (mineral salts) are precipitated in a crystal lattice structure similar to that of the naturally occurring mineral *hydroxyapatite*. The organic matrix consists largely of *collagen* in a gel of cementing substance. The mineral fraction is composed largely of calcium phosphate, carbonate, fluoride, and citrate. Specialized bone cells (osteoblasts, osteoclasts, and osteocytes) control the relationship between the organic matrix and the bone salts. See also *Nutrition, bone* and *Peak bone mass*.

Bone fracture. A break in the continuity of the bone tissue. May be simple, compounded, or comminuted, depending on what bone is involved and the nature of the injury. *Nutrition therapy:* the location of the fracture and the surrounding tissues affected will determine how the diet is modified. Generally, provide an adequate diet supplemented with vitamins and minerals for bone repair. A liberal protein intake will promote healing; give foods as tolerated.

Bone marrow transplantation. Treatment procedure for certain hematologic malignancies, such as acute leukemia, some forms of lymphoma, and solid tumors such as breast cancer. The marrow transplantation procedure presents severe nutritional consequences. Prior to transplantation, high doses of chemotherapy to prevent possible graft rejection produce acute nausea, vomiting, and diarrhea. Delayed effects during the first 2 months after the transplant include varying degrees of mucositis, xerostomia, dysgeusia, stomatitis, taste and salivary changes, anorexia, esophagitis, diarrhea, and steatorrhea. *Nutrition therapy: Pretransplant:* kilocalories at basal energy expenditure (BEE) × 1.1 to 1.2 and protein at 1.0 g/kg body weight per day or higher if nutritionally compromised. Nutritional support by way of enteral or parenteral feeding may be necessary if oral intake is poor for an extended period. Begin TPN after the final dose of chemotherapy before the marrow transplant. *Post-transplant for 2 months:* kilocalories at BEE × 1.2 and protein at 1.2 to 1.5 g/kg/day. Changes in taste acuity persist and food intake may remain low for 30 to 50 days post-transplant. Bland liquids and soft foods with sauces and gravies are best tolerated and provide relief of dry mouth. *Long-term post-transplant:* kilocalories at BEE × 1.2 to 1.5 and protein at 1 g/kg/day. A low-microbial diet prior to and post-transplant is recommended until absolute neutrophil count reaches 500. May use washed fresh fruits and vegetables when the neutrophil count becomes adequate. It is important to follow food safety measures. See *Diet, low-microbial*.

Bone mass. Bone undergoes a continual process of formation and resorption.

Bone mass accumulates during the growth periods and into young adulthood, when bone formation exceeds bone resorption. The long bones stop growing by the age of 20 years, but the cortical bone continues to be formed. Peak bone mass is reached around the age of 25 to 35 years. Around the age of 40, bone mass begins to gradually diminish at the rate of 1.2% per year. There is an accelerated bone loss of 2% to 3% per year in women after the menopause, which continues for 5 to 10 years. Bone loss in men occurs 10 years later than in women. Individuals with a higher *peak bone mass* take longer to reach the critically reduced mass at which bone breaks easily. Estrogen replacement combined with adequate calcium intake (1.0 g/d) is effective in preventing or slowing the rate of bone loss. About 300 mg calcium/day excreted from the bone loss must be replaced.

Boost®. Brand name of a nutritional oral supplement available in vanilla, chocolate, strawberry, and mocha flavors. Lactose-free and low in sodium. See Appendix 39.

Boron (B). A mineral found in most tissues, especially the brain, liver, and body fat. It is generally accepted as essential for plants. There is also accumulating evidence that boron may be an essential nutrient for animals and humans. Boron complexes with many biologic substances, including sugars and polysaccharides, pyridoxine, riboflavin, dehydroascorbic acid, and adenosine-5-phosphate. Boron may influence parathormone action; metabolism of calcium, phosphorus, and magnesium; and the formation of the active form of cholecalciferol. Boron seems to be beneficial in the prevention of calcium loss and bone demineralization in nutritional disorders characterized by secondary hyperparathyroidism and osteoporosis. The daily intake of boron by humans varies widely. Foods of plant origin, especially fruits, vegetables, and nuts, are rich sources; wine, cider, and beer are also high in boron. Meat and fish are poor sources. Studies on the signs of boron deficiency are not complete; the most consistent manifestation is depressed growth. Boron is toxic in large amounts; signs of toxicity include nausea, vomiting, diarrhea, dermatitis, and lethargy. High boron intake also induces riboflavinuria. Boric acid was formerly used as a food preservative but has been declared unsafe as a food additive by a Food and Agriculture Organization/World Health Organization Expert Committee.

Bowel resection. Complete or partial removal of a section of the intestines. The absorption of nutrients occurs throughout the small bowel, so that there is little disturbance of bowel function when a short length of the bowel is resected. If intestine is resected in lengths that compromise absorption, adaptation of the remaining bowel increases function and restores the ability to absorb normally. However, when over 50% of the bowel has been resected, impairment of absorption tends to be permanent and more pronounced. Ileal resection of less than 100 cm, with the colon intact, usually causes steatorrhea. The degree of malabsorption increases with increasing length of resection. The absorption of fat and carbohydrate can be reduced by 50% to 75% of the intake: nitrogen absorption is reduced to a lesser extent. A vitamin B_{12} study should be done and, if abnormal, monthly intravenous administration of 100 μg of the vitamin should be given to prevent pernicious anemia. *Nutrition therapy:* depends on the extent of bowel resected. A normal oral diet should be tried if more than 60 to 80 cm of the bowel still remains. Start by giving small feedings of dry solids, with isotonic fluids 1 hour after the meal. The separation of solids from liquids is important because of the increased speed of gastric emptying, resulting in diarrhea. Electrolyte replacement may be required parenterally in order to meet requirements even when oral intake is sufficient. Depending on the patient's tolerance and on the degree of diarrhea, gradually increase the volume of feeding to reach the goal of 1.5 to 2 times the basal energy expenditure for calories and 1.2 to 1.3 g protein/kg body weight. Total parenteral nutrition may be necessary in resections leaving less than 60 cm of the small bowel and those leaving only the duodenum intact.

Brain injury. Common feeding difficulties encountered in injury of the brain are reduced and ineffective sucking; tongue dysfunction, which affects chewing and swallowing; and a hyperactive or hypoactive gag reflex. A hyperactive gag reflex may lead to vomiting during feeding, and a hypoactive gag reflex may lead to choking. Often a patient may be able to swallow solid foods but will aspirate liquids. *Nutrition therapy:* careful assessment of the patient's chewing and swallowing ability, using a team approach; modification of food consistency; and small, frequent feeding of meals consisting of foods with contrasting colors and flavors to encourage consumption. See *Dysphagia*. See also *Traumatic brain injury*.

BranchAmin®. Brand name of a 4% solution of branched-chain amino acids (isoleucine, leucine, and valine) for intravenous administration; does not contain electrolytes and is not intended to be used alone, but rather as an admixture with other amino acid solutions. See Appendix 40.

Branched-chain amino acid (BCAA). The three BCAAs are leucine, isoleucine, and valine. The metabolism of these amino acids occurs mainly in the periphery, particularly skeletal muscle, rather than in the liver. Thus, BCAA-enriched solutions may be beneficial in the treatment of hepatic failure with hepatic encephalopathy and in chronic renal failure. However, its use in improving nitrogen balance in injured or septic patients has not been clearly established.

Breath hydrogen test. Measurement of the amount of hydrogen in the expired air which is a useful indicator of lactose maldigestion or malabsorption. It is performed after an overnight fast. Dietary sources of breath hydrogen are omitted in the meal preceding the fast. Breath samples are collected and measured for hydrogen after an oral dose of lactose. See *Diet, breath hydrogen test*.

BRFSS. Abbreviation for *Behavioral Risk Factor Surveillance System*.

Bronchial asthma. An obstructive respiratory disorder characterized by recurring attacks of paroxysmal dyspnea, wheezing, and irritable cough with prolonged expiration. It is more common among young children, 3 to 8 years old, than in older individuals. Severe cases need immediate medical attention. Those with chronic asthma receive daily dosage of prescribed medicine and follow a regimen of frequent rest and exercise. *Nutrition therapy:* serve meals consisting of nutrient-dense foods in small, frequent feedings with rest periods. Monitor fluid intake. Adequate fluids help liquify secretions; asthmatics are also at risk for dehydration. Observe for nutritional side effects of drugs, especially of corticosteroids and bronchodilators. Adjust calories, protein, and other nutrients as necessary.

Bronchitis. Acute or chronic inflammation of the membrane lining the bronchial tubes. Acute bronchitis may be due to an extension of infection from the upper respiratory tract. Chronic bronchitis may be caused by irritants in polluted air, particularly smoke or gas fumes. *Nutrition therapy:* provide adequate to liberal calorie and protein intakes to help reduce infection. Avoid milk if it tends to produce mucus. Give small feedings and allow frequent rest periods while eating when there is difficulty in breathing. See also *Chronic obstructive pulmonary disease.*

Bronchopulmonary dysplasia (BPD). Chronic lung disease of infancy which occurs most frequently in premature infants following *respiratory distress syndrome.* Infants with BPD have special nutritional needs. BPD affects growth, energy needs, nutrient metabolism, and the development of feeding skills; infants tire easily and feeding difficulties occur frequently. *Nutrition therapy:* increase basal energy expenditure (BEE) by 25% to 50% or higher (120 to 130 kcal/kg/d). Protein intake should be within RDA standards. Fluid and sodium restriction may be indicated. Monitor electrolytes and supplement with chloride and potassium, if needed. Optimize calcium and phosphorus intake. Meeting high energy needs is difficult because of anorexia, fluid restriction, fatigue, poor coordination, and other feeding problems. Give small frequent feedings and use calorie-dense formulas (>20 to 24 kcal/oz). Feed in a prone position and use thickened formula (½ to 1 tbsp infant cereal/oz formula) if there is gastroesophageal reflux. Introduce solids and food textures gradually.

Brown Bag Program. Administered by the U.S. Department of Agriculture, this program distributes free surplus foods to low-income senior citizens. Each participant receives two grocery bags of about 18 lbs of food, such as breads, dairy products, canned goods, and frozen foods.

BSA. Abbreviation for *body surface area.*

BSL. Abbreviation for *blood sugar level.*

Bulimarexia. An eating disorder in which the individual switches between *anorexia nervosa* and *bulimia.*

Bulimia nervosa. An eating disorder also called the "binge–purge syndrome." The person has an insatiable appetite but is afraid to gain weight; binge eating is followed by purging caused by induced vomiting, diuretics, or laxatives. A typical bulimic pattern of eating involves several secretive binges a day for a few days when the individual consumes several thousand kilocalories in an hour or two. The binging ends when abdominal discomfort or sleepiness occurs, and purging is used. Induced vomiting causes complications, such as erosion and decay of teeth, esophagitis, metabolic alkalosis, dehydration, hypokalemia, and other electrolyte imbalances. Protein-energy malnutrition (PEM) in some cases becomes a serious problem that needs immediate nutritional support and hospitalization. *Nutrition therapy:* correct PEM and fluid and electrolyte imbalances. Use foods of high nutrient density given in small, frequent feedings. Identify factors that trigger the bulimic behavior and the kinds of foods usually eaten during an eating binge. Set a weight goal the patient can accept and establish a meal plan that will result in weight stabilization. Emphasis is not on gaining weight but on developing acceptable eating patterns.

Bulk. The indigestible portion of carbohydrates that is not hydrolyzed by enzymes of the human gastrointestinal tract. See *Fiber.*

Bulking agent. A metabolically inert substance, such as nonfibrous cellulose, added to food to increase its volume without any calorie contribution.

Bumetanide. A sulfonamide loop diuretic used for the management of edema associated with congestive heart failure or hepatic or renal disease. It is a potent diuretic and can produce profound diuresis with loss of sodium, potassium, magnesium, and calcium; it may also cause nausea, vomiting, and anorexia. The brand name is Bumex®.

BUN. Abbreviation for *blood urea nitrogen*. See *Urea* and Appendix 30.

BUN/Creatinine ratio. Useful in evaluating hydration status. The normal ratio is about 10:1. A ratio less than 10:1 may indicate hypovolemia or increased protein metabolism; a ratio greater than 10:1 may indicate hypervolemia, low protein intake, or hepatic insufficiency.

Burn. Tissue injury or destruction caused by excessive heat, caustics (acids or alkalis), friction, electricity, or radiation. On the basis of the extent of injury, burns are divided into three degrees: *first-degree,* with simple redness of the affected parts; *second-degree,* with the appearance of blisters in addition to redness; and *third-degree,* with actual destruction of the skin and underlying tissues. A person with severe burns is in a hypercatabolic state, with an extensive nitrogen deficit, fluid and electrolyte imbalances, and rapid weight loss, unless given nutritional support immediately. It is estimated that a square meter area of a third-degree burn wound allows an evaporation of as much as 4 liters of water and a loss of 580 kcal in 24 hours. One gram of nitrogen is equal to a loss of 30 g of lean body mass, and a weight loss of 40% could lead to mortality. Infection or sepsis is a major cause of death. *Nutrition therapy:* immediately correct the massive losses in fluids, electrolytes, protein, nitrogen, vitamins, and minerals in order to avoid shock and promote healing and restoration of tissues. In minor burns, nutrition support can usually begin in the first 24 hours. In moderate to major burns, feeding usually begins 1 to 2 days after the burn occurs; the first 24 hours are used to restore fluids. Burns, unlike other disorders, result in a highly accelerated rate of hypercatabolism. The basal metabolic rate (BMR) is often twice the normal BMR and protein requirement is about 3 times the RDA. Provide at least 50 kcal/kg body weight per day, with carbohydrate as the main source. Supplement with vitamins and minerals: vitamin C at 3 times the RDA; vitamins A, D, the B-complex, zinc, and iron at 2 times the RDA; and vitamins K and B_{12} once a week. Provide small frequent meals and nutrient-dense snacks. Supplement with tube feeding at night if intake is consistently below 75% of protein and calorie requirements. TPN is indicated for patients with persistent or recurrent paralytic ileus or intractable diarrhea, and patients who are not able to receive enteral nutrition for 3 days or longer. Other complications that need nutrition intervention are infections or sepsis with its accompanying hyperpyrexia, stress hyperglycemia, respiratory problems, and renal shutdown. The goals of the diet are to prevent protein-calorie malnutrition, restore electrolyte and water balances to avoid shock, promote wound healing, and correct stress hyperglycemia. The BMR of a burned patient is often twice his or her normal BMR. The protein requirement is usually three times his or her normal recommended dietary allowance (RDA). The caloric supply should be at least 50 kcal/kg body weight per day. Consider selecting commercial preparations of special formulas to meet the patient's increased nutrient needs. Carefully calculate and monitor the daily diet intake. Burns, unlike other disorders, result in a highly accelerated rate of hypercatabolism. Most body stores are used up; therefore, immediately correct massive losses in fluids, protein, nitrogen, vitamins, and minerals in order to hasten the processes of healing and restoration. For details in calculating nutrient needs, see Appendix 24.

Burning feet syndrome. Set of symptoms consisting of listlessness, fatigue, postural hypotension, rapid heart rate on exertion, and numbness and tingling of the hands and feet. As the condition becomes worse, the throbbing sensation intensifies, with sharp burning and stabbing pains. It is due to a deficiency of *pantothenic acid,* a water-soluble vitamin, and has been induced by administering methylpantothenic acid as an antagonist and by feeding a semisynthetic diet virtually free of pantothenic acid. Burning feet syndrome has been reported by prisoners of World War II and by chronic alcoholics whose diets were deficient in protein and the B complex vitamins.

Busulfan. An alkylating agent used in the treatment of leukemia and other forms of cancer. It interferes with DNA replication or RNA transcription and ultimately results in the disruption of nucleic acid formation. Busulfan may cause nausea, vomiting, diarrhea, cheilosis, glossitis, anemia, and weight loss. The brand name is Myleran®.

Butyric acid. Short-chain saturated fatty acid containing 4 carbons. It is present as triglyceride in butterfat and milk; other fats contain very small amounts of butyric acid.

BV. Abbreviation for *biologic value*.

C

Ca. Chemical symbol for calcium.

Cachectin. Or tumor necrosis factor. A hormone-like protein that induces anorexia and a state of cachexia. See *Tumor necrosis factor (TNF).*

Cachexia. Weakness, extreme weight loss, and severe wasting of tissues due to long-standing chronic diseases, malnutrition, or terminal illnesses. It is often associated with cancer, cardiac disease, rheumatoid arthritis, tuberculosis, and AIDS. See *Cancer cachexia, Cardiac cachexia,* and *Rheumatoid cachexia.*

Cadmium (Cd). A trace element found in the body in minute amounts, mainly in the kidneys and liver. The essentiality of cadmium in humans remains to be established. Nutritional requirements, if they exist, are very low and easily met by the levels in food and drink. Laboratory animals fed diets low in cadmium have impaired reproductive performance and depressed growth. Cadmium is widely distributed in nature. Oysters, seafoods, and grains are rich sources; appreciable amounts may also be obtained from the air and the water supply. Daily intakes by human adults have been estimated to be 25 to 60 mg/day. About 5% of dietary cadmium is absorbed. In the body, it accumulates mainly in the kidney and liver, and to a lesser extent in bones and teeth. The half-life of cadmium is 15 to 30 years; hence, its level increases with age. Cadmium poisoning is an industrial hazard. Chronic intoxication leads to growth retardation, impaired reproduction, hypertension, and renal dysfunction.

Caffeine. Trimethyl xanthine or methyltheobromine; an alkaloidal purine found in coffee, tea, cola drinks, chocolate, and some drugs. It is a cardiac and renal stimulant producing varying pharmacologic effects in humans. To some, caffeine is effective in counteracting drowsiness and mental fatigue; others experience some gastrointestinal distress, increased gastric secretion, insomnia, restlessness, and diuresis. At a rate of 1 g/day, it may cause cardiac palpitations, tremors, and anorexia. If it is over 5 g/day, it can lead to convulsions, coma, respiratory failure, and heart failure. It is contraindicated in certain heart diseases, in peptic ulcer patients who find caffeine irritating, in persons who are undergoing therapy for chemical dependency, and in mothers who are breast-feeding. Caffeine is absorbed by a healthy adult almost completely (99%), and it may take about 5 hours to eliminate half of this amount through the urine. For the caffeine content of selected foods and beverages, see Appendix 35.

Cal. Abbreviation for *Calorie.* If the letter *c* is not capitalized, it refers to a small calorie; if capitalized, it means a large calorie or kilocalorie (kcal).

Calcidiol. 25-Hydroxyvitamin D_3. Also known as 25-hydroxycholecalciferol; previously called calcifediol. A metabolite of vitamin D formed in the liver and further hydroxylated in the kidney to yield 1,25-dihydroxyvitamin D. It is the major circulating form of vitamin D and is the form measured in assessing the vitamin D status of patients.

Calcifediol. Former name for 25-hydroxyvitamin D_3; now called *calcidiol.* Commercially available calcifediol is synthesized for use in the management of hypocalcemia in patients with chronic renal failure undergoing dialysis. The brand name is Calderol®.

Calciferol. Former name given to vitamin D_2 or ergocalciferol. Synthetic preparations are used for prophylaxis and treatment of vitamin D deficiency, rickets, and hypocalcemic tetany. Brand names are Calciferol®, Deltalin®, and Drisdol®.

Calcification. Deposition of calcium salts within the tissues of the body. It is a normal process in bone formation or may be abnormal, as in pathologic calcification of soft tissues, particularly arteries, kidneys, lungs, pancreas, and stomach. Deposition of calcium salts also occurs in areas of fatty degeneration and in dead or chronically inflamed tissues. Areas of necrosis, infarcts, scar tissues, and caseous tuberculous areas have calcium deposits. See *Bone mass* and *Peak bone mass*. See also *Nutrition, bone*.

Calcilo XD®. A low-calcium, vitamin D-free infant formula with iron for use in the dietary management of idiopathic hypercalcemia and osteopetrosis.

Calcinosis. A condition characterized by abnormal deposition of calcium salts in various tissues of the body, including the cartilage and tendons.

Calciol. Formerly called cholecalciferol; vitamin D_3 occurring in animal cells and formed in the skin on exposure of 7-dehydrocholesterol to ultraviolet light from the sun.

Calcitonin (CT). Hormone secreted by the thyroid gland that functions in the regulation of calcium ions in the blood. Its action is opposite that of *parathormone*. When the blood calcium level is high, calcitonin is secreted and "shuts off" the release of calcium from bones. A synthetic preparation of salmon calcitonin can be given by injection to treat osteoporosis and other bone disorders.

Calcitriol. 1,25-Dihydroxyvitamin D_3 (1,25-$(OH)_2D_3$), also known as "1,25-dihydroxycholecalciferol (1,25-DHCC)." It is the most active metabolite of vitamin D and is formed in the kidney from 25-hydroxycholecalciferol. In addition to ensuring adequate absorption of calcium, calcitriol plays a role in the regulation of plasma calcium by increasing bone resorption synergistically with parathyroid hormone and stimulating the reabsorption of calcium by the kidney.

Calcium (Ca). A major mineral constituent of the body that makes up 1.5% to 2% of body weight. Of this amount, 99% is present in bones and teeth; the remaining 1% is found in soft tissues and body fluids and serves a number of functions not related to bone structure. Calcium is important in blood coagulation, transmission of nerve impulses, contraction of muscle fibers, myocardial function, and activation of enzymes. A low calcium intake may be associated with increased incidence of hypertension. Calcium supplements have been reported to lower blood pressure in people with or without hypertension. Calcium deficiency results in *rickets* in children and *osteomalacia* in adults. Good food sources are milk, cheese, and other milk products (except butter) and some leafy green vegetables (such as broccoli, collards, and kale), lime-processed tortillas, and calcium-fortified foods. Intestinal absorption of calcium is influenced by several nutritional and physiologic factors. It is increased during periods of high physiologic requirement, and a higher percentage of calcium is absorbed at low intakes than at high intakes. Vitamin D, ascorbic acid, lactose, and an acid medium favor absorption. Absorption is impaired in the elderly and in the presence of phytate and oxalate, and certain fiber fractions bind calcium and interfere with absorption. The amount of protein and phosphorus in the diet can affect calcium metabolism and requirement. A high protein intake increases urinary calcium excretion, and if the level of phosphorus is high it can combine with calcium to form insoluble compounds. Calcium-to-phosphorus ratios between 1:1 and 2:1 are generally recommended. Calcium supplementation much higher than the recommended dietary allowance is not advisable. A high calcium intake inhibits the intestinal absorption of iron, zinc, and other essential minerals. Ingestion of very large amounts may result in hypercalciuria, hypercalcemia, and urinary stone formation. See Appendix 1 for the recommended dietary allowances. See also *Hypercalcemia* and *Hypocalcemia*.

Calculus. Pl. calculi. An abnormal stony mass in the body, found principally in ducts, passages, hollow organs, and cysts. It is more commonly called a "stone," such as a kidney stone or a gallbladder stone. Seldom "pure," it is usually a mixture of several substances, such as uric acid, cystine, calcium oxalate, calcium carbonate, and calcium phosphate. Once the stone is formed, no diet is effective in bringing about its dissolution. However, for the predisposed individual, dietary management may help prevent or retard the growth or recurrence of the stones. The type of diet depends on the chief component of the stone. Restrict calcium and phosphorus intake with calcium phosphate stones; maintain an acid urine with magnesium phosphate stones; restrict sulfur-containing amino acids and maintain an alkaline urine with cystine stones; restrict oxalate and calcium intakes with calcium oxalate stones; and maintain an alkaline urine to keep urate stones in solution. A high fluid intake is recommended in all types of stone formation. See *Diet, ash; Diet, calcium-phosphorus-restricted;* and *Diet, oxalate-restricted*.

Calf circumference. An anthropometric measurement for estimating body weight. With the person lying in a supine position and the left knee bent at a 90 degree angle, measure the calf circumference (CC) to the nearest 0.1 cm. This value and the other measurements for knee height (KH), mid-arm circumference (MAC), and subscapular skinfold (SSF) are used to derive body weight from the following formula:

$$\text{Weight (women)} = (1.27 \times \text{CC}) + (0.87 \times \text{KH}) + (0.98 \times \text{MAC}) + (0.4 \times \text{SSF}) - 62.35.$$

$$\text{Weight (men)} = (0.98 \times \text{CC}) + (1.16 \times \text{KH}) + (1.73 \times \text{MAC}) + (0.37 \times \text{SSF}) - 81.69$$

Caliper. Instrument for measuring linear dimensions. Calipers such as the Lange and Harpenden may be fixed, adjustable, or movable. The three types that are often used in nutrition surveys are skin-fold, sliding, and spreading calipers.

Calorie (Cal). A unit of heat. The amount of heat required to raise the temperature of 1 kg of water 1°C. This is the large Calorie, or kilocalorie (spelled with a capital *C*), which is 1000 times as large as the small calorie (spelled with a small *c*). In nutrition, the kilocalorie (kcal) is used, and this is generally understood whether the word is written with a capital *C* or a small *c*. The small calorie is used in physics. See also *Joule*.

Calorie/nitrogen (C/N) ratio. The relationship between calories and protein calculated as follows:

$$\text{C/N ratio} = \frac{\text{kcal/day}}{\text{gN/day}}$$

where kcal is the total energy intake and N is nitrogen intake. Because protein is 16% nitrogen, grams of protein divided by 6.25 = grams of nitrogen/day. The desirable calorie-to-nitrogen ratio for normal tissue maintenance is 250 to 300 kcal:1 g of nitrogen. To promote anabolism, the desirable ratio is 150 to 200 kcal:1 g of nitrogen for moderate stress, 100 to 150 kcal:1 g of nitrogen for severe stress, and 80 kcal:1 g of nitrogen for severe stress such as burns and sepsis.

Calorie-protein malnutrition (CPM). Also called protein-energy malnutrition (PEM) or protein-calorie malnutrition (PCM) or energy-protein deficits (EPD). For further details, see *Protein-energy malnutrition*.

Calorimetry. Measurement of heat absorbed or given off by a system or body. The measurement is carried out in either of two ways: *direct* or *indirect* calorimetry. Direct calorimetry is carried out by measuring the heat produced by a subject or food enclosed in a chamber that has water outside the chamber walls. An example is the bomb calorimeter. Food is weighed and placed on a dish inside a container shaped like a bomb (hence the name), which is filled with oxygen. The container is then immersed in water. The amount of heat produced when food is burned which is absorbed by the surrounding water is then measured. Indirect calorimetry is calculated from the amount of oxygen consumed and the amount of carbon dioxide or nitrogen excreted. An example is the respiratory calorimeter, an apparatus that measures the exchange of gases between a living organism or subject and the atmosphere around it, with the simultaneous measurement of the amount of heat produced by that organism or subject.

Cancer. Common term for a malignant cellular growth that tends to spread due to the inability of the DNA to respond to normal physiologic stimuli. The abnormal (cancerous) cells reproduce without control, crowding out the healthy cells and using up nutrients needed by these cells. There are many kinds of cancer, designated by the organ or part of the body affected. The exact cause(s) of human cancer is not definitely established, although contributing or predisposing factors are known in many cases. These include genetics, environmental and occupational hazards, and personal habits such as smoking, alcoholism, and dietary factors. *Nutrition therapy:* Studies on dietary factors in relation to cancer prevention include the following: vitamins A, C, E, and selenium, which are antioxidants, protect the DNA from electron-seeking compounds so that it is not altered; dietary fibers are recommended, since they bind carcinogens in the feces; cruciferous vegetables such as cabbage, cauliflower, and Brussels sprout contain indoles and phenols, which are believed to reduce cell division; nitrate-cured and smoked foods are to be eaten in moderation; and total fat intake should be limited to 30% or less of total calories. The most common problems in patients with metastatic cancer that affect food intake are anorexia, nausea and vomiting, constipation, abdominal pain, mucositis, easy fatigability, and weight loss. The adverse effects of chemotherapy and radiation therapy also contribute to malnutrition and cancer cachexia. Malnutrition adversely affects not only tissue function and repair but also humoral and cellular immunocompetence. A weight loss of more than 50% of body weight is life threatening. *Nutrition therapy:* diet is individualized based on weight and nutritional status, food intolerances, appetite, and nutritional side effects of treatment. In the terminal stage of cancer, emphasis must be placed on the palatability and presentation of food to encourage consumption. Discontinue any dietary restriction and allow the patient to eat as desired. Small but frequent feedings are recommended. Supplemental tube feeding may be necessary if oral feeding is inadequate due to anorexia, dysphagia, or inability to tolerate food. For more specific details, see *Cancer cachexia, Chemotherapy, Dysphagia,* and *Radiotherapy*. See also *Nutrition, cancer*.

Cancer cachexia. Clinical syndrome that develops as a consequence of the nutritional and metabolic abnormalities in advanced malignancies. Characterized by anorexia, weight loss, severe tissue wasting, asthenia, anemia, and organ dysfunction. Severe imbalances in fluid and electrolyte status promote excessive diarrhea or vomiting. Hypoalbuminemia and hypocalcemia are common. Serum levels of ascorbic acid, thiamin, folate, vitamin A, iron, and zinc are often depressed. Alterations in taste and smell sensations lead to anorexia and meat aversion. Food intake is inhibited by mucositis, cheilosis, glossitis, stomatitis, and esophagitis caused by chemotherapeutic drugs. Multiple factors influence cancer cachexia, such as appetite-suppressing toxins produced by the cancer cells, alterations in taste and smell perception, increased resting energy expenditure, ineffective utilization of nutrients, impaired sensitivity to endogenous insulin, increased metabolism of muscle protein, increased mobilization of fatty acids from adipose tissue, and cytokines and regulatory hormones. Specific cytokines implicated in cancer cachexia include tumor necrosis factor (TNF, cachectin), interleukin, and interferon. *Nutrition therapy:* nutrient modification depends on the extent of depletion and specific feeding problem. Patients who have altered taste acuity may benefit from increased use of flavorings and seasonings. Meat aversion requires eliminating red meats. Soft foods moistened with sauce or gravy are useful when there is diminished salivation, dry mouth, or oral and esophageal lesions. Modify food textures and consistency as tolerated. Give frequent small feedings served at moderate or room temperature. Patients with intestinal damage may require modifications in lactose, fat, and fiber content as well as in texture. Oral feeding is preferred if the gut is functional. More aggressive feeding methods may be required if efforts to encourage oral intake fail or are inadequate. Tube feeding or parenteral nutrition may be considered, although the beneficial effect of aggressive nutritional intervention on survival has not been demonstrated. Specific nutrients such as glutamine, arginine, and omega-3 fatty acids may play a more direct role in preventing morbidity. See also *Chemotherapy* and *Radiotherapy*.

Cancer Prevention Awareness Program. Administered by the National Institutes of Health and National Cancer Institute since 1984, the program provides ongoing cancer prevention education for adults in the United States.

Cancrum oris. An infective gangrene of the mouth that erodes the lips and cheeks, giving the appearance of harelip. This has been reported chiefly in South Africa and is presumably caused by a combination of malnutrition and infection. See *Synergism*.

CAPD. Abbreviation for *continuous ambulatory peritoneal dialysis*.

Capillary resistance test. Test to determine the tendency of blood capillaries to break down and produce petechial hemorrhages. This is done by applying enough pressure to obstruct venous blood return on an arm and noting the number of *petechiae* produced after 5 minutes. Capillary fragility is a clinical sign of vitamin C deficiency which is measured by this test.

Caprenin. A fat substitute used in soft candy and confectionery coatings. It is a reduced-calorie triglyceride made with three naturally occurring fatty acids: capric, caprylic, and behenic acid which is only partially absorbed. Provides 5 kcal/g.

Captopril. An angiotensin-converting enzyme inhibitor used in the treatment of severe hypertension and congestive heart failure. It acts by lowering the concentration in the blood of angiotensin II, which is one of the factors responsible for high blood pressure. Captopril has a persistent metallic or salty taste and may decrease taste acuity. It may also cause hyponatremia, hyperkalemia, hypoalbuminemia, proteinuria, anorexia, and weight loss. The brand name is Capoten®.

Carbamazepine. An anticonvulsant drug structurally related to the tricyclic antidepressants; used for control of seizures and relief of pain associated with neuralgia. It may cause anorexia, glossitis, stomatitis, dry mouth, and gastric distress. The brand names are Epitol® and Tegretol®.

Carbamyl phosphate synthetase (CPS) deficiency. A metabolic disorder caused by an enzyme deficiency and characterized by episodes of hyperammonemia. Laboratory findings include elevated plasma glutamine and normal or low orotic acid in urine. The onset is usually in the early neonatal period with vomiting, hypothermia, irritability, respiratory distress, altered muscle tone, and lethargy. *Nutrition therapy:* same as for *citrullinemia* except that arginine in high doses is not indicated.

Carbohydrate. Polyhydroxy aldehyde, ketone, or any substance that yields one of these compounds. The term was originally designated for compounds of hydrates of carbon having the general formula $C_x(H_2O)_y$. Now it includes other compounds having the properties of carbohydrate even though they do not have the required 2:1 ratio of hydrogen to oxygen. Some carbohydrates contain nitrogen and sulfur in addition to carbon, hydrogen, and oxygen. The most important carbohydrates in foods are the digestible sugars and starches and the indigestible cellulose and other dietary fibers. The digestible carbohydrate is the major source of energy, providing about 50% of the calories in the American diet. In the developing countries where protein foods are scarce and expensive, carbohydrate foods may be as much as 80% of the total caloric intake. The chief carbohydrates in the body are glucose ("physiologic sugar") and glycogen ("animal starch"). One gram of digestible carbohydrate yields 4 kcal. A recommended distribution of calories in the daily diet is to allow 50% to 60% of total calories from carbohydrate sources, emphasizing complex carbohydrates and reducing simple sugars. For a classification of carbohydrates and food sources, see Appendix 8.

Carbohydrate counting. One method of meal management for diabetes mellitus based on the assumption that carbohydrate intake is the main consideration in determining insulin requirement. The carbohydrate: insulin ratio often used is 10 to 15 g of carbohydrate to one unit of regular insulin. This method has been most useful for individuals on intensive insulin therapy. See also *Total available glucose*.

Carbohydrate by difference. In the *proximate analysis* of food, this is the difference obtained by subtracting from 100 the sum of the percentages of water, protein, fat, and ash content. Included in this value, in addition to the sugars and starches, which the body can utilize almost completely, are the *crude fiber* and some organic acids that the body cannot utilize.

Carbohydrate functions. Carbohydrate is the primary source of heat and energy (1 g yields 4 kcal). Next to water, carbohydrate is the single largest nutrient in the diet. The sugars are valued for their sweetening power. Carbohydrate has a protein-sparing effect and serves as a carbon skeleton for the synthesis of nonessential amino acids. Glucose is the major energy source of the brain and nervous tissues. As the "physiologic sugar," glucose is the immediate source of energy for all body cells and is stored as glycogen in the muscles and liver. Plants store carbohydrates as starch, which in turn become an inexpensive and readily available energy supply for humans. Lactose increases absorption of calcium from the intestinal tract and provides a medium for the growth of favorable intestinal bacteria. The indigestible carbohydrates (dietary fibers) stimulate peristaltic movement and help prevent constipation, regulate absorption of simple sugars, and control the level of blood cholesterol by preventing or delaying its reabsorption into the bloodstream. Mucopolysaccharides and mucoproteins are compounds related to carbohydrates which are normal constituents of certain body fluids and tissues, such as mucous membrane linings and lubricating fluids. Other compounds in the body that contain carbohydrate are heparin (for blood clotting), galactopins (part of nervous tissues), dermatan sulfate (found in collagen and skin), glycosides (sugar component of steroid and adrenal hormones), and immunopolysaccharides (part of immune bodies). There is no recommended dietary allowance for carbohydrate, but it is a dietary item essential to prevent ketosis. At least 50 g a day is the minimum amount needed to avoid ketosis, assuming normal homeostasis in the body. Allow 125 to 150 g carbohydrate/1000 kcal per day. Of current interest, especially in public health nutrition, is *alcohol,* which is produced by fermentation of glucose in foods (fruit sugars and cereal grains).

Carbohydrate intolerance (malabsorption). A group of conditions in which the absorption of one or more nutritionally important carbohydrates is caused by a deficiency in one or more of the intestinal disaccharidases (lactase, maltase, isomaltase, invertase, and trehalase) or by blockage in the transport mechanism across the gut. Disaccharidase deficiency may be a congenital defect, or it may be acquired in association with certain diseases (celiac disease, enteritis, kwashiorkor, and malnutrition) due to unspecific lesions in the intestinal mucosa. The chief clinical manifestation is diarrhea, which occurs when the sugar that cannot be absorbed is introduced into the diet.

Carbohydrate loading. More appropriately called glycogen loading or glycogen supercompensation. It is a 7-day diet and exercise routine prior to an athletic event lasting 90 minutes or more, for the purpose of increasing glycogen stores in the muscles two to three times the norm. A diet supplying 350 g of carbohydrate/day is given the first 4 days, followed by 550 g/day during the next 72 hours or 3 days prior to the competition. Heavy physical activity is avoided 2 days before the competition, with a tapered rest regimen to maximize glycogen loading. Due to some adverse effects, however, the practice is limited to two times a year. It is not recommended for young athletes (e.g., children and teenagers), diabetics, persons with muscle enzyme deficiencies and renal disorders, and those prone to heart disease.

Carbohydrate utilization. See digestion, absorption, and metabolism of carbohydrate under *Absorption, nutrients,* and Appendices 11, 12, and 15.

Carbon dioxide-combining power. Carbon dioxide capacity of the plasma. Normal values range from 53 to 75 vol%. It is increased in alkalosis and decreased in acidosis.

Carbon dioxide fixation. Process of utilizing carbon dioxide to synthesize more complex molecules. It takes place in photosynthesis, whereby plants, in the presence of solar energy, use carbon dioxide from the atmosphere to build carbohydrates and other organic compounds. The ability to fix carbon dioxide is now known to be possessed also by animal tissues even without radiant energy but with chemical energy and the aid of the vitamin *biotin*. This carbon dioxide fixation in the body is referred to as dark-reaction photosynthesis.

Carbonic anhydrase. Zinc-containing enzyme that catalyzes the reversible hydration of carbon dioxide. It is found in the tissues and erythrocytes, and facilitates the transfer of carbon dioxide from the tissues to the blood and then to the lungs for elimination.

Carboplatin. An antineoplastic agent used in the treatment of ovarian cancer. Adverse side effects include sores in the mouth and on lips, nausea and vomiting, loss of appetite, and anemia. Breast feeding is not recommended while carboplatin is being administered. Brand name is Paraplatin®.

Carboxyhemoglobin (HbCO). Hemoglobin combined with carbon monoxide, which has a stronger affinity for hemoglobin than does oxygen. Carbon monoxide displaces oxygen, causing asphyxia and carbon monoxide poisoning.

Carcinogen. Cancer-producing agent or substance. Carcinogenic compounds such as aflatoxin and cycasin have been identified in some plants. A variety of chemical agents have been used to induce malignancy in animals, but not all of them show the same capability in humans. Some dietary components implicated to be carcinogenic are *N*-nitroso compounds, cyclamates, saccharin, sassafras tea, coffee, some food colors, and smoked products containing significant levels of polycyclic aromatic hydrocarbons (PAHs).

Cardiac cachexia. An abnormal condition of progressive and severe malnutrition resulting from chronic heart failure; characterized by weakness, anorexia, emaciation, malabsorption, increased metabolic rate, and fluid retention. Anorexia may be aggravated by dietary restrictions of salt and fat, and medications may adversely affect the appetite. Dyspnea related to heart failure can increase oxygen consumption and requirement for energy. *Nutrition therapy:* provide for the hypermetabolism [basal energy expenditure (BEE)] × 1.3 to 1.5 plus additional calories for activity). Depending on the extent of malnutrition, give 1.2 to 1.5 g of protein/kg body weight to replenish tissue stores. Restrict sodium (1 to 2 g/day) as necessary if there is fluid retention. Fluid intake of about 1000 to 1500 mL/day (or 0.5 to 0.7 mL/kcal/day) should be sufficient; fluid restriction may not be indicated if the person is compliant with sodium restriction. Supplemental vitamins and minerals may be needed, especially folic acid, thiamin, magnesium, zinc, and iron. Small, frequent feedings are better tolerated. Tube feeding may be necessary if oral intake is not adequate. Monitor fluid and sodium balance; a high-calorie formula (1.5 to 2.0 kcal/mL) may be indicated to reduce formula volume. Consider parenteral nutrition if the tube feeding causes intra-abdominal pressure and increased work of breathing.

Cardiac disease. Also called heart disease. Includes cardiac insufficiency, heart failure, myocardial infarction, pericarditis, and any cardiovascular disease (CVD). *Nutrition therapy:* caloric intake is adjusted to bring about weight loss and consequent lowering of blood pressure, slowing of the heart rate, and reduction in the work of the heart. Rest is the primary consideration in acute heart diseases such as heart failure and coronary occlusion. Fluids are restricted during the first few days. With improvement, soft, easily digested foods are gradually introduced in small amounts as tolerated. Sodium intake is restricted to 500 mg with severe edema and then maintained at 1000 to 1500 mg/day once the edema disappears. In ischemic heart disease involving hypercholesterolemia and atherosclerosis, the fat in the diet should be predominantly of the polyunsaturated type; cholesterol and saturated fats are restricted. In chronic heart conditions, three small meals with between-meal feedings are recommended to avoid strain on the heart. Constipation should be avoided, and maintenance of normal or slightly below-normal weight is desirable. Sodium may be restricted (2000 mg/day) to prevent edema. See also *Diet, prudent.*

Cardiac surgery. Nutrition management before and after cardiac surgery depends on the type of surgery, nutritional status, and medical condition of the patient. Prior to surgery, obesity requires weight reduction whereas undernutrition, as in cardiac cachexia, requires nutritional repletion. Other nutritional considerations include sodium restriction for the management of hypertension and edema, cholesterol and saturated fat restriction to control elevated serum lipid levels, and fluid restriction to manage fluid retention and congestive heart failure. The hypermetabolism following surgery increases the energy requirement and may be 30% to 50% above the basal energy expenditure (BEE). Protein requirement is about 1.0 g/kg of body weight after bypass surgery and 1.2 to 1.5 g/kg body weight after heart transplantation. For other details, see *Coronary artery bypass* and *Heart transplantation.*

Cardiac transplantation. See *Heart transplantation.*

Cardiovascular disease (CVD). Collective term denoting a large group of diseases affecting the heart and blood vessels. The most important of these diseases from the public health point of view are coronary heart disease, cerebrovascular disease, hypertensive disease, and coronary heart disease. The latest statistics in 1993 revealed that about 44% of deaths in the United States are due to cardiovascular diseases, with atherosclerosis as the majority of cases. Cardiac diseases resulted in 500,000 deaths in a year and coronary bypass surgery was done on about 390,000 cases. Reduction of major risk factors such as high blood pressure, high blood cholesterol, and overweight have a significant impact on mortality due to cardiovascular disease. Dietary measures to reduce risk for developing cardiovascular diseases include reducing overweight and reducing the intake of fat, cholesterol, and sodium. For more specific nutrition therapy, see *Cardiac disease, Cerebrovascular disease, Coronary heart disease, Hypertension,* and *Myocardial infarction.* See also *Hypercholesterolemia* and Appendix 7.

Caries. Molecular decay of bones and teeth, making them soft and porous. See *Dental caries* and *Nutrition, dental health.*

Carmustine. A cytotoxic agent used in the treatment of brain tumor, multiple myeloma, and malignant lymphoma. It may cause nausea and vomiting within periods ranging from a few minutes to 2 hours and may persist for up to 6 hours after administration. It may also cause anorexia, dysphagia, esophagitis, diarrhea, glycosuria, and hypophosphatemia. The brand name is BiCNU.®

Carnitine. Designated as vitamin B_t; a dietary essential nutrient for the mealworm *Tenebrio molitor*. Under normal conditions, humans and higher animals can synthesize carnitine from lysine and methionine; hence, it is not necessary to supply this in food. However, recent studies show that the synthesis of body carnitine may be inadequate for some individuals, especially premature infants, and that a number of diseases alter levels of carnitine in body fluids and tissues. Because carnitine is formed from amino acids, protein malnutrition will lower carnitine production. Carnitine functions in lipid metabolism as a carrier of long-chain fatty acids into the mitochondria for beta-oxidation. Abnormalities seen in patients with genetic disturbances in carnitine metabolism include muscle weakness, hypoglycemia, and lipid accumulation between muscle fibers. Evidence of carnitine deficiency has been reported in premature infants maintained on intravenous feeding, and low serum carnitine levels have been found in infants fed soy formula and in certain lipid storage diseases of the muscles.

Carnosine. Beta-alanylhistidine, a dipeptide found in vertebrate muscle tissue. Its biochemical function is not known. A protease, *carnosinase,* that attacks the peptide bond of carnosine is present in the liver, pancreas, and kidneys.

Carnosinemia. Inborn error of amino acid metabolism characterized by excretion of large amounts of *carnosine* in the urine, even when all dietary sources of this dipeptide are excluded. The condition is also associated with unusually high concentrations of homocarnosine in the cerebrospinal fluid and with a progressive neurologic disorder characterized by severe mental defects and myoclonic seizures. The defect may be due to a deficiency in the enzyme *carnosinase*.

Carotene. A *carotenoid* present in green leafy and yellow vegetables. It exists in several forms, of which alpha-, beta-, and gamma-carotene are provitamins A. Beta-carotene is the most active of these three forms. In the body, carotene is converted to vitamin A in mucosal cells. See *Retinol equivalent*.

Carotenemia. Presence of large amounts of carotene in the blood, resulting in a yellowish discoloration of the skin. The condition is harmless and should not be mistaken for jaundice. The conjunctivae and urine are not discolored in carotenemia.

Carotenoids. Group of fat-soluble yellow to red pigments occurring widely in plants. Many different carotenoids exist in nature, but only a few are converted to vitamin A in the body. Beta-carotene is the most active provitamin A carotenoid in food. Cryptoxanthin and alpha-carotene yield only half the vitamin A activity of beta-carotene. Other carotenoids, such as lycopene, cantaxanthin, and zeaxanthin, do not have provitamin A activity.

Carrier. 1. A person who, without showing symptoms of a communicable disease, harbors and transmits disease-producing germs. 2. A substance that transports another substance or compound; for example, fat is a carrier of fat-soluble vitamins. 3. A naturally occurring element added to a pure substance in minute quantity for ease in handling. 4. In physiologic oxidation, a compound that can accept hydrogen or electrons from a substrate and transfer them to another compound in the transport system.

Casal's necklace. Type of dermatitis seen in *pellagra* as a result of niacin deficiency. The lesions on the face and areas of the neck exposed to the sun are distributed to resemble a necklace.

Casec®. Brand name of a powdered protein supplement for infant feeding; may be mixed with cereals, mashed potatoes, and casseroles or blended into milk drinks. It is made of calcium caseinate from skim milk curd; it is low in fat and has only a trace of lactose. Each tablespoon provides 4 g of protein and 75 mg of calcium.

Casein. A phosphoprotein; the principal protein of milk. It is converted to calcium caseinate or milk curd by the enzyme *rennin* in the presence of calcium, leaving a residual clear fluid called whey.

Catabolism. Destructive metabolism; the breakdown of complex substances by living cells into simpler compounds with the liberation of energy. It is the opposite of *anabolism*. Together catabolism and anabolism constitute *metabolism*.

Catalyst. A substance that hastens the speed of a chemical reaction without itself undergoing a change. Catalysts in living cells (biocatalysts) are called *enzymes*.

Catecholamines. Substituted diorthophenols synthesized in the brain, sympathetic nerve endings, peripheral tissues, and adrenal medulla. They are discharged into the circulation under conditions of stress, anger, and fear. Catecholamines (chiefly epinephrine and norepinephrine) are pressor substances and can mobilize sources of rapidly utilizable energy from the body's storage depot to prepare the animal for flight or fight. See *Epinephrine* and *Norepinephrine*.

Cd. Chemical symbol for *cadmium*.

CDC. Abbreviation for *Centers for Disease Control*.

CDP. Abbreviation for cytidine diphosphate. See *Cytidine phosphates*.

Celiac. Abdominal; pertaining to the abdomen.

Celiac disease. More appropriately called "gluten-sensitive enteropathy"; also called "celiac sprue" and "nontropical sprue." See *Gluten-sensitive enteropathy*.

Cellobiose. Disaccharide formed by the partial hydrolysis of *cellulose*. It consists of two glucose units linked in a beta-1,4 configuration.

Cellulose. Polysaccharide that acts as a supporting structure for plant tissues. It yields glucose on complete hydrolysis; partial hydrolysis yields the disaccharide *cellobiose*. It is generally not digested by humans and merely provides bulk or roughage. The glucose units of cellulose are held together by beta-linkages which can be digested by the microorganisms that inhabit the alimentary tract of ruminants. The human gastrointestinal tract has no enzymes capable of hydrolyzing the beta-linkages.

Centers for Disease Control (CDC). Government agency in charge of developing the techniques of nutritional assessment for others to use. Two divisions of the CDC are involved in establishing the scientific base for nutrition policy: the National Center for Health Statistics (NCHS), which provides the database for nutritional status in the United States, and the Division of Nutrition (DON), which aims to prevent nutrition-related diseases by promoting health through good nutrition.

Cephalin. Phosphatide (phospholipid) that, on hydrolysis, yields phosphoric acid, glycerol, a mixture of saturated and unsaturated fatty acids, and either ethanolamine or serine. It is found in the brain, in nerve tissues, and in the lipid portion of glandular organs. It participates in *blood clotting* as thromboplastin, a cephalin–protein complex, and in the manufacture of the protoplasm and the cell membrane.

Cephalosporin. A broad-spectrum antibiotic that resembles penicillin in its action. It may cause nausea, vomiting, diarrhea, dyspepsia, and glossitis. Prolonged use may cause hypokalemia and vitamin K deficiency. Some brand names are Ancef®, Ceclor®, Duricef®, Keflex®, and Ultracef®.

Cerebral palsy. Variety of neurologic dysfunctions secondary to brain damage as a result of birth injury, cerebral hemorrhage, or prematurity. Two general types of motor disability are known: *athetosis,* which is characterized by constant, uncontrollable movements, and *spastic paralysis,* which is characterized by limited activity. The motor dysfunction varies in severity and distribution, affecting one or more extremities or the trunk, head, and neck. Descriptive terms such as monoplegia, hemiplegia, and paraplegia are used to specify the distribution of the dysfunction. *Nutrition therapy:* special feeding devices are needed to ensure adequate food intake and prevent malnutrition. Avoid stringy, hard-to-chew foods. Provide extra fluids and fiber for normal bowel movement. Adjust calorie needs, reducing intake for spastic patients who tend to gain weight, and increasing intake for the athethoid type. Some of the latter patients need as much as 4000 kcal/day during adolescence [1.5 to 2.0 × basal energy expenditure (BEE) or 45 to 50 kcal/kg body weight]. Tube feeding may be indicated when there is difficulty in swallowing. See also *Dysphagia*.

Cerebroside. A glycolipid that, on complete hydrolysis, yields one molecule each of fatty acid, 4-sphingenine, and galactose. It occurs in the brain and myelin sheaths of nerves.

Cerebrovascular accident (CVA). Commonly known as "stroke." Partial brain damage as a result of a constricted blood supply caused either by ruptures, clots, or blood vessel spasms. It may lead to impaired hearing, sight loss, or speech defects, depending on which brain hemisphere is affected. *Nutrition therapy:* immediately restore fluid and electrolyte balances. Give pureed, ground, chopped, or soft foods if the patient has chewing and swallowing difficulties. Avoid milk if the patient is unable to clear salivary and mucus secretions. Tube feeding may be necessary if the gag reflex is lacking. Ensure adequate nutrition, especially calories, protein, and fluid. Monitor weekly weights and adjust caloric intake accordingly. Nutritional intervention after a stroke must address predisposing risk factors such as obesity, hypertension, hyperlipoproteinemia, and diabetes mellitus. A reduction of cholesterol, sodium, and fat intake may be necessary. Weight control and a reduction of dietary cholesterol (300 mg/day or less), fat (30% or less of total calories, with saturated fat under 10%), and sodium (2 to 3 g/day) may be necessary. Avoid excess vitamin K if on warfarin therapy and monitor potassium, magnesium, and calcium for effects of antihypertensives. Increasing physical activity and reducing tobacco use are beneficial.

Ceroid. An insoluble brown substance found in atheromatous plaques and in fat deposits of certain forms of liver disease. It is associated with disorders in lipid metabolism.

Ceruloplasmin. The transport form of plasma copper bound to alpha-globulin.

It is important in the regulation of copper absorption by reversibly binding and releasing copper at various sites of the body. A low plasma ceruloplasmin concentration is associated with *Wilson's disease*. Normal range in plasma is 27 to 37 mg/dL (270 to 370 mg/L).

CF. Abbreviation for *citrovorum factor* and *cystic fibrosis*.

CHD. Abbreviation for *coronary heart disease*.

Cheilosis. Cracks and fissures at the corners of the mouth characteristic of *riboflavin* deficiency. The lesions of the lips begin with redness and denudation along the line of closure or may appear as pale macerations at the angles of the mouth. The lips look dry and chapped, and shallow ulcerations or crusting may occur in severe deficiency. Nonnutritional factors such as cold and wind may also cause cheilosis. Patients receiving long-term antibiotic therapy, especially chloramphenicol, develop cheilosis, which responds readily to riboflavin supplementation.

Chelate. A chemical compound capable of binding a metallic ion into its molecule and thus removing it from a tissue or from the circulating blood.

Chemical dependence. Also called "substance abuse." Addiction to alcohol, nicotine, street drugs (marijuana, cocaine, and the like), and/or certain prescription drugs to the extent of jeopardizing general health and wellness. Nutritional consequences observed with chemical dependence include loss of taste acuity, anorexia, malabsorption, increased nutrient requirement to detoxify and metabolize the drug, poor utilization of nutrients, inactivation of vitamins and coenzymes, increased losses of nutrients through diuresis and diarrhea, and impaired nutrient storage. *Nutrition therapy:* the objectives are to prevent or restore nutritional deficiencies, to correct eating problems, and to ensure the individual consumes a nutritionally adequate intake. Improved nutritional status speeds up treatment and rehabilitation. Establish a regular eating pattern. Dietary guidelines for a recovery diet are the following: eat balanced meals at regular times with nourishing snacks between meals; avoid all drugs including caffeine in coffee, tea, colas, and chocolate; avoid alcoholic beverages; eat more complex carbohydrates and reduce the intake of concentrated sweets; and ensure a liberal supply of good quality protein foods. Supplemental vitamins and minerals, especially thiamin, vitamin B_6, vitamin B_{12}, folate, pyridoxine, vitamin C, calcium, magnesium, and zinc are recommended.

Chemical score. A measurement of protein quality. Calculated by the following ratio:

$$\frac{\text{mg of amino acid per gram of test protein} \times 100}{\text{mg of amino acid per gram of the reference protein}}$$

The reference protein established by the Food and Agriculture Organization contains the essential amino acids in an "ideal" protein pattern. The disadvantage of the chemical score method compared to the biologic value (BV) and the protein efficiency ratio (PER) is that it cannot exhibit any toxicity of the test protein, because the last two methods use animal (rat) feeding.

Chemotherapy. Adverse effects of chemotherapy often lead to malnutrition. Nausea and vomiting occur with use of almost all antineoplastic drugs; taste abnormalities lead to anorexia; food intake is inhibited by cheilosis, glossitis, mucositis, esophagitis, stomatitis, and xerostomia; the intestinal mucosa is affected, altering digestion and absorption; and some drugs cause tissue breakdown and urinary loss of protein, potassium, and calcium. *Nutrition therapy:* the following are general recommendations. *Esophagitis and stomatitis:* liquid and soft diet, fruit drinks, and broth-based soups; avoid acidic juices, spicy foods, tough meats and granular foods, and extremely hot or cold foods. Alter texture and temperature as tolerated. *Xerostomia (mouth dryness) and viscous mucus production:* liquid diet (juices, fruit ades, tea with lemon, popsicles, broth-based soups, thinned hot cereal); avoid thick liquids and nectars, thick cream soups, thick hot cereals, bread products, gelatin, and oily foods. *Hypogeusia (mouth blindness):* regular diet with strongly flavored foods and spicy foods with emphasis on aroma and texture; avoid bland foods, plain meats, unsalted foods. *Decreased salivation:* regular diet with high-moisture foods, gravies, sauces, casseroles, chicken, fish, beverages with foods, citric-acid-containing foods, sherbet, vegetables with sauces; avoid dry foods, bread products, meats, crackers, excessively hot foods, and alcohol. *Dysgeusia (taste alteration):* regular diet with cold foods, milk products; avoid red meats, chocolate, coffee, and tea. *Acute gastrointestinal toxicity:* clear, cold liquids and light, low-fat foods; avoid poorly tolerated foods, such as milk products, cream soups, fried foods, sandwiches, and sweet desserts. *Early satiety:* high-calorie diet with nutrient-dense foods; avoid low or nonfat milk products, broth soups, green salads, steamed or plain vegetables, and low-calorie beverages.

Child Care Food Program. Administered by the U.S. Department of Agriculture (USDA), it provides nutritious meals to children enrolled in child- or day-care centers. See also *Child nutrition programs*.

Child Nutrition Act. A legal measure enacted in 1966 that appropriated funds for schools to start and expand school lunch programs and to feed preschool children. A pilot school breakfast program was also initiated. It provides cash assistance to state educational agencies to help schools operate nonprofit breakfast programs that meet established nutritional standards.

Child nutrition programs. National programs concerned with improving the nutrition of children. These include School Lunch Program, School Breakfast Program, Summer School Program, Child Care Food Program, Summer Food Service Program, and Special Milk Program. Plans are ongoing to streamline the administration of all child nutrition programs as one comprehensive program. See also *Advocates for Better Children's Diets*.

Chinese restaurant syndrome. A disorder usually caused by a reaction to monosodium glutamate (MSG), characterized by a burning sensation, chest tightness, face flushing, and throbbing headaches. These symptoms may persist for about 20 minutes to an hour. *Nutrition therapy:* avoid MSG in seasoning foods and in commercially prepared foods. Drink plenty of fluids. See *Diet, MSG-restricted*.

Chloral hydrate. A sedative and hypnotic used principally in the treatment of insomnia. It may cause an unpleasant taste in the mouth, nausea, vomiting, gastritis, flatulence, and diarrhea. It is excreted in breast milk; infants must be observed for sedation. Brand names are Noctec® and Aquachloral Supprettes®.

Chlorambucil. An antineoplastic agent used in the treatment of lymphocytic leukemia and malignant lymphomas. Adverse gastrointestinal effects, including nausea, vomiting, anorexia, diarrhea, and abdominal discomfort, are usually mild and last for less than 24 hours. However, nausea and weakness may persist for up to 7 days following a single high dose of the drug. The brand name is Leukeran®.

Chloramphenicol. A broad-spectrum antibiotic used for the treatment of serious infections caused by susceptible strains of *Escherichia coli,* pneumococci, and other susceptible organisms. It inhibits protein synthesis and may increase the need for riboflavin, pyridoxine, folic acid, vitamin B_{12}, and iron. It may also cause nausea, diarrhea, glossitis, stomatitis, and impaired taste. Chloramphenicol is excreted in breast milk and can also affect infants adversely. Brand names are Chloromycetin® and Mychel.®

Chloride. An essential mineral found largely in the extracellular fluids. It is a constituent of gastric juice as hydrochloric acid and is essential in maintaining osmotic pressure and acid–base balance. Chloride ions readily pass into and out of red blood cells (chloride shift), which is important in buffering action and maintenance of blood pH. Dietary chloride is supplied mainly by table salt; meats, seafoods, milk, eggs, and processed foods are other sources. It is readily absorbed and is excreted in the urine and sweat. Chloride loss parallels the loss of sodium; hence, chloride is concurrently lost in conditions associated with sodium depletion, as in excessive sweating, chronic diarrhea, and persistent vomiting. Chloride deficiency is characterized by loss of appetite, muscle weakness, lethargy, and metabolic alkalosis. See also *Hyperchloremia* and *Hypochloremia*.

Chloride shift. Exchange of chloride ion for bicarbonate ion in the intracellular fluid without a corresponding movement in cations. Bicarbonate ions diffuse into the plasma when its concentration of red blood cells is high. To maintain ionic equilibrium, chloride ions diffuse from the plasma into the red cells.

Chlorine (Cl). Element universally found in biologic tissues as the *chloride* ion. As free chlorine, it is a disinfecting, bleaching, and purifying agent. Chlorinated water contains 1 part chlorine per 1 million parts of water.

Chlorosis. Form of hypochromic microcytic anemia common in young women. It is characterized by a greenish tinge to the skin.

Chlorothiazide. A thiazide diuretic used in the treatment of edema and hypertension. It increases the excretion of riboflavin, potassium, magnesium, zinc, and sodium. It may also decrease carbohydrate tolerance and increase blood glucose, and may cause dry mouth, anorexia, gastric irritation, diarrhea, and constipation. It is excreted in milk. Brand names are Aldoclor®, Diupres®, and Diuril®.

Chlorpromazine. A phenothiazine antipsychotic agent. It may induce riboflavin depletion, increase serum cholesterol, and alter glucose tolerance, causing hyper- or hypoglycemia. It may also cause fluid retention, with resultant weight gain. It is excreted in breast milk. Brand names are Ormazine® and Thorazine®.

Chlorpropamide. A sulfonylurea antidiabetic agent used for the management of non-insulin-dependent diabetes mellitus. It acts by stimulating the release of insulin from the pancreas. It has a metallic taste. Adverse effects are dose related and may include nausea, vomiting, abdominal cramps, constipation, and diarrhea. Weight gain may occur due to fluid retention or increased appetite and food intake. There is a higher risk of severe hypoglycemia with this drug than with other oral hypoglycemic agents. Alcohol should be avoided; it may cause headache and a flushing reaction. It is excreted in the breast milk. Brand names are Diabenese® and Glucamide®.

CHO. Abbreviation for *carbohydrate*.

Cholecalciferol. Vitamin D_3, now called *calciol*. The form of vitamin D occurring in animal cells and produced by irradiation of the provitamin 7-dehydrocholesterol underneath the skin. See *Vitamin D*.

Cholecystectomy. The surgical removal of the gallbladder to treat cholecystitis and cholelithiasis. *Nutrition therapy:* a low-fat diet (30 to 40 g/day) may be indicated for a few days. Fat stimulates gallbladder contraction which can cause pain in the surgical area. Gradually increase fat to individual tolerance level.

Cholecystitis. Acute or chronic inflammation of the gallbladder. It is commonly due to bacterial infection and obstruction of the cystic duct by stones, tumor, fibrosis, or adhesions. Acute cholecystitis is characterized by epigastric pain that radiates to the shoulder and lower abdominal region, nausea and vomiting, chills and fever, and jaundice. Sensitivity to fatty foods, colicky pain, belching, and flatulence are the general features of the chronic type. *Nutrition therapy:* in acute cases, nothing is given orally for 24 hours or more. Then a clear liquid diet is given for 2 to 3 days, followed by a low-fat diet (30 g/day). The diet progresses to a moderately low fat intake (about 50 to 60 g/day) to promote the flow of bile and induce drainage of the biliary tract. Some patients may not tolerate spices and gas-forming vegetables. For chronic cholecystitis, provide a moderate fat intake, as previously mentioned. The protein level is kept at 1 g/kg body weight. Adjust kilocalories to achieve and maintain desirable weight. See low-fat diet and moderate-fat diet under *Diet, fat modified*.

Cholecystokinin. Hormone that regulates the contraction of the gallbladder. It is secreted in the upper portion of the small intestine when fat enters the duodenum and is carried by the bloodstream to the gallbladder, causing it to contract.

Cholelithiasis. Formation of stones in the gallbladder; the stones are made of cholesterol, bile acids, calcium and other inorganic salts, and bilirubin. An increased incidence of gallstones is associated with obesity, diabetes, use of oral contraceptives, hypercholesterolemia, and cholecystitis. Most patients have no symptoms. Cholecystectomy is recommended in severe attacks or for biliary pain associated with cholelithiasis. *Nutrition therapy:* in an acute gallstone attack, use a low-fat diet to decrease gallbladder contraction and lessen the pain. Unless fat induces symptoms, a low-fat diet is not necessary. If the gallbladder is sluggish, a moderate fat intake is desirable to stimulate its contraction and prevent stagnation of bile. A high fiber intake is also beneficial. It is unlikely that restriction of cholesterol in the diet has any appreciable effect on reducing cholesterol stones.

Cholestanol. Compound formed by the reduction of the double bond of cholesterol. It is a minor constituent of blood sterols and some tissues. It is found in greater concentration in the feces.

Cholestatic liver disease. This disorder disrupts excretion of bile salts, affecting the absorption of fats and fat-soluble vitamins. There is a deficiency in fat-soluble vitamins, as well as C and B complex vitamins, protein, and iron. *Nutrition therapy:* supplement a moderate-fat diet (50 to 60 g/day) with the fat-soluble vitamins A, D, E, and K. Monitor malabsorption and adjust the diet according to the patient's food tolerance. If malabsorption is severe, reduce fat to 30 to 40 g/day and use the water-miscible form of fat-soluble vitamins. Supplement the diet with medium-chain triglycerides for added calories.

Cholesterol. A fatlike compound with a complex ring structure; the chief sterol in the body found in all tissues, especially the brain, nerves, adrenal cortex, and liver. It is synthesized in the liver and other organs (endogenous cholesterol) and is found only in animal products (exogenous source) such as egg yolk, liver and other glandular organs, brain, milk fat and butter, and meats. Biosynthesis of cholesterol is regulated by the amount in the body; as total body cholesterol increases, synthesis tends to decrease. Endogenous cholesterol is also influenced by caloric intake, certain hormones, bile acids, and the degree of saturation of fatty acids in the diet. The average American diet provides at least 450 mg cholesterol per day, but the recommended intake is 300 mg or less per day. Cholesterol circulates in the blood as a component of lipoproteins. The acceptable total blood cholesterol level is <200 mg/dL (<5.2 mmol/L); borderline high level is 200 to 239 mg/dL (5.2 to 6.2 mmol/L), and high total cholesterol is ≥240 mg/dL (>6.2 mmol/L). The current interest in cholesterol is due to its role in cardiovascular diseases, particularly coronary heart disease (CHD). There is strong and consistent evidence for the relationship between high blood cholesterol and increased risk for CHD. See *Coronary heart disease* and *Hypercholesterolemia*. A plasma cholesterol level higher than 200 mg/dL is a risk factor associated with atherosclerotic disease. For the cholesterol content of selected foods, see Appendix 34.

Cholestyramine. Antilipemic agent used as an adjunct to diet therapy to decrease elevated serum cholesterol and low-density lipoprotein levels in Type II hyperlipoproteinemia and to reduce the risks of atherosclerotic artery disease and myocardial infarction. It has a gritty texture and an unpleasant taste. Long-term use may impair the absorption of fat-soluble vitamins, carotene, folic acid, vitamin B_{12}, iron, calcium, and potassium. It may also cause bloating, flatulence, steatorrhea, indigestion, and constipation. The brand names are Cholybar® and Questran®.

Choline. A methyl group donor that occurs in some phospholipids and is a component of sphingomyelin, lecithin (phosphatidylcholine), and the neurotransmitter acetylcholine. It is an essential nutrient for several animal species, including the dog, cat, and rat, but has not been shown to be a dietary essential for humans. Choline can be readily synthesized in the body from ethanolamine and methyl groups derived from methionine. It is found in a wide range of foods; eggs, liver, organ meats, milk, legumes, and nuts are good sources. Choline is not a typical vitamin and has no known coenzyme function. However, some authorities classify it with the B vitamins because it appears to be required in the diet under certain conditions. The demand for choline-containing compounds is high during growth and development and may exceed the synthetic capacity in the human newborn. The American Academy of Pediatrics has therefore recommended that infant formulas should contain 7 mg of choline per 100 kcal, based on the amount of choline in human milk. Choline may also be deficient in certain neurologic diseases, especially in the elderly. Fatty liver is the most common manifestation of choline deficiency in experimental animals. Other manifestations of deficiency include liver cirrhosis and hemorrhagic degeneration of the kidneys, adrenals, heart, and lungs. A pharmacologic dose of choline seems to alleviate symptoms of tardive dyskinesia and Huntington's disease.

Chromium (Cr). An essential trace mineral occurring in minute amounts in the blood and various tissues. Chromium is required to maintain normal glucose metabolism and is a part of the glucose tolerance factor which potentiates the effect of insulin on peripheral tissues. It is also involved in amino acid transport and breakdown of glycogen and lipids. Chromium deficiency results in impaired glucose tolerance in the presence of normal concentrations of circulating insulin, impaired amino acid metabolism, and elevated serum cholesterol and triglycerides. The primary deficiency usually results from inadequate chromium supplementation in patients on long-term parenteral nutrition and with high intakes of fiber and phytate in a usual diet low or marginal in chromium. The usual daily intake is about 25 to 90 μg/day with less than 1% absorbed in the gastrointestinal tract. It is transported to the liver bound to transferrin and is distributed throughout the human body. Its concentration in various tissues declines with age, depending on the dietary intake. This decline may contribute to the glucose intolerance of the elderly. Good food sources are brewer's yeast, liver, meats, and whole grains. Trivalent chromium, the chemical form that occurs in foods, is not toxic in amounts normally consumed. Other chromium compounds, however, are known to be corrosive to the skin and mucous membranes of the respiratory and intestinal tracts. Chronic exposure and inhalation of chromium, as among mine workers, has been associated with an increased incidence of bronchial cancer. Excessive chromium results in allergic and eczematous dermatitis and in systematic effects on the liver and kidneys. See Appendix 2 for the estimated safe and adequate daily dietary intake of chromium.

Chronic obstructive pulmonary disease (COPD). A progressive disorder that is usually a result of asthma, chronic bronchitis, or emphysema and aggravated by cigarette smoking and air pollution. The bronchial air flow is blocked and the person suffering from COPD constantly feels weak, which adds to the problem of not getting enough to eat. Malnutrition usually accompanies the later stages of COPD. *Nutrition therapy:* the goal is to balance the need for oxygen and the elimination of carbon dioxide. Carbohydrate metabolism results in more carbon dioxide production than fat or protein. Fat is a preferred energy source due to its lower respiratory quotient. Thus, carbohydrate should not supply more than 45% of total calories; fat intake should be maintained at about 40% and the rest in good-quality proteins, but not over 15% because excessive protein may increase the ventilatory drive in patients with limited alveolar reserve. Provide adequate kilocalories [basal energy expenditure (BEE) × 1.5] and protein (1.2 to 1.5 g/kg body weight) to restore lean body mass but avoid hypercaloric feeding. Give six to eight small feedings of easily digested foods with supplemental vitamins and minerals. When abdominal bloating is a problem, limit foods associated with gas formation, such as apples, cabbage, cauliflower, legumes, melons, nuts, onions, and sauerkraut. Some patients may require supplemental enteral nutrition by mouth or tube feeding if food consumption is inadequate. Monitor fluid intake because there is a tendency to retain water. For patient with *edema,* sodium restriction (2 to 3 g/day) is indicated. Some patients may need enteral feedings or supplemental formulas for weight maintenance.

Chronic renal failure (CRF). See under *Renal failure*.

Chyle. The milky white emulsion of fat globules formed in the small intestine and transported into the lymph.

Chylomicron. The largest and lightest (least dense) of the *blood lipids,* approximately 1 μm in diameter and consisting chiefly of triglyceride and smaller amounts of cholesterol, phospholipid, and protein. It is therefore classified as a lipoprotein. Chylomicrons are normally synthesized in the intestines and serve to transport absorbed dietary triglycerides to sites of utilization in the tissues passing through the lymphatic system and then into the bloodstream. They are responsible for the turbid, milky appearance of normal plasma after a fatty meal. About 90% of the triglycerides from absorbed fats are transported as chylomicrons. Their presence in the plasma 12 to 16 hours after eating is abnormal. In Type I hyperlipidemia, the body cannot clear the chylomicrons at a normal rate.

Chyme. Thick, semifluid mass into which food is converted after gastric digestion. It is in this form that food passes into the small intestine.

Chymotrypsin. Endopeptidase secreted by the pancreas as an inactive proenzyme, chymotrypsinogen. Administration of chymotrypsin has a marked therapeutic effect on the control of certain cases of insulin-resistant diabetes mellitus.

Cimetidine. A histamine H_2 inhibitor used for the treatment of confirmed duodenal ulcer. It reduces basal gastric acid secretion by 80% and meal-stimulated acid secretion by 50%. Cimetidine can induce vitamin B_{12} deficiency, especially if taken by vegans for an extended period (over 1 year), and can increase the risk of bleeding associated with warfarin-induced vitamin K deficiency. It has a bitter taste and may also cause iron malabsorption, constipation, and transient diarrhea. The brand names are Peptol® and Tagamet®.

Cirrhosis. Chronic, progressive disease of the liver in which fibrous connective tissue replaces the functioning liver cells. There are many types of cirrhosis caused by a number of conditions, such as chronic alcoholism (alcoholic cirrhosis), chronic hepatitis, various infections such as syphilis and malaria, obstruction of the bile duct, and nutritional deficiency. There is loss of appetite, polyneuritis, cheilosis, a low serum albumin level, and edema due to protein deficiency. In advanced cases, ascites and esophageal varices complicate the condition, often ending in liver failure and hepatic coma. *Nutrition therapy:* a liberal calorie and protein diet (about 50% more than the recommended dietary allowance), with a moderate amount of fat, is recommended. The objectives of dietary treatment are to promote healing and regeneration of liver tissue and to prevent fat stasis and formation of fibrous tissues. Provide 1.5 to 2.0 g of protein/kg/day and 300 to 350 g of carbohydrate/day to spare protein. Protein should be immediately curtailed to individual tolerance in impending coma. Restrict sodium to 500 mg/day if ascites and peripheral edema are present. In steatorrhea, reduce dietary fat and substitute with medium-chain triglycerides (MCT). Use the water-soluble forms of vitamins A, D, and E if steatorrhea is severe and supplement with vitamin K if hypoprothrombinemia exists. Restrict sodium to 1000 mg to reduce fluid retention and reduce further to 500 mg if ascites is present. Avoid fibrous and coarse foods in cases of esophageal varices. Alcohol is not allowed. Supplemental vitamin C, the B-complex vitamins, magnesium, potassium, and zinc may be needed.

Cisplatin. A potent antineoplastic agent used for the treatment of metastatic tumors and other neoplasms. The drug is generally considered the most emetogenic of all the antineoplastic agents. It causes marked nausea, vomiting, and anorexia 1 to 6 hours after administration, which may persist for 5 to 10 days. It may also cause depletion of magnesium, calcium, potassium, zinc, and phosphate, which may result in low serum levels of these minerals. The brand name is Platinol®.

Citrisource®. Brand name of an oral supplement that can be used to fortify clear liquid diets. Because it is fat-free and low in sodium and potassium, it can be used for renal and pre- or postoperative patients. See Appendix 39.

Citrotein®. Brand name of a high-protein, low-fat, flavored supplement for oral feeding. It comes in powder form that readily dissolves in cold water and is appropriate for clear liquids. See also Appendix 39.

Citrovorum factor (CF). Growth factor for the organism *Leuconostoc citrovorum;* has been given the name *folinic acid,* a biologically active form of the vitamin folic acid.

Citrulline. An amino acid first isolated from watermelon, although it does not appear to be a constituent of common proteins. It is closely related to arginine and is involved in the *urea cycle.*

Citrullinemia. An inherited metabolic disorder of the urea cycle due to lack of the enzyme argininosuccinic acid synthetase. Characteristic symptoms are elevated levels of citrulline in the blood, urine, and cerebrospinal fluid; hyperammonemia; persistent vomiting; ataxia; seizures; and progressive mental retardation. Until recently, this disorder was fatal within a few weeks. *Nutrition therapy:* protein restriction at 1 to 2 g/kg per day, depending on individual tolerance. Give the highest level of protein tolerated and provide sufficient kilocalories to meet energy needs. Use modular formulas or dilute infant formula to meet the prescribed protein level. Supplement with L-arginine (1 g/day for infants; 2 g/day for older children) to prevent arginine deficiency, and sodium benzoate (0.25 g/kg/day) to aid in ammonia excretion.

Cl. Chemical symbol for chlorine.

Clearance test. Test for kidney function. It measures the excretory efficiency of the kidney in "clearing" blood of a substance over a given period of time. It may also be used as a test for liver function to determine the ability of the liver to remove a substance from the blood.

Clearing factor. A lipoprotein lipase present in various tissues, notably the heart, lungs, and adipose tissue. It appears in the plasma following a meal and "clears" the blood of its turbid, milky appearance by hydrolyzing the triglycerides present in the chylomicrons and in lipoproteins. It also mobilizes fatty acids from the fat depots. This enzyme is enhanced by *heparin* and activated by calcium and magnesium ions.

Cleft palate. Congenital deformity characterized by incomplete closure of the lateral halves of the palate or roof of the mouth. This presents feeding difficulties, as food passes through the roof of the mouth into the nasal cavity. *Nutrition therapy:* inability to suck adequately presents a problem. Food may back up in the nose and can cause choking. In the newborn, a medicine dropper or a plastic bottle and a soft nipple with an enlarged hole may be used. Milk can be squeezed a little at a time in coordination with the infant's chewing motions. Pureed baby foods may be diluted with milk and spoon-fed. Feed slowly in a more upright position (90 degree angle) and with frequent burping. For young children, special feeding devices are needed. Most patients with cleft palate usually undergo surgical repair, after which a normal diet is gradually introduced. Start with pureed or very soft foods, then progress to other textures. Avoid foods that are hard and sticky.

Clindamycin. An antibiotic used in the treatment of serious respiratory tract and skin infections caused by susceptible organisms. The drug has a persistent bitter taste. It may cause nausea, vomiting, bloating, esophagitis, anorexia, weight loss, and hypokalemia. In addition, nonspecific colitis characterized by severe diarrhea and abdominal cramps may develop 2 to 9 days after initiation of treatment or even several weeks after the drug has been discontinued. The brand names are Cleocin® and Dalacin®.

Clofibrate. An antilipemic drug used as an adjunct to diet in the management of Type III hyperlipoproteinemia that does not respond to diet therapy alone. The drug has an unpleasant aftertaste and decreases taste acuity. It decreases the absorption of carotene, vitamin B_{12}, iron, and electrolytes; increases the urinary excretion of protein; and may decrease the activity of intestinal disaccharidases. Adverse gastrointestinal effects include nausea, dyspepsia, diarrhea, stomatitis, and gastritis. The brand name is Atromid-S®.

Clonidine. An imidazole derivative antihypertensive agent. It may cause constipation, dry mouth, loss of appetite, and weight gain due to sodium and fluid retention. A mild sodium-restricted diet may be necessary. Brand names are Catapres® and Combipres®.

Clotting time. The time it takes for blood to clot or coagulate. Normally, it is 4½ minutes. A delay in blood clotting time indicates vitamin K deficiency.

CMP. Abbreviation for cytidine monophosphate. See *Cytidine phosphates*.

Co. Chemical symbol for cobalt.

Co I, Co II, Co III. Abbreviations for *coenzyme I, coenzyme II,* and *coenzyme III,* respectively.

CoA. Abbreviation for *coenzyme A*.

Cobalamin. Collective term given to the many forms in which vitamin B_{12} may appear in animal tissues, all of which contain cobalt as an integral part of the molecule. The predominant forms are methylcobalamin, adenosylcobalamin, and hydroxocobalamin. Cyanocobalamin is the commercially available form.

Cobalt (Co). A trace mineral normally present in animal tissues as an integral part of vitamin B_{12}. The need for cobalt other than its role in the synthesis of vitamin B_{12} is not known. Only 10% of the cobalt in the body is in vitamin form; inorganic cobalt is found in the plasma attached to albumin and deposited in the bone, muscle, and other tissues. Whether inorganic cobalt has a role in human metabolism has not been demonstrated. There are no documented cases of what might be considered cobalt deficiency. The average American consumes 5 to 10 μg/day cobalt, mainly as inorganic cobalt from vegetables and whole grains and as cobalamin in animal foods. It is readily absorbed and is excreted mainly in the urine, with little body storage. Cobalt salts have been used pharmacologically in the treatment of anemias that are refractory to iron, folate, and vitamin B_{12}. When taken in large doses, cobalt can produce toxic effects, including goiter, hypothyroidism, hypotension, and heart failure. Toxicity from dietary intake has not been a problem.

Cobamide. Generic name given to vitamin B_{12}-containing coenzymes. The two cobamides active in human metabolism are methylcobalamin and 5-deoxyadenosylcobalamin.

Codex Alimentarius Commission. Committee created by the Food and Agriculture Organization and the World Health Organization to develop international food standards on a worldwide, regional, or group-of-countries basis and to publish these standards in a food code called *Codex Alimentarius*. These food standards aim at protecting consumers' health and ensuring fair practices in the food trade.

Coefficient of digestibility. Percentage portion of the ingested food constituent retained in the body and not excreted in the feces. The *average* coefficients of digestibility of foods are 98% for carbohydrate, 95% for fat, and 92% for protein.

Coenzyme. An organic dialyzable and heat stable enzyme cofactor. It is required for the activity of many enzymes. Coenzymes usually contain vitamins as part of their structure, and they function as acceptors of electrons and functional groups.

Coenzymes I and II (Co I and Co II). Hydrogen and electron transfer agents known to be a complex of nicotinamide, D-ribose, phosphoric acid, and adenine. They play a vital role in metabolism and participate in many forms of biologic oxidation. Their two major functions are the removal of hydrogen from certain substrates in cooperation with dehydrogenases and the transfer of hydrogen (or electrons) to another coenzyme in the hydrogen transport series. Coenzyme I is also called nicotinamide adenine dinucleotide (NAD) or diphosphopyridine nucleotide (DPN). Coenzyme II is also called nicotinamide adenine dinucleotide phosphate (NADP) or triphosphopyridine nucleotide (TPN). It is a complex of nicotinamide, adenine, two molecules of D-ribose, and three molecules of phosphoric acid.

Coenzyme III (Co III). Cysteine sulfinic acid dehydrogenase, an enzyme that catalyzes the oxidation of cysteine sulfinic acid to cysteic acid. This dehydrogenase has been named coenzyme III because it requires as coenzyme a pyridine nucleotide similar in structure to *coenzymes I* and *II*. It is found in microorganisms and animal tissues, particularly in the liver, kidneys, and heart.

Coenzyme A (CoA). Pantothenic acid joined to adenosine phosphate by a pyrophosphate bridge and to beta-mercaptoethylamine by a peptide bridge. It functions in acetylation and acylation reactions, oxidation of ketoacids and fatty acids, formation of acetylcholine, and synthesis of triglycerides, phospholipids, steroids, and porphyrins.

Coenzyme Q (CoQ). A lipidlike substance, similar to vitamins K and E in its chemical makeup, that belongs to a group of compounds known as *ubiquinones*. It is found in practically all living cells and appears to be concentrated in the mitochondria. The form that appears to be biologically active has long carbon side chains (about 30 carbon atoms). It helps in the release of energy in the oxidation–reduction system and the respiratory chain.

Cofactor. General term given to the nonprotein fraction of an enzyme necessary for its full activation. Cofactors are divided loosely into three groups: a *prosthetic group,* which is firmly attached to the protein portion of the enzyme; a *coenzyme,* which is easily dissociated from the protein enzyme; and *metal activators,* which are mono- and divalent cations such as K, Mn, Mg, Ca, and Zn and which may be either firmly or loosely bound to enzyme protein. See also *Enzyme*.

Colchicine. Alkaloid from the root of a lily plant, *colchicum autumnale,* that is of value in the treatment of gout. It may decrease the absorption of vitamin B_{12}, fat, carotene, sodium, potassium, and iron and may decrease lactase activity. It may also cause nausea, vomiting, diarrhea, abdominal cramping, loss of appetite, and weight loss. Brand names are Colbenemid® and Proben-C®.

Colectomy. Surgical excision of the colon to treat cancer or severe ulcerative colitis. *Nutrition therapy:* provide a low-residue diet several days before surgery. Postsurgical feeding is by parenteral means initially, followed by per os feeding with the same dietary modifications as for *rectal surgery*.

Colestipol. Basic anion-exchange resin that binds bile acids in the intestine, forming a nonabsorbable complex that is excreted in the feces. It is antilipemic and is used for the treatment of primary Type IIa hyperlipoproteinemia. High long-term use causes folate deficiency and depletion of fat-soluble vitamins due to malabsorption, especially of vitamins A and E. It may also cause abdominal discomfort and distention, constipation, and occasionally fecal impaction. The brand name is Colestid®.

Colitis. Acute or chronic inflammation of the colon or large bowel due to one or more causes. *Mucous colitis* is characterized by abdominal distress, constipation or diarrhea, and the passage of mucous or membranous masses in the stools. *Spastic colitis,* also called irritable colon or "unstable colon," is associated with increased tonus or abnormal activity of the colon. *Ulcerative colitis* is a chronic condition characterized by an ulceration of the mucosa and passage of pus and blood in the stools. For *Nutrition therapy:* see *Crohn's disease*.

Collagen. An albuminoid; the insoluble protein of connective tissue, bones, tendons, and skin. It is resistant to animal digestive enzymes but is hydrolyzed to soluble gelatin by boiling in water, dilute acids, or alkalis. Collagen constitutes one-third of the body proteins and possesses remarkable qualities. In some tissues, collagen is dispersed as a gel to stiffen structures, as in vitreous humor of the eye. Collagen may be bundles in tight parallel fibers to provide strength, as in tendons. In bones, collagen occurs as fibers arranged at right angles to resist mechanical shear in any direction.

Colon. The large intestine, which extends from the cecum to the rectum. It has four parts: the ascending colon, transverse colon, descending colon, and sigmoid colon. For its function, see *Absorption, nutrients*.

Colostomy. Formation of an artificial outlet from the colon through the wall of the abdomen. The need arises when part of the colon becomes obstructed or has to be removed because of a diseased condition. The colostomy opening may be in the ascending, transverse, or descending portion of the colon. *Nutrition therapy:* start with a clear liquid diet and progress gradually to one low in residue. Then give a soft or low-fiber diet as tolerated. A liberal supply of calories and protein (at least 1.5 times the recommended dietary allowance) will speed up recovery and prevent weight loss. During the first 6 to 8 weeks after surgery, encourage the intake of soluble fiber (oatmeal, applesauce, bananas, rice) to bind the stool; gradually introduce insoluble fiber (wheat, bran, corn, nuts) after ostomy output has normalized. Most patients with colostomies can return to a normal diet. Dietary modifications depend on individual tolerance. Restrict only those foods that are not tolerated and that produce excessive gas or loose stools. Avoid foods that may cause stoma obstruction, such as raw celery, raw cabbage, nuts, seeds, and foods with kernels; also avoid gas-forming foods, including asparagus, beans, broccoli, brussels sprouts, cabbage, cauliflower, and onions. Take plenty of fluids, at least 8 to 10 cups per day or more if the ostomy output is excessive. If diarrhea occurs, omit insoluble fibers because they tend to reduce intestinal transit time; reintroduce gradually when the diarrhea subsides. Reinstitute a soft diet with adequate fluids if there is abdominal pain, distention, or decreased ostomy output. Fat restriction may be necessary if steatorrhea occurs.

Colostrum. First milk secreted by the mammary gland a few days after parturition. Compared to later milk secretion, it is higher in protein content and immunoglobulins that contain antibodies responsible for the immunity of the newborn. Colostrum is also higher in beta-carotene, riboflavin, and niacinamide content but has less fat and carbohydrate than later milk.

Commodity Supplemental Food Program (CSFP). Administered by the U.S. Department of Agriculture (USDA), this program provides commodities to infants from low-income families, children up to age 6, and pregnant or lactating mothers. Also distributes commodities to senior centers, school breakfast and lunch programs, and food banks.

Community Nutrition Institute (CNI). A private, nonprofit organization that provides nutritional and technical assistance to individuals, groups, and programs. Its major activity is to publish the *CNI Weekly Report,* a newsletter that provides information on legislative and bureaucratic actions that affect all aspects of nutrition.

Compleat®. Brand name of a blenderized tube feeding formula made with meat, vegetables, and fruit. Each liter contains 4.2 g of fiber. Also available as Compleat® Regular, which contains milk, and Compleat® Modified, which is lactose-free. See Appendix 39.

Comply®. Brand name of a high-calorie, lactose-free nutritional supplement containing sodium and calcium caseinates, maltodextrin, sucrose, and corn oil. See Appendix 39.

Congestive heart failure (CHF). Reduced efficiency of heart pumping, with less circulation of blood to the body tissues. It is caused by a coronary or other cardiac disease, lung disorder, or severe anemia. Clinical signs include edema, dyspnea, cachexia, and decreased renal function. *Nutrition therapy:* the overall objective is to supply nutrients adequately without overworking the heart. Provide small, frequent meals of soft-textured and easy-to-eat foods. Restrict calorie intake to reduce weight if the person is obese or overweight and maintain desirable body weight. In severe failure and with hypermetabolism, increase calories by 30% to 50% above basal energy requirements. Protein intake may be increased from 1.0 to 1.5 g/kg body weight if malnutrition is present. Restrict sodium to 1 to 2 g/day if there is fluid retention; restrict further to 0.5 to 1.0 g/day, if necessary, and liberalize to 2 to 3 g/day when the patient improves. Fluid restriction may not be needed if the patient is compliant with sodium restriction; if necessary for some patients, restrict fluids to 0.5 to 0.7 mL/kcal or about 1000 to 1500 mL/day. Drug–nutrient interactions primarily associated with diuretics cause loss of nutrients which may require supplementation, especially of potassium, magnesium, zinc, folic acid, and vitamin B_6.

Congregate Meal Program. Administered by the Department of Health and Human Services (DHHS), this program provides senior citizens with at least one hot meal a day, supplying one third of the recommended dietary allowances. Meals are served in churches, community centers, and schools. The gathering also aims to provide educational sessions, social interactions, and companionship for the elderly.

Conjunctival xerosis. Condition characterized by dryness, thickening, pigmentation, lack of luster, and diminished transparency of the conjunctiva of the exposed part of the eyeball. It is due to keratinization of the epithelial cells of the conjunctiva and is seen in *vitamin A* deficiency.

Conservation of energy. Energy cannot be created or destroyed, although it can be changed from one form to another. Thus the sum of all forms of energy remains constant. For example, the amount of heat (or energy) obtained from food is equal to the amount of energy expended in the forms of heat, work done, and energy in waste products (assuming that there is no gain or loss of weight). Energy intake greater or less than its expenditure would result, respectively, in storage or utilization of potential energy in the form of depot fat.

Constipation. Infrequent or difficult bowel movement; retention of the feces in the colon beyond the normal length of emptying time. There are many causes of constipation, including organic and functional impairments, inactivity, and less than adequate intake of fluids and dietary fiber. Chronic or habitual constipation often causes headache, a feeling of malaise, loss of the power to concentrate, foul breath, and dullness of special senses. There are two kinds of constipation: *atonic,* which indicates a lack of muscle tonus of the intestines, resulting in stasis of the colon, and *spastic* which pertains to narrowing and uncontrolled contractions of the colon. *Nutrition therapy:* diet is not a cure but provides relief or comfort to the patient. In atonic constipation, a high-fiber diet (35 to 40 g/day) will stimulate peristalsis, provide bulk to the intestinal contents, and help retain water in the feces to facilitate bowel movement. Good sources of fiber are whole grain cereals, legumes, corn, raw and dried fruits, and leafy vegetables. (See Appendix 32.) Extra fiber can be incorporated by adding 1 to 2 tablespoons of bran each day to the usual foods. Increase fluid intake to 8 to 10 cups/day. The use of prunes or prune juice, which contain a natural laxative, may be helpful. Use fiber-containing formulas for those on tube feeding. (See Appendix 39.) In spastic constipation, a low-fiber diet will prevent undue distention and stimulation of the bowel. See *Diet fiber-modified*.

Continuous ambulatory peritoneal dialysis (CAPD). A self-dialysis technique that does not require a machine and allows the patient to be ambulatory and complete the dialysis at home. The dialysate is left in the peritoneum for 4 to 6 hours and then drained out manually. This process of instillation, dwelling, and drainage is repeated four to five times a day, making it a 24-hour treatment. See also *Peritoneal dialysis*.

Continuous arteriovenous hemofiltration (CAVH). Method of removing fluid from the blood by ultrafiltration, using the force of the patient's own blood pressure to remove electrolytes, urea, and other solutes. It is indicated for acute renal failure in critically ill patients who are hemodynamically unstable and cannot be dialyzed, have severe fluid overload, or have multiple organ failure. The fluid volume removed may be replaced by infusion of parenteral nutrients, thus providing calories and protein at the same time.

Continuous peritoneal dialysis (CCPD). A dialysis technique applied at home using automated devices for night therapy. It provides three or four peritoneal exchanges during the night, leaving 1 to 2 liters of dialysate solution in the peritoneal cavity for 12 to 15 hours during the day or until the start of the night automated exchanges. This procedure allows daytime freedom from exchanges. See *Peritoneal dialysis*.

Copper (Cu). Essential trace mineral in humans, all vertebrates, and some lower animal species. It is a component of several enzymes involved in various functions, including the release of energy in the cytochrome system; melanin production in skin; absorption and transport of iron; formation of hemoglobin and transferrin; production of catecholamine in the brain and adrenal gland; metabolism of glucose, cholesterol and phospholipid; synthesis of collagen and elastin; and detoxification of superoxide radicals. Copper deficiency in humans is not common but has been observed in long-term use of total parenteral nutrition inadequate in copper, and in association with hypoproteinemia, malabsorption syndromes, jejunoileal bypass, nephrotic syndrome, high zinc intake, and *Menke's disease,* a rare genetic disorder of copper metabolism. Excess loss of copper, as through kidney dialysis, can also cause a deficiency. Symptoms of deficiency include decreased serum copper and *ceruloplasmin,* anemia similar to that caused by iron deficiency, impaired glucose tolerance, poor wound healing, immune defects, and central nervous system and cardiovascular disorders. Copper is widely distributed in foods; good food sources are liver, shellfish, legumes, nuts, and whole grains. High intakes of zinc, ascorbic acid, iron, and fiber interfere with the bioavailability of copper. The body's copper content is homeostatically controlled, and there is little storage of the excess, except in certain disease states such as *Wilson's disease*. Copper toxicity has been seen in renal dialysis in which an excess of copper was transferred from the dialysis bath to the patient or when copper tubing was used in hemodialysis machines. Acute copper poisoning has occurred from accidental ingestion of copper salts by children, soft drinks served from defective equipment, and consumption of foods and beverages cooked or stored in copper-lined vessels. Symptoms of acute poisoning are vomiting, dizziness, and diarrhea with bleeding, and, in severe cases, circulatory collapse, severe hemolysis, and liver and kidney failure. See Appendix 2 for the estimated safe and adequate daily dietary intake.

CoQ. Abbreviation for *coenzyme Q*.

Cori cycle. Series of reactions that make possible the conversion of muscle glycogen to blood glucose. Muscle glycogen cannot contribute directly to blood glucose because of lack of glucose-6-phosphatase, which is present only in the liver. In the absence of oxygen, lactic acid formed in the muscles diffuses into the blood and is taken up by the liver, which is able to convert lactic acid to glucose and, eventually, to liver glycogen.

Corneal vascularization. Formation of fine capillary blood vessels on the periphery of the cornea due to congestion of the normal limbal plexus. It may be nonspecific and may occur in any inflammatory or irritating process affecting the cornea. It is also seen in *riboflavin deficiency*.

Corneal xerosis. Hazy, milky, or opaque appearance of the cornea, usually most marked in the lower central area. It is due in part to cellular infiltration of the corneal stroma; it is seen in *vitamin A deficiency*.

Coronary artery bypass. An open-heart surgery to bypass a blockage or narrowing in a coronary artery; a prosthesis or section of a blood vessel is grafted onto one of the coronary arteries and connected to the aorta. *Nutrition therapy:* immediately after bypass surgery, the diet is advanced from clear to full liquids, followed by solid foods as tolerated. Supplemental tube feeding may be necessary if oral intake is inadequate for 10 or more days. Parenteral nutrition may be necessary if the gut is not functional. On discharge, follow the Step-One diet of the National Cholesterol Education Program (NCEP) and monitor serum cholesterol and triglyceride levels. If serum cholesterol is not within acceptable level after 3 months, follow the NCEP Step-Two diet for 6 months before initiating drug therapy. Limit sodium intake to 2 to 3 g/day. Also limit consumption of alcohol and caffeine. Emphasize foods rich in potassium if the patient is on thiazide diuretics; potassium supplement may be necessary if dietary intake is poor. See *Diet, cholesterol-restricted, fat-controlled*.

Coronary heart disease (CHD). Condition characterized by inadequate coronary circulation because of narrowing of the lumen or complete occlusion of the coronary arteries due to atherosclerosis, thrombus formation, or embolism. As a consequence, the heart is deprived of its oxygen and its nutrient supply. Epidemiologic and clinical investigations have identified several risk factors associated with susceptibility to coronary heart disease. An individual is considered to have high risk status if he or she has borderline high cholesterol and two of the following CHD risk factors: hypertension, high-density lipoprotein-cholesterol below 35 mg/dL (< 0.9 mmol/L), diabetes mellitus, $\geq 30\%$ overweight, family history of premature CHD, cigarette smoking, and history of cerebrovascular or peripheral vascular disease. *Nutrition therapy:* weight reduction and/or maintenance of healthy weight by an appropriate combination of physical activity and caloric intake is recommended. Follow the NCEP diet to lower blood cholesterol. Increase the intake of complex carbohydrates and soluble fiber, such as legumes, oats, oat bran, fruits, and tubers. The consumption of ocean fish and shellfish two or three times per week may be beneficial. A moderate intake of alcohol (1 to 2 oz/day) is recommended; high alcohol consumption can increase triglycerides and high-density lipoprotein cholesterol. There is no specific recommendation for the consumption of coffee in the management of hyperlipidemia; avoid or limit caffeine if it causes tachycardia, arrhythmia, and palpitation. See also *Diet, cholesterol-restricted, fat-controlled*.

Cor pulmonale. A disorder due to hypertension of the pulmonary circulation; characterized by hypertrophy of the right ventricle of the heart. Symptoms and clinical signs include coughing, fatigue, cyanosis, and hypoxia. About 80% of patients with cor pulmonale have chronic obstructive pulmonary disease. *Nutrition therapy:* give foods of high nutrient density in small, frequent meals. Monitor fluid and potassium intake and restrict sodium if edema is present. See also *Chronic obstructive pulmonary disease*.

Corticosteroid. Term applied to the steroid hormones secreted by the adrenal cortex and other natural or synthetic compounds having the same activity as the *adrenocortical hormones*. These hormones are used for their anti-inflammatory and immunosuppressant properties. Cortisone, hydrocortisone, and prednisone are among the corticosteroids available. Long-term use of these drugs can cause generalized protein depletion and can have a variety of nutritional implications, including increased need for ascorbic acid, folic acid, vitamin B_6, and vitamin D; decreased wound healing; decreased bone formation; decreased absorption of calcium and phosphorus; and increased urinary excretion of ascorbic acid, potassium, zinc, and nitrogen. Some brand names are Decadron®, Deltasone®, Medrol®, and Merticorten®.

Corticosterone. Steroid hormone found in the adrenal cortex. It influences carbohydrate metabolism and electrolyte balance.

Cortisol. 17-Hydroxycorticosterone, the major adrenal cortical steroid influencing carbohydrate metabolism. It increases the release of glucose from the liver, stimulates gluconeogenesis from amino acids, and decreases the peripheral utilization of glucose. Cortisol is released into the blood and transported to the tissues in combination with a globulin as transcortin.

Coumarin. An anticoagulant with a hypoprothrombinemic effect, especially when the dietary intake of vitamin K is inadequate. Prescribed for treatment of embolism and thrombosis, but not in cases of hemorrhagic risks.

Cow's milk allergy. Syndrome of malabsorption, diarrhea, and vomiting associated with variable villus atrophy attributable to an allergy to cow's milk protein. Exclusion of cow's milk results in clinical and histologic remission, while challenge produces inflammatory changes and villus atrophy within 24 hours. Additional symptoms include urticaria, asthma, and anaphylaxis after milk intake. Some patients also exhibit a low-grade colitis that responds to elimination of cow's milk. It is a disease of neonates; after the age of 2 years, most children overcome their intolerance. *Nutrition therapy:* exclude cow's milk and all products made with cow's milk. If breast feeding or use of human milk is not possible in feeding the infant, the American Academy of Pediatrics recommends the use of hypoallergenic formulas based on casein hydrolysates, such as Nutramigen® and Pregestimil®. Goat's milk and soy protein or whey protein hydrolysate formulas may not be acceptable alternatives for some infants allergic to cow's milk. See *Diet, milk-restricted*.

C. Peptide. A part of the endogenous insulin, but not present in exogenous (manufactured) insulin. Its measurement indicates functional pancreatic beta cells. This helps in determining whether a person is Type I (insulin-dependent) or Type II (non-insulin-dependent) diabetic.

Cr. Chemical symbol for chromium.

Creatine. Methyl guanidine derivative of acetic acid. As creatine phosphate, it acts as a source of *high-energy phosphate* and plays an essential part in the release of energy in muscular contraction. It is present in the muscle, brain, and blood; trace amounts are also normally present in the urine. It is excreted in abnormally large amounts in conditions accompanied by failure to burn carbohydrate (as in starvation, diabetes mellitus, and severe liver disease), in diseases of muscles (as in myasthenia gravis and muscular dystrophy), and in conditions accompanied by excessive tissue breakdown (as in fevers and wasting diseases).

Creatine index. Measure of the ability of the body to retain ingested creatine, as determined under standard conditions. The percentage retained is high in hypothyroidism and low in hyperthyroidism and other conditions accompanied by muscle wasting.

Creatinine. A waste product of *creatine*, formed largely in the muscles by irreversible nonenzymatic removal of water from creatine phosphate. It is excreted in the urine through the glomeruli of the kidney. The excretion of creatinine in the urine is constant from day to day for a given individual. Thus urinary creatinine is an index of muscle mass when kidney function is normal. It is increased in body protein breakdown (catabolism), as in trauma and surgery. One gram of urinary creatinine is equivalent to about 17 to 20 kg body mass. The serum creatinine is a useful index of renal function; the normal level is 0.6 to 1.5 mg/dL (53 to 132 μmol/L). It may be slightly elevated in severe volume depletion and in individuals with acromegaly or a large muscle mass. Serum creatinine is increased in diseases of the kidney when 50% or more of the glomeruli is damaged. A value over 5 mg/dL (442 μmol/L) is of serious significance. Values below 0.7 mg are of no known significance. See *Creatinine-height index*.

Creatinine clearance test. Test for renal function based on the rate at which ingested creatinine is filtered through the glomeruli. Normally this is 70 to 130 mL/min. In severe renal disease, creatinine clearance may fall to 5 mL/min. The optimum protein intake per kilogram of body weight may be determined from the creatinine clearance (CC), as follows:

CC (mL/min)	PROTEIN (g/kg/day)
20 to 30	0.75 to 0.8
10 to 20	0.5 to 0.75
<10	0.5

If the urine collection has not been done, the following formula may be used to estimate creatine clearance (CC) from lean body mass, serum creatinine, and age:

$$\text{In men: CC} = \frac{(140 - \text{age}) \times \text{weight (kg)}}{\text{serum creatinine} \times 72}$$

In women: Use the same equation and multiply the result by 0.85. The creatinine clearance parallels the glomerular filtration rate. See also *Glomerular filtration rate*.

Creatinine coefficient. Amount of urinary creatinine excreted in 24 hours/kg body weight. Average excretion is 23 mg/kg body weight for men and 18 mg/kg body weight for women. When expressed in terms of body size, the creatinine excretion of different individuals of the same age and sex is constant from day to day. The creatinine coefficient may thus be regarded as an index of muscle mass. However, creatinine excretion decreases with age, and a consensus is lacking on the effects of exercise, stress, steroids, and amount of meat intake. Creatinine excretion can be expressed in terms of an expected value for body weight. See *Creatinine-height index*.

Creatinine-height index (CHI). An indirect measurement of somatic protein or lean body mass if renal function is normal and fluid intake is adequate. The amount of *creatinine* in a 24-hour urine output is measured, and compared with the expected value for a normal individual of the same sex and height. The index is calculated from the following formula:

$$\text{CHI} = \frac{\text{actual 24-hour creatinine excretion (mg)}}{\text{expected 24-hour creatinine excretion (mg)}} \times 100$$

See Appendix 29.

Cretinism. Chronic condition occurring in fetal life or early infancy due to deficient thyroid activity. It is characterized by arrested physical and mental development, dry skin, chubby hands, large protruding tongue and abdomen, and low basal metabolic rate. The disorder usually occurs in areas where goiter is endemic and the diet is deficient in iodine. *Nutrition therapy:* the use of iodized salt (76 μg iodine/g salt) reduces the incidence of cretinism. Seafoods and saltwater fish are good sources of iodine. See also *Myxedema*.

Criticare HN®. Brand name of a monomeric (elemental) formula for oral or tube feeding; contains free amino acids and small peptides. It is lactose-free, low in fat, and has minimum residue. See also Appendix 39.

Crohn's disease. Also known as "regional enteritis." A disorder, usually chronic and inflammatory, that affects the digestive tract. Ordinarily, it involves the colon and the ileum, but it may also affect areas anywhere from the mouth to the anus. Symptoms include ulcerative lesions, chronic diarrhea, and upper abdominal pain. Scarring and thickening of the intestinal wall result in intestinal obstruction and malabsorption. Of unknown cause, it has no effective medical treatment as yet. Corticosteroids, immunosuppressive agents, and sulfasalazine are used to suppress inflammation, but long term use of these medications promotes catabolism and poor utilization of nutrients. One or more intestinal resections may also be required, resulting in reduced absorptive capacity. *Nutrition therapy:* protein-energy malnutrition (PEM) is a common problem. Increase caloric and protein intake about 50% more of required needs. Due to the occurrence of certain food intolerances and lack of appetite, the patient has reduced intake of vitamins and minerals, particularly vitamin A, B_1, B_2, B_6, B_{12}, folate, calcium, magnesium, potassium, selenium, zinc, and iron. Supplemental multivitamins and minerals are therefore beneficial, especially if the patient is on anti-inflammatory drugs. Dietary fat should be limited (40 to 50 g/day) if steatorrhea is evident. Give medium-chain triglycerides (MCT) and supplemental calcium, zinc, and magnesium if steatorrhea is severe. If hyperoxaluria is present, avoid grapefruit, tea, and cola drinks. Restrict lactose if there is intolerance. Stop oral feeding to rest the bowel during acute flare-ups, then progress from an elemental or minimal residue formula to a moderate-residue diet. Parenteral nutrition may be indicated in severe cases if nutritional needs cannot be met orally. Gradually introduce fiber and regular foods as tolerated during periods of remission. See *Diet, residue-modified*. See also *Bowel resection*. For severe cases of Crohn's disease an elemental diet is given, progressing to a low-residue diet in small, frequent feedings.

Crucial™. Brand name of a complete elemental solution for critically ill patients with head injury, burns, trauma, and hypermetabolism. It is peptide-based, high in protein (94 g/L), and contains arginine and glutamine. See Appendix 39.

Crude fat. The fat, lipids, and other fat-soluble materials in food that are extractable by fat solvents, such as petroleum ether, ethyl ether, chloroform, or benzene. In the *proximate analysis* of foods, this is sometimes referred to as ether extract or ether-soluble fraction.

Crude fiber. The ash-free, insoluble residue left after boiling a food sample first with dilute acid and then with dilute alkali to simulate gastric and intestinal digestion. It is commonly called "indigestible carbohydrate." See also *Fiber*.

Crude protein. Food nitrogen content, as derived from protein and nonprotein nitrogenous materials. Crude protein is obtained by multiplying the nitrogen content of foods (as determined by the Kjeldahl process) by the factor 6.25. See *Kjeldahl method*.

Cryptoxanthin. A yellow pigment belonging to the *carotenoid* group that can be converted into *vitamin A* in the body. It is one of the chief pigments in yellow corn, paprika, and oranges.

CT. Abbreviation for *calcitonin*.

CTP. Abbreviation for cytidine triphosphate. See *Cytidine phosphates*.

Cu. Chemical symbol for copper.

Cultural foods. Refers to the food practices of diversed population groups. Coming from different ancestry, race, and ethnic backgrounds, these people may be newly arrived immigrants or may have settled in the United States for generations. Many practice their native culture while others adapt the American diet or use foods available locally. For the unique food practices of these groups, see Appendix 47.

Curling's ulcers. Deep ulcers that usually develop in the duodenum of patients suffering from extensive burns, causing major bleeding. *Nutrition therapy:* initially, total parenteral nutrition may be required, because the patient cannot be fed orally. When oral feeding can be resumed, follow dietary modifications for *burns*.

Cushing's syndrome. Condition due to hypersecretion of the adrenocortical hormones because of hyperplasia or tumor of the adrenal cortex, basophilic adenoma of the pituitary gland, or prolonged use of large doses of adrenocorticotropic hormone (ACTH). The chief features of the syndrome include obesity of the trunk, face, and buttocks; purplish striae over the abdomen; pigmentation of the skin; hypertension; hyperglycemia; excessive growth of hair; and loss of sexual function. *Nutrition therapy:* correct *hyperglycemia, hypertension,* and *obesity*.

CVD. Abbreviation for *cardiovascular disease*.

Cyanocobalamin. Synthetic form of vitamin B_{12} used in vitamin pills and pharmaceuticals. It is a *cobalamin* with a cyanide group attached to the cobalt atom.

Cyclamate. An artificial, nonnutritive sweetener which is the sodium or calcium salt of cyclohexylsulfamic acid. It is 30 times sweeter than sucrose. Used widely in the 1960s, it was banned in 1970 for commercial use in the United States pending additional studies to resolve some problems regarding its safety for human consumption. On June 10, 1985, the National Academy of Sciences (NAS) panel of scientists reported that studies did not indicate any carcinogenicity of cyclamates in human beings. Petition for reapproval by the Food and Drug Administration (FDA) is pending, based on data from about 75 new studies in favor of cyclamate use. At least 40 other countries have approved the use of cyclamates.

Cyclinex®. Brand name of a nonessential amino acid-free medical food for infants and toddlers (Cyclinex®-1) and children and adults (Clyclinex®-2) with a defect in a urea cycle enzyme or with gyrate atrophy of the choroid and retina. See Appendix 37.

Cyclophosphamide. An alkylating agent used alone or as a component of various chemotherapeutic regimens in the treatment of lymphomas and other malignant diseases. Anorexia, nausea, and vomiting occur commonly, especially at high doses. Occasionally, diarrhea, hemorrhagic colitis, mucosal irritation, and oral ulceration may occur. The agent is excreted in breast milk. The brand names are Cytoxan® and Neosar®.

Cycloserine. An antibiotic used in treatment of tuberculosis and urinary tract infections caused by susceptible bacteria. It acts as a pyridoxine antagonist and decreases protein synthesis. It may also decrease the absorption of calcium, magnesium, and folate. The brand name is Seromycin®.

Cyclosporine. An immunosuppressant drug used with glucocorticoids to prevent rejection in kidney, heart, and liver transplantation. Adverse side effects include anorexia, gingival hyperplasia, nausea, vomiting, diarrhea, and abdominal discomfort. It may also cause elevated blood lipids, glucose, potassium, and uric acid, and decreased magnesium. Excreted in breast milk and is contraindicated in breast feeding. Brand name is Sandimmune®.

Cystathionine. An intermediate product in the conversion of methionine to cysteine.

Cystathioninuria. Inborn error of metabolism characterized by mental retardation and elevated *cystathionine* in blood and urine. It is probably a defect in the methionine–cysteine pathway. In some instances, the biochemical abnormality can be corrected by administration of pyridoxine.

Cysteine. Alpha-amino-beta-mercaptopropionic acid, a sulfur-containing amino acid formed by the reduction of cystine. Cysteine is a reducing agent, a precursor of taurine, and also part of glutathione. It is considered an essential amino acid in the preterm infant.

Cystic fibrosis (CF). Also called mucoviscidosis due to the highly viscous secretions of the mucus-producing exocrine glands. It is a life-threatening congenital disease occurring in infants and young children. CF is characterized by generalized dysfunction of the exocrine glands involving the pancreas, respiratory system, salivary glands, gastrointestinal tract, biliary system, and paranasal glands. The sweat contains large amounts of sodium chloride and, to a lesser extent, potassium. Abnormal mucus secretion causes obstruction of mucus-producing cells or organ passages, resulting in asthma, recurrent pneumonia, and sinusitis as common symptoms of the disease. Lack of pancreatic enzymes interferes with the utilization of protein, fat, and carbohydrate. Vitamin deficiencies occur because of digestive defects, and death usually results from malnutrition or bronchopneumonia. *Nutrition therapy:* provide a high-calorie, high-protein, liberal-fat, and high-sodium diet, supplemented with fat-soluble vitamins, considering the age and food tolerances of the child. Dietary intakes are increased to 120% to 150% of required needs for calories and 150% to 200% for protein. Supplementation with fat-soluble vitamins (150% of RDA) may be necessary if there is fat malabsorption. Water-miscible preparations of fat-soluble vitamins may be indicated. Monitor for iron, copper, magnesium, selenium, and zinc status and supplement if necessary. An additional salt intake per day (⅛ to ¼ tsp for infants and ½ to 1 tsp for children) is necessary when there is excessive loss during strenuous physical activity, extremely hot weather, and febrile illness. Besides sodium and chloride, other minerals that need monitoring are potassium, calcium, iron, zinc, and selenium. Fat restriction is not necessary if a pancreatic enzyme preparation (e.g., Cotazym®, Pancrease®, and Viokase®) is taken at each meal and with snacks. The capsule-encased enteric coated pancreatic enzyme should be swallowed intact without chewing and should not be crushed or given with foods with an alkaline pH because the enteric coating would be destroyed. For infants, the enzyme powder can be added directly to the conventional infant formula or placed in the infant's mouth if the infant is on breast-milk feeding; enzyme beads can also be added to a small amount of soft baby food, such as applesauce. Infants who fail to gain weight in spite of adequate caloric intake need formulas containing casein hydrolysate and medium-chain triglycerides

(MCTs). Nutritional modules and supplements (e.g., MCT oil, Polycose®, Pediasure®, Sustacal®) can be given orally or by tube feeding at night if oral intake is not adequate to meet nutritional requirements. Older children usually tolerate a regular diet without difficulty with pancreatic enzyme therapy. Younger children absorb only 50% to 60% of their food intake.

Cystine. A sulfur-containing nonessential amino acid found in large amounts in *keratins*. It is relatively insoluble at the pH of normal urine and can crystallize or precipitate to form cystine stones.

Cystinosis. A rare hereditary disorder characterized by accumulation of cystine in the lysosomes of many cells, including those of the bone marrow, cornea, and kidneys. Progressive accumulation of cystine leads to renal tubular damage, aminoaciduria, hypophosphatemic rickets, and potassium-wasting acidosis. *Nutrition therapy:* adequate fluid intake, supplemented by potassium in the form of potassium citrate, to correct the acidosis and vitamin D with or without phosphate to correct the rickets. Salt restriction may be beneficial because a high sodium intake increases the urinary excretion of cystine. Excessive dietary protein (more than 100 g/day) should be avoided.

Cystinuria. A hereditary disorder characterized by a defect in the transepithelial transport of cystine, lysine, ornithine, and arginine in the intestine and kidney. The renal tubules fail to reabsorb these acids, which pass in large amounts into the urine, where cystine, the least soluble amino acid, tends to precipitate out and form stones. The simplest therapeutic measure is to ensure a good flow of dilute urine by maintaining a high fluid intake (about 3 liters/day), especially during the night, when the most concentrated urine is produced. An alkaline-ash diet can help alkalinize the urine to prevent precipitation of cystine, although the use of alkalinizing agents and penicillamine is more effective. See *Diet, alkaline-ash*.

Cytarabine. A pyrimidine antagonist; an antimetabolite used with other chemotherapeutic agents for remission induction in nonlymphocytic leukemia. Adverse effects are dose-related and may include severe nausea, vomiting, diarrhea, oral inflammation or ulceration, and anorexia. These may result in hypokalemia, hypocalcemia, protein-losing enteropathy, and weight loss. Breast feeding is not recommended. The brand names are Cytosar-U® and Tarabine PFS®.

Cytidine phosphates. Mono-, di-, and triphosphoric esters of cytidine as cytidine monophosphate (CMP), cytidine diphosphate (CDP), and cytidine triphosphate (CTP). As cytidine phosphate derivatives, they are involved in the biosynthesis of lecithin and cephalins.

Cytochrome oxidase. An iron-containing oxidase that accepts electrons from the other *cytochromes* and passes them on to oxygen, which is the ultimate hydrogen acceptor.

Cytochromes. Group of oxidation–reduction enzymes that are primarily concerned with the transfer of electrons from flavoproteins and other substrates to oxygen or their electron acceptors.

Cytokines. Inflammatory mediators produced by cells of the immune system in response to toxins and viral particles. Cytokines produced by lymphocytes (both B and T cells) are called lymphokines, and those produced by monocytes and macrophages are called monokines. Cy-

D

Dactinomycin. An antineoplastic antibiotic; used as a component of various chemotherapeutic regimens in conjunction with surgery and/or radiation therapy. Nausea and vomiting occur within a few hours after administration and can last for up to 24 hours. The antibiotic also causes diarrhea, stomatitis, oral lesions, and anorexia, and may decrease the absorption of calcium, iron, and fat. The brand name is Cosmegen®

Daily reference value (DRV). Reference value on food labels for nutrients and food components, such as fat, cholesterol, fiber, and sodium, for which the RDAs (recommended dietary allowances) have not been established but are important to health. See Appendix 6.

Daily value (DV). Collective term for the *Reference Daily Intakes (RDIs)* and the *Daily Reference Values (DRVs),* which are the two sets of reference values used in nutrition labeling of food. See Appendix 6. See also *Nutrition labeling.*

Dairy Ease™. Brand name for a *lactose* enzyme, may be added to milk 24 hours in advance or taken just before eating a meal that contains lactose.

Dark adaptation. Speed with which the eye adjusts to a change in intensity of light. This depends on the amount of vitamin A in the body. In humans, maximal dark adaptation of the rods requires about 25 minutes; it is longer if vitamin A deficiency is present.

Daunorubicin. An anthracycline antibiotic that has cytotoxic and immunosuppressive effects; used for treatment of leukemia. Nausea and vomiting may occur soon after administration and last for 1 to 2 days. The drug may also cause stomatitis as early as 5 to 7 days after administration, which begins as a burning sensation with erythema of the oral mucosa, leading to ulceration. The brand name is Cerubidine®.

DBW. Abbreviation for desirable body weight. See *Weight.*

DCH. Abbreviation for *delayed cutaneous hypersensitivity.*

DCCT. Abbreviation for *Diabetes Control and Complications Trial.*

Decalcification. Demineralization; the loss of calcium salts from bones and teeth. It may be due to a number of dietary or physiologic factors, including lack of calcium in the diet; deficiency of vitamin D, which is needed for the absorption of calcium from the intestines; loss of excessive dietary fats (steatorrhea) excreted as insoluble soaps with calcium; the presence of oxalates that bind dietary calcium; a low ratio of calcium to phosphorus (less than 1.0) in the blood; inadequate *parathormone;* and *immobilization,* as in bedridden persons.

Decubitus ulcer. See *Pressure sore.*

Deficiency disease. Condition arising from a deficiency or lack of one or more of the essential nutrients because of *primary* or dietary inadequacy or as a result of *secondary* or conditioned inadequacy. The condition may be a progressive, continuous process that, if uncorrected, eventually leads to depletion of body nutrient reserves. Biochemical changes or "lesions" occur in selected tissues or in the body at large; these eventually result in functional changes such as loss of appetite, easy fatigability, and gastrointestinal disturbances. As the nutritional deficiency continues, anatomic lesions develop and gross clinical signs and symptoms such as glossitis, cheilosis, and dermatitis become manifest. In summary, the development of a deficiency disease may be envisioned to occur in five stages: nutritional inadequacy, tissue depletion, biochemical changes, functional changes, and anatomic lesions. There are two categories of deficiency disease:

Primary. Also called dietary deficiency disease; a condition due to the failure to ingest an essential nutritional factor in amounts sufficient to meet existing requirements of the body. This may, in turn, be due to poor food habits, poverty, ignorance, lack of food, or excess consumption of highly refined foods.

Secondary. Also called conditioned deficiency disease; a condition due to failure to absorb or utilize an essential nutrient because of an environmental condition or a bodily state and not because of dietary lack or failure of ingestion. The conditioning factors may be grouped into six categories: interference with ingestion, interference with absorption, interference with utilization, increased nutrient requirement, increased nutrient destruction, and increased nutrient excretion.

Dehydration. 1. Drying; removal of water from food, tissue, or substrate. 2. Condition resulting from excessive loss of water, often accompanied by losses of electrolytes, particularly sodium, potassium, and chloride ions. A 3% weight loss from dehydration in 24 hours may impair physical performance and decrease blood volume; 5% weight loss may cause nausea and difficulty in concentration; 8% weight loss may cause dizziness, failure to regulate body temperature, and increasing weakness; 10% weight loss may cause muscle spasms, delirium, and wakefulness; and a loss of more than 10% may result in severely reduced blood circulation to the kidneys, resulting in renal failure. Other signs of dehydration include dry oral membranes, decreased skin turgor, increased urine specific gravity (>1.030), decreased urine output (<1 to 2 mL/kg/hour), and depressed anterior fontanel in an infant. There are three types of dehydration: hypertonic, isotonic, and hypotonic. *Hypertonic dehydration,* also called hypernatremic dehydration, occurs when the body loses more water than sodium, as in fever and profuse perspiration. The body fluid loss results in hemoconcentration, with elevated serum sodium level (>145 mEq/L) and blood osmolality (>295 to 300 mOsm/kg). *Isotonic dehydration* occurs when the body loses equal amounts of water and sodium, as when body fluid and electrolytes are lost through vomiting and diarrhea. Serum sodium and serum osmolality may remain within the normal range, but there may be acute weight loss, tachycardia, and orthostatic hypotension. *Hypotonic dehydration* is characterized by hyponatremia, which occurs when body sodium loss exceeds water loss. It can develop when isotonic dehydration is treated with clear water or solution of unsuitable electrolyte concentration, or when diuretics are used concomitantly with a sodium-restricted diet. *Nutrition therapy:* in hypertonic dehydration, provide additional 1 to 2 liters of water above maintenance needs and adjust total volume as needed. In isotonic dehydration, replace fluid and electrolyte deficits with 2 to 3 liters/day of an oral rehydration solution (ORS) or with salty broths and fruit juices. In hypotonic dehydration, replace lost fluid and electrolytes with ORS. Monitor hydration status and adjust the volume according to symptoms. See *Oral rehydration solution* and Appendix 38.

Dehydroascorbic acid. Oxidized form of vitamin C; the reduced form is ascorbic acid. These two compounds are readily and reversibly oxidized and reduced. Both are biologically active, although dehydroascorbic acid is somewhat less stable than ascorbic acid.

7-Dehydrocholesterol. Cholesterol derivative in the skin that is converted to vitamin D by irradiation or exposure to ultraviolet light. See *Vitamin D*.

Dehydroretinol. Vitamin A_2; the form of vitamin A found in the retina and liver of freshwater fishes. It differs from retinol (vitamin A_1) in having one more conjugated double bond and has about one-third the biologic activity of retinol. See *Vitamin A*.

Delayed cutaneous hypersensitivity (DCH). A skin test used to assess immune function prior to and following institution of nutritional support. A negative reaction or induration less than 5 mm in diameter to the intradermal injection of three to five antigens is indicative of *anergy* or lack of immunocompetence. DCH is reduced in malnutrition, although infection, hepatitis, trauma and other critical conditions, and certain drugs such as steroids and immunosuppresants also alter DCH in the absence of malnutrition. See *Hypersensitivity skin test*.

Deliver®. Brand name of a high-calorie, high-nitrogen formula for oral or tube feeding. It is specifically formulated to meet the needs of patients who are hypermetabolic or fluid-restricted. See Appendix 39.

Denaturation. Any alteration in the structure of a native protein, giving rise to definite changes in chemical, physical, or biologic properties such as decrease in solubility at the isoelectric point, loss of biologic specificity, loss of ability to crystallize, increase in viscosity and digestibility, and changes in molecular shape. Denaturation may be brought about by heating, freezing, irradiation, pressure, and treatment with organic solvents.

Dental caries. Tooth decay; demineralization of the inorganic portion and dissolution of the organic substance of the teeth. Three factors are involved in the development of dental caries: the host's teeth, microflora, and the substrate in the mouth. Dental caries will not develop if the teeth are caries resistant and the mouth is kept clean and free of food particles that will sustain the growth of cariogenic bacteria. Evidence indicates that sucrose, especially if taken between meals, is the chief dietary component promoting dental caries. The production of acid from the bacterial fermentation of carbohydrates is the immediate cause of the breakdown of the enamel and dentin. Fluoride and other nutrients in foods can also accelerate or reduce the development of dental caries. The following are recommended dietary measures to prevent tooth decay: fluoridation of drinking water where the supply is deficient in natural fluorine; restricted consumption of sticky starches and sugary confections; avoidance of between-meal consumption of sweets; and eating a balanced diet and including starch and dietary fiber with meals. Raw fruits and vegetables have a cleansing effect if sticky starches and sweets are eaten with the meal. See also *Fluoride* and *Nutrition, dental health*.

Dental (oral) problems. Difficulties associated with chewing or eating, including lack of teeth, broken teeth, dental caries, mouth sores, broken or wired jaw, and periodontal disease. *Nutrition ther-*

apy: provide the proper consistency and type of food. Give a "dental" soft or mechanical soft diet and include foods that are easy to chew; some foods may need to be chopped, ground, or pureed as necessary. Avoid citrus juices and acidic foods in mouth ulcers. In individuals with jaw wiring, give blenderized foods thin enough to pass through a straw. Restrict sweets and sticky foods in dental caries.

Deoxycorticosterone (DOC). Also called "desoxycorticosterone"; a hormone produced by the adrenal cortex. It is corticosterone without the oxygen atom at carbon atom 11. Except for aldosterone, it has a greater electrolyte-regulating effect than the other corticoids. It is prepared synthetically as the acetate (DOCA) and is useful in the treatment of Addison's disease.

Deoxypyridoxine. Synthetic structural analog and antagonist of vitamin B_6; it competes for binding with pyridoxal phosphate-dependent enzymes but cannot function as a coenzyme. It is used in the experimental induction of vitamin B_6 deficiency.

Deoxyribonucleic acid (DNA). Also called deoxyribose nucleic acid; the main carrier of genetic information necessary for the synthesis of specific proteins. It is found in the nuclei of cells as part of the chromosome structure and consists of phosphoric acid, purines (adenine and guanine), pyrimidines (cytosine and thymine), and the sugar deoxyribose. The structure of DNA is envisioned as a double helix in which the purine and pyrimidine bases are inside the helix, with the sugar and phosphate backbone outside the helix. The chains are held together by hydrogen bonding.

Depression. An abnormal emotional state characterized by an excessive, unrealistic, inappropriate reaction to events (environmental changes) or inner (endogenous) conflicts. There are many factors leading to intrapsychic conflicts. Visible signs related to nutrition include mood swings in eating (lack of appetite or increased appetite and irregular mealtimes), poor choices of food, skipping meals, and weight control problems. *Nutrition therapy:* adjust caloric intake and maintain optimum body weight. Assess eating habits and identify nutritional inadequacies; modify the diet accordingly. Consider the effects of certain drugs prescribed for depression on specific nutrients, i.e., observe dietary guidelines for drug–nutrient interactions.

Dermatitis. Inflammation of the skin. There are many forms and causes, e.g., prolonged exposure to the sun's rays, allergy, and infection. Dermatitis is also seen in association with deficiency states due to lack of biotin, niacin, vitamin B_6, and essential fatty acids. See also *Nutrition, dermatology.*

Dermatitis herpetiformis. A chronic skin disorder associated with a mild mucosal lesion similar to that of celiac sprue, suggesting that the disorder is gluten-sensitive. It is characterized by itching and vesicular lesions that leave hyperpigmented spots. The onset occurs in childhood and continues with varying severity for life. *Nutrition therapy:* follow a gluten-free diet. It may take several months to 2 years on this diet to produce a normal-appearing skin. The diet must be followed for life. See *Diet, gluten-free.*

Detoxification. Also called detoxication; the chemical changes in the body that serve to convert toxic substances into forms that are less toxic and more readily excretable. The detoxification mechanism involves oxidation, reduction, hydrolysis, or conjugation with a compound occurring normally in the body, such as glycine, cysteine, or glutamine.

Developmental disabilities. Conditions and diseases characterized by handicaps, both motor and cognitive, and delays in growth and development. Malnutrition is common in developmentally disabled children. Their nutritional needs vary and depend on energy expenditure, physical impairment, growth failure, and medications. Evaluation is difficult because standard methods of measurement do not apply. Obtaining an accurate height is a problem due to physical conditions such as contractures, scoliosis, and inability to stand. Other measurements such as arm span, knee–heel height, and sitting height can be used to estimate actual height. Recumbent length can also be taken, but it must be noted that it is 2 cm greater than standing length until the age of 4 years. Weight-for-height ratio is helpful in evaluating children's weight below the normal height for age. Conditions that may affect eating include oral–motor dysfunction, behavioral problems, gastrointestinal reflux with choking and aspiration, and delayed gastric emptying. *Nutrition therapy:* the diet is individualized, taking into account the developmental level, physical problems, specific feeding difficulties, and drug–nutrient interactions. A multidisciplinary approach is necessary to maximize the outcome of nutritional support. Energy intake must balance with expenditure, which may vary from hypotonia and spasticity to ataxia and athetosis. Feeding may require modifications in consistency, use of nutritional supplements, adaptive equipment and supportive devices, or assistance with eating. Monitor drugs for nutritional side effects. Vitamin/mineral supplementation may be necessary, especially vitamins C, D, B_6, and B_{12}, folic acid, calcium; iron; phosphorus; magnesium; and zinc. Particular attention must be given to the bone-mineral status of children taking anticonvulsants. Problems with aspiration may require gastrostomy feedings. For nutrition therapy in specific developmental disabilities, see *Cerebral palsy, Down's syndrome, Prader–Willi syndrome,* and *Myelomeningocele.* See also *Self-feeding problems.*

Dextrimaltose. Preparation of dextrin and maltose; used as a carbohydrate modifier in infant milk formulas.

Dextroamphetamine. The dextrorotatory isomer of amphetamine; used as an anorexiant and for treatment of narcolepsy and hyperactivity. Its anorexigenic effect is temporary, lasting for only a few hours. When it is used for the treatment of obesity, a low-calorie diet is recommended. The drug has a metallic taste and may cause cramps, constipation, and a decreased tissue ascorbic acid level. Brand names are Dexedrine® and Biphetamine®.

Diabetes. Condition characterized by excessive thirst and urination. If used without qualification, the term refers to *diabetes mellitus.*

Diabetes Control and Complications Trial. A 10-year study that ended in June, 1993, involving more than 1400 subjects. It proved that blood glucose control or intensive therapy prevents or delays complications in patients with Type I diabetes mellitus. Multiple daily injections of insulin and the use of insulin pumps allowed people with diabetes to deliver insulin more like a normal pancreas would. The result of the study was so dramatic that it was concluded early at the end of nine years. In the study, damage to the eyes was reduced by 76%, damage to the kidneys by 35% to 56%, and damage to the nerves by 60%.

Diabetes insipidus. A condition due to lack of *vasopressin,* causing a secondary deficiency in *antidiuretic hormone.* Excessive fluid is lost through the kidneys, leading to dehydration if fluids are not replaced. Urine output may be as much as 10 to 20 liters/day. *Nutrition therapy:* salt restriction (2 to 3 g sodium/day) may be beneficial with diuretic therapy. Monitor fluid intake and output, especially in patients who are confused or immobilized and are unable to drink adequately. Encourage foods high in potassium to prevent depletion with potassium-losing diuretics.

Diabetes mellitus (DM). A group of diseases characterized by abnormal glucose metabolism. Insulin is not manufactured by the pancreas or is produced in too small quantities, and, if present, is not utilized by the body. Characteristic symptoms include elevated blood glucose (hyperglycemia), increased appetite (polyphagia), increased urination (polyuria), increased thirst (polydipsia), nocturia and unplanned weight loss. DM is classified as follows: Type I or insulin-dependent diabetes mellitus (IDDM), which requires insulin to prevent ketosis; was previously called "brittle diabetes" or juvenile-onset diabetes mellitus and commonly appears in youth and young adulthood but can occur at any age; Type II or non-insulin-dependent diabetes mellitus (NIDDM) and subdivided into obese NIDDM and non-obese NIDDM, which usually appears in middle age and was previously called adult-onset diabetes or maturity-onset diabetes; Type III or gestational diabetes mellitus (GDM), which occurs during pregnancy but the blood glucose level reverts to normal after parturition; Type IV or secondary DM, which includes other types of diabetes associated with a pancreatic disease, adverse effects of drugs, hormonal changes, and other causes; and Type V or impaired glucose tolerance (IGT), in which plasma glucose levels are abnormal but not sufficiently beyond the normal range to be diagnosed as diabetic and was previously called "borderline diabetes." *Nutrition therapy:* consider age, body weight, lifestyle as well as goals for blood glucose and lipid levels. Determine the total energy needs and adjust calories to bring about a gradual weight loss in obese patients and maintenance of desirable weight, especially for those with Type II or NIDDM. Distribute total calories as follows: 60% to 70% from carbohydrate and monounsaturated fat, 12% to 15% from protein, <10% from saturated fat, and 6% to 8% from polyunsaturated fat. A high-carbohydrate, high-fiber diet reduces insulin requirement and improves glycemic control. Generally, about 55% to 60% of total calories should come from carbohydrates consisting of complex polysaccharides, fiber, and modest amounts of sucrose or other sugars (10% to 15% of total calories), depending on blood glucose control. "Simple sugars" need not be avoided and may be included in the total carbohydrate content of the diet; alternate sweeteners such as honey and fructose are also acceptable in limited amounts. For fiber, 25 g/1000 kcal up to 40 to 50 g/day is recommended, with emphasis on water-soluble fibers. For protein, the RDA allowance of 0.8 g/kg body weight is adequate for adults; elderly persons may need slightly more or 1.0 g/kg body weight, unless renal disease is present which may necessitate a reduced protein intake. Ideally, fat should provide less than 30% of energy intake, of which less than 10% comes from polyunsaturated fatty acids, less than 10% from saturated fatty acids, and 10% to 15% from monounsaturated fatty acids. Limit cholesterol intake to under 300 mg/day and sodium to 3000 mg/day. Avoid alcohol or limit to 1 to 2 alcohol equivalents once or twice a week if allowed by the physician. Snacks are not necessary for Type II or NIDDM but may be included as part of the total calories if desired, depending on the person's glucose control. However, for Type I or IDDM and Type III GDM, snacks should be coordinated with insulin schedule and the kind and/or dosage used. Timing and regularity of meals are important, especially for those with IDDM, to avoid swings in glycemic levels. Alcohol may be taken occasionally (no more than 2 drinks per week) without altering the meal plan; calories from alcohol in excess of 2 drinks per week should be included as part of the daily meal plan. See Appendix 33 for the alcohol and caloric content of alcoholic beverages.

Diabetic ketoacidosis (DKA). One of the acute complications of *diabetes mellitus*. It occurs when blood glucose cannot get into the body cells or when not enough insulin is available to utilize glucose. Instead, *ketones* are used to provide energy. The warning signs of DKA are: the appearance of ketones in the urine, especially when the blood glucose levels are 500 mg/dL or higher, excessive thirst, polyuria, nausea and vomiting, dehydration, and abdominal pain. Immediate treatment consisting of insulin, electrolytes, and fluids is essential. *Nutrition therapy:* in severe ketoacidosis, fluid and electrolytes are administered intravenously. A 5% glucose solution is given as hyperglycemia and glycosuria subside. If there is no vomiting, clear broth and tea may be given, followed by fruit juices and other liquids. Eventually return to the usual diet but avoid alcohol. Dietary adjustments may be necessary if insulin coverage is altered. See also *Ketoacidosis* and *Ketone test*.

Dialysis. Process of separating substances in solution by selective diffusion through a semipermeable membrane. Excess fluid, electrolytes, waste products, and toxic substances in blood can pass through an external semipermeable membrane or the peritoneum for removal into the dialysate solution. See *Renal dialysis*. See also *Hemodialysis* and *Peritoneal dialysis*.

Diarrhea. Condition characterized by frequent passage of loose, watery, and unformed stools. Kinds of watery diarrhea are the passive movement of water and electrolytes, with increased hydrostatic pressure, as in inflammatory bowel disease, salmonellosis, and fungal infections; and the active movement of water into the mucosa, as in enterotoxic infections. Diarrhea may be functional or organic and may be due to a number of causes, e.g., intestinal infection, ingestion of poison, overeating, nervous disorder, endocrine disturbance, and food malabsorption and sensitivity. Diarrhea may also be associated with fecal impaction, side effect of medications, and nutritional disorders such as pellagra and protein-calorie malnutrition. Diarrhea may be acute (less than 2 weeks duration) or chronic (longer than 2 weeks). Two major consequences are dehydration and malabsorption due to mucosal injury or inflammation of the bowel. *Nutrition therapy:* immediately replace fluid and electrolyte losses and provide adequate nutrition to prevent a state of starvation. Commercial or home *oral rehydration solutions (ORSs)* are effective in rehydrating even severe cases of diarrrhea. The oral route should be used whenever possible, but intravenous fluid and electrolytes may be necessary for patients who are in shock or severely dehydrated. The bowel should not be "rested" by omitting oral intake. Fasting has not been shown to be beneficial in acute gastroenteritis. The Subcommittee on Nutrition and Diarrheal Disease Control of the Food and Nutrition Board supports the continuation of feeding during acute diarrhea, and the American Academy of Pediatrics recommends reintroduction of feeding during the first 24 hours of the diarrheal episode unless there is a specific contraindication. Start refeeding after 4 to 8 hours of oral rehydration therapy. In infants, continue with breast feeding or full-strength formula feeding, preferably lactose-free; supplement with water or maintenance ORSs to maintain hydration. In children and adults, reintroduce solids previously tolerated over the next 3 days. Begin with cereals and starches, followed by meat, poultry, fish, and eggs, then cooked vegetables, and finally fruits and milk. If not tolerated, dilute regular milk to half strength or use a lactose-free formula; gradually reintroduce regular milk later. Avoid full-strength juices and regular (sweetened) beverages and gelatin because their high sugar content makes them hyperosmolar. Fats are not restricted, although medium-chain triglyceride (MCT) is preferable to long-chain triglyceride (LCT) in steatorrhea. Frequent small meals are generally better tolerated. Give nutrient-dense foods to prevent weight loss. Chronic diarrhea that lasts longer than 2 weeks may require dietary changes, such as lactose-restricted diet for lactase deficiency, high fiber diet for irritable bowel syndrome, MCT fat diet for fat malabsorption, and gluten-free diet in celiac sprue. In protracted diarrhea of infancy, give a lactose-free, hydrolyzed protein (casein or whey) formula via nasogastric feeding or parenteral feeding if the enteral route is not tolerated. Supplemental copper, iron, phosphorus, potassium, sodium, zinc, folic acid, and thiamin may be needed.

Diazepam. A benzodiazepine tranquilizer used primarily for relief of anxiety and agitation; also used as a muscle relaxant, as an anticonvulsant, and in the management of alcohol withdrawal. It can stimulate appetite, increasing food intake and weight, but when given in high doses it may cause dry mouth and reduced food intake. Brand names include Valium®, Valrelease®, and Zetran®.

Diazoxide. A rapid-acting hypotensive and hyperglycemic agent that is structurally related to the thiazide diuretics but has no diuretic activity; used in severe hypertension and hypoglycemia due to hyperinsulinism. Severe hyperglycemia may develop, especially in patients with renal disease and carbohydrate metabolic disorders; weight gain may also occur due to retention of sodium and water. Other side effects include nausea and vomiting, taste alteration, abdominal discomfort, anorexia, and constipation. Brand names are Hyperstat® and Proglycem®.

Dicumarol®. Brand name of bishydroxycoumarin; an anticoagulant present in spoiled sweet clover that is structurally similar to vitamin K. It counteracts the formation of clotting factors, notably prothrombin, whose production requires the presence of vitamin K. Dicumarol and its various derivatives are prepared synthetically and used in the treatment of thrombosis.

Didanosine. Antiretroviral drug used for the treatment of symptomatic HIV infection and AIDS. It may cause dry mouth, anorexia, weight loss, stomatitis, anemia, pancreatitis, diarrhea, nausea, vomiting, abdominal pain, anemia, hepatic failure, and edema. Contraindicated in lactation and phenylketonuria. Food may decrease absorption by 50%. Brand name is Videx®.

Diet. 1. The usual foods and drinks regularly consumed. 2. To take food according to a regimen. 3. Food prescribed, regulated, or restricted in kind and amount for therapeutic or other purposes. Dietary guidelines for all healthy Americans over 2 years old are as follows: eat a variety of foods; achieve and maintain a healthy weight; choose a diet low in fat, saturated fat, and cholesterol; choose a diet with plenty of vegetables, fruits, and grain products; use sugars in moderation; use salt and sodium in moderation; and if you drink alcoholic beverages, do so in moderation.

Diet, adequate. Diet that meets all the nutritional needs of an individual. A dietary pattern based on the Food Guide Pyramid is a practical way of planning for dietary adequacy to meet the recommended dietary allowances for specific nutrients. See Appendices 1 and 4.

Diet, antiesophageal. See *Reflux esophagitis*.

Diet, ash. The mineral elements in foods (ash) form a residue that is excreted in the urine. Foods are said to be acid-ash (acid-forming) or alkaline-ash (basic-forming) on the basis of their influence on the pH of the urine. The acid-forming elements in foods, such as sulfur, phosphorus, and chloride, decrease the pH of the urine; the alkaline-forming elements, such as sodium, potassium, magnesium, and calcium, increase the pH of the urine. Foods that have no effect on urine pH are neutral-ash foods. Thus, by changing the composition of the diet, the urine may be made either acidic or alkaline. In general, cereals and protein-rich foods have a predominance of the acid-forming elements. Fats, oils, and sugars give a neutral ash. Milk, fruits, and vegetables have a preponderance of the basic-forming elements. An acid-ash or alkaline-ash diet is used as an adjunct to acidifying or alkalinizing drugs.

Acid-ash d. Diet emphasizing the use of large amounts of acid-forming foods while restricting the intake of alkaline-forming foods such as fruits (except cranberries, plums, and prunes), vegetables (except corn and lentils), and milk. Acid-forming foods include meat, fish, poultry, shellfish, and eggs; all types of breads, cereals, and grain products; corn and lentils; cranberries, plums, and prunes; and cakes, and cookies. An acid urine favors the excretion of kidney stones consisting of calcium and magnesium phosphates, carbonates, and oxalates; it also enhances the effect of some medications used for urinary tract infection.

Alkaline-ash d. Diet emphasizing the intake of large amounts of alkaline-forming foods, including all fruits (except cranberries, plums, and prunes), all vegetables (except corn and lentils), and milk and milk products. while limiting the intake of acid-forming foods such as meat, fish, eggs, and cereals. The diet is prescribed for patients with uric acid and cystine stones in the kidneys. An alkaline urine is beneficial in keeping these stones in solution.

Diet, aspartame-restricted. Diet restricting the artificial sweetener, aspartame, because of sensitivity or adverse reaction. The artificial sweeteners Nutrasweet and Equal contain aspartame. Read labels to identify foods containing aspartame. One can of diet soda, for example, contains about 200 mg. The maximum acceptable level of intake set by the FDA is 40 to 50 mg aspartame per kg body weight.

Diet, balanced. Diet containing all the required nutrients in *proper proportion* with respect to one another for optimum nutrition. A more appropriate term to use is "adequate diet," since a diet that is quantitatively balanced for optimum nutrition is rather difficult to attain.

Diet, basic-ash. See *Alkaline-ash d* under *Diet, ash*.

Diet, bland. Diet previously used in treating gastric ulcers; eliminates or restricts the intake of substances known to cause gastric irritation and excessive gastric acid secretion. These substances include black pepper, chili powder, and red pepper; coffee, both regular and decaffeinated; alcohol; soft drinks with caffeine; and any food that is not tolerated. The current dietary recommendation is to encourage the patient to eat anything that does not produce symptoms, to restrict only those foods that cause discomfort, and to have three well-balanced meals a day but no late-evening snack, since it stimulates nocturnal secretion of acid. Spices are restricted by individual tolerances. Avoidance of regular and decaffeinated coffee and caffeinated soft drinks that induce acid secretion may prevent heartburn and discomfort due to reflux. Avoid beer and high-proof alcoholic drinks. Less concentrated alcoholic beverages, such as wine taken in moderate amounts with food, are generally well tolerated. Also avoid drinking a lot of milk; a 12-oz serving stimulates as much gastric acid production as 12 oz of beer which is a potent gastric acid stimulant. The diet is highly individualized.

Diet, brat. Diet consisting of banana, rice, unsweetened applesauce, and toast. It is called "bratty" diet when served with tea and yogurt. It is prescribed for diarrhea in infants and children but is not recommended for the treatment of diarrheal illness. The beneficial effect of this diet is not supported by controlled scientific studies. Its use should be limited to 1 day when prescribed. The diet is nutritionally inadequate.

Diet, breath hydrogen test. A preparatory diet given in the evening meal, followed by an overnight fast preceding the *breath hydrogen test* for lactose intolerance. Food sources of breath hydrogen, including wheat and products containing wheat such as breads and breaded items, cereals, crackers, cakes, cookies, and pastas; beans (baked, butter, kidney, mung, navy, pinto, soya, and string); peanuts; lentils; and peas (chick, garden, and split), are omitted 12 hours prior to the test. A high-lactose evening meal is given. A rise in exhaled hydrogen (above that obtained during fasting) indicates lactose intolerance following an oral dose of lactose.

Diet, calcium-modified. Calcium intake varies widely among individuals, depending upon the consumption of dairy products. The calcium content of a diet that includes a variety of foods, but no milk and milk products, can be estimated as approximately 200 mg/1000 kcal/day. For most individuals, this is equivalent to 300 to 500 mg of calcium per day. The intake of milk (300 mg calcium per cup), cheese, and other milk products (average of 150 to 200 mg of calcium per serving) contributes more than 55% of the total daily calcium intake.

High-calcium d. Diet offered to meet the calcium needs of persons depleted by disease, traumatic stresses, or prior dietary inadequacies. The diet is basically a well-balanced regular diet supplemented with additional high-calcium foods (about 400 to 600 mg Ca/day), derived mainly from milk and dairy products. Other food sources of calcium are green leafy vegetables (broccoli, collards, dandelion greens, endive, escarole, kale, mustard greens, spinach, turnip greens), tofu, dried fruits, lime-processed tortillas, soft bones of canned fish, quick-cooking cereals, and calcium-fortified foods.

Low-calcium d. Diet that generally restricts the intake of milk, milk beverages, cheese, yogurt, and ice cream and other products made with milk to the equivalent of 1 cup of milk per day. Other foods containing calcium are allowed in moderate amounts but should not be used frequently. These include vegetables high in calcium, such as spinach, rhubarb and mustard greens, canned fish with bones, tofu, dried beans, quick-cooking cereals, and quick breads. It usually consists of 500 to 600 mg of calcium, but the diet should be supplemented with multivitamins and minerals if the regimen is to be continued for an extended period of time. A low-calcium diet is recommended for patients with renal hypercalciuria and Type II absorptive hypercalciuria.

Moderate-calcium d. Diet that limits the daily calcium intake to approximately 800 mg for men, and 1000 mg for premenopausal and 1200 mg for postmenopausal and pregnant or lactating women with Type I absorptive hypercalciuria. Calcium restriction below these levels provides no additional benefit. An excessive calcium intake is avoided because it may produce significant hypercalciuria. Milk and dairy products are limited to the equivalent of 2 cups of milk per day. Other food sources of calcium are not restricted, unless used frequently and in large amounts. However, foods high in oxalate should be avoided to prevent an increase in urinary oxalate excretion.

Diet, calcium test. A calculated diet used in the study of calcium metabolism under conditions of minimum intake. Calcium in the diet is limited to 200 mg or less, and phosphorus is limited to 600 mg. Fats and sugars are used to maintain caloric intake. The calcium content of drinking water should be checked; if it is unusually high, distilled water should be used for cooking and drinking. When screening for hypercalciuria, a 1000-mg calcium diet is given, with food sources providing 400 mg and the remaining 600 mg provided by calcium gluconate. A 24-hour urinary calcium excretion that exceeds 250 mg for a female and 300 mg for a male, or a calcium excretion that exceeds 4 mg/kg body weight, is diagnostic for hypercalciuria.

Diet, calorie-modified. Diet in which the total intake of calories is either increased or decreased from normal to allow for a gain or loss in body weight.

Low-calorie d. Diet planned to permit loss of weight while maintaining health. A reduction of 500 kcal/day from the usual intake, while keeping activity constant, should bring about a loss in body weight of about 1 lb/week. It is best to arrive at a caloric allowance that is acceptable to the patient. A weekly weight loss of ½ to 1 lb is considered satisfactory. Weight loss of more than 2 lb/week is not advisable except under the close supervision of a physician. In general, the caloric deficit should not exceed 500 to 1000 kcal per day and total calories should not be less than 1000 kcal/day. Weight loss is greater when the low-calorie diet is combined with regular exercise. Care should be observed in planning the diet; caloric levels below 1200 are marginal in nutrient content and may necessitate vitamin supplementation. A liberal protein intake is essential for its satiety value and to prevent negative nitrogen balance; carbohydrate and fat are restricted to the caloric level desired. Foods to avoid include sauces, gravies, nuts, sweets, desserts, and fried foods. Extra fiber is recommended to reduce caloric density and promote satiety.

Moderate-calorie d. A well-balanced mixed diet that is most widely prescribed for weight reduction or to maintain the weight loss that has been achieved. The energy level varies with the individual's size and activities but it is usually 30% less than the usual intake. About 20% of total calories come from protein, with emphasis on lean and well-trimmed lean meats; 50% from carbohydrate, with emphasis on complex forms and extra fiber; and 30% or less from fat, with emphasis on low amounts of animal fats. Alcohol, simple sugars, and calorie-dense foods are limited or avoided. When combined with exercise and physical activity, the diet is decreased to the point where the body's fat stores must be utilized to meet daily energy needs. It is nutritionally adequate and can be followed for months without supplementation in most individuals, except pregnant or lactating women and children under 10 years old.

High-calorie d. Diet with a prescribed caloric intake above normal to meet increased energy requirements and to provide weight gain. It is indicated in febrile conditions, hyperthyroidism, athetosis, undernutrition, and other conditions that result in loss of weight. The caloric increase may vary from 30% to 100% above the usual intake. It is best to individualize the diet. Observations with patients of both sexes and different ages show that men seem to prefer ingesting the additional calories through extra portions of the usual foods served at meals, children and adolescents prefer between-meal nourishment, and women seem to favor more concentrated foods.

Very-low calorie d (VLCD). Reducing diet that severely limits calories to fewer than 800 kcal/day to promote a large, rapid weight loss averaging 2 to 4 lb/week for women and 3 to 5 lb/week for men. The diet is very high in protein and low in carbohydrate, which is usually 30 to 50 g/day; fat is restricted to that present in the protein source. Supplementation with vitamins and minerals up to 100% of the recommended dietary allowance is recommended. A VLCD is safe when administered properly under medical supervision. It is indicated for individuals who are at least 30% overweight and have a minimum *body mass index* of 32. Because of the potentially life-threatening side effects of the diet, it is contraindicated in children, the elderly, and in certain conditions: pregnancy and lactation, cardiac dysfunction, renal failure, hepatic disease, protein-wasting diseases, and severe psychologic disturbances. A VLCD may be planned with conventional foods or with the use of semisynthetic formula preparations. The formula is either a liquid or a powder to be mixed with liquid. Some formulas contain soluble fiber. Generally, the dieter is limited to the formula alone or in combination with very limited quantities of low-calorie foods. The diet is usually preceded by 2 to 4 weeks on a well-balanced 1200-kcal diet to allow the body to adjust to the very low calorie level and to promote gradual diuresis, instead of the rapid sodium and water loss seen with abrupt introduction of the VLCD. At the end of the weight loss period of about 12 weeks, a period of gradual refeeding for 2 to 4 weeks follows. The VLCD is the basis of commercial weight-reduction programs such as Optifast, Medifast, and the HMR Fasting Program. Throughout the weight reduction program, sessions are held on behavior modification, nutrition education, and exercise. Potential side effects associated with VLCD include alopecia, dry skin, diarrhea, fatigue, irritability, depression, and nausea. See also *Diet, protein-sparing modified fast.*

Diet, carbohydrate-modified. Diet that provides a specified level of carbohydrate or is restricted in the amount of a particular type of carbohydrate such as glucose, sucrose, fructose, galactose, or lactose. There is no recommended dietary allowance for carbohydrate. The National Research Council recommends that at least 50% of the energy requirement after infancy should be provided by carbohydrate, especially complex carbohydrate. Thus in a 2000-kcal diet, carbohydrate would be at least 250 g/day. A carbohydrate intake of less than 100 g per day will lead to a state of ketosis and excessive breakdown of tissue protein.

High-carbohydrate d. Diet high in available carbohydrate to allow for glycogen formation, to ensure sufficient calories to meet needs, to spare protein, and to minimize tissue catabolism. It is indicated in liver diseases, Addison's disease, fasting hypoglycemia, acute glomerulonephritis, uremia, pernicious vomiting, and toxemias of pregnancy.

The diet is modified in consistency or other nutrient content to suit specific disease conditions. Emphasis is placed on easily available carbohydrates such as sugar, syrups, jellies, and jams. A high-carbohydrate diet is also used by athletes in pre- and postgame meals, especially in endurance events lasting 1 hour or longer.

Restricted-carbohydrate d. Diet limited in carbohydrate content to reduce available glucose when carbohydrate metabolism is impaired, as in spontaneous hypoglycemia. It is also indicated for the dumping syndrome, epilepsy, and following a vagotomy and partial or total gastrectomy. Simple carbohydrates are restricted, including sugar, cookies, pies, pastries, and other concentrated sweets. Restriction of specific types of carbohydrate, such as sucrose, lactose, maltose, and galactose, is also indicated in cases of intolerance of these sugars. See *Diet, fructose-restricted, Diet, galactose-restricted; Diet, glucose-restricted; Diet, lactose-restricted; Diet, maltose-restricted;* and *Diet, sucrose-restricted.*

Diet, chemically defined. See *Diet, elemental.*

Diet, cholesterol-restricted. Diet in which the intake of dietary cholesterol is restricted to a prescribed level. It is indicated for hypercholesterolemia, atherosclerosis, and gallbladder stones with cholesterol esters. Individuals who eat one or more eggs a day and use organ meats at regular intervals ingest 1000 mg/day or more of cholesterol. Limiting eggs to 3 yolks/week and avoiding organ meats will bring down the intake of cholesterol to 300 mg/day. This amount can be reduced further if butter is not used and if skim milk and dairy products made from skim milk are substituted for whole milk. Omission of eggs from the diet will bring cholesterol intake down to about 200 mg/day. See Appendix 34 for the cholesterol content of foods.

Diet, cholesterol-restricted, fat-controlled. Diet recommended by the National Cholesterol Education Program (NCEP) of the National Institutes of Health (NIH) as the primary treatment for lowering high blood total cholesterol and low-density lipoprotein-cholesterol in individuals considered at high risk for coronary heart disease, including adolescents and children over 2 years. The diet is a two-step program. The step-one diet is tried for at least 3 months; if there is no significant decrease in blood cholesterol, the individual is placed on the step-two diet. In the step 1 diet, cholesterol is restricted to 300 mg or less per day; total fat is 30% or less of total calories, with saturated fats less than 10%, polyunsaturated fats up to 10%, and the remaining total fat calories from monounsaturated fats up to 15%. In the step-two diet, cholesterol is restricted further to 200 mg or less and saturated fats to less than 7% of total calories. Protein intake is about 12% to 15% of total calories. Step-one diet is accomplished by limiting the intake of meat, fish, and poultry to two average servings per day (2 to 3 oz per serving); restricting the intake of eggs (i.e., egg yolk) to three per week; limiting the intake of organ meats (liver, kidney, sweetbreads, and heart) to a 2-oz portion as a substitute for an egg; restricting the intake of saturated fats by trimming visible fat and using only lean cuts of meat and low-fat milk and dairy products; using only liquid vegetable oils rich in MUFA and PUFA such as corn, cottonseed, soybean, safflower, canola, and olive oils; avoiding commercially prepared, packaged foods containing whole egg, whole milk, and saturated fats; and eliminating cholesterol-rich foods such as fish roe, caviar, and brain. In the step-two diet, omit egg yolk (contains 215 mg of cholesterol/yolk), substitute fish and poultry (white meat, no skin) for red meats, and use only the leanest cuts of meat, skim milk, and nonfat products. Include good sources of soluble fiber, such as oats, legumes, barley, dried beans, pectin, psyllium, and guar gum for their cholesterol-lowering effect. Physical activity, weight control, and maintenance of desirable weight are important. See Appendix 34.

Diet, copper-restricted. Diet prescribed in conjunction with oral administration of copper chelating agents used in the treatment of Wilson's disease and other disorders associated with elevated copper storage in the liver. Large quantities of food rich in copper are avoided. These include liver and organ meats, shellfish, mushrooms, nuts, oysters, dried beans, lentils and peas, dried fruits, bran breads and cereals, sweet potatoes, avocado, tofu, soybeans and soymilk, and chocolate. At the discretion of the physician, low to moderate amounts of these foods may be allowed.

Diet, corn-free. Diet prescribed for corn allergy. In addition to corn, all products made from corn are omitted. Corn is present in baking powder, corn flour, cornstarch, cornmeal, corn sweetener or corn syrup solids, grits, and hominy. Check the labels closely. Corn or corn products are used in baked goods, beverages, candy, canned fruits, cereals, cookies, jams, jellies, luncheon meats, snack foods, and syrups. Corn oil is not restricted because the protein in corn that causes the allergy is removed during processing.

Diet, dental soft. See *Diet, mechanical soft*.

Diet, desensitizing. Diet aimed at decreasing the sensitivity of a person to a given food allergen. The food allergen is first excluded from the diet for an indefinite period. Then it is gradually added to the diet in small amounts until an average portion is tolerated.

Diet, egg-free. Allergy diet restricting the intake of egg and products that contain egg. These include French toast, pancakes, waffles, quick breads, custards, puddings, cakes, cookies, mayonnaise, egg noodles, marshmallows, souffles, and others. Egg may be present in beverages such as root beer (foaming agent), wine, and coffee (as clarifier). Baked goods may not only contain egg but also may be glazed with eggwhite. The fat substitute Simplesse contains egg. Common egg substitutes contain egg (without the cholesterol), although "egg-free" egg substitutes are available. Look for terms in food labels to indicate the presence of egg, such as albumin, globulin, livetin, ovalbumin, ovoglobulin, ovomucin, ovomucoid, ovovitellin, Simplesse, and vitellin.

Diet, elemental. Term indicating use of monomeric solutions for enteral feeding, which contain protein as peptides and/or amino acids, carbohydrate as partially hydrolyzed maltodextrins and glucose oligosaccharides, and fat as medium chain triglycerides (MCT) or a mixture of MCT and long chain triglycerides (LCT). It requires very little digestion, is almost completely absorbed by the upper small intestine, and is very low in residue. Elemental diets are indicated in inflammatory bowel disease, short bowel syndrome, malabsorption, pancreatitis, and preoperative nutrition.

Diet, elimination. A normal diet that excludes the intake of a specific food or food group suspected to produce allergic manifestations. Attention should be given to commercially processed or packaged foods that may contain the offending substance in disguised form. It is important to read the list of ingredients on all labels and packages. The elimination diet is maintained for 1 to 2 weeks. If symptoms do not clear and food allergy is still suspected, more restrictive diets can be implemented to omit cow's milk, eggs, wheat, and other common allergens. If symptoms persist causes other than the foods eliminated should be investigated. If improvement occurs, eliminated foods are reintroduced one at a time at 2- to 3-day intervals. If a food provokes a reaction, a food challenge is performed, either single or double blind, which should be done in a hospital or physician's office equipped with emergency equipment.

Diet, elimination test. A series of allergy test diets beginning with a basic diet that consists of a few carefully selected foods least likely to cause allergic reactions. If the person on the basic diet is asymptomatic for 2 weeks, other foods are gradually added at 4-day intervals. Milk, eggs, and wheat are added last because these items are the most notorious food allergens. If relief of symptoms is not obtained even on the basic diet, it is probable that an agent other than food is the offending allergen. The most widely used test diet is the one designed by Rowe. See *Diet, Rowe's elimination.*

Diet, faddist. A craze diet based on the purported "magic quality" in a particular kind of food to treat or cure certain conditions. Common fad diets are the *Zen macrobiotic diet,* molasses diet, fruit diet, and various formula diets that have caught the fancy of overweight persons.

Diet, fat-modified. Diet that prescribes a specified level of fat in the diet, fatty acids ratio, or percentage of calories from fat. Modifications in the fat content of the diet are necessary in weight reduction; in diseases of the gallbladder, pancreas, and cardiovascular system; and in disturbances in fat absorption in association with such diseases as sprue, cystic fibrosis, and pancreatitis.

Fat-controlled d. Also called proportioned-fat diet; both the *amount* and the *kind* of fat are regulated. The diet is generally planned to provide 30% of the total calories/day from fat. The total amount of fat is then "controlled" or proportioned into less than 10% from saturated fatty acids, 10% to 15% from monounsaturated fatty acids, and up to 10% from polyunsaturated fatty acids. The objective of the diet is to increase the intake of the unsaturated fatty acids and reduce the intake of saturated fatty acids to bring about a decrease in serum cholesterol and triglycerides. Polyunsaturated fatty acids are supplied by such oils as corn, soybean, cottonseed, and safflower oils. Animal products, palm oil, and hydrogenated fats are high in saturated fatty acids. Monounsaturated fatty acids are present in most vegetable oils, particularly olive oil and canola oil. It is recommended that people trim visible fats from meats and use lean meats and poultry without the skin, skim milk, and low-fat or nonfat yogurt and nondairy creamers.

High monounsaturated fat (HMF) d. Diet that limits the amount of saturated fat and cholesterol. Protein-rich foods that contain saturated fats are also limited; energy intake is at recommended level. Monounsaturated fat-rich foods, such as canola oil, olive oil, nuts, and avocado are used as replacements for a portion of the starches and other carbohydrate-rich foods in the diet. The HMF diet is used in the management of non-insulin-dependent diabetes mellitus. The diet may improve glycemic control and lipoprotein profile.

Liberal-fat d. A liberal fat allowance is about 35% to 40% of caloric intake. A liberal fat intake often accompanies a high-protein, high-calorie diet prescribed for certain conditions that do not require any restriction in the fat content of the diet, such as undernutrition, burns, nephrosis, and ulcerative colitis. A high-fat diet is indicated in conditions that restrict the intake of carbohydrate and/or protein, as in dumping syndrome, functional hyperinsulinism, and uremia. In such cases, it is best to supply fat in the form of butter, margarine, cream, salad dressing, and vegetable oils.

Low-fat d. Reduction in the fat content of the diet to supply about 15% to 20% of caloric intake. Indicated for acute attacks of pancreatitis and cholecystitis, and in chylomicronemia and plasma triglycerides above 500 mg/dL. This amount of fat is supplied by about 5 to 6 oz of lean meat, poultry, or fish per day. No foods rich in fat are allowed. Visible fats are trimmed from meat, and foods are prepared simply by broiling, baking, or boiling. Avoid fried, fatty, or heavily marbled meat; cold cuts; sausages; canned fish in oil; nuts; creamed sauces; gravies; and all fats including butter, margarine, mayonnaise, vegetable oils, and cream. Use of MCT oil as source of kilocalories may be necessary. The use of sugar, sweets, fruits, cereals, and starchy vegetables will increase caloric intake, and protein may be increased through the use of skim milk, egg white, and gelatin.

MCT fat d. Diet in which MCT oil is used in place of ordinary cooking fats and oils. MCT is a unique oil containing triglycerides of medium-chain fatty acids, which are more easily hydrolyzed and absorbed than the long-chain triglycerides present in conventional dietary fats. The diet is indicated for conditions in which ordinary dietary fats are poorly digested and absorbed, such as pancreatitis, cystic fibrosis, chyluria, celiac sprue, intestinal resection, pancreatic insufficiency, deficient bile secretion, and biliary obstruction. Three to four tablespoons of MCT oil (about 40 to 55 g of fat) or an amount recommended by the physician may be used in cooking, in salad dressings, sauces, or marinades for meats and vegetables, or simply mixed with fruit juices. Add MCT oil gradually to the desired level and use no more than 1 tablespoon each time to avoid unpleasant side effects. MCT oil does not provide essential fatty acids; use at least 10 g/day (2 tsp) of vegetable oil, such as corn, safflower, and sunflower oil, or margarine made with these oils.

Moderate-fat d. Diet with a fat allowance of 25 to 30% of caloric intake. A moderate fat intake is indicated in hepatitis, cirrhosis of the liver, and chronic gallbladder and pancreatic diseases. It is also recommended in weight reduction to lend palatability and satiety value to the diet. Fried foods, nuts, sauces, gravies, and other fatty foods should be avoided. It is best to get the fat allowance from lean meats, eggs, milk, butter, and other highly emulsified fats. Whenever necessary, additional calories may be supplied by sugar, sweets, and other carbohydrate-rich foods.

Very-high-fat d. One in which fat is approximately 80% or more of caloric intake. Carbohydrate intake is severely restricted to no more than 30 g/day, and the protein allowance is at the RDA level. The aim of the diet is to maintain a ketogenic/antiketogenic ratio that will produce a state of ketosis. See *Diet, ketogenic*.

Diet, fat test. Two test diets are used in the diagnosis of gallbladder disease and the determination of fat absorption.

Fat-free test d. A test meal given the night before a radiologic examination of the gallbladder. The meal may consist of fat-free broth, fruits, plain gelatin, skim milk, and black coffee or tea. Nothing is given orally after midnight, and breakfast is withheld until after the examination is made. In some cases the examination is repeated after a *fatty meal,* which may consist of two fried eggs, buttered toast, whole milk, and coffee with cream.

High-fat test d. Calculated 100-g fat test diet used to diagnose steatorrhea. It is usually given for 3 days before the stool specimen is collected for fat analysis. The fat allowance should come mainly from eggs, meats, whole milk, and highly emulsified fats such as butter and cream. It is important to choose foods that are simply prepared to facilitate the fat intake calculation. The amount of fecal fat is analyzed after the high-fat diet is ingested. See *Fecal fat test.*

Diet, fiber-modified. Diet in which the amount of *dietary fiber* is higher or lower than the usual intake of about 15 g or less per day.

High-fiber d. Diet providing about 20 to 35 or more of fiber per day, or approximately 10 to 13 g fiber/1000 kcal but not less than 20 g/day. Excellent sources are whole grain cereals and unrefined breads, high-fiber vegetables, raw fruits, and legumes. See Appendix 32. A high-fiber diet is a regular diet with an additional two to three servings of foods high in dietary fiber. It is indicated in atonic constipation, diverticulosis, irritable bowel syndrome, diabetes mellitus, and hypercholesterolemia. A high fiber intake, especially from cruciferous vegetables, may also reduce the risk of colon cancer.

Low-fiber d. Diet that contains a minimal amount of indigestible carbohydrates or dietary fiber. The fiber content of the diet may be reduced by removing gristle and tough connective tissue in meats, removing seeds and skins from fruits and vegetables, omitting high-fiber foods, and using refined cereals and breads. It is indicated in narrowing of the intestine, gastroparesis, small bowel obstruction, and acute diverticulitis or inflammatory bowel disease.

Diet, Feingold. Diet postulated to improve behavior and reduce hyperkinetic activity by eliminating foods containing salicylates, artificial colors (especially FD&C yellow no. 5), and the preservatives butylated hydroxyanisole (BHA) and butylated hydroxytoluene (BHT), monosodium glutamate (MSG), and sodium benzoate. No convincing experiments have been done to confirm the effectiveness of the diet.

Diet, fish-restricted. Allergy diet eliminating or restricting fish; may also include various shellfish, crustacea (crab, crawfish, lobster, and shrimp), and mollusks (abalone, clam, mussel, oyster, scallop). Fish may be found in some sauces that contain anchovies, Caesar salad, caviar, and roe.

Diet, fructose-restricted. Diet that restricts the intake of fructose from fruits and other foods high in sucrose content (on hydrolysis, sucrose yields glucose and fructose). It is prescribed for fructose intolerance. The degree of fructose restriction is highly individualized to fit the tolerance level. Table sugar, syrups, honey, sweets, and processed foods with added sugar are avoided or restricted. Fructose is found in all fruits; it is also present in sugar beets, green peas, and sweet potatoes.

Diet, full hospital. Also called regular diet or general diet. See *Diet, regular.*

Diet, galactose-restricted. Diet that eliminates foods containing galactose and lactose for the treatment of galactosemia. Milk and all products made with milk are not allowed because galactose is a component of lactose in milk. If a galactose-free diet is required, liver, pancreas, brain, and other organ meats that store galactose are also not allowed. Package labels of commercially prepared products should be carefully checked. Those with added milk, lactose, casein, whey, dry milk solids, and curd should be avoided. Lactose is also used as a filler in certain artificial sweeteners and in the pharmaceutical industry. Some liberalization of the diet at about 12 years of age has been recommended, but this may not be wise because of the damaging effects of accumulated galactitol in the lens, liver, and kidney. Proprietary formulas low in galactose for infant feeding include Nursoy®, Nutramigen®, Isomil®, Prosobee®, and Pregestimil®. See Appendix 39 for lactose-free formulas for children and adults.

Diet, general. See *Diet, regular*.

Diet, gliadin-restricted. More commonly called gluten-restricted diet; it excludes gliadin, which is a component of gluten found in wheat and other cereals. See *Diet, gluten-free*.

Diet, glucose-restricted. Diet prescribed in rare cases of glucose intolerance. This diet is difficult to plan, as glucose is present, either as free glucose or as a disaccharide component in milk, fruits, vegetables, and cereals. Newborn infants require a special proprietary formula (Product 3232A and RCF®) based on calcium caseinate, fructose, and vegetable oil, with added vitamins and minerals. This regimen is usually followed until 6 months of age. Semisolid foods low in starch content are gradually introduced by the seventh month. At the age of 1 year, the infant should be taking egg, fish, meat, and restricted amounts of fruits and vegetables. Milk and starchy foods are not introduced until about 2 to 3 years of age.

Diet, glucose tolerance test. A preparatory diet given before a *glucose tolerance test* when the carbohydrate intake is known to be inadequate or if the person is on a weight reduction diet. A carbohydrate intake of at least 150 g/day must be consumed for 3 days prior to the test. This is not necessary if the person has been consuming an adequate diet prior to the test.

Diet, gluten-free. Also called gliadin-free diet; used in the treatment of gluten-sensitive enteropathy (celiac disease or nontropical sprue) and dermatitis herpetiformis. The diet eliminates wheat, oats, rye, and barley and products containing these cereals. Products made from corn, potato, rice, arrowroot, tapioca, and soybean can be used as substitutes. Rice, corn, and buckwheat contain a smaller amount of gluten, but they are usually tolerated. Milk and dairy products, meat, fish, poultry, eggs, fruits, vegetables, potato, yam, soybeans, fats, sweets, and gluten-free wheat starch can be taken as desired. It is important to read labels carefully, as many processed foods contain the restricted cereals in disguised form, as in fillers, stabilizers, or emulsifiers. Examples of such foods are malted milk, Ovaltine®, hot cocoa mixes, cold cuts, soy protein meat substitutes, nondairy creamer, commercial salad dressing and mayonnaise, puddings and fruit pie fillings, and commercially prepared meat sauces.

Diet, high-altitude. High-carbohydrate (65%), low-fat (20%) liquid diet recommended prior to rapid ascent to high altitudes. This diet has been found beneficial in reducing the clinical symptoms observed at high altitudes.

Diet, histamine test. Foods high in histamine are eliminated from the diet the afternoon before the 24-hour urine collection until the collection is completed. These include cheeses (Blue, Monterey, Parmesan, and Roquefort), chicken liver, beef, spinach, eggplant, tomatoes, and red wines. Urinary excretion of histamine is used in the diagnosis of systemic mastocytosis or carcinoid syndrome.

Diet, hospital. Diet used for hospital patients. The routine hospital diets are the *regular*, *soft*, and *liquid* diets. These may be modified to suit individual requirements for certain therapeutic purposes.

Diet, house. See *Diet, regular*.

Diet, 5-hydroxyindole acetic acid (5-HIAA) test. See *Diet, serotonin test*.

Diet, hydroxyproline test. Test diet eliminating meat and meat products, fish, gelatin, ice cream, marshmallows, salad dressings, and puddings. Used during the test period for urinary excretion of hydroxyproline to study bone collagen turnover in hyperparathyroidism and Paget's disease.

Diet, hyperlipoproteinemia. See *Hyperlipoproteinemia*.

Diet, ketogenic. Diet that is very high in fat and severely restricted in carbohydrates; it is prescribed to control certain types of seizures when drug therapy alone is not fully effective or has undesirable side effects if given in large doses. The diet is calculated to have a fatty acid/available glucose ratio exceeding 3:1. Such a ratio causes the accumulation of ketones (produced from fat oxidation) and the inhibition of convulsive seizures. Per 100 g, fat yields 90% fatty acids and 10% glucose; protein yields 54% glucose and 46% fatty acids, and carbohydrate yields 100% glucose. For clinical purposes, the following simplified ratio may be used:

$$\frac{\text{Ketogenic}}{\text{Antiketogenic}} = \frac{(0.9\text{ F}) + (0.5\text{ P})}{(0.1\text{ F}) + (0.5\text{ P}) + (1.0\text{ C})}$$

About 80% to 90% of total calories come from fat; protein is at the RDA level and carbohydrate should not be reduced below 10 g. Increase fat gradually as carbohydrate is decreased. A 4-day progression to obtain the desired ketogenic-to-antiketogenic ratio would be: 1.1:1.0 on day 1, 1.6:1 on day 2, 2.2:1 on day 3, and 3:1 on day 4. Foods high in carbohydrate such as breads, cereals, fruits, desserts, sweets, and beverages containing sugar are excluded from the diet; concentrated fat sources such as butter, cream, bacon, mayonnaise, salad dressings, and oils are taken in generous amounts. The use of medium-chain triglycerides in place of the usual dietary fats makes the diet more effective in inducing ketosis. It also allows more carbohydrate. In the MCT ketogenic diet, distribute the calories as follows: 60% MCT oil and 12% dietary fat, 10% protein, and 18% carbohydrate. Introduce MCT oil gradually to prevent vomiting, abdominal pain, and diarrhea. The ketogenic diet is inadequate in B-complex vitamins, vitamin C, folate, calcium, iron, and zinc.

Diet, lactose-restricted. Diet designed to limit the intake of lactose to the tolerance level in individuals who experience bloating, cramping, and diarrhea after the ingestion of products containing lactose. The restriction is highly individualized. Many lactose-intolerant individuals usually have no difficulty with small amounts of lactose. Most can manage 4 to 8 oz/day of milk if it is taken with meals, while others need to eliminate milk as a beverage. Cottage cheese, aged cheddar cheese, and fermented milk products such as yogurt are tolerated by most persons. A few individuals highly intolerant of even a small amount of lactose may have to eliminate all products containing milk and lactose. It is important to examine labels carefully to detect hidden lactose. For example, lactose is added to some nondairy creamers, artificial sweeteners, and even some pharmaceutical products as a filler. The diet may be low in calcium, riboflavin, and vitamin D, depending on the degree of milk restriction. Calcium supplementation may be indicated, especially in growing children, postmenopausal women, and women at risk for developing osteoporosis. A commercial lactase enzyme, when added to milk or taken with meals, can sufficiently hydrolyze lactose to allow the use of milk in the diet. Lactose-free formulas for infant feeding include Alimentum®, Isomil®, Nursoy®, Nutramigen®, Pregestimil®, and ProSobee®. See also Appendix 39 for lactose-free enteral products.

Diet, leucine-restricted. Diet restricting the intake of leucine to the minimum requirement of 150 to 230 mg/kg body weight. For infants, a proprietary product low in leucine is available (MSUD Diet Powder). Protein-rich foods, particularly milk and eggs, are restricted until tolerance to these foods is established. Fruits and vegetables are added to the diet according to the infant's normal feeding schedule. Small amounts of carbohydrate feeding, equivalent to 10 g, are given 30 to 40 minutes after each meal to help counteract the hypoglycemic effect of leucine. The child is usually able to tolerate a normal diet by the age of 5 to 6 years.

Diet, light. Diet that consists of foods that are easily digested and readily emptied from the stomach. It is often prescribed prior to surgery or gastric analysis; it is also indicated for patients, especially older ones, who are quite sick and cannot tolerate rich and heavy foods. The diet is given in three small meals, with between-meal feedings. Foods are prepared simply; fatty foods, rich pastries, concentrated desserts, and fibrous fruits and vegetables are restricted or given as tolerated.

Diet, liquid. Diet consisting of a variety of foods that are liquid, can be liquefied, or can easily melt in the mouth or at body temperature.

Blenderized liquid d. Diet consisting of fluids and foods blenderized to a liquid puree consistency. It is used in patients with dysphagia, wired jaw and other oral surgery, and inadequate oral control. The blenderized mixture may vary in consistency, ranging from the thickness of a fruit nectar to that of cream soup. The diet is highly individualized based on patient tolerance. Patients who are unable to use straw or open their mouths may need a syringe or container with a long spout to assist in feeding.

Clear liquid d. Diet of clear liquids that leaves little or no residue. It is used in pre- and postoperative bowel surgery, partial paralytic ileus, acute inflammatory conditions of the gastrointestinal tract, and as a brief transition from intravenous or parenteral nutrition to oral feeding. The primary purpose of the diet is to relieve thirst and help maintain water balance. Plain tea, black coffee, fat-free broth, ginger ale, plain gelatin, and glucose solution are the usual liquids given. Other liquids, such as Popsicles®, fruit ices, fruit drinks, carbonated beverages, and clear fruit juices such as apple, grape, and cranberry, are often allowed to contribute additional calories. The diet is nutritionally inadequate and must not be used for more than 24 hours. The use of high-protein broth, high-protein gelatin, and commercially prepared low-residue liquid formulas should be considered if needed longer to provide more calories, protein, and other nutrients.

Cold semiliquid d. Diet prescribed after tonsillectomy and other mouth and throat surgery. Only cold liquids and cold soft, bland foods are given to avoid irritation and to prevent bleeding. Foods included are gingerale and other flavored drinks, iced tea, plain flavored gelatin, Popsicles®, plain yogurt, soft custard and pudding, fruit purees, and other cold foods tolerated. Chocolate products and red-colored beverages are not given because they may mask bleeding; milk beverages are omitted or restricted if excessive mucus is produced; acidic fruit juices may be irritating and should be avoided.

Full liquid d. Diet consisting of foods that are liquid or easily become liquid in the mouth or at body temperature. This diet bridges the gap between the clear liquid diet and the soft diet. It is used in acute conditions, for patients with fractured jaws, after oral and other types of surgery, and for patients too ill to eat solid foods. When properly planned, the diet can be made nutritionally adequate and can be used for relatively long periods of time. Six or more feedings per day are recommended. All liquids and foods that easily become liquid, such as plain ice cream, plain gelatin, strained cream soups, strained cereal gruel, soft custard, and puddings, are allowed in the diet. Supplemental liquid vitamins and minerals or liquid enteral products may be necessary if the diet is used longer than 2 weeks.

Restricted liquid d. The diet order specifies the volume of liquid allowed in 24 hours, such as 500, 1000, 1200, or 1500 cc. The total fluid intake used by the nursing service for medication should be added to the total daily fluid intake. In addition to the usual liquid beverages, there are semisolid foods or foods that liquefy in the mouth which should be counted for fluid equivalents. For example, ½ cup gelatin yields 60 cc liquid; ½ cup ice cream or sherbet has 90 cc liquid; and thin cooked cereals have 50% fluid.

Diet, low-microbial. Also called "low-bacteria" diet. Diet used before and after organ transplantation to reduce the risk of foodborne infection in the immunosuppressed patient. Safety precautions in implementing the diet vary. It may just avoid foods that are inherently contaminated with microbes, such as raw eggs; unpasteurized milk; and raw or undercooked meat, fish, and seafood. All raw vegetables and unpeeled fresh fruits are also excluded. A more restrictive diet avoids all foods that may harbor other pathogens, including yeasts and molds. Only well cooked foods are allowed and the following are excluded: pasteurized milk, buttermilk, whipped cream, sour cream, yogurt, cheese, except American, ice cream and sherbet, cottage cheese, fresh fruits and juices, raisins and other dried fruits, potato and macaroni salad, sweet rolls, deli meats and luncheon meats, and dried meats. Food sanitation guidelines must be followed in the handling, preparation, and serving of foods. The scientific evidence for the low microbial diet is lacking; however, most dietitians strongly support the food sanitation guidelines. *See also, Diet, neutropenic.*

Diet, macrobiotic. See *Diet, Zen macrobiotic.*

Diet, maltose-restricted. Diet that eliminates foods containing maltose or available maltose; prescribed for maltose intolerance. Excluded from the diet are corn syrup, corn sugar, beets, malted cereals, and other malted products. Because maltose is an intermediate product of starch digestion, the intake of starchy foods such as wheat, rice, corn, potato, and sweet potato is limited.

Diet, MCT. See under *Diet, fat-modified.*

Diet, meat-free test. Diet taken prior to a fecal occult blood testing with Hemoccult® and other guaiac tests. Meat, fish, and poultry are not given for 3 days prior to the test. These foods contain hemoglobin and myoglobin, which can give a false-positive result. Also avoid horseradish, beets, bananas, tomatoes, iron, vitamin C, and aspirin-containing compounds. Less dietary restriction is required with the new Hemoquant® test which requires only the avoidance of red meat. Moderate amounts of chicken, fish, ham, or bacon are allowed.

Diet, mechanical soft. Also called the "dental soft diet." It is used for patients who have difficulty in chewing due to poor dental condition or lack of teeth, or to the presence of sores and lesions in the mouth, following head and neck surgery, and for those who are debilitated and are too ill to eat the regular diet. It consists of foods that are soft, well cooked and easy to chew, and, if necessary, chopped, ground, or minced. Foods are best served moist, as in casseroles, or with gravy or sauce. The diet must be highly individualized according to each patient's chewing tolerance. All beverages are allowed, although patients with lesions in the mouth may not be able to take tart fruit juices.

Diet, methionine-restricted. Diet prescribed for homocystinuria. Products low in methionine and methionine-free chemically defined formulas are commercially available for infant feeding (e.g., Low Methionine Diet Powder, Product 3200K, Maxamaid, and Maxamum XMET). The methionine level is controlled by adding specified amounts of milk or proprietary infant formula to these products. Small amounts of solid foods are introduced at the usual ages. Considerations in planning the diet are similar to those in phenylketonuria. This involves calculating the amount of formula needed to meet protein and calorie requirements and then determining the amounts of other foods permitted. As an adjunct to this diet, supplementary betaine or choline may be given to promote remethylation of homocysteine.

Diet, milk-restricted. Diet that omits foods containing cow's milk proteins. Avoid all forms of cow's milk, dairy products, and foods prepared with added milk or milk solids. Check food labels and avoid those that contain terms indicating the presence of milk protein, such as casein, caseinate, curd, milk solids, lactalbumin, lactoglobulin, Simplesse, and whey. Infants and young children are given special formulas or nonmilk substitutes, such as Nutramigen®, Pregestimil®, and Alimentum®. Infants with milk allergy may also be sensitive to soy. Soy-based formulas and those made with whey protein hydrolysates should not be used.

Diet, modified. A *regular diet* altered to meet specific body requirements under different conditions of health or disease. The diet may be modified in consistency, content (calories, carbohydrates, protein, fat, or specific nutrient), flavor, methods of preparation or service, and frequency of feeding.

Diet, mold-restricted. Diet restricting foods containing molds due to allergy or sensitivity to molds. Mushrooms and food sources of molds, such as bacon, beer, buttermilk, cheeses, ham, sauerkraut, sausage, sour cream, soured milk, and wine are eliminated or restricted. Because molds grow on food, eat foods that are freshly prepared and use canned foods immediately. Do not store foods for an extended period of time.

Diet, motor test meal. Test diet to determine the emptying time of the stomach. It consists of rice and raisins or berries with seeds, or a meat sandwich with 2 tablespoons of raisins, or a meal with stewed prunes given 12 hours before gastric analysis. The presence of fibers in the gastric contents indicates decreased stomach motility.

Diet, MSG-restricted. Diet restricting the use of MSG (monosodium glutamate) in cooking. Ingestion of MSG is believed to cause the "Chinese restaurant syndrome" and exacerbation of asthma in individuals sensitive to monosodium glutamate which is used as a flavor enhancer in Chinese and other Asian foods. MSG is on the generally recognized as safe

(GRAS) list of food additives and is often added in bouillon cubes, commercial spice mixtures, and canned, packaged, or prepared foods. Read food labels and avoid those containing the following terms: glutamate, monosodium glutamate, and MSG.

Diet, neutropenic. Diet that is prepared and served under strict sanitary conditions to minimize the microbial count, especially pathogens. Useful for immunocompromised patients who have neutrophil counts of less than $500/mm^3$. The following measures in handling and serving foods are observed: restrict or avoid fresh fruits and vegetables; cook foods adequately; cover all food items properly; serve immediately after preparation; keep hot foods hot and cold foods cold, observing recommended temperatures for food safety; avoid cross-contamination; thaw fish, meat, and poultry in the refrigerator; and cook thoroughly. Add spices while at least 5 minutes of cooking time remains; properly cool and refrigerate foods quickly; and maintain adequate storage temperatures. See also *Diet, low microbial*.

Diet, nickel-restricted. Diet prescribed for individuals sensitive to nickel and who experience a flare-up of dermatitis following ingestion of a 2.5-mg nickel challenge. Foods high in nickel are avoided, such as beans, bran products, buckwheat, chocolate and cocoa drinks, figs, kale, leeks, lentils, lettuce, multigrain bread, oatmeal, peanuts, peas, pineapple, prunes, raspberries, shellfish, spinach, and mineral supplements containing nickel. Avoid using stainless steel or nickel-plated utensils. The nickel-restricted diet is generally recommended for 1 to 2 months, then small amounts of foods containing nickel are gradually introduced as tolerated.

Diet, normal. Diet that supplies all the nutritional needs of a normal, healthy individual, with due consideration for age, sex, activity, and physiologic needs. It contains enough calories for energy, adequate protein for growth, and sufficient minerals, vitamins, and water for the proper functioning of the body.

Diet, no sugars and concentrated sweets. A liberalized diabetic diet prescribed for individuals with mildly impaired glucose tolerance who are maintaining an acceptable weight; those who need a diabetic diet but have poor intake; and diabetics who cannot or will not follow the food exchange system. "Simple sugars" and foods high in sugar need not be avoided; small amounts, up to 20% of total carbohydrate, are included in the diet. Alternate sweeteners such as honey and fructose are also allowed. Generous intake of fiber (30 to 50 g) is recommended to improve glycemic control. The diet does not use measurements or food exchanges.

Diet, optimal. The *best possible* diet that will supply all the essential nutrients at the *highest possible* level to achieve the *ultimate* goal of nutritional intake. The optimal diet is difficult to define in precise quantitative terms, as the optimum intake for each nutrient for every individual has not yet been established.

Diet, oxalate-restricted. Diet prescribed for urinary calcium oxalate stones. Foods high in oxalic acid such as spinach, rhubarb, endive, beets, green beans, sweet potato, swiss chard, collard, plums, all types of berries, nuts, tea, and cocoa are avoided. Plenty of fluids (3 to 4 liters) should be taken throughout the day to ensure a constantly high and diluted urine output.

Diet, penicillin-restricted. Diet aimed at eliminating foods containing penicillin; prescribed in the treatment of some types of chronic urticaria or other allergic reactions to penicillin. Milk and other dairy products that may contain penicillin as a contaminant are avoided. A calcium supplement or an alternative source should be taken daily to meet the RDA for calcium.

Diet, phenylalanine-restricted. Diet restricting the intake of phenylalanine to approximately 15 to 25 mg/kg body weight, depending on the tolerance of amino acid and the age of the patient. Because natural protein foods contain about 5% phenylalanine, high-quality protein foods such as meats, fish, poultry, milk and milk products, and eggs are not allowed. Special formulas low in phenylalanine are used, such as: Lofenalac® and Analog XP for infants; Maxamaid XP, PKU-1, and PKU-2 for young children; and Phenyl-Free®, PKU-3, and Maxamum XP for older children and adults. Controlled amounts of phenylalanine and low-protein foods such as fruits, vegetables, and starches are gradually added as the child grows older. Foods and beverages containing aspartame (Equal® or NutraSweet®) are not allowed because the sweetener is derived from phenylalanine. Phenylalanine restriction is usually continued until the age of 5 years or later, although there is no safe age when dietary restrictions can be discontinued. The current recommendation is to continue the PKU diet indefinitely. Blood phenylalanine levels maintained between 2 and 10 mg/dL are used as guidelines for allowing limited amounts of phenylalanine in the diet. Women with PKU should maintain blood phenylalanine levels between 2 and 6 mg/dL prior to conception and during pregnancy.

Diet, phosphorus-restricted. Diet restricting the phosphorus intake to the minimal level. The intake of foods rich in phosphorus is limited. These include milk and milk products, legumes, fish, meat, leafy green vegetables, nuts, and whole-grain cereals. Calcium intake is also limited if the diet is prescribed for urinary calcium phosphate stones.

Diet, phytanic acid-restricted. Diet restricting the intake of phytanic acid and phytol, which are found in a wide variety of foods. Eliminated are rich sources of phytanic acid (primarily milk and dairy products) and phytol (primarily green vegetables, nuts, and legumes). The diet is prescribed for Refsum's disease.

Diet, potassium-modified. Potassium intakes vary widely depending on food selection. The richest dietary sources of potassium are unprocessed foods, fruits (especially bananas, berries, citrus fruits and juices, guava, kiwi, figs, melons, and dried fruits), and some vegetables (tomatoes, potato, sweet potato, asparagus, broccoli, carrots, corn, mushroom, spinach, and summer squash). Individuals who eat plenty of fruits and vegetables can have a high potassium intake of up to 8000 mg/day. The normal daily intake is about 2500 to 3500 mg (65 to 90 mEq). The minimum requirement is approximately 1600 to 2000 mg (40 to 50 mEq) per day.

High potassium d. Diet that provides about 4000 to 6000 mg of potassium (100 to 155 mEq K) per day. The diet is designed to prevent hypokalemia as a result of drug therapy with potassium-wasting diuretics and steroids, or from a disease state. A high potassium intake may also be beneficial in mild to moderate essential hypertension. Increased amounts of fruits and vegetables that are good sources of potassium are included daily in the diet. Salt substitutes that contain potassium chloride provide, on the average, 400 to 500 mg (10 to 13 mEq) of potassium per gram. However, dietary measures alone may not be able to reverse a preexisting potassium deficiency; prescription for a potassium supplement is often necessary.

Restricted potassium d. Diet prescribed in hyperkalemia restricting potassium intake to 1500 to 2000 mg/day (37 to 50 mEq potassium), which is about half the average potassium content of a regular mixed diet. Since potassium is widely distributed in foods, its restriction also limits the intake of other essential nutrients, particularly protein and vitamins. Milk, meats, legumes, whole grains, leafy vegetables, and some fruits (bananas, prunes, melons, and citrus fruits) supply considerable amounts of potassium. Many other foods are supplementary sources. The diet thus requires careful planning and selection of foods. Avoid low-sodium baking powder and potassium-based salt substitutes.

Diet, protein modified. Diet that prescribes a specified level of protein or restricts the amount of a protein fraction or amino acid. An increase in protein intake is necessary in any of the following conditions: excessive metabolism of protein, as in fevers and hyperthyroidism; loss of protein from the body, as in severe burns and nephritis; failure of protein synthesis, as in liver disease; failure of protein absorption, as in sprue and celiac disease; and inadequate intake of protein, as in starvation and kwashiorkor. A reduction in protein intake is necessary whenever the body's ability to excrete waste products of protein metabolism is impaired, as in hepatic coma and acute glomerulonephritis. Restriction of a specific amino acid is called for in certain conditions, such as phenylketonuria and other inborn errors of amino acid metabolism. The suggested terminology for the different levels in protein content of the diet is listed below.

High-protein d. An allowance of 1.5 to 2.0 g/kg protein for adults. Indicated in severe stress, depleted protein stores, hepatitis, and long bone fractures.

Liberal-protein d. A protein allowance of 1.2 to 1.4 g/kg body weight or about 75 to 100 g/day. A liberal protein intake is indicated in moderate stress, surgery, chronic obstructive pulmonary disease, and peritoneal dialysis.

Low-protein d. A protein allowance of 0.5 to 0.7 g/kg/day for adults, but at least 40 g/day. Indicated in chronic glomerulonephritis and chronic uremia. The protein in the diet is supplied by 1 egg, ½ cup of milk, 2 oz meat, 3 slices of bread or the equivalent, fruits, and low-protein vegetables. Additional calories are supplied by sugar and other sweets, fats, carbonated beverages, and baked products made from low-protein wheat starches.

Minimal-protein d. A protein allowance of 0.2 to 0.3 g/kg or approximately 20 to 25 g/day. It is prescribed for patients with acute renal failure, acute glomerulonephritis, and hepatic coma. The protein allowance is supplied by 1 egg, ½ cup of milk, 3 slices of bread or substitute, fruits, and low-protein vegetables. Extra calories are provided by liberal use of sugars, fat-rich foods, and protein-free starches. This dietary regimen will supply the essential amino acids for tissue synthesis; intake of nonessential amino acids is minimized. Because of their relatively greater content of nonessential amino acids in comparison with egg and milk proteins, meat, fish, and poultry are not recommended at this low level of protein intake.

Normal-protein d. A protein allowance of 0.8 to 1.0 g/kg/day or about 50 to 65 g/day for adults. Higher protein allowances are required during pregnancy, lactation, and growth periods, which may range from 1.5 to 2.0 g/kg body weight.

Very-high-protein d. A protein allowance of 2.5 to 3.0 g/kg/day for adults. Indicated for severe burns, severe sepsis, multiple fractures, and head injury. It is also prescribed for celiac disease and premature infants. Eggs, milk, cheese, meat, fish, and poultry are excellent protein sources of high biologic value. It is best to divide the protein allowance into three meals and three between-meal feedings.

SUGGESTED TERMINOLOGY	ALLOWANCE (g/kg body weight)	APPROX. LEVEL (g/day)
Minimal protein	0.2–0.3	20–25
Low protein	0.5–0.7	30–40
Normal protein	0.8–1.0	50–65
Liberal protein	1.2–1.4	75–85
High protein	1.5–2.0	90–110
Very high protein	2.5–3.0	120–150

Diet, protein-redistribution. Diet suggested for patients with Parkinson's disease who do not respond well to levodopa. The usual protein intake of 0.8 g/kg body weight is redistributed so that only 7 to 10 g of protein are given during the day (divided between breakfast and lunch), and the remaining protein is given after 5 P.M. Large neural amino acids (LNAAs) from protein in the diet compete with levodopa in crossing the blood–brain barrier, making the drug less effective. By restricting the daytime protein, the diet aims to improve the effectiveness of the drug and thus improve daytime mobility and motor performance. However, consuming a larger protein meal in the evening can result in suboptimal levodopa effect with consequent increase in rigidity and trembling at night. See also *Diet, seven:one (7:1)*.

Diet, protein-sparing modified fast (PSMF). A form of very low calorie diet that provides 1.5 to 2.0 g of high-quality protein/kg desirable body weight per day and 600 to 800 kcal. The diet consists of lean meat, fish, poultry without the skin, and no carbohydrate, and fat is restricted to that present in the protein sources. Four cups of raw or two cups of cooked nonstarchy vegetables daily are recommended. Vitamin and mineral supplements, as well as essential fatty acids, are required.

Diet, provocative. Allergy test diet containing the most allergenic foods, such as milk, egg, and wheat, unless the patient's history definitely contraindicates their use. Foods are introduced (challenged) one at a time to provoke a reaction. About ½ teaspoon to 1 tablespoon of fresh food (or 1 to 2 g of dried food) is given for the initial challenge. The amount is gradually increased until the amount approximates the usual serving size (or 8 to 10 g of dried food). If an allergy is due to food, some manifestations will show up within a week; otherwise, the allergy is due to a nonfood allergen.

Diet, prudent. Diet recommended by the American Heart Association for all healthy Americans over 2 years old as safe and prudent to prevent or reduce the incidence of coronary heart disease (CHD) and other atherosclerotic diseases. The general guidelines are as follows: reduce the amount of fat to less than 30%, with up to 10% polyunsaturated fatty acids, 10% to 15% monounsaturated fats, and saturated fats not exceeding 10% of the total kcal/day intake. A wide variety of foods should be consumed. Choose margarines made from liquid vegetable oils and salad dressings made with oils high in linoleic acid. Limit cholesterol intake to no more than 300 mg/day. Egg consumption should not exceed three per week for adults and four to six per week for children. Limit low-fat milk to 2 cups, include 1 oz of hard cheese (as a substitute for 2 oz of lean meat), and limit portion sizes to 2 or 3 oz of cooked lean meat, poultry, and fish twice a day. Increase carbohydrate intake to 55% to 60% of the total calories/day, including one or more whole grain breads, cereals, rice, oats, pasta, dried beans, and other legumes in each meal. Include at least four servings per day with as many fresh fruits and vegetables as possible to provide more fiber, vitamins (especially A and C), and other nutrients. Limit the use of refined sugar, salt, coffee, and alcohol. Sodium intake should not exceed 3 g/day; alcohol should not exceed 1 to 2 fluid oz of ethanol per day. Two oz of 100 proof whisky, 8 oz wine, or 24 oz of beer each contain about 1 oz of ethanol. Total calories should be sufficient to maintain the desirable body weight. See Appendix 33 for the alcohol content of various alcoholic beverages.

Diet, pureed. Also called the "blenderized diet"; consists of a normal variety of foods that have been strained or put in a blender or osterizer. It is indicated for patients who have difficulty in chewing and/or swallowing, as in cases of stroke, neurologic disorders, jaw wiring, and oral surgery. Liquids may be added to achieve the desired consistency. Some individuals may need a pureed diet thinned in order to be fed via a straw or syringe; others may need a thicker consistency that will form a bolus or not break apart in the mouth. Use milk, cream, fruit and vegetable juices, broths, gravies, and sauces to enhance the flavor and nutrient value. The addition of butter or margarine, sugars, powdered milk, and grated cheese will provide extra calories and protein.

Diet, purine-restricted. Diet restricting the daily intake of purine to approximately 120 to 150 mg compared to a normal intake of 600 to 1000 mg/day. The diet is prescribed as an adjunct to drug therapy for gout and other disorders affecting purine metabolism; it is designed to lower the uric acid level in the body. Good food sources of purine such as liver and glandular organs, anchovies, sardines, and meat extractives are excluded from the diet. Moderate purine sources such as fish and seafoods, meats, fowl, beans, peas, asparagus, mushrooms, cauliflower, and spinach are restricted, depending on the patient's condition. These foods are not allowed during acute gouty attacks. A 2-oz portion is permitted when the acute stage subsides, and two moderate servings per day may be taken in chronic conditions. Foods that are essentially free of purines may be taken as desired. These include breads and cereals, milk and milk products, eggs, fruits, vegetables (except those previously listed), sugars and other sweets, and beverages, except for alcohol which should be taken only in moderation. Encourage liberal fluid intake (2 to 3 liters/day) to dilute the urine and to promote uric acid excretion.

Diet, reducing. Low-calorie diet designed to reduce weight. The amount of caloric reduction from the normal intake varies, depending on the rate at which one expects to lose weight. Ideally, this should be about ½ to 1 lb/week, although a greater reduction in weight is necessary in extremely obese individuals. Reducing the number of calories from the normal intake by 3500 kcal/week or 500 kcal/day theoretically would reduce body weight by 1 lb/week. Several reducing regimens have been proposed, ranging from starvation diets to nibbling and from formula diets to calculated food intake. In order to be effective, a reducing diet must be nutritionally adequate (except for calories), acceptable to the patient, compatible with the food pattern to which the patient is accustomed, economically feasible, palatable, varied, and capable of providing a sense of well-being. Common types of reducing diets are a low-carbohydrate diet (reduction in total carbohydrate intake, both sugar and complex carbohydrates); a low-carbohydrate, high-protein, high-fat diet (high fat content risks increasing the serum cholesterol, high protein aggravates any kidney disorder, and vitamin B deficiency may develop if diet is prolonged); low-carbohydrate, limited-fat, limited-cholesterol diet (attempts to prevent an increase in serum cholesterol); a high-fat diet (high fat intake leads to atherosclerosis; fatigue and lassitude may occur); and a high-carbohydrate, low-fat, low-cholesterol, high-fiber diet (an increased intake of complex carbohydrates and vegetables, with limited intake of meat, refined sugar, and processed foods, is difficult to follow for prolonged periods). See also *Diet, calorie-restricted.*

Diet, regular. Also called "general," "house," or "full hospital diet"; it is a normal diet planned to provide the recommended daily allowances for essential nutrients but designed to meet the caloric needs of a bedridden or ambulatory patient whose condition does not require any dietary modification for therapeutic purposes. It also serves as a basis for the modification of therapeutic diets in the hospital. While there is no restriction as to the amounts and type of foods allowed, the diet calls for careful planning of menus, wise selection and proper preparation of foods, and attractive service so that it will appeal to patients with relatively poor appetites. The Food Guide Pyramid and the Dietary Guidelines for Americans are good guides to follow in planning meals. (See Appendix 4 and Appendix 7B.) Other recommendations from various professional, federal, and health organizations may also be utilized.

Diet, residue-modified. Diet that limits or eliminates the intake of foods that leave a high amount of residue in the colon. Foods with decreasing amounts of residue are carbohydrates with indigestible material, digestible carbohydrates, milk, fats, and protein.

Low-residue d. Diet often prescribed in the treatment of diarrhea and diseases involving the bowel, particularly in association with obstruction, distention, edema, and inflammation. A diet low in residue is desirable in these conditions, where the presence of bulky fecal masses would strain the colon. Foods recommended are fish, tender cuts of meat, chicken, hard-cooked egg, liver, gelatin, refined cereals, and nonfibrous cooked or canned fruits and vegetables. Limit milk or milk products and foods that contain milk to the equivalent of 2 cups of milk per day.

Minimal-residue d. Diet designed to leave the least amount of residue in the lower bowel after digestion and absorption have taken place proximally. The diet is indicated prior to and following intestinal surgery, particularly of the colon, and during radiation therapy of the pelvic area. It may also be used initially during the acute stage of diarrhea, ileitis, colitis, and diverticulitis. Eliminated are milk, tough cuts of meat, fish or poultry with skin, fibrous fruits and vegetables unless canned or cooked and strained, excessive fats, excessive sweets, spices, and condiments. As a substitute for milk, use a lactose-free, low-residue enteral product; several are available commercially. An elemental formula may be indicated during the acute stage of inflammatory bowel disease. See Appendix 39. The diet is intended for short-term use only. Supplemental vitamins and minerals may be necessary.

Diet, rotation. Series of meal plans for 4 to 5 days to test for relief from food allergies. Closely related foods that are not tolerated or suspected to cause allergy are eaten one at a time on the first day of the 4- to 5-day cycle.

Diet, routine, hospital. Term referring to the *regular, soft,* and *liquid* diets commonly used in hospitals. They differ in the consistency and type of foods allowed.

Diet, Rowe's elimination. Series of four test diets for the diagnosis of food allergy.

Each of the first three diets contains a cereal or a starch, one or two meats, a group of fruits and vegetables, and condiments and seasonings. The fourth diet consists only of milk, tapioca, and sugar. The patient is placed on each diet for a period of 1 week unless relief of symptoms is obtained. If the patient shows no improvement on any of the diets, the allergy is probably not caused by food.

Diet, salicylate-restricted. Diet recommended for some individuals who are sensitive to salicylates and develop urticaria (hives), itching, angioedema, and rhinitis or bronchial asthma after ingestion of aspirin or foods containing salicylate, such as apples, apricots, berries, cherries, currants, grapefruit, grapes, lemons, melons, nectarines, oranges, peaches, plums, prunes, raisins, cucumbers, green peppers, potatoes, tomatoes, and almonds.

Diet, serotonin test. Also called the "5-hydroxyindoleacetic acid (5-HIAA) test diet." Foods high in serotonin are eliminated from the diet 3 days prior to and during the urinary collection for the diagnosis of carcinoid tumors of the intestinal tract. These tumors produce excessive serotonin, which is metabolized and excreted in the urine as 5-HIAA. Intake of foods rich in serotonin alters the result of the test. Foods excluded are bananas, plantains, avocados, pineapples, passion fruit, plums, tomatoes, walnuts, eggplants, alcohol, and vanilla.

Diet, seven:one (7:1). A new diet that aims to control the amount of large neural amino acids (LNAAs) in the blood to allow for better levodopa utilization in patients with Parkinson's disease. High blood levels of LNAAs, components of protein in the diet, compete with levodopa for the same transport mechanism into the brain. When blood levels of LNAA are lowered, levodopa moves across the blood–brain barrier more effectively. The diet consists of balancing the carbohydrates and protein in a 7:1 ratio to produce what is believed to be the most stable level of LNAAs for optimal levodopa effect. Daytime meals are allowed as long as the 7:1 carbohydrate to protein ratio is maintained. A convenience food product (Hearty Balance™) has been formulated to provide this ratio.

Diet, sodium-restricted. Diet in which the sodium content is limited to a specified level, which may range from mild restriction to severe restriction. Sodium restriction is used primarily for the elimination, control, and prevention of edema accompanying congestive heart failure, cirrhosis of the liver, nephritis, nephrosis, toxemias of pregnancy, and adrenocorticotropic hormone therapy. It is also beneficial in the treatment of some cases of hypertension sensitive to sodium. The average American diet contains approximately 3 to 6 g of sodium (about 7.5 to 15 g salt or sodium chloride) per day. Sodium in the diet comes from two sources: sodium *naturally* present in foods and sodium *added* during cooking and food processing. Table salt is 40% sodium by weight; 1 tsp of table salt contains approximately 2000 mg of sodium (87 mEq). Foods differ widely in their natural sodium content. In general, animal foods are relatively high in sodium; plant foods are generally low in sodium. Most of the sodium added to foods comes from table salt (or sodium chloride) and monosodium glutamate. The other forms of sodium commonly used in food processing are sodium bicarbonate, sodium citrate, sodium alginate, sodium benzoate, sodium hydroxide, and sodium sulfite. In addition, there is sodium in water and in some medicines. It is important to read labels on packaged foods. (See Appendix 5.) In the United States, sodium added during food processing provides 75% of sodium in the diet; 15% is added during cooking and at the table, and 10% is from the natural sodium content of foods. The estimated minimum requirement for sodium is 0.5 g/day. The usual American diet contains approximately 3 to 6 g of sodium (7.5 to 15 g salt or sodium chloride) per day. The National Research Council has recommended limiting the intake of salt to 6 g/day (2400 mg of sodium). The national dietary goal is an intake of less than 5 g of salt/day, and the American Heart Association and other health agencies recommend limiting sodium intake to 3 g or less. The levels of sodium restriction given below are based on these recommendation.

3000 mg sodium d (130 mEq sodium). Also called "no added salt" diet (N.A.S. diet). The preferred name is "no extra salt" diet. It is essentially a normal diet with moderate use of salt in cooking (no more than ½ tsp/day; sodium chloride or table salt contains 400 mg sodium per ¼ tsp). No additional (extra) salt or salty condiments is allowed at the table.

2000 mg sodium d (90 mEq sodium). Mild sodium restriction. A small amount of salt is allowed during cooking (¼ tsp/day). No further addition of salt or salty condiments is allowed at the table. Eat sparingly or limit to one small serving/day in place of ¼ tsp salt, foods processed in salt such as ham, cold cuts, and sausage. Avoid foods extremely high in sodium, such as bouillon cubes, salt pork, sauerkraut, pickles, soy sauce, and other high-sodium sauces. Read food labels.

1000 mg sodium d (45 mEq sodium). Moderate sodium restriction. Foods are prepared without added salt or sodium compound, and no processed foods high in sodium are allowed. Avoid or exclude all salted and processed foods allowed in limited amounts in the 2000-mg diet. Also avoid bread and crackers with salted tops, commercial bread stuffing, instant soup mixes, salty condiments and sauces, pickled foods, and other foods high in sodium. Up to 4 servings of regular bread, 3 cups of regular milk, 7 oz of unsalted meat or substitute, 3 servings of fresh or unsalted frozen and low-sodium canned vegetables, and 2 or more servings of fruits are allowed. There is no limit on the amount of low-sodium bread, unsalted cooked cereals, potatoes, and pastas, unsalted margarine or butter, and cooking oils.

500 mg sodium d (22 mEq sodium). Strict sodium restriction. It restricts the intake of foods naturally high in sodium content, such as meat, fish, eggs, milk, and vegetables high in sodium (i.e., beets, carrots, celery, and spinach). Prepared foods with added salt are not allowed. The diet thus requires the use of unsalted bread, unsalted butter, and other low-sodium dietetic foods. Regular milk is limited to 2 cups/day; unsalted meats are limited to 6 oz/day. One egg may be used in place of 1 oz of meat. Omit the following vegetables high in sodium: beets, beet greens, carrots, celery, chard, kale, mustard greens, peas, rutabagas, and spinach. This level of restriction is not recommended for long-term use.

250 mg sodium d(11 mEq sodium). Lowest level of sodium restriction prescribed for the acutely ill patient. At this level of sodium restriction, the intake of meats is limited to 5 oz/day and low-sodium milk is substituted for regular milk. Only those foods low in natural sodium content are allowed. The diet is therefore unpalatable and monotonous. This severely restricted sodium diet is seldom prescribed now because of available drug therapy.

Diet, soft. Diet consisting of foods that are soft in texture, easily digested, without any harsh or coarse fibers, and prepared simply and not highly seasoned. Foods generally included are milk and dairy products; eggs, tender cooked meat, fish, and poultry; well-cooked vegetables; crisp lettuce and salad tomatoes; cooked or canned fruits; bananas and citrus sections; whole-grain or enriched breads and cereals; plain desserts; all beverages; and other foods as tolerated. The soft diet is traditionally used as a progression from the *full liquid* to a *regular* diet after surgery, although it has no proven physiologic advantage over the regular diet. Smaller but more frequent feedings might be better tolerated by patients who are still unable to handle all the foods on the regular diet served in three meals. A liberal interpretation of a soft diet is to allow the patient's tolerance guide the selection of foods to include in the diet, except for specific food items that should be restricted for therapeutic purposes.

Diet, soy-restricted. An allergy diet restricting the intake of soybeans totally or at a level tolerated. Although soybean is the most allergenic legume, other foods classified as legumes, such as beans, chickpeas, lentils, peanuts, and peas may also have to be eliminated. Soybeans and soy products include soybean curd (tofu), miso, tempeh, textured vegetable protein, soy extenders, soy sauce, soy flour, soy milk, soy oil, and hydrolyzed soy protein. Check labels for ingredients containing soy.

Diet, space. Bulk-free diet designed for the use of astronauts in space travel. It requires little or no digestion and supplies calories and other nutritional requirements. Foods developed for space travel must conform to weight and volume restrictions and must be protected against chemical and biological deterioration. Freeze-dried finger foods and beverage powders that easily rehydrate have been developed.

Diet, starch-restricted. Diet limiting the intake of starch from bread, cereals, cereal products, and root crops. It is prescribed in cases of starch intolerance due to lack of pancreatic amylase. Foods allowed are milk, eggs, meats, fruits, and vegetables low in starch content. Carbohydrate in the diet is obtained from sugar, syrups, sweets, fruits, and enzymatically hydrolyzed malted starches such as dextrimaltose and glucose polymers.

Diet, starvation. Calorie-free diet designated to effect rapid weight reduction in a short period of time. Vitamins and minerals are given to meet specific nutrient requirements, and water intake is liberal to prevent dehydration. Weight loss of 4 to 8 lb/day in the early days of starvation is not rare. How long one should stay on this diet depends on several factors. Since severe complications may develop, starvation or fasting as a treatment for extreme obesity should be done in a hospital under strict medical supervision. With fat loss, the individual also loses nitrogen. No method has been successful in effecting loss of fat without loss of some nitrogen. See also *Nutrition, starvation*.

Diet, sterile. Diet aimed at eliminating foodborne pathogens in foods served to severely immunocompromised patients; used in conjunction with other measures such as immunosuppressive drugs, reverse isolation, and protected isolation environment. Steam autoclaving, prolonged oven baking, and gamma irradiation are methods that are employed to sterilize foods. Food preparation and tray assembly are done aseptically in a protected environment, such as a laminar air flow hood, to prevent contamination from airborne microorganisms. All tray service items, utensils, cookware, and paper goods are also sterilized. The protective advantage of a sterile diet over a low-microbial diet has not been established. See *Diet, low-microbial*.

Diet, sucrose-restricted. Diet that limits the intake of sucrose because of intolerance to this disaccharide. All forms of sucrose are excluded, such as table sugar, jellies, jams, marmalades, preserves, corn syrup, maple syrup, other syrups, sweetened condensed milk, and other foods with added sugar and sorbitol. Most fruits, except for berries, lemons, and grapes, contain sucrose and are given in limited amounts. Vegetables high in sucrose content, such as beets, green peas, sweet potatoes, navy beans, and soybeans, are also restricted. Cereals, milk, eggs, meat, fish, poultry, and vegetables low in sucrose are allowed in normal amounts. Sucrose-free formulas for infant feeding include Enfamil®, Isomil® SF, ProSobee®, Similac® and SMA®. See also *Product 3232A* and *RCF®*.

Diet, sulfite-restricted. Diet restricting the intake of foods that contain sulfiting agents; prescribed for individuals who experience a variety of adverse reactions after ingesting sulfites, including asthmatic attacks, bronchospasm, flushing, hives, gastrointestinal disturbance, and anaphylactic shock. Sulfites are used in the food industry to prevent browning, bleach certain foods, modify dough texture, and control microbial growth. Foods that may contain sulfites include cookies and crackers; pie and pizza crust; beer, cocktail mixes, wine, and wine coolers; canned clams, dried cod, frozen lobster, scallops, and shrimp (fresh, frozen, canned, or dried); dried and glazed fruits; fruit juices (canned, bottled, frozen, or "dietetic"); maraschino cherries; fresh and precut potatoes; condiments and relishes (horseradish, olives, onion and pickle relish, pickles, salad dressing mixes, and wine vinegar); and confections and frostings containing brown, powdered, or white sugar obtained from sugar beets. Read labels on packaged foods and avoid those containing sulfites. Restaurants are required to post signs if they have sulfite-containing items on the menu. The Food and Drug Administration (FDA) has banned the use of sulfite on fresh fruits and vegetables served raw.

Diet, supraglottic. Diet prescribed for patients with difficulty in swallowing even liquids. Thin liquids are usually more difficult to swallow without aspirating or spilling into the laryngeal inlet, as compared to thick liquids and semisolid foods that form a cohesive bolus. Provide smooth, moist, blenderized foods to facilitate swallowing. These include mashed potatoes with gravy and cereals of medium consistency such as oatmeal. Give puddings, custards, solid ice cream, and milkshakes. For fruits, bananas, pureed apples, and strained pears are recommended. Finely chopped meats with large amounts of gravy or egg salad with mayonnaise should be considered.

Diet, synthetic. Diet used for the diagnosis of food allergy. It contains amino acids, sugar, water, vitamin concentrates, salt mixtures, and sometimes emulsified fats. The mixture is given orally or by tube feeding. Persons who are allergic to food should show marked improvement with the synthetic diet; those who are allergic to substances other than food do not show any relief of symptoms. If allergy to food is ascertained, other foods may be added to the synthetic diet to determine which ones are allergenic.

Diet, tartrazine-restricted. Diet recommended for some patients with asthma who develop severe asthmatic attacks following the ingestion of tartrazine, a coloring dye (FD & C Yellow No. 5) used in foods and drugs. Check food labels and avoid those products that contain "FD & C Yellow No. 5" or "Yellow No. 5" in the ingredients.

Diet, test. Simple meal or specific food items taken prior to or during a test procedure for diagnosing a disorder. See *Diet,* *calcium test, Diet, elimination test, Diet, fat test, Diet, glucose tolerance test, Diet, histamine test, Diet, hydroxyproline test, Diet, meat-free test, Diet, motor test, Diet, serotonin test,* and *Diet, VMA test.*

Diet, taurine-restricted. Protein-modified diet prescribed for the treatment of psoriasis. Certain high-protein foods are entirely eliminated; others are restricted as to amounts allowed per day. The daily food intake should not exceed ½ to 1 can of evaporated milk diluted with an equal amount of water; 3 oz of chicken, turkey, beef, or veal; and ½ cup of cottage cheese or 1 slice of American or Swiss cheese. Fruits, vegetables, fats, sugars, and breads may be taken as desired. Foods that are to be avoided are eggs and dishes containing eggs, pasteurized and homogenized milk, ice cream, fish, cold cuts, organ meats, meat extractives, meat gravies, soups, and broths.

Diet, therapeutic. A normal diet adapted or modified to suit specific disease conditions; one designed to treat or cure diseases. See *Diet, modified.*

Diet therapy. The branch of dietetics that is concerned with the use of food for therapeutic purposes. Its goals are to maintain good nutritional status, correct deficiencies that may have occurred, afford rest to the whole body or to certain organs that may be affected by disease, adjust the food intake to the body's ability to metabolize the nutrients, and bring about changes in body weight whenever necessary. A therapeutic diet is planned on the basis of a normal or regular diet, with amounts of nutrients adjusted to meet the requirements imposed by disease or injury; essential nutrients should be provided as generously as the limitation of the diet allows. The diet must be flexible and adapted to the patient's food preferences, eating habits, economic status, religious beliefs, and social customs. Foods included should be acceptable to the patient and should emphasize natural and commonly used items that are available and easily prepared at home. A correctly planned diet is successful only if the food is eaten. Compare with *Nutrition therapy.* See also *Medical nutrition therapy.*

Diet, tyramine-restricted. Diet eliminating the intake of foods containing high amounts of tyramines and other pressor amines, such as dopamine and histidine. It is prescribed for patients receiving monoamine oxidase (MAO) inhibitory drugs in order to prevent a sudden hypertensive reaction to the amine. The diet should continue for 4 weeks after the drug is discontinued. Foods with high tyramine content are alcoholic beverages such as ale, beer, liqueurs, and red wines such as burgundy, Chianti, sherry, sauterne, and vermouth; caviar; aged and processed cheeses; pickled herring and salted dried fish; dry and fermented sausages such as salami, pepperoni, bologna, and corned beef; broad beans, lima beans, lentils, snow peas, and soybeans; liver; meat extracts; brewer's yeast; and soy sauce in large amounts. A limited amount of beer (12 to 24 oz/day) and white wine (4 to 8 oz/day) may be taken if approved by the physician. The individual should keep the tyramine intake below 5 mg/day by consuming only fresh or freshly prepared foods. Avoid any protein food that has been aged, stored, or refrigerated. Perishable refrigerated items should be consumed within 48 hours of purchase.

Diet, vanillylmandelic acid (VMA) test. Diet given 3 days before and during the urinary collection for vanillylmandelic acid (VMA) to diagnose pheochromocytoma in persons with unexplained hypertension. Foods that give rise to phenoxy acid in the urine and give false-positive results are restricted. These include bananas, pineapples, oranges, cheese, nuts, chocolate, coffee, tea, and foods containing vanilla extract. (Newer test procedures, such as urinary metanephrine measurement, do not require any dietary restriction.)

Diet, vegetarian. Diet that is predominantly from plant foods, and may or may not include milk, eggs, and certain animal products depending on the individual practice of vegetarianism. The diet is used for religious, health, economic, environmental, or other reasons. The nutritional adequacy of the diet is dependent on the type of vegetarian diet chosen. See *Vegetarian* and *Nutrition, vegetarianism*.

Diet, wheat-free. An allergy diet that excludes wheat and foods containing wheat products. Omit breads prepared with wheat flour, bran, wheat germ, wheat, gluten, or cracker meal; cereals containing wheat products; salad dressings thickened with wheat products; foods prepared with flour, bread, or other wheat products; and cream soups, gravies, and sauces thickened with wheat or wheat products. Ingredients on food labels that may indicate the presence of wheat include bulgur, durum, farina, flour, graham flour, malt, modified food starch, semolina, wheat bran, wheat germ, wheat starch, vegetable gum, and vegetable starch. Use flour made from barley, oat, potato, rice, and rye as a substitute for wheat flour.

Diet, Zen macrobiotic. Diet based on the belief that one's health and happiness depend on a proper balance between the "ying" and "yang" foods. The macrobiotic dietary pattern progresses in ten stages; eliminates desserts, fruits and salads, animal foods, soup, and vegetables, in that order; and replaces these with increasing amounts of cereal grains. The highest stage contains 100% cereals. All dietary stages encourage the restriction of fluid intake.

Dietary allowances. See *Recommended dietary allowances*.

Dietary antioxidant. See *Antioxidant*.

Dietary counseling. See *Nutrition Counseling*.

Dietary fiber. See under *Fiber*.

Dietary guidelines. Dietary recommendations to promote health and prevent or delay the onset of chronic diseases and cancer. At least eight different organizations, mostly federal, have issued dietary guidelines within the last 10 years. Their recommendations are similar and include the following: adjust energy intake and activity level to achieve and maintain appropriate body weight; reduce the intake of total fat, especially saturated fat; and increase the consumption of carbohydrates, especially complex carbohydrate. Other recommendations included in most guidelines are: eat a variety of foods; eat less sodium and cholesterol; eat more fruits, vegetables, grain products, and fiber; reduce the intake of refined sugar; and drink alcohol in moderation or not at all.

Dietary Guidelines Advisory Committee (DGAC). A committee, appointed by the U.S. Department of Agriculture (USDA) and the Department of Health and Human Services (DHHS), in charge of discussing revisions to the "Dietary Guidelines for Americans" formulated in 1990. The USDA and DHHS are required by law to publish dietary guidelines every 5 years; thus, the next revision is due in 1995. See Appendix 7 for the current guidelines.

Dietary history. Dietary study method used in evaluating or assessing the dietary intakes of individuals. It is taken by 24-hour recall or repeated food records to provide information on the subject's past and present dietary habits, food likes and dislikes, usual food pattern, and type of meals normally eaten over a relatively long period of time. The dietary history is useful in food habit studies and furnishes data for classifying individuals into certain groups. See also *Dietary study*.

Dietary requirement. Minimum amount of a specific nutrient that is needed by the body to attain a specified state of health. Unlike the *recommended dietary allowance,* it has no added margin of safety and is stated for *definite* (not average) conditions of age, weight, activity, and food intake, as well as physiologic status and pathologic state. There is variability and lack of precision in the assessment of nutritional requirements for different nutrients. For this reason, the *minimum daily requirements* for certain nutrients are often stated as ranges, and average minimum dietary requirements should be considered only as close approximates and should not be interpreted as final and accurate.

Dietary standard. Quantitative summary or compilation of nutrient allowances or requirements for various groups of people. It is used to formulate and evaluate food intakes of large population groups; it also serves as a rationale or yardstick for planning adequate nutrition and scheduling agricultural production. The establishment of a dietary standard is not easy because of the lack of available information about certain nutrients, the wide range of individual variation in nutrient requirements, and the lack of agreement among authorities setting the standard. The nutrients included in the standard are those that are apt to be absent from or inadequate in the usual diet. Other nutrients, although required by the body, are not included in the tabulation. Either they are present in adequate amounts in the usual diet, or they are trace elements or vitamins for which there are insufficient data to serve as a basis for recommendation. Several countries have established dietary standards for their populations. One must recognize the philosophy behind the standard set in each country and must distinguish carefully between the use of such terms as "standard," "requirement," and "allowance" in the interpretation for purposes of comparison or evaluation. See Appendices 1 and 4.

Dietary study (survey). Method of determining or evaluating the dietary intake of an individual, group, or population. The adequacy of a given diet is determined by *qualitative* comparison with the basic food groups or by *quantitative* comparison with the recommended dietary standard of a particular country. A dietary study is used to detect the adequacy or inadequacy of diets in order to give valuable information concerning food habits, menu preparation, and food procurement, availability, and distribution. See also *Nutrition assessment*.

Dietary study, methodology. There are several methods of obtaining dietary information. These are generally classified into those applicable to individuals and those applicable to groups. Methods applicable to individuals include *estimation by recall,* with the subject or food provider for the subject recalling the food intake of the previous 24 hours or longer; the *food intake record,* which is a listing of all foods eaten (including between-meal intakes) for varying lengths of time, usually 3 to 7 days; the *food frequency,* which obtains the frequency of use of common food items or food groups on a per day, per week, or per month basis; the *dietary history* taken by recall, repeated food records, or both to discover the usual food pattern over relatively long periods of time; and the *weighed food intake* of subject taken by a trained person, a parent of the subject, or the subject personally. Methods applicable to groups include a *food account* or running reports of foods purchased or produced for household use; a *food list* or recall of estimated amounts of various foods consumed during a previous period, usually the past 7 days; and a *food record* or weighed inventory of foods at the beginning and end of the study, with or without records of kitchen and plate wastes.

Dietetic fellow. A certification award by the American Dietetic Association (ADA) to an active member who meets certain criteria specified by the ADA, such as minimum educational attainment of a master's degree, work experience as a registered dietitian for at least 8 years, professional achievement, and community services. Details about criteria are undergoing revision for 1995.

Dietetic foods. Processed foods for therapeutic purposes. In the United States, a wide array of dietetic foods are commercially available. The most common are products containing nonnutritive (artificial) sweeteners such as low-calorie soft drinks, canned fruits and juices, puddings and gelatin desserts, confectionery, and baked goods. Other dietetic foods may be low in sodium, cholesterol, total fat or saturated fat, and calories. Specific nutrition information is required on the food labels for these products. See Appendix 5. See also *Nutrition labeling*.

Dietetics. Combined science and art of regulating the planning, preparing, and serving of meals to individuals or groups under *various conditions of health and disease* according to the principles of nutrition and management, with due consideration for economic, social, cultural, and psychologic factors. The science consists of knowledge of nutrition, food, and the dietary constituents needed in different states of health and disease. The art consists of knowledge of the practical planning and preparation of meals at various economic levels, as well as attractive and pleasing service of food so that an individual, well or ill, will be encouraged to eat the food and adhere to the diet.

Dietetic technician, registered (DTR). An individual who has completed a minimum of an associate degree in dietetics or a related area at a regionally accredited U.S. college or university, completed a supervised clinical experience, and passed a national examination administered by the Commission on Dietetic Registration. A dietetic technician, registered, must earn 50 hours of continuing education every 5 years. DTRs are qualified to perform nutrition screening and other nutrition services under the direction of a registered dietitian.

Dietitian, registered (RD). An individual who has completed a baccalaureate degree in dietetics or a related area at a regionally accredited U.S. college or university, completed a supervised clinical experience, and passed a national examination administered by the Commission on Dietetic Registration, which is recognized by the National Commission for Certifying Agencies. To retain registered dietitian status, 75 hours of continuing education activities are required every 5 years. Registered dietitians are qualified to perform nutrition screening, assessment, and treatment. A *licensed* dietitian is credentialed (licensed) by individual states and may have the same or similar qualifications as a registered dietitian.

Diffusion. Redistribution of material by random movement; spreading out. *Simple diffusion* is movement of solutes from higher to lower concentration. *Facilitated diffusion* is carrier-mediated movement of solutes down their electrochemical potential, but the rate of movement is faster than can be accounted for by simple diffusion.

Diflunisal. A nonsteroidal antiinflammatory agent used for relief of pain and inflammatory diseases; a derivative of salicylic acid. It imposes a risk of gastrointestinal blood loss leading to anemia; it may also cause gastric ulceration, stomatitis, dyspepsia, anorexia, nausea ,and vomiting. The brand name is Dolobid®.

Digestibility. Extent to which a foodstuff is digested and absorbed from the digestive tract and not excreted in the feces. The fecal residue excreted is primarily indigestible materials, secretions, linings shed from the digestive tract, and microoganisms with their end products.

Apparent d. Measure of the difference between food intake and output in the feces, without consideration of fecal excretion not due to food eaten.

True d. Measure of the difference between food intake and fecal output, with allowances for linings shed from the intestinal tract, bacteria, and residues of digestive juices that are not part of indigestible food fecal output.

Digestion. The mechanical and chemical breakdown of complex substances into their constituent parts; the conversion of food into smaller and simpler units that can be absorbed by the body. See Appendix 11.

Digoxin. A cardiac glycoside obtained from the leaves of *Digitalis lanata;* used in the treatment of congestive heart failure. Low body potassium causes increased levels of digoxin in the myocardium and potential toxicity; ensure a high potassium intake, especially if digoxin is taken with a potassium-wasting diuretic. Digoxin may increase urinary excretion of calcium, magnesium, and potassium; it may also cause anorexia, nausea, and vomiting. Brand names are Lanoxin® and Lanoxicap®.

1,25-Dihydroxycholecalciferol. Also called "calcitriol" and "calcifetriol." It is the active form of vitamin D in humans; it is a steroid derivative made by the combined action of the skin, liver, and kidneys. See *Vitamin D*.

Diphenoxylate. A synthetic narcotic used in the management of diarrhea. It may cause abdominal pain, distention, dry mouth, nausea, and vomiting when taken in large amounts. Brand names are Lomotil®, Diphenatol®, Lomanate®, Lofene®, and Lonox®.

Diphosphopyridine nucleotide (DPN). See Coenzymes I and II.

Disaccharide. Sugar containing two monosaccharides joined in glycosidic linkage, with the elimination of a molecule of water. The most common are lactose, maltose, and sucrose.

Disaccharide intolerance. Inability to absorb certain disaccharides because of lack of certain specific disaccharidases, such as maltase, lactase, sucrase, and isomaltase. The condition may also be acquired from secondary surgical operations, infectious enteropathies, celiac disease, and other malabsorption states. Diarrhea is the principal symptom. Other clinical features include flatulence, abdominal pain, vomiting, and excretion of large amounts of volatile fatty acids. *Nutrition therapy:* exclusion of the poorly tolerated disaccharide from the diet results in disappearance of symptoms. Replace it with utilizable carbohydrate. The deficiency of the enzyme is apparently compensated for in later years. See *Diet, lactose-restricted; Diet, maltose-restricted; and Diet, sucrose-restricted.*

Disaster feeding. See *Nutrition, disaster feeding*.

Disulfiram. Drug used as an alcohol deterrent to aid in the treatment of alcohol dependence. Avoid alcohol; a severe reaction occurs if it is ingested while receiving disulfiram. Symptoms of this reaction include flushing, throbbing headache, nausea, vomiting, sweating, chest pain, hyperventilation, and hypotension. Trade name is Antabuse®.

Diverticula. Herniations or outpouchings of the mucous membrane through gaps or weak spots in the circular muscle of the colon. Diverticula may be single or multiple and occur most frequently in the distal descending portion of the colon. They are formed because of unusually high intraluminal colonic pressures.

Diverticulitis. Inflammatory condition of a diverticulum (or diverticula) characterized by nausea, vomiting, fever, abdominal tenderness, distention, pain, and intestinal spasm. The inflammatory process may eventually lead to intestinal obstruction or perforation, necessitating surgery. *Nutrition therapy:* conservative treatment in the acute stage includes intravenous feeding and NPO. Start oral feeding with an elemental formula and progress to a minimal, then low-residue diet, followed by increasing amounts of fiber and eventual return to a regular diet high in fiber. Avoid excessive intake of raw fruits and vegetables and hot spices such as chili pepper. See *Diet, residue-modified.* See also Appendix 39.

Diverticulosis. This condition is characterized by many small mucosal sacs, called *diverticula,* protruding through the intestinal wall. It occurs mostly in the sigmoid colon but is also evident in the gastrointestinal tract. There is apparently a defect in the muscle layers of the sigmoid colon, causing the lumen to narrow and the intraluminal pressure to increase. It may be the result of a diet lacking in fiber. Diverticulosis may lead to diverticulitis. *Nutrition therapy:* the main goal is to increase the bulk of the stools and distend the bowel wall by increasing dietary fiber. Increase the intake of whole grain cereals and other foods rich in insoluble fiber. Ensure adequate fluid intake. Two teaspoons of bran three times a day have been found effective in maintaining normal colonic function and regular bowel action. Bulk-forming agents such as methylcellulose and psyllium are also beneficial. See high-fiber diet under *Diet, fiber-modified.*

DMF. Refers to the total number of decayed, missing, and filled teeth. It is often used in surveys to determine the amount of *dental caries* present in a community or given population. However, teeth may be missing for other reasons, including removal because of local custom.

DNA. Abbreviation for *deoxyribonucleic acid.*

D/N ratio. Abbreviation for dextrose/nitrogen ratio. See *Glucose/nitrogen ratio.*

Double-blind, placebo-controlled food challenges (DBPCFCs). Considered to be the "gold standard" for diagnosing food allergy. Subjective bias is excluded by this method but it should be done in a hospital or physician's office equipped with appropriate emergency equipment because of the risk of an anaphylactic reaction. After a probable food allergen has been eliminated from the diet for at least 2 weeks, a small amount of the fresh or dried food is reintroduced in a food challenge to provoke symptoms. If no reaction occurs within 24 hours, the amount is doubled daily up to a standard portion of the fresh food or 8 to 10 g of dried food. This amount of dried food taken without problems usually implies that the food can be included in the diet.

DOPA. 3,4-Dihydroxyphenylalanine, an intermediate product in the formation of *melanin,* epinephrine, and norepinephrine from tyrosine.

Dopamine. 3,4-Dihydroxyphenylethylamine, an intermediate product in tyrosine metabolism and the precursor of norepinephrine and epinephrine. It is present in the central nervous system and is localized in the basal ganglia.

Down's syndrome (mongolism). Children suffering from this congenital defect are usually short and overweight, with signs of mental retardation. The condition is caused by trisomy of chromosome 21, directly correlated with the advanced reproductive age of the mother, particularly if she is over 35 years of age. *Nutrition therapy:* aided feeding may be required if there is poor lip and tongue control and problems with sucking or swallowing. Use adaptive feeding devices, adjust food texture and consistency, and provide oral stimulation. Prevent obesity by monitoring calorie intake and encouraging increased activity. Calorie recommendations for children aged 2 to 14 years are 16.1 kcal/cm height and 14.3 kcal/cm height for boys and girls, respectively. Avoid caloric and fat restrictions before age 2. Give a liberal protein intake and supplement the diet with vitamins and minerals, particularly vitamins A, B_6, and C.

Doxepin. A tricyclic antidepressant used for the treatment of depression. It may cause altered taste acuity, anorexia, and weight loss, as well as nausea, vomiting, diarrhea and stomatitis. Brand names are Adapin® and Sinequan®.

Doxorubicin. An antineoplastic antibiotic used for the treatment of tumors, including breast and ovarian carcinomas. It causes nausea and vomiting on the day of therapy, which can be severe. Stomatitis and esophagitis may occur, especially if doxorubicin is administered daily for several days; stomatitis usually begins with a burning sensation, accompanied by erythema of the oral mucosa, and in 2 to 3 days may progress to ulceration which is sometimes severe enough to result in difficulty in swallowing. The anorexia and reduced food intake may cause weight loss. The brand name is Adriamycin®.

DPN. Abbreviation for diphosphopyridine nucleotide. See *Coenzyme I* under *Coenzymes I* and *II*.

DRV. Abbreviation for *daily reference value*.

Dulcin. Nonnutritive sweetening agent 250 times as sweet as sugar. Not approved for food use in the United States but permitted in some European countries. Also called Sucrol and Valzin.

Dumping syndrome. Also known as "jejunal hyperosmotic syndrome." A gastrointestinal disorder characterized by physical signs that develop when the stomach contents are emptied into the jejunum at an abnormally fast rate. Symptoms include nausea, vomiting, weakness, syncope, and diarrhea. It is experienced by individuals who have had a partial or total gastrectomy. The syndrome is aggravated by eating meals high in sugar or simple carbohydrates, which increases the osmolality of the gastric contents, and/or by taking liquid with meals, which enhances the rapid emptying of food from the stomach. *Nutrition therapy:* a diet relatively low in simple sugars but high in fiber and complex carbohydrates, given in five or six small, dry meals should prevent discomfort. Take liquids between meals but avoid fluids for at least 1 hour before and after meals. Initially, limit fluids to 4 oz per serving. Fluid restriction retards gastric emptying of solids. Also avoid sugar, sweets or desserts, alcohol, caffeine, and carbonated beverages. Artificial sweeteners may be used. Restrict or omit milk and milk products if there is lactose intolerance. Lying down for 30 mintes after meals may reduce the symptoms of dumping. Liberalize the diet as symptoms improve. See *Gastrectomy*.

DV. Abbreviation for *daily value*.

Dynamic state. Continuous metabolism or turnover (i.e., synthesis, degradation, and replacement) of body constituents even at constant composition; a state of flux.

Dysgeusia. Taste perversion associated with a generalized decrease in taste acuity. This is a common complaint of cancer patients whose decreased appetite is partly due to dysgeusia, directly causing weight loss. One way to diminish this problem is to let the patients eat foods they like before chemotherapy. Taste acuity is also affected by drugs such as phenytoin, potassium chloride, lincomycin, antilipemics, and digoxin.

Dyspepsia. Gastric indigestion; a wide variety of complaints referable to the upper gastrointestinal tract following the ingestion of food. Typical symptoms are heartburn, nausea, epigastric pain, abdominal discomfort, belching, distention, and flatulence. It may be the result of an organic disease of the gastrointestinal tract, such as esophagitis, gastritis, or peptic ulcer, or it may be functional in nature and occur in the absence of any demonstrable organic lesion. Dyspepsia is also due to rapid eating, inadequate chewing, swallowing of air, or emotional stress. *Nutrition therapy:* no simple dietary rule can be set down. Foods should be adequate, well cooked, not too spicy, and served in a relaxed atmosphere. It is best to eat small meals. In the majority of cases, dyspepsia is of nervous origin and disappears once the psychoneurotic cause is removed. If it is due to organic causes, a soft diet low in fat may be beneficial.

Dysphagia. Difficulty in swallowing. It may be temporary or more serious if caused by a stroke or head injury, structural or mechanical impairment, and diseases involving the nervous system. Symptoms include facial drop, drooling, pouching or pocketing of food in the mouth, choking or coughing when eating or drinking, piecemeal swallow, abnormal gag reflex, and frequent throat clearing. Dysphagia may result in aspiration if food enters the respiratory tract below the level of the true vocal cords. *Nutrition therapy:* provide the appropriate food and fluid consistencies, and monitor the patient's overall nutritional status. Foods that form a cohesive bolus within the mouth are well tolerated. Those that break apart, such as dry chopped or ground meat or plain rice, are not recommended. Moisten foods by adding gravies and sauces to form a bolus. Avoid sticky foods that adhere to the roof of the mouth. Liquid consistencies cause the greatest problem; thicken liquids to avoid this difficulty. Depending on the severity of the swallowing dysfunction, the patient is placed in any of the four levels of dysphagic consistency: Level 1, all pureed foods with thick liquids, including smooth hot cereals, smooth yogurt, creamed cottage cheese, puddings, and strained soups thickened to pureed consistency; Level 2, lumpy pureed foods with thick liquids, soft moist cake or pancake and breads, finely chopped tender meats and cooked leafy greens, eggs, and fish moistened with sauce, soft cheeses, noodles and pasta, and sliced ripe banana; Level 3, finely ground meat bound with thick sauce, soft fish, eggs any style, soft chopped vegetables, and drained canned fruits; and Level 4, all regular chopped (bite-size) foods with thin liquids, except hard and particulate foods such as dry breads, tough meat, corn, rice, and apples. Pureed food must be thick enough to hold its shape on the plate. Thin liquids include water, all juices thinner than pineapple, Italian ice, and other clear liquids except gelatin desserts. Thick liquids include milk, fruit nectars, sherbet, and ice cream. It may be necessary to thicken liquids to a puree consistency for patients who cannot tolerate any kind of liquid. Foods that are mildly spiced or moderately sweet and served at room temperature are usually well accepted. Feed in small spoonfuls, allow adequate time for complete swallowing, and get the patient in the best possible position for feeding.

Dyssebacia. Disorder of the sebaceous follicles; plugging of the sebaceous glands. Nasolabial dyssebacia (seborrhea) is often seen in *riboflavin* deficiency. It is characterized by the appearance of enlarged follicles around the sides of the nose which may extend over the cheeks and forehead. The follicles are plugged with dry sebaceous material which often has a yellow color.

E

EAA. Abbreviation for *essential amino acid.*

Eating disorder. Collective term for anorexia nervosa, bulimia nervosa, and binge eating. Basically, these are emotional disorders: anorexia nervosa is compulsive self-starvation with extreme fear of weight gain, resulting in 25% or more body weight loss; bulimia nervosa is compulsive binging, eating up to 10,000 kcal in 8 hours, and then purging by forced vomiting. About half of anorexics are also bulimics, hence the coined term "bulimarexia." The health risks of these eating disorders include malnutrition, osteoporosis, menstrual cycle shutdown, heart problems with dangerously low blood pressure, circulatory collapse, and cardiac arrest in the advanced stage. Behavioral problems include depression, lack of self-esteem, suicidal tendencies, impulsiveness, and inability to cope with problems, especially family relationships. Compulsive eating or binge eating without purging is the consumption of abnormally large amounts of food, possibly caused by psychologic factors, leading to *obesity*. Compulsive eaters respond more readily to external cues than to internal factors that control appetite. See *Anorexia nervosa* and *Bulimia nervosa*. See also *Binge eating*.

Eclampsia. See *Pregnancy-induced hypertension*.

Edema. Presence of an abnormally large amount of fluid in the tissue spaces of the body due to a disturbance in the mechanisms involved in fluid exchange. Factors that tend to increase the volume of interstitial fluid include reduction in plasma osmotic pressure, rise in capillary blood pressure, increase in permeability of the capillary membrane, and obstruction of the lymph channels. Edema is seen in congestive heart failure, nephrotic syndrome, cirrhosis, and idiopathic edema. It is also seen in association with malnutrition, such as in beriberi, protein deficiency, and chronic starvation. Hypoalbuminemia is a major factor in the production of edema in patients with hepatic cirrhosis and nephrotic syndrome. Edema observed in burns is due to loss of plasma proteins and increased capillary permeability. *Nutrition therapy:* depending on the cause of the edematous condition, protein, fluid, and/or electrolyte intakes are modified. Fluid retention is relieved with sodium restriction and increased potassium intake. Because edema is a sign of a disease or disorder, refer to the nutritional modifications for a particular disease for further details on dietary guidelines.

EFA. Abbreviation for *essential fatty acid.*

EFAD. Abbreviation for *European Federation of the Associations of Dietitians.*

EFNEP. Abbreviation for *Expanded Food and Nutrition Education Program.*

Egg white injury. Condition characterized by exfoliative dermatitis, muscle pains, anorexia, and emaciation as a result of prolonged, excessive intake of raw egg white. This is actually a *biotin* deficiency caused by the presence of *avidin* in raw egg white, which combines with biotin, rendering it unavailable for use. Cooking the egg destroys avidin. Biotin deficiency can be produced by eating abnormally large amounts of raw egg whites for a long time.

Eicosanoic acid. A fatty acid with 20 carbons (eicosa) in its straight chain. An example is arachidic acid, found in butter, peanut oil, and other fats.

Eicosanoids. Hormonelike compounds synthesized in the body from polyunsaturated fatty acids. The primary substrate is arachidonic acid which is derived from linoleic acid. Eicosanoids include prostaglandins, thromboxanes, prostacyclins, and leukotrienes. They are believed to play a role in platelet aggregation, vasoconstriction, immunologic and allergic reactions, and inflammatory conditions.

Eicosapentaenoic acid (EPA). An omega-3 fatty acid with 20 carbons in its chain and five double bonds. Fish oils from salmon, black cod, mackerel, brook trout, herring, and sardines are good sources of EPA. The ingestion of EPA results in the lowering of blood pressure and reduction of very-low-density lipoprotein (VLDL) synthesis and platelet aggregation. Supplementation with EPA may also be beneficial in septic shock syndrome and rheumatoid arthritis.

Elastin. An albuminoid, the characteristic protein of yellow elastic fibers abundant in ligaments, lung matrix, and blood vessel walls.

Elbow breadth. An estimate of frame size based on the measurement of elbow breadth is done in the following manner: with the arm extended and the forearm bent upward to form a 90 degree angle, measure the distance (cm) between the two prominent bones on either side of the elbow. See Appendix 21 for the interpretation of results.

Electrolyte balance. Condition of electroneutrality in the body: the number of positively charged ions equals the number of negatively charged ions. Compensating shifts and losses/gains occur biochemically to maintain electrolyte balance. Electrolytes regulate water balance in the body. Failure to maintain the proper kind and amount of fluid in body compartments, i.e., fluid and electrolyte imbalance, is a medical problem. A typical electrolyte balance of cations and anions in the body fluids is as follows:

INTRACELLULAR FLUID (ICF) (mEq/liter)		EXTRACELLULAR FLUID (ECF) (mEq/liter)	
Na^+	35	Na^+	142
K^+	123	K^+	5
Ca^{2+}	15	Ca^{2+}	5
Mg^{2+}	2	Mg^{2+}	3
	175		155
Cl^-	5	Cl^-	104
$HPO_4^=$	80	$HPO_4^=$	2
$SO_4^=$	10	$SO_4^=$	1
Protein	70	Protein	16
HCO_3^-	10	HCO_3^-	27
	175	Organic acids	5
			155

The principal cation in the plasma and interstitial fluids is sodium, and the principal anion is chloride. In the intracellular fluid, the principal cation is potassium and the main anion is phosphate. The electrolytes, particularly Na^+ and K^+ ions, control the amount of water in the body compartments. Chloride (Cl^-), which is the main anion in the ECF, is the "balancer" for sodium; phosphate ion ($HPO_4^{=1}$) is the main anion in the ICF. There is a larger concentration of protein in the blood plasma than in the interstitial fluid. If body water is decreased, serum electrolytes will be increased. Water can move freely across cell membranes, but salts cannot. *Osmotic pressure* moves water from one side of the semipermeable membrane consisting of a dilute solution to the other side of a strong solution. When water in the cell is lost due to hypertonicity of the surrounding ECF, this is designated as *hypertonic dehydration*. When water in the cell is increased due to decreased solutes (hence, decreased osmotic pressure), this results in *hypotonic dehydration*, which causes reduced blood volume. Cellular edema occurs, and renal blood flow is impaired. Clinically, dehydration is a loss of both water and electrolytes. See also *Water balance* and *Dehydration*.

Elementra(TM). Brand name of a powdered elemental protein module made of hydrolyzed whey peptides. Provides 2.6 g of protein per scoop. For tube or oral feeding of patients with elevated protein needs. See Appendix 39.

Emaciation. Extreme leanness; wasting away of body flesh.

Emergency feeding. Also called disaster feeding. See *Nutrition, disaster feeding*.

Emphysema. Abnormal condition of the lung characterized by destructive changes and overdistention of the pulmonary alveoli. May be caused by respiratory infections or air pollution. Characterized by labored breathing with wheezing and chronic cough. Dyspnea and fatigue are accentuated if the patient is obese. *Nutrition therapy:* because of the shortness of breath, the patient frequently does not eat enough and finds it difficult to chew and swallow. A soft diet high in nutrient density should be given in small, frequent feedings. Provide adequate kilocalories [basal energy expenditure (BEE) × 1.5] and protein (1.2 to 1.5 g/kg body weight) to correct any tissue wasting but avoid hypercaloric feeding of foods high in carbohydrate. Promote weight loss if the patient is obese. Avoid gas-forming vegetables and encourage intake of plenty of fluids (2 to 3 liters/day). Restrict sodium (2 to 3 g/day) if there is pulmonary edema and monitor for effects of nutrient–drug interactions. Supplementation with vitamins and minerals may be necessary if intake is inadequate. Provide high-protein supplements for snacks to correct any tissue wasting.

"Empty" calories. Refer to carbohydrate-rich foods such as sugars, syrups, jellies, and other sweets that contribute mainly calories, with either no or insignificant amounts of the other nutrients. See also *Junk food*.

Emulsification. Process of lowering surface tension or breaking up large particles of an immiscible liquid into smaller ones that remain suspended in another liquid. An agent that has this ability is called an emulsifier, e.g., bile salts and lecithin. Emulsification of fat by the action of bile salts facilitates its digestion.

Enalapril. An angiotensin-converting enzyme inhibitor used in the management of hypertension. It may cause loss of taste perception, dry mouth, anorexia, and gastritis; it may also cause hypoalbuminemia and hyperkalemia. Brand names are Vasotec® and Vaseretic®.

Encephalomyelitis. Inflammation of the brain and spinal cord; characterized by fever; head, neck, and back pains; and nausea and vomiting. Depending on the extent of the damage to the central nervous system, more serious manifestations include personality changes, seizures, paralysis, coma, and even death. *Nutrition therapy:* feeding problems are related to symptoms such as fever, nausea and vomiting, and paralysis, which vary with each case. The ability to eat and swallow, self-feeding problems, loss of appetite, and reduced digestive processes need individualized monitoring. Enteral tube feeding is used for comatose patients whose gastrointestinal tract is functional.

Encephalopathy. Any abnormal function or structure of tissues of the brain, especially chronic, destructive, or degenerative conditions. See *Hepatic encephalopathy, Septic encephalopathy,* and *Wernicke's encephalopathy.*

Endemic. Prevalent in a particular region or locality. An endemic disease is one that has a low incidence but is more or less constantly present in a given population, e.g., endemic goiter.

Endergonic reaction. Reaction requiring an input or supply of energy to push the reaction; usually associated with anabolism.

Endocrine. Secreting internally or into the bloodstream. The *endocrine glands* are ductless glands producing internal secretions called "hormones" that are discharged directly into the bloodstream. These glands include the adrenals, thyroid, parathyroid, pituitary, gonads, pancreas, pineal gland, and thymus. The last gland is now classified as part of the lymphatic system. The endocrine system affects growth and development, digestion and metabolism, water and electrolyte balance, and many other vital processes in the body. See also *Hormone.*

Endopeptidase. Proteolytic enzyme that splits centrally located peptide bonds of protein. Examples are pepsin, trypsin, and chymotrypsin.

End-stage renal disease (ESRD). Or end-stage renal failure (ESRF). See under *Renal failure.*

Energy. Capacity to do work. It exists in six forms: kinetic, potential, thermal, nuclear, radiant or solar, and chemical. Chemical energy in the human body is released by the metabolism of foods. Carbohydrates, fats, proteins, and alcohol are chemical sources of energy. Energy is needed by the body for muscular activity, to maintain body temperature, and to carry out metabolic processes. Energy comes from the oxidation of foods and is measured in terms of *calories* or *joules.* See *Basal metabolism; Food, energy value;* and *Metabolizable energy.* See also *Energy balance, Energy requirement,* and Appendix 15.

Energy balance. Also called caloric balance: the equilibrium between energy intake and energy output. When energy intake exceeds energy needs, the extra energy is stored, leading to an increase in body weight (positive balance). When energy intake is less than energy needs, the body utilizes its own reserves, resulting in loss of weight (negative energy balance). See *Nutrition, weight control.*

Energy measurement. The potential energy available in a food is measured directly by using a bomb calorimeter or indirectly by calculation using these values: 4 kcal/g carbohydrate, 4 kcal/g protein, 9 kcal/g fat, and 7 kcal/g alcohol. For dextrose, 3.4 kcal/g is used. Energy values in foods and beverages as reported in food composition tables and the food exchange lists use these physiologic fuel values expressed in kcal or Calories. For measuring energy expenditures of the body, see *Calorimetry.*

Energy requirement. For a normal adult, the energy needs are estimated by adding the basal energy expenditure (BEE), plus the energy used for physical activity (PA), plus the thermal effect of food, formerly called "specific dynamic action of food (SDA)." The thermal effect of food is usually only 5% to 10% of total energy needs, and for practical purposes, some practitioners do not add this factor. Energy needs vary among individuals due to the effects of factors such as age, body composition and size, genetic makeup, growth, lactation, pregnancy, nutritional status, and climate or environmental temperature. In certain pathologic conditions or disorders (e.g., endocrine abnormalities, fevers, infections, burns, trauma), energy requirements are altered. For details on calculating energy requirements in various conditions, see Appendix 24. Recommended dietary allowances (RDAs) for energy are average values for groups of individuals according to age, sex, and physiologic state, as given in Appendix 1. See also *Basal metabolism, Nutrition, weight control,* and *Resting energy expenditure*.

Enfamil®. Brand name of a milk formula for routine infant feeding. Provides 20 kcal/oz and is also available in 13 kcal/oz and 24 kcal/oz formulations, with or without iron. Enfamil® Premature is a modified formula for premature infants with a limited intake or for infants recovering from illness with increased needs. See also Appendix 37.

Enrichment. 1. Addition of vitamins and minerals (thiamin, niacin, riboflavin, and iron) to flours and cereal products to restore those lost in milling and processing. Minimum and maximum levels for addition have been set. 2. In countries other than the United States, this refers to the addition of vitamins, minerals, amino acids, or protein concentrates to foods to improve their nutrient content. See *Fortification*.

Ensure®. Brand name of a polymeric, lactose-free low-residue product for tube feeding or oral supplementation. Also available in higher-nitrogen (Ensure® HN), high-calorie (Ensure Plus®), high-calorie, high-nitrogen (Ensure Plus® HN), high protein, normocaloric (Ensure® High Protein) and with 14.4 g of dietary fiber per liter (Ensure® with Fiber). See Appendix 39.

Enteral feeding (nutrition). Ingestion of food by mouth or by means of a tube via the gastrointestinal tract. Enteral nutrition products may be administered orally, via nasogastric tube, gastrostomy tube, or needle catheter jejunostomy. Enteral products may be *monomeric* or *oligomeric* (i.e., chemically defined formula made up of amino acids or short peptides and simple carbohydrates) or *polymeric* (i.e., more complex protein and carbohydrate sources) in composition. *Modular* supplements are used for individual supplementation of protein, carbohydrate, or fat when formulas do not offer sufficient flexibility. *Specialized* formulas are indicated for specific disease states, such as hepatic encephalopathy, renal failure, and trauma or high stress. Enteral products are used for patients who are unable to meet their nutritional requirements through solid foods. Conditions in which enteral nutrition is indicated include: protein-energy malnutrition; severe dysphagia due to stroke, brain tumor, or head injury; major trauma; hypercatabolic states; liver failure and severe renal dysfunction; radiation and chemotherapy regimen. Conditions in which enteral nutrition may be contraindicated include: complete or small bowel obstruction; ileus or intestinal hypomotility; severe pancreatitis; high output (>500 mL per day) external fistulas; and shock. See Appendix 39. See also *Tube feeding*.

Enteric. Pertaining to the intestines.

Enteritis. Acute or chronic inflammation or irritation of the intestinal mucosa, chiefly the small intestine. It may result from overeating, food or chemical poisoning, ingestion of irritants, or bacterial or protozoan invasion. Often accompanied by diarrhea and abdominal pain. *Nutritional therapy:* see *Gastritis*.

Enterocolitis. Inflammation of the small intestine and colon. *Nutrition therapy:* monitor the ability to digest food, giving foods tolerated by the patient. As the condition improves, give a diet low in fiber and moderately restricted in fats to 25% of the total caloric needs. See also *Enteritis*.

Enterocrinin. Hormone produced by the intestinal mucosa that stimulates the glands of the small intestine to secrete digestive fluid.

Enterogastrone. Hormone produced by the duodenum stimulated by the ingestion of fat. It inhibits gastric secretion and motility.

Enterostomies. Term for tube placement, usually by a surgical procedure, for the purpose of delivering nutrient formulas for enteric feeding via the pharynx, cervical esophagus, stomach, or jejunum. See also *Gastrostomy, Jejunostomy,* and *Ostomy*.

Entrition™. Brand name of a lactose-free tube feeding formula for delivery in a closed system. Entrition™ 0.5 is half-strength for initial tube feeding; Entrition™ HN is high in nitrogen for patients with increased protein needs. See Appendix 39.

Enzyme. Organic catalyst produced by living cells that is responsible for most of the chemical reactions and energy transformation in both plants and animals. Many enzymes are simple proteins, often existing as inactive *proenzymes or zymogens;* other enzymes require, in addition to the protein molecule, another factor in order to exhibit full activity. This *cofactor* may be an inorganic element such as zinc, or it may be an organic molecule such as a vitamin or its derivative. Some enzymes are thus described in terms of their protein portion, or *apoenzyme,* and their cofactor portion is designated a *coenzyme, prosthetic group,* or *activator*.

Epilepsy. Nervous disorder characterized by episodes of motor, sensory, or psychic dysfunction, with or without unconsciousness and/or convulsions. The most common episodes are *grand mal, Jacksonian, petit mal* and *psychomotor epilepsy*. *Nutrition therapy:* a regular diet is sufficient for those who respond to drug therapy. Monitor the side effects of anticonvulsant drugs, especially those that interfere with vitamin D and the B complex. Some drugs cause gastric irritation and weight loss. If drug therapy is not effective, ketogenic diets could be beneficial. See *Diet, ketogenic*.

Epinephrine. Adrenaline, the major hormone of the adrenal medulla. It is secreted in response to stimulation of nerve fibers by a variety of factors, including fear, anger, pain, hypoglycemia, hemorrhage, muscular activity, and anesthetic drugs. Its action is varied, causing dilatation of the skeletal muscles as well as coronary and visceral vessels and resulting in *increased* blood flow in these regions, *constriction* of the capillaries of the skin and arterioles of the kidney, elevation of the respiratory quotient with an increase in oxygen consumption and carbon dioxide production, and acceleration of glycogenolysis with formation of glucose in the liver and lactic acid in the muscles. See also *Norepinephrine*.

Epithelium. Covering of the skin and internal mucous membrane lining the body cavities, including vessels and passages. It consists of cells joined by small amounts of cementing substances. See *Nutrition, dermatology*.

EquaLYTE™. Brand name of an enteral rehydration solution for the prevention and treatment of mild-to-moderate isotonic dehydration. It is designed to replace fluids and electrolytes when gastrointestinal fluids are lost in such conditions as diarrhea, vomiting, malabsorption, or fistula drainage. See Appendix 38.

Ercalcidiol. New name for 25-hydroxyvitamin D_2, also known as "25-hydroxyergocalciferol." See *Vitamin D*.

Ercalciol. New name for ergocalciferol; vitamin D_2 or irradiated ergosterol. A compound that has vitamin D activity and that is obtained by ultraviolet irradiation of ergosterol.

Ercalcitriol. New name for 1,25-hydroxy vitamin D_2, also known as 1,25-hydroxyergocalciferol. See *Vitamin D*.

Ergocalciferol. Also called ercalciol; irradiated ergosterol or vitamin D_2. Formerly called calciferol or viosterol. See *Vitamin D*.

Ergosterol. Provitamin D_2, a sterol found chiefly in yeasts and widely distributed in plants. See *Vitamin D*.

Erucic acid. *Monounsaturated fatty acid* containing 22 carbon atoms; found in rapeseed, and other vegetable seed oils.

Erythrocuprein. A copper-containing protein found in the red blood cells. It contains two atoms of copper per mole and accounts for most, if not all, of the copper in the red blood cells.

Erythrocyte. Or red blood cell (RBC). The pigmented biconcave and nonnucleated cell that transports oxygen to the tissues. See *Hemoglobin* and Appendix 30.

Erythromycin. A broad-spectrum antibiotic effective against a wide variety of organisms, including gram-negative and gram-positive bacteria. Gastric acidity inactivates the drug, and the rate of drug absorption is decreased when given with food; it should not be taken with fruit juice or other acidic juices. Gastrointestinal side effects are dose-related. Abdominal pain and cramping occur frequently; nausea, vomiting, and diarrhea may also occur. Brand names are E-Mycin®, Erythrocin®, Ilosone®, and Pediazole®.

Erythropoiesis. Formation or development of erythrocytes from the primitive cells to the mature erythrocytes. See *Hemopoiesis*.

Erythropoietin. Hormone produced primarily in the kidneys and released into the bloodstream in response to tissue hypoxia, as in anemia. It stimulates the bone marrow to produce red blood cells (RBCs). In chronic renal failure, the kidneys are unable to produce adequate amounts of erythropoietin which results in a condition of chronic anemia. The anemia of chronic renal failure is reversed by administration of *recombinant human erythropoietin (r-HuEPO)*. See *Hemopoiesis*.

ESADDI. Abbreviation for *estimated safe and adequate daily dietary intake*.

Esophageal cancer. A neoplastic disease usually affecting the middle or lower third of the esophagus; causative factors are prolonged, heavy smoking and alcohol consumption, achalasia, hiatal hernia, aflatoxins, and betel-nut chewing seen in some Asian and African regions. Symptoms include anorexia, weight loss, persistent coughing, dysphagia, anemia, dehydration, malnutrition, and general malaise. In advanced cases, there is vocal cord paralysis and hemoptysis. *Nutrition therapy:* in the early stages when oral feeding is still adequate, provide a liberal intake of calories and protein in small, frequent feedings. Medium-chain triglycerides are helpful in malabsorption cases, especially when esophageal resection has been done. In advanced stages when per os feeding is not sufficient to correct weight loss and malnutrition, or dysphagia has worsened, specialized nutrition support is provided with gastrostomy feeding or total parenteral nutrition.

Esophageal reflux. See *Gastroesophageal reflux disease*.

Esophagitis. Inflammation of the mucosal lining of the esophagus. May be due to irritations, infections, or backflow of gastric contents from the stomach, as in gastroesophageal reflux. Drug therapy is the treatment of choice to neutralize acidity. *Nutrition therapy:* a soft diet, as tolerated by the patient, is recommended to prevent irritation of the esophagus. Many patients avoid acidic fruits and juices. Some find fatty and spicy meals irritating. Avoid alcohol and caffeine-containing beverages.

Esophagus. A hollow muscular tube measuring about 2 cm in diameter and 25 to 30 cm in length extending from the pharynx to the stomach. See *Digestion.*

ESRD. Abbreviation for *end-stage renal disease.*

Essential amino acid (EAA). See under *Amino acid.*

Essential fatty acid (EFA). See under *Fatty acid.*

Estimated safe and adequate daily dietary intake (ESADDI). Amount recommended for a nutrient listed as tentative by the Food and Nutrition Board, but the given value or range is considered safe and adequate until more information is gathered to establish the recommended dietary allowance (RDA). Upper levels in the "safe and adequate range" should not be habitually exceeded because of the toxic levels of many trace elements. See Appendix 2.

Estrogen. Collective term for the natural or synthetic female sex hormones. The naturally occurring estrogens, such as estradiol, estrone, and estriol, are produced principally by the maturing follicles in the ovary, although they are also formed in the placenta and adrenal cortex. Two synthetic estrogenic compounds of therapeutic importance are ethynylestradiol and diethylstilbestrol (DES). The estrogens are responsible for the development of the female sex organs, growth of the genitalia and mammary glands, and development of secondary sex characteristics; they also affect calcium and phosphorus metabolism and probably are related to bone metabolism, lipid metabolism, and skin or related structures. Synthetic estrogens are used in treating a variety of conditions associated with a deficiency of estrogenic hormones. A high dose impairs folate absorption and may cause folate deficiency; it may also increase serum triglycerides, decrease glucose tolerance, and increase weight due to salt and fluid retention. Estrogen is excreted in breast milk. Brand names are Estrace®, Estrone®, Estratab®, and Premarin®.

Ethacrynic acid. A loop diuretic used in the treatment of hypertension and edema associated with congestive heart failure. Diuresis results in increased excretion of potassium, sodium, chloride, calcium, magnesium, and zinc. Potassium depletion occurs frequently, and dehydration is most likely to occur in geriatric patients and those on restricted salt intake. Other side effects may cause carbohydrate intolerance, anorexia, dysphagia, and abdominal discomfort. The brand name is Edecrin®.

Ethanol. Ethyl alcohol, the form of alcohol that is ingested in alcoholic drinks. See *Alcohol* and Appendix 33.

Ethanolamine. Beta-aminoethyl alcohol, a basic component of certain cephalins. It may be formed by reduction of glycine or decarboxylation of serine and forms choline when methylated by methionine. A lipotropic agent, it can prevent the formation of fatty livers.

Etoposide. Antineoplastic drug used in treatment of leukemias, lung and testicular cancer, lymphomas, and ovarian cancer. Its side effects include nausea, vomiting, anorexia, and anemia. Brand name is VePesid®.

Etretinate. Systemic retinol derivative used for treatment of severe recalcitrant psoriasis. It may cause anorexia, abdominal pain, nausea, hepatitis, constipation, and hematuria. May also elevate potassium, calcium, phosphate, and sodium blood levels. Brand name is Tegison®.

European Federation of the Associations of Dietitians (EFAD). An organization of national dietetic associations in European countries. Membership is approved if the national dietetic association has more than 50% legally qualified dietitians. The aims of the federation are to encourage a better nutritional status for the populations of the member countries of the Council of Europe; to develop dietetics on a scientific and professional level; to promote the development of the dietetic profession; to improve the teaching of dietetics; and to pursue these objectives with the help of international organizations.

Exceed®. Brand name of a fluid replacement and energy drink; provides 460 kcal/16 oz from 100% carbohydrate calories.

Exchange list. Grouping of foods based on similarity in nutrient composition. Each food within the list has approximately the same food value and may be used interchangeably.

Exergonic reaction. Reaction accompanied by a release of energy; usually associated with catabolism.

Expanded Food and Nutrition Education Program (EFNEP). A program designed by the U.S. Department of Agriculture to teach low-income families, particularly those with small children, the skills to develop and consume nutritionally adequate meals. Trained nutrition aides, who are often members of the local community, work on a one-to-one basis in the homes of the low-income homemakers, using demonstration techniques reinforced with newsletters and printed educational handouts.

Expert Panel on Nutrition Monitoring. Established in 1990 by the Life Sciences Research Office of the Federation of American Societies for Experimental Biology (FASEB) to review the dietary and nutritional status of the American population. The panel report summarized the survey results from NHANES II, Hispanic HANES, and the NFCS and Continuing Survey of Food Intake of Individuals (CSF II).

Extension, USDA. Also called the Cooperative Extension Service. Established to teach nutrition directly to rural residents in the United States. Extension nutrition program activities include the Expanded Food and Nutrition Education Program (EFNEP), group meetings with youth clubs, homemakers' clubs, agricultural producers, and other special interest groups in both urban and rural areas.

Extrinsic factor. Literally, a substance originating from the outside. In nutrition, this refers to the extrinsic factor discovered by Castle or vitamin B_{12} obtained from food. See *Vitamin B_{12}*.

Exudative enteropathy. Diarrhea seen in inflammatory intestinal diseases, such as ulcerative colitis and Crohn's disease. It is accompanied by mucus, cellular components, and intestinal loss of serum protein. *Nutrition therapy:* increase protein intake by about 50% of the RDA and treat dehydration.

F

F. Chemical symbol for fluorine. See also *Fl.*

Factor. Any constituent that tends to produce a result. In nutrition, it refers to an essential or desirable element in the diet that has some effect on the growth, reproduction, or health of organisms. It may be a vitamin, a mineral, or any other nutrient. The structure of the factor may be identified or it may remain unidentified.

FAD. Abbreviation for flavin adenine dinucleotide. See *Flavin nucleotide.*

Failure to thrive (FTT). Also called "failure to grow"; a rate of gain in length and/or weight less than two standard deviations below the mean. Other criteria used to define FTT are: weight less than 80% of the 50th percentile for age, height for chronologic age less than 80%, and weight less than the 3rd percentile of the National Center for Health Statistics (NCHS) growth curve. Several factors may be involved in FTT, such as endocrine diseases, gastrointestinal disorders or obstruction, central nervous system disturbances, unusually high nutrient needs or losses, inadequate food intake, and psychosocial problems in the family that result in emotional deprivation of the child, causing erratic appetite, lack of interest in food, and conflicts concerning meal patterns. Nutritional support and guidance are needed to rehabilitate these infants and children. Nutrient intakes, particularly for calories and protein, should be assessed.

Fair Packaging and Labeling Act. This act, passed in 1966, set regulations for labeling and packaging consumer commodities such as foods, drugs, medical devices, and cosmetics as defined in the Food, Drug, and Cosmetic Act. Meat products, poultry, poultry products that are not canned, and prescription drugs do not come under this law. See *Nutrition labeling*.

Famotidine. A histamine H_2-receptor antagonist effective in reducing gastric acid secretion; used for the short-term management of confirmed active ulcer. There is a risk of vitamin B_{12} depletion with famotidine; it may be excreted in breast milk and may also cause anorexia, dry mouth, dysgeusia, and abdominal discomfort. The brand name is Pepcid®.

Fanconi syndrome. An inherited or acquired condition characterized by a low renal threshold for amino acids. A generalized aminoaciduria occurs even with normal levels of amino acids in the blood; there is also acidosis, hypokalemia, and loss of glucose and phosphates in the urine. This leads to withdrawal of calcium from the bones to neutralize the acids. As a result, conditioned rickets or osteomalacia may occur. *Nutrition therapy:* give plenty of liquids and dietary supplements of bicarbonate, potassium, phosphate, calcium, and vitamin D.

FAO. Abbreviation for *Food and Agriculture Organization*.

Farnoquinone. Vitamin K_2, a naphthoquinone derivative. Originally isolated from putrid fish meal; it is synthesized by a large number of intestinal bacteria. See *Vitamin K*.

Fat. In the strictest sense, the term means *neutral* or *true fat*. It is a *triglyceride*, which is an organic ester of three molecules of fatty acid combined with one molecule of glycerol. This is the form in which fats occur chiefly in foodstuffs and in fat depots of most animals. Neutral fat may be *soft* or *hard*, depending on the length of the fatty acid chain; or it may be *liquid* or *solid*, depending on the degree of saturation or unsaturation of the fatty acids. See also *Fatty acid* and *Lipid*.

Fat, body. There are two types of body fat: protoplasmic and depot fat. Protoplasmic fat is part of the essential structure of cells. In addition to neutral fat, it contains other lipids, such as phospholipid, cholesterol, and cerebroside. Protoplasmic fat is of constant composition and is not altered by variations in food intake; hence it is not reduced during starvation. Depot fat or *adipose tissue* is the fuel store of the body, and is found mainly in the subcutaneous tissue and around visceral organs. This fat store is either filled up or depleted, depending on the balance between the energy value of food eaten and expended. The adipose cell can store as much as 50 times its weight, and if filled up to its maximum, new adipose cells can be formed; hence the body has a very large capacity for storing fats. See also *Obesity*.

Fat functions. Fat is a concentrated source of energy, yielding about 9 kcal/g. It is a carrier of fat-soluble vitamins, adds palatability and satiety value to the diet, and has protein-sparing action. In addition, fat acts as a shock absorber, serves as a padding around vital organs, and insulates the body against the loss of heat. It stores energy efficiently, using the least amount of water among the fuel nutrients. When the person is at rest or engaging in light activities, about 40% of the energy needed by the body comes from fatty acids.

Fat intake, recommended. A satisfactory fat intake is 30% of the total caloric needs per day. For example, an adult requiring 1800 kcal/day needs 60 g of fat (540 kcal from fat). Fatty acid intake should be proportioned as follows: up to 10% of calorie needs from polyunsaturated fatty acids, 10% to 15% from monounsaturated fatty acids, and 10% or less from other fatty acids. Linoleic acid at a level of 1% to 2% of the total kcal/day is recommended to prevent essential fatty acid deficiency. Infants consuming 100 kcal/kg body weight per day need 0.2 g of dietary linoleic acid per kilogram of body weight. An adequate linoleic acid intake for adults is 3 to 6 g/day.

Fat-soluble. Generally refers to substances that cannot be dissolved in water but can be dissolved in fats and oils or in fat solvents such as ether and chloroform. The fat-soluble vitamins are vitamins A, D, E, and K.

Fat substitutes. Dietary replacements for fat in an effort to reduce fat and calories in food. About 37% of total calories in an average American diet come from fat. Food manufacturers and consumers aim to meet the "Healthy People 2000" objective of reducing fat intake to 30% by using fat substitutes.

Fatty acid (FA). An organic acid containing carbon, hydrogen, and oxygen; generally consists of a carbon chain that terminates in a —COOH group. It is found abundantly in ester linkage in several lipid compounds, or it may exist as a free or nonesterified fatty acid. With a few exceptions, fatty acids occurring in natural fats contain an even number of carbon atoms (4 to 24). Fatty acids are generally classified according to dietary essentiality, the number of carbon atoms, or the degree of saturation between the carbon atoms. See Appendix 10.

cis-FA (CFA). *Cis-* refers to the structural configuration of the fatty acid molecule where the hydrogens next to the carbon-to-carbon double bonds are on the same side, assuming a U-shape. The *cis-*configuration occurs naturally in foods, but becomes *trans-* in food processing. See also *trans-FA*.

Essential FAs (EFAs). Polyunsaturated fatty acids that are necessary for growth, reproduction, health of the skin, and proper utilization of fats. According to this definition, three FAs may be considered *physiologically essential:* linoleic, alpha-linolenic, and arachidonic acids. Linoleic acid and alpha-linolenic acid must be present in the diet. Linoleic acid serves as a precursor for the biosynthesis of arachidonic acid. EFAs are important for membrane structure and transport processes. Clinical symptoms of EFA deficiency include mild diarrhea; dryness, thickening, and desquamation of the skin; coarsening of the hair and hair loss; impaired wound healing; brittle and osteoporotic bones, and growth retardation in infants. Blood changes include decreased cholesterol levels, increased red blood cell fragility and anemia, and increased capillary permeability. Hepatomegaly, increased serum glutamic-oxaloacetic transaminase, serum glutamic-pyruvic transaminase, and lactic dehydrogenase, and fatty liver have been reported after 4 to 6 weeks of fat-free parenteral nutrition. Newborns, especially premature infants, with limited body fat stores of EFA are also susceptible to deficiency.

Free FA (FFA). Also called "nonesterified fatty acid"; plasma FFAs, such as oleic, palmitic, stearic, and linoleic acids, are bound to serum albumin as part of the lipoproteins. They are believed to aid in the transport of fat, both from alimentary sources and from fat depots, to be oxidized in various tissues. During fasting, FFA from depot fat increases, whereas glucose and insulin administration decrease the movement of depot fatty acids to plasma FFA. In fats and oils, the amount of free FA is a measure of the degree of hydrolytic rancidity.

Long-chain FAs (LCFAs). Those containing 12 to 22 carbon atoms. The most prevalent one in foods is stearic acid (18 carbon atoms). Food sources include cocoa butter, dairy fats, lard, tallow, and palm oil. Long-chain triglycerides contain essential fatty acids.

Medium-chain FAs (MCFAs). Those containing 8 to 10 carbon atoms. Although not prevalent in foods, MCFAs are more readily absorbed than LCFAs. Examples are caprylic acid (8 carbon atoms) and capric acid (10 carbon atoms). Commercial preparations of triglycerides made from these FAs (MCT) are used to treat patients with gastrointestinal tract disorders.

Monounsaturated FAs (MUFAs). Also called monoenoic fatty acids; those with only one unsaturated linkage or double bond, having two hydrogen atoms fewer than the saturated form. The most abundant monounsaturated FA in fats and oils are oleic and palmitoleic acids. Olive oil and canola oil are excellent sources of monounsaturated FA. Of animal fats, lard, beef suet, and chicken fat have about 40% to 45% of total FA as monounsaturates. See Appendix 34 for MUFA content of selected foods.

Nonesterified FAs (NEFAs). See *Free FA* under *Fatty acid*.

Omega FAs (OFAs). FAs designated by the position of the double bond starting from the omega end or the methyl (CH_3—) carbon. The symbol ω for omega is used, followed by a number indicating the location of the nearest double bond. The three classes of omega FA of dietary significance are omega-3 (ω3) such as *linolenic acid, eicosapentaenoic acid,* and *docosahexaenoic acid;* omega-6 (ω6) such as *linoleic acid* and *arachidonic acid;* and omega-9 (ω9) such as *oleic acid*. The omega-3 FAs are abundant in fish oils. Good sources are salmon, mackerel, tuna, and anchovy; plant sources are limited and include linseed, rapeseed (canola), walnut, and wheat germ. The omega-6 FAs are found in corn, cottonseed, safflower, soybean, and sunflower oil; they are also present in dairy products and organ meats. Omega-9 is a monounsaturated fatty acid. Recent studies support the role of omega-3 FAs in reducing the risk of heart attack. Omega-3 FA and omega-6 FA strengthen the immune system of the body and produce *eicosanoids*.

Polyunsaturated FAs (PUFAs). Those having two or more unsaturated linkages or double bonds and classed as dienoic, trienoic, and tetraenoic. Of greatest interest are linoleic acid (two double bonds), linolenic acid (three double bonds), and arachidonic acid (four double bonds). PUFAs play a role in immune function, fat transport and metabolism, and maintenance of the integrity of cell membranes. Fat sources high in PUFAs include the following oils and products made with these oils: corn, cottonseed, safflower, sesame, soybean, and sunflower oil. See Appendix 34 for PUFA content of selected foods.

Saturated FAs (SFAs). Those having all the carbon atoms of the molecule linked to hydrogen so that only single bonds exist. Saturation of FAs accounts for the firmness of fats at room temperature. The most common ones in animal fats are palmitic and stearic acids. Primary food sources are animal fats (butter, bacon, cream, lard, salt pork, meat and poultry fat, etc.) and some plants (palm oil and coconut oil). See Appendix 34 for SFA content of selected foods.

Short-chain FAs (SCFAs). Those containing fewer than 6 carbon atoms. These are not abundant in food fats and yield only about 5 kcal/g compared to 9 calories/g from the LCFAs. Examples are caproic acid (6 carbon atoms) and butyric acid (4 carbon atoms). Also known as "volatile" fatty acids, SCFAs are primarily the products of bacterial fermentation of indigestible carbohydrate and fiber polysaccharides. The three principal SCFAs produced are acetate, propionate, and butyrate. In addition to their caloric contribution, SCFAs promote sodium and water absorption in the colon and stimulate pancreatic enzyme secretion and gut mucosal proliferation. SCFAs are the major energy source of the colonocyte. Thus, conditions such as short-bowel syndrome, ulcerative colitis, and disuse atrophy may benefit from the effects of SCFAs. Short-chain fatty acids may also promote wound healing after bowel resection.

trans-FA (TFA). *Trans-* refers to the structural configuration of the fatty acid molecule where the hydrogens next to the carbon-to-carbon double bonds are on the opposite sides, assuming a linear form. Recent epidemiologic and metabolic studies provide evidence that *trans*-fatty acids raise the plasma levels of low-density lipoprotein-cholesterol and lower the high-density lipoprotein-cholesterol to the same extent as saturated fatty acids do. The process of hydrogenation converts some *cis*-form of fatty acid to *trans*-FA. It is suggested that in food labeling, information about *trans*-fatty acid content be included.

Unsaturated FAs (UFAs). Those containing one or more double bonds between one or more of the carbon atoms in the chain. Unsaturation alters certain properties of FAs. In general, the melting point is greatly lowered and the solubility in nonpolar solvents is enhanced. All the common UFAs in nature are liquid at room temperature. They are abundant in vegetable oils such as olive oil, corn oil, cottonseed oil, and soybean oil. See also

Monounsaturated FA and *Polyunsaturated FA* under *Fatty acid*.

Very-long-chain FAs (VLFAs). FAs with a carbon chain consisting of 24 or more carbon atoms. An example is hexacosanoic FA (C26:0), which is associated with *adrenoleukodystrophy*.

Fatty liver. Accumulation of fatty deposits rich in cholesterol esters in the liver. This may be the result of a number of causes, including lack of *lipotropic factors;* liver poisoning by phosphorus, chloroform, and other chlorinated compounds; and following chronic infectious diseases such as tuberculosis, metabolic disorders such as diabetes, or various nutritional disorders such as kwashiorkor, chronic alcoholism, and vitamin E deficiency. *Nutrition therapy:* restrict fats to 50 g/day and cholesterol to 200 mg/day. Use of *MCT* is beneficial.

Favism. An inherited metabolic disorder characterized by vomiting, dizziness, prostration, jaundice, and hemolytic anemia (which is sometimes fatal in children) following the ingestion of fava beans or even the inhalation of its pollens by susceptible individuals. It is due to deficiency of the enzyme glucose-6-phosphate dehydrogenase. Children are more prone to hemolytic episodes than adults. A singular feature of favism is the increasing tolerance to fava beans with age in affected individuals.

FDA. Abbreviation for *Food and Drug Administration*.

Fe. Chemical symbol for iron.

Fecal fat test. Diagnostic test for malabsorption by determining the amount of fat in the stools. Fecal fat greater than 7 g is indicative of steatorrhea or fat malabsorption. Values of 10 to 25 g indicate mild, 25 to 30 g moderate, and over 40 g severe fat malabsorption. A diet containing at least 100 g/day of fat is given for 3 days before the stool specimen is collected. The amount of fat may be reduced to 70 g per day for individuals who cannot consume the standard 100-g fat diet. The test result for the lower fat intake is interpreted as follows:

$$\text{Grams fecal fat/day} = 2.93 + (0.02 \times \text{grams dietary fat/day})$$

In children, steatorrhea is defined as fecal fat excretion of 5 g or more per day on a diet containing 40 to 65 g of fat.

Fenoprofen. A nonsteroidal anti-inflammatory agent used for treatment of rheumatoid arthritis and osteoarthritis; it also has analgesic and antipyretic activity. It has a metallic taste and can cause gastric mucosal damage, which may result in ulceration and/or bleeding. It can also induce fluid retention and weight gain, and may cause dyspepsia, dry mouth, aphthous lesions, and abdominal discomfort. Brand name is Nalfon®.

Fermentation. Also called glycolysis; the enzymatic oxidation of carbohydrate under anaerobic or partially anaerobic conditions.

Ferriprotoporphyrin. See *Hemin*.

Ferritin. An iron–protein complex found chiefly in the liver, spleen, bone marrow, and reticuloendothelial cells; it contains 23% iron. Ferritin functions in the absorption of iron through the intestinal mucosa and serves as a storage form of iron in the body. The normal serum ferritin range is 20 to 200 ng/mL (20 to 200 μg/L). Levels below 10 ng/mL indicate iron deficiency and levels above 400 ng/mL indicate iron excess. Increases in serum ferritin concentrations are associated with several diseases, including infections, liver disease, and cancer. When the storage capacity of ferritin is exceeded, iron accumulates in the liver as *hemosiderin*. See also *Hemosiderosis*.

Ferroprotoporphyrin. See *Heme.*

Fetal alcohol syndrome (FAS). Congenital defect in an infant delivered by an alcoholic mother who consumed at least 3 oz of alcohol per day during pregnancy. It has also been observed that ingestion of alcohol even occasionally during the gestation period may be harmful. The exact amount of alcohol that will result in FAS is not known; therefore, pregnant mothers should be advised to avoid alcohol throughout pregnancy. It is best to educate and warn expectant mothers of the serious consequences of alcohol on the fetus. Clinical signs seen in the infant are low birth weight, limb and facial/head malformations, cardiovascular defects, and retarded physical growth and mental development. *Nutrition therapy:* in serious cases, tube feeding or total parenteral nutrition may be needed. Oral formulas are resumed as soon as possible. See nutritional principles for *Low-birth-weight infants.*

Fever. 1. Elevation in body temperature above normal (98.6°F or 37°C); pyrexia. 2. Any disease characterized by a marked increase in temperature; acceleration of basal metabolism; increase in tissue destruction; and loss of body water, sodium, and potassium. Fevers may be *acute,* as in influenza, chickenpox, and pneumonia; *chronic,* as in tuberculosis; or *intermittent,* as in malaria. When fever rises sharply, it may cause convulsions and delirium. *Nutrition therapy:* The diet should be high in calories [1.5 × basal energy expenditure (BEE)] due to the elevated metabolic rate, which increases 7% for every 1°F (13% for every 1°C) rise in body temperature. Protein is increased (1.3 to 1.5 g/kg/day) to replace nitrogen losses from tissue destruction characteristic of febrile conditions. A liberal supply of carbohydrates will spare protein and replenish glycogen stores. Fluid intake will depend on losses from excretion (urination, perspiration, insensible losses, etc.), which may be as much as 2.5 to 4 liters/day. Severe dehydration may require intravenous treatment. For oral liquids, salty broths, nourishing drinks such as milk, vitamin C-rich beverages, and nutrient-dense liquids are preferred. In acute fevers, at least eight feedings are provided, progressing from a full liquid diet to a soft diet. Eventually, a regular regimen is resumed as tolerated. If fever is a symptom of a disorder that requires other dietary modifications, follow the guidelines as needed for the primary cause of the febrile condition.

FFA. Abbreviation for *free fatty acid.*

FH_4. Abbreviation for *tetrahydrofolic acid.*

Fiber. 1. Threadlike, elongated structure of organic tissue such as muscle fiber or nerve fiber. 2. In nutrition and diet therapy, fiber consists of nondigestible materials which include *plant fibers:* mostly cellulose, hemicellulose, lignin, agar, pectins, gums, and mucilages; animal tissues such as ligaments and gristle in meats; and undigested pharmaceutical products. Formerly called bulk or roughage. The term *dietary fiber* is now used in dietetics.

Crude fiber. Complex polysaccharide that cannot be hydrolyzed by human digestion or by acids and alkalis in the laboratory. Earlier food composition tables that reported *fiber* referred to crude fiber, which is mainly cellulose and does not include hemicellulose and other forms of fiber. Because of this limitation, the crude fiber values in foods are no longer used in clinical nutrition. Newer chemical analyses now include all food fibers.

Dietary fibers. Plant polysaccharides and lignin that are not digested in the human intestine. These include the *insoluble* dietary fibers: cellulose, hemicellulose, and lignin, which do not dissolve in water but instead absorb water, contributing bulk to the stools and preventing constipation and diverticulosis; and the *soluble* dietary fibers: pectins, gums, mucilages, and algal substances. Soluble dietary fibers are useful in lowering blood cholesterol, managing obesity, cardiovascular disease, and diabetes mellitus, and preventing colon cancer. Insoluble fibers are only 10% to 15% digested; good food sources are bran, whole wheat and other grains, cabbage family, green leafy vegetables, peas and beans, and mature vegetables. Soluble fibers are 90% to 99% digested and are also fermented by bacteria in the colon, producing short-chain fatty acids (SCFAs); good food sources are fruits, rolled oats and oat bran, tubers, dried beans, peas, and other legumes. Recommended daily intake of dietary fiber is 20 to 35 g/day for adults. This is about 10 to 13 g of fiber per 1000 calories, which can be obtained from five or more servings of fruits and vegetables and six or more servings of whole grain breads, cereals, and legumes. No recommendations exist for children, the elderly, or persons consuming special diets. Too much fiber may cause diarrhea in persons not accustomed to a high fiber intake and may interfere with the absorption of calcium and zinc. See also Appendix 32.

Fiberlan™. Brand name of an isotonic, lactose-free formula containing 14 g of dietary fiber per liter. See Appendix 39.

Fibersource™. Brand name of a formula with soy fiber (10 g/L); used for patients with maintenance protein needs. Also available as Fibersource™ HN for patients with increased protein needs. See Appendix 39.

Fibronectin. A plasma glycoprotein and *opsonin* of the reticuloendothelial system. It is not synthesized in the liver and has a short half-life (less than 1 day). As a nutritional marker, plasma fibronectin is sensitive to starvation and repletion; a decrease in plasma concentration is seen in burns, sepsis, surgery, and shock. The normal plasma fibronectin level is 2.92 ± 0.2 g/dL.

FIGLU. Abbreviation for *formiminoglutamic acid.*

Fish skin. Dry skin that comes off in fine or rough scales. See *Xeroderma.*

Fistula. 1. Deep ulcer, often leading to an internal hollow organ, as a result of incomplete healing of a wound, an abscess, or other disease conditions. 2. An abnormal passage or communication between two organs, often from an internal organ to the surface of the body. Most bowel fistulas occur secondary to abdominal surgery; other causes include inflammatory bowel disease, cancer, radiation therapy, and trauma. Patients become malnourished because of loss of nutrients, especially albumin, through drainage, increased metabolism associated with infection, and poor nutritional intake secondary to decreased appetite and mechanical obstruction. *Nutrition therapy:* provide a high-calorie (up to twice the basal energy expenditure) and high-protein (1.5 to 2.0 g/kg) diet. The anatomic location of the fistula and the volume of drainage determine the optimal means of nutritional support. Patients with high upper gastrointestinal fistula (stomach and duodenum) can be fed a chemically defined or polymeric product by jejunostomy feeding or a nasogastric tube placed 30 to 40 cm beyond the fistula to avoid reflux of food through the fistula. Gradually progress to a low-residue, followed by soft and regular diet as tolerated. Other patients with distal fistulas or high fistula output should be evaluated for parenteral nutrition support. Total parenteral nutrition should be reserved for patients with high-output (drainage) fistulas in the upper small intestine and with distal ileal or colonic fistulas.

Fl. *Fluoride* ion.

Flaky-paint dermatosis. Also called crazy-pavement dermatosis: extensive, often bilateral hyperpigmentation of the skin that peels off, leaving a hypopigmented skin with superficial ulceration. It is characteristic of protein-energy malnutrition (PEM), which occurs in patches, usually on the buttocks, thighs, and arms.

Flatulence. Gastrointestinal discomfort, with or without pain, due to the presence of excessive amounts of air or gas in the stomach and intestinal tract. The gases in these organs are nitrogen, hydrogen, oxygen, carbon dioxide, methane, and traces of others. They are produced in the gastrointestinal tract or are swallowed. Ingested air accounts for most of the gas in the esophagus and stomach, which may escape by belching. Hydrogen, carbon dioxide, and methane are produced in the intestine and constitute the bulk of the flatus, most of which passes through the rectum. Complaints of "too much gas" by patients may be due to an abnormality of intestinal motility associated with some diseases. *Nutrition therapy:* flatulence among individuals varies according to the foods they eat and their swallowing and eating habits. Flatulence can be relieved by avoiding foods that an individual finds "gassy." The foods usually implicated are legumes, the cabbage family, onions, prunes, raisins, bananas, sugar alcohols, bran and high-fiber grains, rich sauces and gravies, lactose in milk and ice cream, fermented foods, and carbonated beverages. Eating practices that are conducive to swallowing air include eating rapidly, drawing on straws, repetitive swallowing as in chewing gum and tobacco, sucking candies, and sipping liquids. Lack of exercise and stress factors may cause gas retention.

Flavin. Any one of a group of yellow pigments widely distributed in plants and animals, including *riboflavin*. They have an intense green fluorescence.

Flavin nucleotide. Derivative of the vitamin *riboflavin* that participates as a coenzyme in many oxidation–reduction reactions.

Flavin adenine dinucleotide (FAD). Riboflavin-5-phosphate attached to adenosine monophosphate. It forms the prosthetic group of certain enzymes and is important in electron transport mitochondria.

Flavin mononucleotide (FMN). Riboflavin-5-phosphate; it arises from riboflavin by reaction with adenosine triphosphate and the enzyme *flavokinase*. FMN acts as a coenzyme for a number of oxidative enzymes.

Flavonoids. More commonly called *bioflavonoids* to indicate having biologic activity; a large group of flavone derivatives, including hesperidin, rutin, and quercitin, that are widely distributed in plants and concentrated in the skin, peel, and outer layers of fruits and vegetables, as well as in tea, coffee, wine, and beer.

Flavoprotein. Flavin-containing protein that constitutes the *yellow enzyme;* has as a prosthetic group either a phosphoric acid ester of riboflavin (flavin mononucleotide, or FMN) or the latter combined with adenylic acid (flavin adenine dinucleotide, or FAD).

Fletcherism. System advocating thorough chewing of food (as many as 30 times) to obtain greater satisfaction from food flavors, to induce more effective secretion of digestive juices, and hence to enhance digestion and utilization of food. It is also claimed that chewing for a long time satisfies the appetite with much less food, thus reducing the total amount of food ingested.

Floxuridine. A pyrimidine antagonist used as an antiviral and antineoplastic agent. Among its adverse effects are nausea, vomiting, anorexia, glossitis, stomatitis, diarrhea, enteritis, and gastric ulceration. The brand name is FUDR.

Fluconazole. Antifungal drug used for treatment of oral or esophageal candidiasis in patients with AIDS. It may cause nausea, vomiting, diarrhea, cramping, and elevated blood BUN. Brand name is Diflucan®.

Fluid. (1) Physiologically, the term refers to intracellular or extracellular body fluids, such as blood, lymph, and cerebrospinal fluid; involved in the transport of electrolytes and other solutes. (2) In foods, the term refers to any liquid or substance that becomes liquid at room temperature or melts in the mouth. Thus, dietary fluid intake includes water and other beverages, gelatin, popsicle, sherbet, and ice cream. See also *Diet, liquid* and *Water*.

Fluid balance. A state of equilibrium in which the amount of liquids ingested equals the amount lost from the body via the urine, feces, perspiration, lungs, etc. See also *Water balance* and *Water requirement*.

Fluoridation. Addition of small amounts of fluoride in public water supplies wherever natural fluoride concentrations are low; an effective and practical means of reducing dental caries. Recommendations approved by national and international organizations call for fluoride concentrations between 0.7 and 1.2 mg/liter. In the United States, the Food and Nutrition Board recommends that water supplies be fluoridated at a level of 1 ppm (1 mg/liter) in areas where natural fluoride levels are substantially below 0.7 mg/liter. This level is considered safe and does not present any known health risk. The topical application of stannous fluoride and the addition of fluoride to mouth rinse and toothpaste are other ways of preventing tooth decay. See also *Dental caries*.

Fluoride. A binary compound of fluorine; found in trace amounts in human tissues, particularly in the teeth, bones, thyroid gland, and skin. There is controversy regarding the status of fluoride as a dietary essential nutrient. However, the beneficial effect of fluoride in preventing dental caries in the teeth of growing children is well established. Fluoride is retained in the teeth and bones, where it forms fluoroapatite, which is important for hardening tooth enamel and contributes to the stability of bone mineral matrix. Fluoride intake in the United States from food, beverages, and water is about 0.9 to 1.7 mg/day. It is present in small but varying concentrations in all soils, water supplies, plants, and animals; it is therefore a constituent of all diets. The chief source is drinking water; tea, seafoods, and marine fish are also good sources. It is readily absorbed (about 90%), and excretion is mainly in the urine. The potential for toxicity is high when it is consumed in excessive amounts. Of all the elements, fluoride has the narrowest range of safe and adequate intake. Mottling of the tooth enamel results from slight exposure of about three to four times the intake necessary to prevent caries. Chronic toxicity, called *fluorosis,* occurs after years of daily exposure. An acute toxicity, resulting in death, has been reported with the ingestion of one dose of 5 to 10 g of sodium fluoride. See Appendix 2 for the estimated safe and adequate daily dietary intake.

Fluorine (F). A nonmetallic gaseous element belonging to the halogen group.

Fluorine, in the form of *fluoride,* is incorporated into the structure of bones and teeth and provides protection against dental caries.

Fluoroapatite. The form in which fluoride, along with calcium and phosphorus, is incorporated into dentin and enamel. It is more resistant to acid erosion than hydroxyapatite.

Fluorosis. Condition arising from the excessive intake of fluoride; seen mainly in localities where the water supply contains over 10 ppm fluoride. It is also an industrial hazard among workers handling fluoride-containing compounds, such as cryolite used in smelting aluminum. Fluorosis is characterized by deformed teeth, erosion of tooth enamel, calcification of tendons and ligaments, increased bone density, neurological disturbances, and generalized weakness.

Fluorouracil. A fluorinated pyrimidine antagonist that interferes with DNA synthesis; used as an antineoplastic for treatment of carcinomas of the breast and gastrointestinal tract. Anorexia, nausea, vomiting, and stomatitis are common side effects, usually during the first week of administration. The stomatitis may be severe enough to limit food intake or preclude intake of solid or dry foods. The drug also increases the need for thiamin and may cause anemia. Brand name is Adrucil®.

Fluphenazine. A phenothiazine antipsychotic agent; a major tranquilizer. It has a metallic taste and may cause dry mouth but can increase appetite and weight; it may also alter glucose tolerance and cause hypo- and hyperglycemia. Brand names are Permitil®, and Prolixin®.

FMN. Abbreviation for flavin mononucleotide. See under *Flavin nucleotide.*

Folacin. Generic descriptor for all *folates* and related compounds that exhibit qualitatively the biological activity of tetrahydropteroylglutamic acid or *folic acid.*

Folate. Generic name for compounds that have nutritional properties and chemical structures similar to those of *folic acid.*

Folic acid. Pteroylglutamic acid (PGA), a water-soluble vitamin; alternative names are folacin and folate. The generic name "folic acid" was originally applied to a number of compounds having the same biologic property as PGA. This vitamin was first recognized as a factor necessary for normal growth and hemopoiesis of certain microorganisms and animals and has been given various names such as vitamin M, vitamin B_c, vitamin B_{10} and B_{11}, rhizopterin, citrovorum factor, *Lactobacillus casei* factor, norite eluate factor, and factor SLR. When the substance was finally isolated from spinach, it was called folic acid because of its great abundance in dark green leaves (foliage). As a coenzyme, folic acid is involved in single carbon metabolism, which is an important cellular synthetic mechanism. It also plays a role in the synthesis of purines, citrulline and aspartic acid, metabolism of fatty acids, and carboxylation and decarboxylation reactions. Folic acid has been shown to reduce the incidence of neural tube defects. It is widely distributed in plant and animal tissues. Good food sources are liver, beef, leafy vegetables, legumes, and whole grains. Folate is absorbed as mono- and polyglutamate in the proximal part of the intestine, and then converted in the liver to tetrahydrofolic acid (folinic acid) which is the active metabolite. This conversion is enhanced by ascorbic acid. Folate deficiency may occur either as a result of inadequate intake, increased requirement (infancy, pregnancy, hyperthyroidism), deficient absorption (malabsorption syndromes, bowel resection, genetic disorders), increased losses (kidney dialysis, liver disease), or drug interference (anticonvulsant, antimalarial, barbiturate, cholestyramine, methotrexate, pyrimethamine, triamterene, trimethoprim, etc). Manifestations of deficiency resemble those of vitamin B_{12} but does not lead to degeneration of the cord; symptoms include glossitis and other oral lesions, gastrointestinal disturbances, and megaloblastic anemia. There have been rare reports of gastrointestinal disturbances with high doses of folic acid. An excessive intake is not recommended because it can mask a diagnosis of pernicious anemia, antagonize anticonvulsant therapy, elicit hypersensitivity reaction, and interfere with zinc utilization. In laboratory animals, large doses of folic acid given parenterally precipitated in the kidney and produced kidney damage. For Recommended Dietary Allowances, see Appendix 1.

Folinic acid. Also called citrovorum factor and leucovorin; a reduced and formylated derivative of folic acid that is more stable to air oxidation than the parent compound. It is the name given to citrovorum factor, a substance in the liver required for the growth of *Leuconostoc citrovorum*. The calcium salt (leucovorin calcium) is used to treat folic acid deficiency and as an antidote to folic acid antagonists.

Follicular keratosis. Also called "follicular hyperkeratosis"; a skin condition seen in vitamin A deficiency. The skin becomes rough, dry, and scaly, and the hair follicles are blocked with plugs of keratin, which appear as prominent projections along the upper forearms and thighs, and also along the shoulders, back, abdomen, and buttocks.

Folliculosis. Condition characterized by dry, rough skin, especially in the area of the shoulder and back of the arm. The follicles are raised above the surface, giving the superficial appearance of chronic gooseflesh. The condition is seen in vitamin A deficiency but should not be confused with *follicular keratosis,* as no horny plugs project from the follicular orifices.

Food. Anything that, when taken into the body, serves to nourish, build, and repair tissues; supply energy; or regulate body processes. Aside from its nutritional function, food is valued for its palatibility and satiety effect, as well as for the varied meanings attached to it (emotional, social, religious, cultural, etc.) by different individuals, groups, or races. The so-called ethnic foods reflect the cultural differences of food habits.

Food account. Method used in dietary studies and surveys involving groups or the population at large. It consists of running reports of foods purchased (or produced) for household use. It is useful in checking trends in purchasing certain foods or food groups.

Food additive. Any substance, other than the basic foodstuff but not including chance contaminants, present in food as a result of any aspect of food production, processing, storage, or packaging. Generally classified as *intentional additives* (those added to perform a specific function, such as improvement in nutrient value, flavor, color, etc.) and *accidental additives* (those that unavoidably become part of the product through some phase of production, processing, or packaging).

Food adulterant. Any substance, such as a toxic organism, filth, pesticide residue, or poisonous substance, found in food that is harmful to health or any substance added to increase the bulk or weight of a product.

Food allergy. Or food hypersensitivity. An altered immune reaction (Type I or

IgE-mediated) to the ingestion of food or a food additive. The most common food allergens are wheat, milk, egg, shellfish, chocolate, nuts, corn, soy, and seafoods. Food allergy, however, may develop with any kind of food. The adverse immunologic reaction to allergen(s) in an individual may be harmless to most persons in similar amounts. Allergic symptoms may appear within a few seconds, a few hours, or 1 to 3 days after ingestion of the food allergen. Symptoms are varied, generally affecting the nasobronchial and cutaneous tissues. Gastrointestinal disturbances include diarrhea, nausea and vomiting, and abdominal pain. *Nutrition therapy:* the initial treatment is to identify the food allergen and avoid it entirely. Methods of diagnosis include skin tests, in vitro tests such as ELISA (enzyme-linked immunosorbent assay) and RAST (radioallergosorbent test), and food elimination or challenge procedures such as DBPCFC (double-blind, placebo-controlled food challenges). A food history is taken, and food records are kept. If symptoms have disappeared when the suspected food allergen has been eliminated for at least 6 weeks, that food is reintroduced and reactions are observed again. Sometimes more than one item is implicated and a series of elimination diets can be useful. During the test period, monitor nutritional adequacy, giving supplementary vitamins and minerals if necessary. Symptoms of food allergy tend to resolve with age; reintroduce foods occasionally in a food challenge to ensure that foods are not being restricted unnecessarily. See *Diet, corn-restricted; Diet, egg-restricted; Diet, gluten-restricted; Diet, milk-restricted; Diet, mold-restricted; Diet, MSG-restricted; Diet, nickel-restricted, Diet, salicylate-restricted; Diet, soy-restricted; Diet, sulfite-restricted;* and *Diet, wheat-restricted.* See also *Diet, elimination test; Diet, provocative; Diet, rotation;* and *Diet, Rowe's elimination.*

Food analysis. Quantitative or qualitative determination of food components using different techniques. The general method of food analysis is by proximate determination of water, nitrogen (protein), ether extract (fat), ash, and carbohydrate (by difference). Specific mineral and vitamin components are determined by different methods. The *Official Methods of Analysis* published by the *Association of Official Agricultural Chemists* is a good reference for accepted methods. See also *Proximate analysis.*

Food and Agriculture Organization (FAO). International organization of the United Nations directly concerned with the production, distribution, and consumption of foods to improve nutrition and raise the standard of living of the people of all countries. It provides technical assistance through five divisions: agriculture, economics, fisheries, forestry, and nutrition.

Food and Drug Administration (FDA). Agency of the U.S. Department of Health and Human Services that enforces legislation concerning food, drugs, and cosmetics. It safeguards the food supply to ensure that it is fit for human consumption; some of its main activities are to monitor food and nutrition labeling, food additives, and food–drug interactions.

Food and Nutrition Board (FNB). Board of the National Academy of Sciences/National Research Council that was established in 1940 to serve as an advisory group on nutrition to U.S. agencies. It promotes needed research and helps interpret nutritional science in the interest of public welfare. Specific activities are carried on by committees composed of experts in each field.

Food and Nutrition Information Center (FNIC). A branch of the U.S. Department of Agriculture that provides serials, monograms, and audiovisual materials including computer information to nutritionists, dietitians, food service managers, and others. It also prepares bibliographies on specific food and nutrition topics, which are also available to consumers, educators, and other professionals.

Food and Nutrition Service (FNS). A branch of the Food and Consumer Services, U.S. Department of Agriculture, responsible for programs that provide food for needy persons, such as the Family Nutrition Program, Food Stamp Program, National School Lunch Program, and Special Milk Program for Children.

Food balance sheet. Measure of the food available per person. It is calculated by dividing the total food for the year by the number of people in the country.

Food Basket/Food Vouchers. Administered by the U.S. Department of Agriculture, this program provides indigent families a box or basket of nutritionally balanced foods, or distributes vouchers used to purchase food at a grocery store.

Foodborne disease. Disease caused by the ingestion of food that contains bacteria, parasites, naturally occurring toxicants, chemical poisons, or radioactive fallout. The most common foodborne diseases are caused by salmonella, campylobacter, and staphylococcal bacteria. Vomiting, diarrhea, and other gastrointestinal disturbances that accompany the disease may be severe and may cause malabsorption of nutrients, dehydration, and electrolyte imbalance. Immunocompromised individuals (including those with diabetes, AIDS, liver disease, rheumatoid arthritis, collagen disease, kidney disease, the very young and elderly, and those on corticosteroid, radiation, and chemotherapy) are at high risk of acquiring foodborne diseases. The use of a low microbial diet in conjunction with other measures that suppress microbial growth has been recommended for these individuals to reduce the number of organisms acquired during hospitalization, especially after organ transplantation. See *Diet, low-microbial*.

Food challenge. A method for establishing an intolerance to a food item. When the symptoms resolve after a trial elimination of suspected food items, the suspected food is reintroduced to "challenge" the recurrence of symptoms. See *Double-blind, placebo-controlled food challenge (DBPCFC)*. See also *Diet, elimination* and *Diet, provocative*.

Food composition tables. Tabulated data on quantities of different nutrients per 100 g or per serving portion of a food item. Most commonly used in the United States is Handbook Number 8, *Composition of Foods—Raw, Processed and Prepared,* published by the U.S. Department of Agriculture. Food values are given in terms of a 100-g edible portion (EP) and 1 lb as purchased (AP) of the foods. Information is given about the energy value and nutrient content of various food items, including water, protein, fat, saturated fatty acids, unsaturated fatty acids (oleic acid and linoleic acid), carbohydrate, calcium, phosphorus, magnesium, iron, sodium, potassium, copper, zinc, manganese, vitamin A, thiamin, riboflavin, niacin, vitamin B_6, folacin, vitamin B_{12}, pantothenic acid, ascorbic acid, and others. There are other food composition tables compiled by others (e.g., Pennington etc.) using U.S. Department of Agriculture data, food industry analyses, university laboratories, and so on that are also useful references, some of which deal with special dietary components such as alcohol, cholesterol, and caffeine. Currently, the use of computer software for food composition data is an important tool for practitioners and researchers.

Food Cooperative/Food Co-op or SHARE. A private, nonprofit self-help organization that allows the community to save money on groceries by doing voluntary community service. A supply of commodities consisting of lean meats, fresh fruits, and staples worth $13 to $15 is given to any individual putting in 2 hours of community service.

Food Distribution Program. The program aims to improve the diets of schoolchildren, the poor, and the elderly. It distributes domestically produced foods acquired under price support operations or surplus removal.

Food elimination. Foods suspected of causing an allergic reaction are omitted from the diet for 2 weeks or until symptoms clear. If symptoms do not clear and food allergy is still suspected, more restrictive diets are used. See *Diet, elimination*.

Food, energy value. Also called "fuel value of food"; oxidation of foodstuffs in the *bomb calorimeter* yields, on the average, 4.15, 9.40, and 5.65 kcal/g of pure carbohydrate, fat, and protein, respectively. However, the *physiologic fuel* values of foods when burned by the body are somewhat lower because of incomplete digestion of the three nutrients and incomplete oxidation of protein. Experiments conducted by Atwater on typical American mixed diets showed that the digestibility of carbohydrate was 98%, of protein 92%, and of fat 95%, and that the urinary energy loss for incomplete oxidation of protein was 1.25 kcal/g of protein. On the basis of these observations, the Atwater values for the available energy of the three foodstuffs were derived. Later experiments showed that each food has a specific coefficient of digestibility, making the fuel value specific for each type of food. The Nutrition Division of the Food and Agricultural Organization therefore proposed the use of *specific fuel factors* for estimating the caloric value of foods. See *Atwater values* and *Specific fuel factor (value)*.

Food fad. Idea associated with food that becomes fashionable for a time to meet the needs of a current trend, usually at the sacrifice of important nutrients. A food fad may be an exaggerated truth about a food or a claim that it is a cure-all. Food fads are usually short-lived and remain popular only until they are replaced by another fad. Many diets for weight reduction are currently based on common fads.

Food fallacy. False belief about food; misrepresentation, misinterpretation, or misinformation about a food fact. A *fact* implies a scientific basis, and facts about food are the result of research.

Food group. Classification of various foods into groups on the basis of similarity in nutrient content of the members of each group. It is a practical guide in planning diets that will satisfy nutrient allowances by merely defining the number of servings to eat from each group. See Appendix 4.

Food Guide Pyramid. A guide to daily food choices graphically presenting the main food groups. It lists a range for the number of servings in each of the five food groups. See Appendix 4.

Food intolerance. Collective term for a wide variety of adverse nonimmune reactions to foods. Food intolerance may be caused by diseases (malabsorption) as in cystic fibrosis and enteropathies; enzyme deficiency, as in lactase deficiency; and inborn errors of metabolism, as in galactosemia and phenylketonuria. Cutaneous, gastrointestinal, and respiratory symptoms of food intolerance are similar to those seen in food allergy. *Nutrition therapy:* avoid or restrict the specific food that is not tolerated, depending on the tolerance level of the individual. Some individuals cannot tolerate any amount, whereas others can take small amounts of the offending item. For specific dietary guidelines, see *Diet, fructose-restricted; Diet, galactose-restricted; Diet, glucose-restricted; Diet, lactose-restricted; Diet, leucine-restricted; Diet, maltose-restricted; Diet-methionine-restricted; Diet, phenylalanine-restricted; Diet, starch-restricted;* and *Diet, sucrose-restricted.*

Food inventory. Method of recording the food intake of a group or family. It consists of a weighed inventory of foods at the beginning and end of the study together with day-to-day records, with or without records of kitchen and plate wastes. Nutrients are calculated from food tables or by laboratory analysis of the foods.

Food labeling. A format on packaged foods that gives nutrition information and a list of ingredients, as required by law. The Food and Drug Administration (FDA) regulates the labeling of most food products, except alcoholic beverages, which are regulated by the Bureau of Alcohol, Tobacco and Firearms. Produce such as fresh fruits and vegetables, and some products with Standards of Identity filed with the FDA, need not comply with the food and nutrition labeling requirements. Food labels must state the product's name, manufacturer and address, the amount of the contents, and the ingredients in descending order by weight. If a manufacturer adds a nutrient to the product or makes some claim about its nutritional value, it must be so stated on its *nutritional labeling*. The Nutrition Labeling and Education Act of 1989 was approved and became effective in May 1994. See Appendices 5 and 6 for details.

Food list. Method used in dietary studies applicable to groups. The subject reports an estimated quantity (by weight, retail unit, or household measure) of various foods consumed during a previous period, usually the past 7 days. A list of foods may be used to aid recall.

Food record. Method used in dietary studies. It consists of taking records of foods (including between-meal snacks) for varying periods of time, usually 3 to 7 days. The accuracy of the method depends on the ability of the subjects to estimate quantities of foods and the correct application of food tables to calculate the nutrient content.

Food Research and Action Center (FRAC). A nonprofit national organization that aims to alleviate hunger and malnutrition in the United States. It publishes a bimonthly newsletter, *Foodlines*.

Food Stamp Program. The primary purpose of this nationwide program is to alleviate hunger and malnutrition among the indigent. It enables households with incomes below the poverty line to receive a specified number of stamps without money transactions, which can be used to purchase food items in normal food outlets. There are restrictions on the food items to be purchased (e.g., no gourmet items or alcoholic beverages) to encourage the purchase of more nutritious foods.

Foot drop. Condition in which the foot hangs or falls down due to paralysis of the lower limbs, as seen in polyneuropathy of *beriberi*.

Formiminoglutamic acid (FIGLU). A metabolite of histidine seen in the urine of individuals deficient in folic acid. It is also seen in patients with acute leukemia who are receiving folic acid antagonist therapy such as methotrexate and aminopterin.

Fortification. Addition of one or more nutrients such as vitamins, minerals, amino acids, or protein concentrates to food so that it contains more of the nutrients than were originally present. Examples include the addition of vitamin A to margarine, vitamin D to milk, lysine to bread, and iodine to salt.

Foscarnet. Antiviral drug against all herpes viruses and for treatment of retinitis in AIDS. Adverse side effects include nausea, vomiting, anorexia, abdominal pain, dysphagia, dry mouth, glossitis, stomatitis, mouth ulcers, pancreatitis, and dehydration. It may also increase glucose and decrease potassium, calcium, and magnesium in blood. Brand name is Foscavir®.

FreAmine®. Brand name of an amino acid solution with electrolytes for total parenteral nutrition; available in 3%, 8.5%, and 10% amino acid concentration. Also available as 6.9% FreAmine®-HBC, with high branched-chain amino acids. See Appendix 40.

Free fatty acid (FFA). Unesterified fatty acid. See *Fatty acid.*

Free radicals. Compounds containing one or more unpaired electrons produced during cellular metabolism. Free radicals are highly reactive and can cause damage to body cells. Damage to cellular components accumulates over time. The *free radical theory* correlates the cellular damage caused by free radicals with degenerative changes and the development of atherosclerosis, cancer, cataract, Parkinsonism, rheumatoid arthritis, and other chronic diseases common among older adults. This process appears to be retarded by antioxidants; however, clinical studies are generally inconclusive. See *Antioxidants.*

Fröhlich's syndrome. Disturbance in fat deposition usually occurring in children before puberty; characterized by the presence of superficial fat in the body and the external genitalia. It may be the result of an inherent defect in the pituitary gland or atrophy of the secretory cells by tumor injury or infectious disease.

Fructose. Levulose or fruit sugar. A 6-carbon monosaccharide found in fruits and honey and obtained from hydrolysis of sucrose to glucose and fructose. It is an essential intermediate in carbohydrate metabolism. Much sweeter than sugar, it is sometimes used as a sweetener in foods. Substitution of fructose for sucrose in the diets of individuals with diabetes mellitus improves glycemic control. Provides 4 kcal per gram. Its use in large amounts can cause osmotic diarrhea and high blood triglyceride levels.

Fructose intolerance. A genetically inherited metabolic defect due to a deficiency of the enzyme fructose-1-phosphoaldolase, which results in hypoglycemia and hypophosphatemia with associated vomiting, lethargy, and coma; the long-term effect may lead to jaundice and enlargement of the liver. Another disorder in enzyme deficiency, involving fructose 1,6-diphosphatase, also results in hypoglycemia with hyperventilation, shock, and convulsions. *Nutrition therapy:* the essential treatment for hereditary fructose intolerance is immediate and lifelong removal of all sources of fructose from the diet, with frequent meals and avoidance of long fasts. The management of crises with severe hypoglycemia and acidosis may require the administration of glucose and bicarbonate. See also *Diet, fructose-restricted.*

Fructosuria. Presence of fructose in the urine; a genetically inherited metabolic defect due to a deficiency of the enzyme fructokinase. The disorder does not cause any clinical symptoms and is usually found incidentally when urine is tested as a screen for diabetes. It does not require any dietary modification.

Fruitarian. Person whose diet consists chiefly of fruits. See also *Vegetarian*.

Fuel factors. Energy values per gram of carbohydrate, protein, and fat. See *Atwater values; Food, energy value;* and *Specific fuel factor (value)*.

Furosemide. A powerful diuretic used in the management of edema associated with congestive heart failure and hypertension. It acts by inhibiting reabsorption in the ascending loop of Henle in the kidney (hence it is a loop diuretic), thus causing diuresis and increased excretion of potassium, sodium, chloride, magnesium, and calcium. It may cause low blood levels of these electrolytes if dietary intake is inadequate. Furosemide is excreted in breast milk. Its side effects include anorexia, dry mouth, constipation, and stomach distress. Brand names are Lasix® and Furomide®.

G

GA. Abbreviation for *glutaric acidemia.*

GABA. Abbreviation for *gamma-amino butyric acid.*

Galactoflavin. An analog of riboflavin that is a potent inhibitor; D-galactose replaces D-ribose in the molecule.

Galactosamine. An amino sugar; galactose containing an amino group in carbon atom 2. It is found in the polysaccharide of cartilage, in chondroitin (as chondrosamine), and in the structural material of the skeletons of arthropods.

Galactose. A 6-carbon monosaccharide differing from glucose only in the position of the hydroxyl group on carbon atom 4. It is seldom found free in nature; it occurs mainly linked with glucose to form lactose (milk sugar). It is also a constituent of cerebrosides, gangliosides, and certain polysaccharides such as agar and flaxseed.

Galactosemia. 1. Presence of galactose in the blood. 2. An inborn error of metabolism characterized by the inability to metabolize galactose due to a deficiency in one of three enzymes: galactokinase, galactose-1-phosphate uridyl transferase, and uridine diphosphate galactose-4-epimerase. Galactose accumulates in the blood and is excreted in the urine; the clinical symptoms are varied and include jaundice, enlarged liver and spleen, anorexia, weight loss, vomiting, diarrhea, ataxia, mental retardation, and cataract formation. *Nutrition therapy:* The key element in the treatment of galactosemia is a diet as low as possible in galactose. All forms of milk and lactose are not allowed. See *Diet, galactose-restricted.*

Galactose tolerance test. A liver function test that measures the ability of the liver to remove galactose from the bloodstream and convert it to glycogen; it is performed in a manner similar to that of a glucose tolerance test.

Galacturonic acid. A sugar acid resulting from the oxidation of the primary alcohol group of galactose to a carboxyl (—COOH) group. It occurs in various vegetable sources such as pectins and certain plant gums.

Gallstones. See *Cholelithiasis.*

Gamma-aminobutyric acid (GABA). An important neurotransmitter synthesized from glutamic acid; found in the brain, heart, lungs, and kidneys.

Gastrectomy. Surgical removal of all or part of the stomach. *Nutrition therapy:* routine pre- and postoperative diets are followed after partial or minor gastrectomy. Special dietary modifications are needed after total gastrectomy because of the resulting nutritional problems such as diarrhea, weight loss, malabsorption, and the dumping syndrome. The absence of intrinsic factor may eventually lead to macrocytic anemia, unless monthly injections of 100 μg vitamin B_{12} are given. Start with clear then full liquids low in carbohydrate (broth, unsweetened gelatin, diluted unsweetened fruit juices), followed by gradual introduction of easily digested, soft, and low-fiber foods in small frequent feedings. As the patient's condition improves, give five to six small meals high in protein, moderate in fat, and relatively low in carbohydrate, especially simple sugars. It is best to take fluids between meals rather than with meals to retard the rapid emptying of hypertonic food into the duodenum and jejunum. Small amounts of milk and lactose-free nutritional products are better tolerated if there is milk intolerance due to a lactase deficiency. Maintain a healthy weight and watch for signs of general malabsorption. Use of medium-chain triglycerides (MCT oil) and supplemental vitamins and minerals, especially folate, iron, calcium, and vitamin D, may be indicated. Vitamin B_{12} should be given parenterally in total gastrectomy.

Gastric acidity. Amount of hydrochloric acid in the stomach, which may be present in two forms—as *free acid* or as *combined acid* (in combination with protein from food, regurgitated duodenal secretion, saliva, and mucus). The sum of the free and combined acids is termed "total acidity." Protein foods stimulate more acid secretion than do carbohydrates and fats. Protein foods have an initial buffering effect, with less free acid to erode tissues.

Gastric bypass (stapling). A surgical procedure designed to restrict or divert normal intake of food as a method of controlling weight in morbid obesity. The stomach is divided by means of several rows of staples into a small proximal pouch of about 50 to 60 mL in capacity and a nonfunctioning distal pouch. The proximal pouch is then attached to the jejunum. This procedure results in rapid filling of the reduced stomach, giving a feeling of early satiety with small meals and consequent decrease in food intake. However, there are side effects, such as dehydration due to persistent vomiting, dumping syndrome, vitamin B_{12} and iron deficiency, and severe weight loss associated with malabsorption. *Nutrition therapy:* postoperative progression from clear to full liquids, then blenderized and soft, semisolid foods in small quantities (30 to 60 mL). Frequent feedings are necessary to meet energy and protein needs. Gradually increase the volume to 120 to 150 mL and introduce soft foods, with gradual progression to regular foods. The individual should eat slowly and chew foods well. Avoid large pieces that could block the pouch opening; also avoid hard foods, tough and gristly meats, fibrous and raw vegetables, nuts, spicy foods, sweets, and carbonated beverages. If there's dumping syndrome, take dry solid meals low in simple sugars and high in complex carbohydrate. Take liquids between meals. Use lactase-treated milk if the individual is intolerant to lactose. In diarrhea, limit milk and milk products and greasy, fibrous foods. Supplemental vitamins and minerals, especially the B-complex vitamins and iron, may be necessary.

Gastric emptying. Nutrient density, osmolality of solutions, and gastric pH affect gastric emptying. The greater the nutrient density, the slower the emptying rate. High fat can significantly delay gastric emptying and cause retention. Hypertonic formulas can induce large fluid shifts into the small bowel, causing diarrhea, nausea, and vomiting. Gastric motility is decreased by foods with a pH of less than 3.5.

Gastric enzymes. Pepsin, rennin, and gastric lipase. See Appendix 11.

Gastric juice. A colorless secretion of the gastric gland containing mucus, hydrochloric acid, enzymes, and intrinsic factor.

Gastric surgery. Gastric surgical procedures include gastrectomy, vagotomy, pyloroplasty, Bilroth I, and Bilroth II.

Dumping syndrome, hypoglycemia, and steatorrhea are common complications after surgery. Postgastrectomy dumping syndrome occurs when food is rapidly "dumped" into the jejunum 10 to 15 minutes after eating, rather than being released gradually from the stomach. In total gastrectomy or vagotomy, hypoglycemia may occur 1 to 2 hours following meals. The rapid digestion and absorption of foods causes an elevation of blood glucose, which results in overproduction of insulin and subsequent hypoglycemia. Steatorrhea occurs most often after a Billroth II procedure. *Nutrition therapy:* NPO then sips of cold fluids or crushed ice, followed by progression from clear to full liquids and gradual introduction of soft, mildly flavored and easily digested foods. Most patients can tolerate regular foods by the fifth to seventh postoperative day. The diet should be low in simple carbohydrates, high in complex carbohydrates and protein, and moderate in fat, given in small, frequent meals. Carbonated beverages and milk are not recommended initially. Introduce small amounts of milk to determine tolerance; a lactose-restricted diet may be necessary. Address postoperative complications as they arise. Avoid sugars and simple carbohydrates if there is alimentary hypoglycemia 1 to 2 hours after a meal. Pectin in fruits and vegetables can reduce glycemic response by delaying gastric emptying and carbohydrate absorption. Use products with medium-chain triglycerides (MCTs) if there is steatorrhea. See also *Dumping syndrome*.

Gastrin. Hormone secreted by the pyloric mucosa that stimulates hydrochloric acid secretion by the parietal cells.

Gastritis. Acute or chronic inflammation of the mucous lining of the stomach. *Acute gastritis* may be due to a variety of causes, including allergy, ingestion of poison or irritants, and dietary indiscretion (i.e., rapid eating, overeating, or large intakes of alcohol and fibrous or highly seasoned foods). It may also accompany severe burns, trauma, radiation therapy, major surgery, and use of aspirin and other nonsteroidal anti-inflammatory drugs. *Chronic gastritis* is usually a sign of an underlying disease, such as stomach cancer, peptic ulcer, and pernicious anemia. Chronic gastritis may lead to atrophy of the gastric mucosa, achlorhydria, and loss of intrinsic factor, resulting in vitamin B_{12} and iron malabsorption. *Nutrition therapy:* correct faulty eating habits and provide nutritionally adequate meals. For chronic gastritis, depending on the patient's tolerance of solid foods, a soft diet is given in six small feedings, with vitamin and mineral supplements, particularly iron and vitamin B_{12}. Avoid foods known to cause discomfort. In acute gastritis, withholding food for 24 hours to allow the stomach to rest may be indicated, followed by clear fluids the next day before a full liquid to soft diet is started. Avoid highly seasoned foods and foods that increase gastric acidity.

Gastroenteritis. Inflammation of the stomach and intestine, usually due to viral infection. Characterized by diarrhea, vomiting, and abdominal pain with or without fever. *Nutrition therapy:* see under *Gastritis*.

Gastroenterostomy. Surgical formation of a communication between the stomach and the small intestine, usually performed to short-circuit the food around a stomach ulcer.

Gastroesophageal reflux disease (GERD). A condition characterized by a decreased pressure or relaxation of the lower esophageal sphincter (LES) which allows the backflow of acidic gastric contents into the esophagus, causing tissue irritation and erosion. Symptoms include heartburn, cramping, pressure sensation, and pain aggravated by lying down or by any increase in abdominal pressure, such as caused by tight clothing. Chronic reflux may lead to reflux esophagitis, esophageal ulcer, esophageal stricture, and swallowing difficulty. *Nutrition therapy:* decrease fat intake to about 45 g/day and limit rich desserts. Avoid foods that are known to decrease LES pressure or irritate the esophageal mucosa. These include tomato products, citrus fruits and juices, chocolate, coffee, carbonated beverages, peppermint, raw onions, spicy foods, high-fat meals, and alcohol. Avoid large meals; eat small, frequent meals but do not eat within 3 hours before bedtime. Take only a small amount of liquid during meals and drink mostly between meals. Weight reduction is essential if the person is obese. Avoid tight clothing around the abdominal area and maintain an upright posture during and after eating. When feeding an infant, hold the infant in an upright position or a prone position with the head elevated to a 30 degree angle. Thickening the formula with cereals may help prevent reflux.

Gastrointestinal tract (GIT). Also called "gut" or the "*alimentary tract.*" Refers to the entire digestive tract from the mouth through the stomach, intestines, and anus. See *Gut barrier* and *Nutrition, gut integrity.*

Gastroparesis. Partial paralysis of the stomach which may be related to local neuritis. It is a result of inadequate or absent contractions of the gastric muscles. There is delayed gastric emptying, which may be acute, as seen in postoperative ileus and diabetic ketoacidosis, or chronic, as seen in muscular dystrophies and diabetic autonomic neuropathy. Anticholinergics, levodopa, tricyclic antidepressants, and other drugs may also cause gastroparesis. The emptying of fluids in the stomach is normal but the emptying of solids is markedly delayed. Symptoms include anorexia, nausea, vomiting, constipation or diarrhea, and abdominal pain and distention. Symptoms are of varying degree and severity, and may lead to weight loss and *bezoar* formation. *Nutrition therapy:* patients with mild to moderate gastroparesis may tolerate oral feedings of easy-to-digest foods given in small amounts. Avoid foods high in fiber and fat. In diabetic gastroparesis, synchronize insulin activity with food intake and emptying from the stomach. If the gastroparesis is severe, pureed foods or an isotonic enteral formula may be tolerated orally in small amounts at frequent intervals, or administered slowly by tube feeding. Liquids high in osmolality have slower emptying rates and should be avoided or diluted in water to reduce osmolality.

Gastroplasty. Surgical procedure reducing the volume of the stomach to only 50 mL, or about 2 liquid ounces. See *Gastric bypass (stapling).*

Gastrostomy. Opening into the stomach from the outside, usually for artificial feeding. The introduction of a nutrient formula through this gastrostomy tube is sometimes called "gastrogavage." A feeding gastrostomy is used when long-term tube feeding is required and the gag reflexes are intact, and there is normal gastric and duodenal emptying. It is contraindicated in severe gastroesophagel reflux, intractable vomiting, inadequate gastric emptying, and high risk for aspiration. The gastrostomy tube can be inserted surgically, usually under general anesthesia, or as a *percutaneous endoscopic gastrostomy (PEG),* a nonsurgical procedure. Under endoscopic guidance, a feeding tube is percutaneously placed in the stomach and secured by rubber "bumpers" or an inflated balloon catheter. The tube can be advanced into the jejunum to create a *percutaneous endo-*

scopic jejunostomy (PEJ) when small bowel feedings are more appropriate.

Gaucher's disease. A rare familial disease caused by an enzyme deficiency. Characterized by disturbance in lipid metabolism; accumulation of *kerasin* in the liver, spleen, lymph nodes, and bone marrow; and abnormal bone growth. Mortality is high in infancy and early childhood. *Nutrition therapy:* milk formula low in fats but otherwise adequate in other nutrients and calories for infants. For children, provide the daily diet meeting the recommended dietary allowances according to age, controlling the fat level to 20% to 25% of total caloric intake.

GDM. Abbreviation for *gestational diabetes mellitus*.

Gemfibrozil. A lipid-lowering drug that decreases serum triglycerides, principally very-low-density lipoprotein triglycerides and to a lesser extent low-density-lipoprotein triglycerides; used in the management of Type IV and mild Type V hyperlipoproteinemia. The most frequent side effects involve the gastrointestinal tract and may cause dry mouth, anorexia, abdominal and epigastric pain, diarrhea, flatulence, nausea, and vomiting. The brand name is Lopid®.

Gentamicin. An aminoglycoside antibiotic; active against many aerobic gram-negative bacteria and some gram-positive bacteria and used for treatment of skin and eye infections. Gentamicin increases the urinary excretion of potassium and magnesium, and may induce hypokalemia and hypomagnesemia. Brand names include Garamycin®, Gentafair®, Genoptic®, Gentacidin®, and G-myticin®.

Geophagia. A form of pica involving the consumption of dirt or clay. See *Pica*.

Gerovital. A buffered solution of procaine hydrochloride; a compound that is being promoted as one that alleviates symptoms of aging and has been designated by some as "vitamin H_3 or "vitamin GH_3." These claims have not been supported, and gerovital is not recognized as a vitamin.

Gestational diabetes mellitus (GDM). High blood glucose level (hyperglycemia) developed during pregnancy which usually disappears after the baby is delivered. One cause is an overproduction of hormones by the placenta that antagonizes the action of insulin. Women who have a family history of diabetes or who are overweight have a greater chance of exhibiting GDM, which may recur with subsequent pregnancies. About 60% of the cases of GDM develop later in life as Type II or adult-onset diabetes. It is important to note that normal glucose levels in pregnancy are lower than in the nonpregnant state. Fasting blood glucose levels during pregnancy should be within 60 to 110 mg/dL (3.3 to 6.1 mmol/L) and 2 hours after a meal the acceptable range is 80 to 140 mg/dL (4.4 to 7.8 mmol/L). Self blood glucose monitoring (SBGM) and food records are very important, along with regular exercise and prescribed medications. The urine must be ketone-free. Total weight gain during pregnancy should be within 24 to 28 lb. *Nutrition therapy:* give 30 kcal/kg/day, of which 50% should be provided by carbohydrates. Distribute the carbohydrates as follows: 10% for breakfast, 30% for lunch, 30% for dinner, and 10% each for in-between snacks. Breakfast is small and low in carbohydrate because of the tendency during pregnancy for peak postprandial glucose levels to occur early in the day. Avoid simple sugars and emphasize complex carbohydrates, which are more slowly absorbed and converted into glucose. Provide about 20% to 25% of the total calories from proteins, emphasizing proteins of high biologic value, supplied by a quart of milk a day, and the rest by lean meats, cheese, fish, and eggs. Encourage liberal intake of fluids. See also *blood sugar level* and *Diabetes mellitus*.

GFR. Abbreviation for *glomerular filtration rate*.

Gingivitis. Inflammation of the gums (gingiva). Gingivitis is seen in association with vitamin C deficiency and is characterized by spongy, swollen gums that bleed readily. This may result in a high susceptibility to infection; teeth may become loose if the condition is not corrected.

GIT. Abbreviation for *gastrointestinal tract*.

Gliadin. Glutamine-bound fraction of protein in wheat, oats, rye, and barley; a simple protein that lacks lysine and is classified as a partially complete protein. See also *Gluten*.

Glipizide. A sulfonylurea oral hypoglycemic agent for management of non-insulin-dependent diabetes mellitus (NIDDM). It has a metallic taste. Side effects are dose related and may cause nausea and vomiting. Alcohol may induce flushing and headache, and hypoglycemia may occur with prolonged exercise without caloric supplementation or with alcohol consumption. The brand name is Glucotrol®.

Globin. A colorless basic protein often found in combination with other proteins; soluble in water, acid, or alkali and coagulable by heat. See Appendix 9.

Globulins. Group of plant and animal proteins slightly soluble in water but extremely soluble in salt solution; an important component of human blood. Fractionation of plasma globulins by electrophoresis gives alpha, beta, and gamma globulins. See Appendix 9.

Glomerular filtration rate (GFR). Number of milliliters of blood that is passed through the glomeruli in 1 minute. The normal GFR is approximately 130 mL/min. A polysaccharide such as inulin can be used to measure the GFR, because it is filtered but not reabsorbed by the kidney tubules. The amount of inulin in the urine corresponds to the amount that has been filtered from the plasma. This test is used to estimate the degree of function of the kidney. Aging and some disease conditions reduce the GFR. In chronic renal failure, the GFR helps determine the daily protein intake as follows:

GFR (mL/min)	PROTEIN (g/kg/day)
>30	0.8
20 to 30	0.7 to 0.75
10 to 20	0.6 to 0.65
5 to 10	0.5 to 0.55
<5	0.4 to 0.45

For every 50% reduction in GFR, the serum creatinine level will double. For example, with a GFR of 100 mL/min, the serum creatinine concentration is 1.0 mg/dL. As GFR decreases to 50 mL/min, the serum creatinine level increases to 2.0 mg/dL.

Glomerulonephritis. Inflammatory disease of the kidneys affecting chiefly the glomeruli. It may be acute or chronic and generally follows streptococcal infections of the respiratory tract such as tonsillitis, sinusitis, pneumonia, and influenza. Symptoms include nitrogen retention, presence of albumin and blood in the urine, and varying degrees of hypertension, edema, and uremia preceding convulsions and death. *Nutrition therapy:* restrict protein to 0.5 g/kg body weight if blood urea nitrogen (BUN) is high (>80 mg/dL) and oliguria is present; gradually increase to 0.8 to 1.0 g/kg body weight as renal function improves. About 75% of protein should be of high biological value, such as egg and meat. Provide sufficient calories from carbohydrates. Salt is usually not restricted unless there is edema, hypertension, or oliguria. Restrict sodium to 1 to 2 g/day if there is edema or hypertension, and restrict further to 0.5 to 1 g/day if oliguric. Potassium is restricted to 2 g/day if hyperkalemia is present. Fluid intake is adjusted to output. Replace fluids for normal diuresis, but if oliguria is present, restrict fluids to 600 mL/day. Fluids may be taken as desired if restriction is not indicated.

Glossitis. Inflammation of the tongue (glossa); usually due to biting, burning, or injuring the tongue. It may also be a symptom of a gastrointestinal disorder or a nutritional deficiency. Deficiencies in niacin, riboflavin, vitamin B_{12}, folic acid, and iron may all give rise to glossitis. It is a feature of pellagra, sprue, and the various types of nutritional anemias. In acute glossitis, the tongue is swollen, the papillae are very prominent, and the color is characteristically red or purplish blue. Deep, irregular fissuring is common and shallow ulcers may occur, especially on the sides or tip. In chronic atrophic glossitis, the tongue is small and the surface appears smooth or has fine fissuring.

Glucagon. A hyperglycemic–glycogenolytic hormone of the pancreas that is protein in nature. Also found in gastric and duodenal mucosa, as well as in some commercial insulin preparations. It increases the blood sugar level by stimulating the breakdown of glycogen into glucose in the liver.

Glucerna®. Brand name of a nutritional product, low in carbohydrate and with fiber, for patients with diabetes mellitus and stress-induced hyperglycemia. Contains 14.3 g fiber from soy polysaccharide/liter. See Appendix 39.

Glucoascorbic acid. An analog of ascorbic acid; acts as an antivitamin C and can cause scurvy even in animals that do not normally require the vitamin in their diet.

Glucocorticoids. Steroid hormones secreted by the adrenal cortex or synthetic analogs of these hormones. These include corticosterone, cortisone, and cortisol. Glucocorticoids act primarily on carbohydrate, lipid, and protein metabolism, and also have anti-allergic and anti-inflammatory effects. Prolonged use may cause protein wasting, decreased bone density, decreased glucose tolerance, increased appetite, weight gain, hypercholesterolemia, sodium and water retention, and peptic ulcer.

Glucogenesis. Formation of glucose; may arise from any of the intermediates in glycolysis or from glucogenic substances such as glycerol and some of the amino acids that can be converted into one of the intermediates in carbohydrate metabolism.

Glucola®. Brand name of a drink that may be used in place of a 100-gram carbohydrate meal for a glucose tolerance test. A 7-oz bottle contains the equivalent of 75 g of glucose; the rest of the carbohydrate is in the form of a partial hydrolysate of cornstarch dissolved in a cola-flavored solution that is also carbonated.

Gluconeogenesis. Formation of glucose or glycogen from noncarbohydrate sources such as glycerol and glucogenic amino acids.

Glucose. Also called "dextrose," "grape sugar," or "blood sugar"; a 6-carbon monosaccharide occurring naturally in plant tissues and obtained from the complete hydrolysis of starch. It is the chief form in which carbohydrate is absorbed into the bloodstream. Physiologically, glucose is considered the most important sugar, and hence is referred to as the "physiologic sugar." It is the only source of energy for the brain which needs about 140 g/day. It is the form in which carbohydrate is circulated in the blood and can be utilized by all cells. Glucose can be converted to glycogen or fat for future energy needs. It provides 4 kcal/g. See also *Blood sugar level*.

Glucose/nitrogen (G/N) ratio. The ratio between the amount of glucose and the amount of nitrogen excreted in the urine following induction of diabetes by phlorizin. It is an indication of the proportion of sugar that can be derived from protein. The ratio varies with the severity of the diabetes, reaching a maximum value of about 3.65.

Glucose tolerance factor (GTF). The biologically active form of chromium bound to an organic compound. It potentiates the action of insulin and increases the uptake of glucose by the cell. GTF is present in the liver, blood plasma, and other cells, and is also found in brewer's yeast, black pepper, and whole grains.

Glucose tolerance test (GTT). Also called the "oral glucose tolerance test (OGTT)." Test that measures the ability of the body to utilize a known amount of glucose. It is performed after a 12-hour fast and a carbohydrate intake of at least 150 g daily for 3 days prior to the test. The subject is given orally 50 to 100 g of glucose or an allowance of 1 g/kg body weight with a maximum of 100 g in adults, and 1.75 g/kg body weight in children. Blood samples for glucose analysis are taken ½ hour and then at hourly intervals for the next 4 or 5 hours. A normal individual shows a rise in blood sugar about ½ hour after ingestion of glucose, but the blood sugar level returns to normal after 2 hours. A diabetic person shows a much higher rise in blood sugar after ½ hour, a rise that continues even after 2 hours and remains higher than normal after 4 hours. The plotted results of blood sugar values form the glucose tolerance curve.

Glutamic acid. Alpha-aminoglutaric acid, a nonessential amino acid that is a constituent of *folic acid*. It is involved in transamination and deamination reactions and in the synthesis of glutathione, *GABA*, and glutamine. The monosodium salt of glutamic acid (MSG) is widely used as a flavoring agent.

Glutamine. Compound formed from glutamic acid and ammonia in the liver, brain, and kidneys. It plays an important role in transamination reactions and serves as a source of ammonia in the kidneys for base conservation. Glutamine can cross the blood–brain barrier, thus providing a source of glutamic acid for brain oxidation. Glutamine is a major fuel for the gastrointestinal tract and plays a role in the maintenance of the immune system. The body's requirements for glutamine appear to increase during stress and catabolic diseases. Traditionally considered a nonessential amino acid, glutamine is now recognized as a conditionally essential nutrient. Glutamine supplementation is beneficial in wound healing, liver injury, impaired immune funtion, and intestine-related infections. Large doses of glutamine have also been shown to prevent gut atrophy, maintain bowel integrity, and prevent *bacterial translocation*.

Glutarex®. Brand name of a lysine-free and tryptophan-free product for nutrition support of infants and toddlers (Glutarex®-1) and children and adults (Glutarex®-2) with glutaric acidemia Type I.

Glutaric acidemia (GA). A rare inborn error of lysine and tryptophan metabolism caused by a defect in the enzyme glutaryl-coenzyme A dehydrogenase.

Symptoms include seizures, vomiting, acidosis, and elevated glutaric acid in the plasma. There may be progressive neurologic damage, if untreated. *Nutrition therapy:* diet restricted in lysine and tryptophan and pharmacologic oral doses of riboflavin (100 to 200 mg/day) and L-carnitine (100 to 300 mg/kg/day) may arrest the neurologic deterioration. Restriction of dietary lysine and tryptophan is achieved by using a formulated product free of these amino acids (e.g., Glutarex®). Maximize energy and protein intakes to promote normal growth. Depending on the age, provide additional whole protein foods in amounts to meet the requirements for protein and the essential amino acids lysine and tryptophan. Use protein-free foods to supply energy. Feeding problems in a neurologically impaired child should be addressed.

Glutathione. A tripeptide composed of glutamic acid, cysteine, and glycine; widely distributed in nature and isolated from yeast, muscle, and liver. It is the prosthetic group of glyceraldehyde phosphate dehydrogenase and is a constituent of most cells. As a sulfhydryl-reducing agent, glutathione takes part in the cell's defense against oxygen radicals.

Glutelin. A simple protein found in cereals; insoluble in water and neutral solutions but soluble in dilute acids and alkalis. Examples are *glutenin* in wheat, *oryzenin* in rice, and *hordein* in barley. See Appendix 9.

Gluten. Protein fraction of wheat and other cereals that gives flour the elastic property which is essential for bread making. It is composed of two fractions: gliadin and glutenin. It is believed that the gliadin portion is responsible for the malabsorption syndrome in susceptible individuals with gluten-sensitive enteropathy. This condition is corrected by eliminating wheat, oats, rye, and barley from the diet. See also *Diet, gluten-free*.

Gluten-sensitive enteropathy. Also called "gluten-induced enteropathy," "celiac disease," and "nontropical sprue"; a form of malabsorption primarily affecting the proximal portion of the small intestines and characterized by damage to villus epithelial cells in response to ingestion of *gliadin,* a constituent of the protein *gluten*. A secondary lactose intolerance sometimes appears. It occurs in both children and adults. The principal abnormality is the failure of the jejunal mucosa to absorb digested substances adequately because of villus atrophy and a resultant reduction in the number of functioning absorptive cells. As a consequence, there is general malabsorption of fats, carbohydrates, proteins, vitamins, and minerals. The condition is characterized by loss of weight, nausea and vomiting, abdominal pain, weakness, and diarrhea consisting of pale, bulky, frothy, foul-smelling stools. There is generalized wasting, and signs of multiple vitamin and mineral deficiencies develop, such as anemia, cheilosis, glossitis, peripheral edema, tetany, rickets, and hypoprothrombinemia with a tendency to bleed. *Nutrition therapy:* the condition is completely relieved if *gluten,* derived chiefly from wheat, oat, and rye, is excluded from the diet. Give a high-calorie, high-protein diet if there is weight loss and malabsorption. Supplemental vitamins and minerals are recommended, especially vitamins A and E to replenish stores depleted by steatorrhea; iron, folate, or vitamin B_{12} depending on the type of anemia; vitamin K if there is bleeding or prolonged prothrombin time; and calcium and vitamin D to correct osteomalacia. Replace fluids and electrolytes if diarrhea is severe. Avoid or restrict milk and milk products if there is a lactose intolerance, which usually disappears after the omission of gliadin and the gastrointestinal mucosa begins to heal. Life-long adherence to a gluten-free diet is necessary to prevent recurrence of symptoms. See *Diet, gluten-free*. See also *Diet, lactose-restricted*. Read food labels carefully.

Glutethimide. A sedative and hypnotic used primarily in short-term treatment of insomnia. It probably induces inactivation of 25-hydroxyvitamin D_3, and thus may increase vitamin D turnover and bone resorption; it may also cause polyneuropathy, dry mouth, and gastric irritation. It is excreted in milk. The brand name is Doriden®.

Glyburide. A sulfonylurea antidiabetic agent used in the management of non-insulin-dependent diabetes mellitus. Hypoglycemia may occur with inadequate food intake and prolonged exercise or with alcohol consumption. Simultaneous alcohol ingestion may cause flushing, headache, nausea, and vomiting. The drug has a metallic taste. Brand names are DiaBeta® and Micronase®.

Glycemic index. Measurement of the 2-hour glucose response to a 50-g available carbohydrate portion of a food. It is expressed as a percentage of the response to the same amount of carbohydrate from a standard food such as white bread. Compared to white bread as 100%, the glycemic index of some foods are the following: puffed rice, 132; cornflakes, 121; instant mashed potato, 120; baked russett potato, 116; corn chips, 99; raisins, 93; sucrose, 89; banana, 84; boiled polished rice, 81; sweet potato, 70; ice cream, 69; canned green peas, 60; whole milk, 44; fructose; 31; canned soybeans, 22; and peanuts, 16.

Glyceraldehyde. Glyceric aldehyde, the simplest form of carbohydrate, containing only 3 carbon atoms. In the form of phosphate ester, it acts as an intermediate product in the anaerobic breakdown of carbohydrate; it also serves as a precursor of glycerol.

Glycerol. Also called "glycerin"; trihydroxypropane, a trihydroxy alcohol. It is a clear, colorless, sweetish, and viscous liquid obtained from the hydrolysis of fats and oils. Esters of glycerol with fatty acids are called "glycerides." Glycerol possesses three hydroxy groups; it can combine with three molecules of fatty acid to form a *triglyceride* or simple fat.

Glycine (Gly). Formerly called "glycocoll"; aminoacetic acid, the simplest amino acid. It is sweet-tasting. It is an essential constituent of body tissues, a precursor of bile acid, and a participant in the detoxication mechanism and synthesis of purine, creatine, glutathione, and porphyrins. It is not considered a dietary essential except for growing chickens.

Glycinuria. Condition characterized by excessive excretion of glycine in the urine and elevated glycine blood levels accompanied by metabolic acidosis, respiratory distress, seizures, hematologic abnormalities, and mental retardation, leading to death. It is considered an inborn error of metabolism, although the specific enzyme defect has not been identified; it is probably due to a failure of conversion of glycine to glyoxalate. Symptoms are controlled by a low-protein diet with arginine and pyridoxine supplementation.

Glycogen. Animal starch; a branched-chain polysaccharide composed of glucose units. It is the chief carbohydrate storage material in animals, especially in the liver and muscles. In a well-nourished adult, glycogen is about 2% to 8% of the weight of the liver. Although muscle contains less glycogen (approximately 1%) than the liver, the greater mass of the muscle accounts for a considerable quantity of this storage form of carbohydrate. The amount of reserve glycogen stored in the liver and muscle depends largely on the nature of the diet and the amount of exercise. Fasting results in a rapid depletion of liver glycogen, and muscle glycogen is depleted during continuous or violent exercise.

Glycogenesis. Formation or synthesis of glycogen. Glycogenic substances are hexoses, as well as a wide variety of other compounds, e.g., glycogenic amino acids, glycerol derived from lipids, intermediates in glycolysis such as lactic acid and pyruvic acid, and products structurally related to hexoses.

Glycogenolysis. Breakdown or splitting up of glycogen, as opposed to glycogenesis. The conversion of glycogen to glucose occurs in several steps involving phosphorylation, rearrangement, dephosphorylation, and condensation or hydrolysis, with each step requiring a specific enzyme.

Glycogenosis. Type II glycogen storage disease characterized by extreme deposition of glycogen in the heart. The heart enlarges to as much as five times its normal weight. Infants affected by this condition usually die of heart failure before the age of 2 years. They resemble cretins or mongoloids, have poor appetites, and fail to grow. The specific enzyme defect is lysosomal α-1,4 glucosidase.

Glycogen storage disease (GSD). One of a group of genetically inherited metabolic disorders characterized by an abnormal accumulation of glycogen in the liver and other tissues. The condition is due to a deficiency in one of several enzymes involved in the interconversion of glycogen and glucose. As a result, normal blood glucose levels are not maintained. Hypoglycemia following only short periods of fasting can be profound and is generally associated with lactic acidosis. The clinical manifestations vary with the enzyme deficiency and may include fasting hypoglycemia, hepatomegaly and liver cirrhosis, failure to thrive, convulsions, and muscular atrophy. *Nutrition therapy* for specific GSD disorders: *GSD I (glucose-6-phosphatase deficiency).* Continuous overnight 12-hour gastric feeding of glucose (4 to 6 mg/kg/min). During the day, give frequent feedings of complex carbohydrate every 2 to 3 hours. After 8 months of age, cornstarch therapy may be used as an alternative to nasogastric feeding. Give uncooked cornstarch (1.8 to 2.5 g/kg body weight) mixed in 120 mL low-fat (2%) milk every 4 to 6 hours. Children 1 to 3 years of age can be managed with 2 g of cornstarch/kg body weight 4 to 5 times per day. (1 tbsp cornstarch = 8.3 g.) Mix the uncooked cornstarch in tap water or sugar-free flavored beverage at room temperature. Maintain blood glucose level above 70 mg/dL (3.9 mmol/L). Introduce supplemental foods at the usual age. Emphasize complex carbohydrates (oatmeal, pastas, rice, legumes) and limit or avoid foods containing sucrose, galactose, and fructose, such as table sugar, milk and milk products, and fruits. *GSD IIB (α-1,4-glucosidase deficiency).* Frequent glucose feeding and increased protein may be beneficial. *GSD III (amylo-1,6-glucosidase deficiency).* In infants, frequent feedings every 3 hours during the day and continuous 12-hour overnight feeding of a standard formula. Introduce supplemental feedings of high-protein strained foods. At age 1 year, replace infant formula with low-fat milk. Give high-protein, low-carbohydrate feedings during the day and a high-protein snack at night. Supplementation with uncooked cornstarch may be beneficial to increase energy intake from carbohydrates. *GSD IV (α-1,4-glycan 6-glycosyl transferase deficiency).* There is no effective treatment; liver transplantation is recommended during early childhood. While the patient is awaiting transplantation, give a high-protein, low-carbohydrate diet similar to the GSD III diet.

GSD VI (hepatic phosphorylase deficiency). Diet high in protein and energy at normal level, distributed as frequent daytime feedings similar to the GSD III diet plus a midnight or early morning high-protein snack. *GSD IX (hepatic phosphorylase kinase deficiency).* High-protein diet similar to that for GSD III.

Glycogen supercompensation. More popularly called *carbohydrate loading.* A high-carbohydrate diet taken for 7 days before an endurance event to build up muscle glycogen reserves for optimal performance during a single endurance event lasting 90 minutes or longer. See *Carbohydrate loading.*

Glycolysis. Also called "Embden–Meyerhof pathway"; a series of reactions involving the anaerobic breakdown of glycogen (or glucose) in the tissues, with lactic or pyruvic acid as the end product. The net result is the synthesis of two molecules of adenosine triphosphate (ATP) per mole of glucose catabolized; oxygen is not consumed in the overall process. In the presence of oxygen, pyruvic acid is completely oxidized to carbon dioxide and water in the mitochondria by way of the Krebs tricarboxylic acid cycle and its associated oxidative phosphorylation. Although the yield of ATP from glycolysis is small compared to the net yield of 38 moles of ATP/mole of glucose with complete oxidation, glycolysis provides a means of rapidly obtaining ATP (hence, energy) in a relatively anaerobic organ such as muscle.

Glycosuria. Presence of sugar in the urine. It may be due to *diabetes mellitus,* a lowered renal threshold for glucose without any accompanying blood glucose elevation, a brain tumor or injury, or temporary emotional tension or worry.

Glycosylated hemoglobin (HbA_{lc}). A compound formed in the red blood cells by the irreversible reaction of hemoglobin A with glucose. The glycosylated hemoglobin concentration indicates the average blood glucose level over the previous 6 to 8 weeks. The rate of formation increases when the blood glucose level is elevated, as in uncontrolled diabetes mellitus. Normal values in nondiabetics are about 5% of the total hemoglobin, whereas uncontrolled diabetics have more than 10%. It is a reliable tool for evaluating long-term management of diabetes mellitus.

Glymidine. Sulfonylpyrimidine derivative that has a hypoglycemic action; used for the oral treatment of diabetes mellitus.

Glytrol®. Brand name of a nutritional product for patients with hyperglycemia. Lower in fat and higher in soluble fiber content than other similar formulas. Available in cans and in UltraPak enteral closed system. Contains 15 g fiber/liter. See Appendix 39.

G/N ratio. Abbreviation for *glucose/nitrogen ratio.*

Goiter. Enlargement of the thyroid gland due to lack of iodine (simple goiter), overproduction of the thyroid hormone in hyperthyroid states (exophthalmic goiter or thyrotoxicosis), or decreased production of the thyroid hormone in hypothyroid states (cretinism and myxedema). A hyperthyroid state is characterized by a high basal metabolic rate (BMR) which leads to loss of weight; a hypothyroid state is accompanied by a low basal metabolic rate and weight gain. *Nutrition therapy:* a calorie-controlled diet based on the age, sex, and activity level and adjusted for the basal metabolic rate, which can be 10% to 40% higher (in hyperthyroid goiter) or lower (in hypothyroid goiter) than the normal BMR. In simple goiter due to iodine deficiency, provide good sources of iodine such as iodized salt, seafoods, and marine fish. Foods containing natural *goitrogens* should be eaten cooked.

Goitrogens. Substances present in some foods that are capable of producing goiter by interfering with the production of the thyroid hormone. These have been identified as *arachidoside,* present in the red skin of peanuts, and *thiooxazolidone,* present in plants of the genus *Brassica* (e.g., brussels sprouts, cabbage, cauliflower, kale, and turnip). Goitrogens are destroyed by heat.

Gooseflesh. Also called "goose pimples"; term given to rough skin characterized by erection of the hair follicles, as from cold or shock. Also seen in vitamin A deficiency. See *Keratosis*.

Gout. A disorder of purine metabolism characterized by abnormally high uric acid levels in the blood and deposits of sodium urate in soft and bony tissues such as joints, cartilage, and tendons. The hyperuricemia of gout may be due to overproduction of uric acid or inadequate excretion by the kidney. *Nutrition therapy:* some limitation in dietary purines may be advisable in the acute stage to avoid adding exogenous purines to the existing high uric acid load. As an adjunct to drug therapy, a purine-restricted diet will reduce the excretion of uric acid by 200 to 400 mg/day and lower the mean serum uric acid level by 1 mg/100 mL. Other dietary considerations include weight control and maintenance of desirable body weight, high fluid intake (up to 3 liters/day) to prevent urate precipitation in the kidneys, and moderate alcohol consumption ($\leq$ 2 oz/day), as alcohol inhibits renal excretion of urates. Limit protein to 0.8 to 1 g/kg body weight; meat should not exceed 3 to 4 oz per meal. Have a high-carbohydrate and low-fat intake to enhance the excretion of urates. Avoid large, heavy meals late in the evening and maintain an alkaline urine to increase the solubility of uric acid in the urine. See *Diet, purine-restricted,* and alkaline-ash diet under *Diet, Ash*.

Graft-versus-host disease (GVHD). Also called "homologous disease." A rejection response to a tissue or organ transplant, especially allogenic bone marrow, commonly associated with inadequate immunosuppresive therapy. GVHD may involve the skin, liver, gastrointestinal tract, and other organs. Gastrointestinal symptoms may be severe and include abdominal pain, nausea, vomiting, diarrhea, bloody stools, malabsorption, steatorrhea, and altered intestinal motility. Frequent problems affecting food intake are anorexia, taste changes, and xerostomia. Weight loss is common for 3 to 12 months. *Nutrition therapy:* in severe diarrhea, NPO until the stool volume is less than 500 mL/day for 2 days or more; nutrition is provided by parenteral nutrition (PN). Gradually introduce oral feedings of isotonic liquids low in residue and lactose, in small frequent feedings while PN support continues. The recommended intake (combined oral and PN) is basal energy expenditure (BEE) $\times$ 1.8 and 2 g protein/kg desirable body weight. When symptoms improve, progress to solid foods low in residue, lactose, and fat (about 30 g/day). Introduce foods one at a time in small amounts; avoid gastric irritants. Taper PN and gradually advance to a normal diet as tolerated.

Griseofulvin. An antifungal antibiotic used for treatment of mycoses of the skin, hair, and nails. It may cause altered taste acuity, dry mouth, epigastric distress, nausea, and vomiting. Concurrent ingestion with alcohol may also cause tachycardia, headache, and flushing. High-fat meals increase drug absorption. Brand names include Fulvicin®, Grifulvin®, and Grisactin®.

Growth factor. Any factor, such as a mineral, a vitamin, or a hormone, that promotes the growth of an organism.

Growth hormone. See *Somatotrophin (somatotropic hormone)*.

GSD. Abbreviation for *glycogen storage disease*.

GTT. Abbreviation for *glucose tolerance test*.

Guanethidine. A postganglionic adrenergic agent used in the treatment of hypertension. Long-term use may cause weight gain due to sodium and fluid retention; it may also cause dry mouth, taste disturbances, diarrhea, and possible anemia. Brand names are Ismelin® and Esimil®.

Guillain-Barré syndrome. A neurologic disorder characterized by an acute postinfectious polyneuritis, involving numbness, increasing fatigue, pain, and paralysis. It is an autoimmune reaction often associated with a viral infection or with immunization. There may be respiratory failure requiring a ventilator, fluctuating blood pressure, coupled with weakening of the lower extremities, and even personality changes. *Nutrition therapy:* enteral or parenteral feeding may be necessary in severe cases during the acute episode. Increase energy and protein by 20% to 30% as necessary. If there is respiratory distress requiring a ventilator, use more fat as energy source rather than carbohydrate to prevent overproduction of carbon dioxide. Weak facial and throat muscles may cause chewing and swallowing problems. Modify food consistency as tolerated and gradually introduce soft, then regular foods as the patient improves.

Gut barrier. The gastrointestinal tract or "gut" serves as an important barrier to infection by preventing the migration of pathogenic microorganisms into the systemic circulation and other organs. The gut barrier includes the intestinal mucosa and submucosa, the lymphatic system of the mesentery, the reticuloendothelial system, and other immunocompetent cells. The gastrointestinal tract has high nutrient needs because of rapid turnover of epithelial cells; the absence of nutrients results in villous atrophy with a loss of gut mass and function. Maintenance of gut barrier function is essential; physical or functional disruption of this barrier can result in the translocation of bacteria. See *Bacterial translocation*. See also *Nutrition, gut integrity*.

GVHD. Abbreviation for *graft-versus-host disease*.

H

Haldane apparatus. Open-circuit type of respiration apparatus that applies the principle of *indirect calorimetry*. A small animal placed in the chamber is weighed before and after the experiment. The difference in weight represents carbon dioxide and water eliminated and oxygen consumed. The latter is indirectly determined by subtracting the sum of carbon dioxide and water losses from the total weight loss of the animal.

Halibut liver oil. Expressed oil from fresh halibut liver standardized to contain approximately 100 times the amount of vitamin A and 10 to 30 times the amount of vitamin D.

Hamwi formula. A quick estimate for ideal body weight (IDW) that is easy to remember:

IBW for adult males (in lb)
= 106 lb for five feet plus 6 lb per inch over 5 feet.
IBW for adult females (in lb)
= 100 lb for five feet plus 5 lb per inch over 5 feet.

Subtract 10% for small-framed individuals and add 10% for large-framed individuals.

HANES. Abbreviation for Health and Nutrition Examination Survey. See *NHANES*.

Haptoglobin (Hp). Serum mucoprotein that binds hemoglobin. Its major role appears to be the conservation of body iron by binding hemoglobin and preventing its loss from the body. Increased levels of Hp are found in individuals with inflammatory or neoplastic disease. Plasma levels are decreased in acute hepatitis and hemolytic conditions.

Harelip. See *Cleft palate*.

Harris–Benedict formula. Formula for estimating basal energy expenditure (BEE) based on body weight and standing height. It is used to calculate the basal metabolic rate (BMR).

$$\text{BEE (men)} = 66.5 + (13.8\ \text{W}) + (5.0\ \text{H}) - (6.8\ \text{A})$$
$$\text{BEE (women)} = 655.1 + (9.6\ \text{W}) + (1.8\ \text{H}) - (4.7\ \text{A})$$

where W is weight in kilograms, H is standing height in centimeters, and A is age in years. The energy expenditures for activity and stress factors are added to the BEE to estimate the total energy expenditure (requirement). See Appendix 24.

Hartnup disease. Also called "H disease"; a genetic abnormality in the renal and intestinal transport of amino acids, especially alanine, threonine, phenylalanine, and tryptophan. The most striking feature is a pellagra-like skin rash sensitive to light, which is most likely due to nicotinamide deficiency secondary to the tryptophan malabsorption. There is also aminoaciduria and some growth retardation and psychological changes. *Nutrition therapy:* the only treatment needed is nicotinamide supplementation (50 to 200 mg/day). A high-protein diet is recommended to counter amino acid loss in the urine.

Hb. Abbreviation for *hemoglobin*.

HbA$_{lc}$. Abbreviation for hemoglobin A$_{lc}$. See *Glycosylated hemoglobin*.

HE. Abbreviation for *Hepatic encephalopathy*.

Head injury. Generally characterized by hypermetabolism, hypercatabolism, hyperglycemia, depressed immunocompetence, and altered gastrointestinal function. *Nutrition therapy:* patients with a minor head injury and a Glasgow Coma scale (GCS) score of 13 to 15 are not hypermetabolic. They may be disoriented but can take adequate nutrition orally. Patients with moderate (GCS of 9 to 12) and severe (GSC of 3 to 8) head injuries require nutrition support by tube feeding or parenteral feeding. Start with a full-strength formula at 25 mL/hour and increase every 4 hours at 25 mL/hour until the kilocalorie goal is achieved. Provide 1.5 g of protein and 30 to 40 kcal per kilogram body weight per day. Monitor prealbumin and adjust protein accordingly. Also monitor gastric contents to detect gastric residual and feeding reflux. When there is a high risk of infection, formulas enriched with arginine, nucleotides, and omega-three fatty acids may be beneficial for immunologic enhancement. Most patients need additional free water to maintain fluid balance, given as enteral flushes throughout the day. Gradually introduce oral feeding as the patient improves.

Head Start Program. National program administered by the Department of Health and Human Services. Preschool children whose ages range from three to five are provided with comprehensive health, education, nutrition, social, and other services if they are the recipients of a federal assistance program. The educational services are extended to their parents, using nutritious meals as teaching tools, with the supervision of a nutritionist.

Health. State of physical, mental, and emotional well-being, and not merely freedom from disease or the absence of any ailment.

Health claim. The food labeling regulation permits the use of symbols, vignettes, and other forms of explicit and implied health claim. A food bearing a health claim must provide at least 10% of the recommended daily intake (RDI) of at least one of the following nutrients: protein, calcium, iron, vitamin A, vitamin C, and fiber. Entree items must provide 10% of the RDI of two of the six nutrients, and meals must provide three of the six. The health claim for food also specifies the levels of total fat, saturated fat, cholesterol, and sodium above which a food will be disqualified from making any health claim. The Food and Drug Administration (FDA) has authorized the following health claims: calcium and osteoporosis; dietary saturated fat and cholesterol and risk of coronary heart disease; dietary fat and cancer; sodium and hypertension; fiber-containing grain products, fruits, and vegetables and risk of cancer and coronary heart disease. Any substance or nutrient for which a health claim is made must be safe.

Healthy People 2000. A broad-based initiative to improve the health of all Americans over the next decade. Led by the U.S. Public Health Service, more than 300 national organizations form the Health Objectives Consortium. Priorities in the development of the objectives include health promotion, health protection, disease prevention, clinical preventive services, and treatment of age-related health problems. National programs on health promotion and disease prevention are coordinated by the Office of Disease Prevention and Health Promotion

(ODPHP), which is responsible for the development and implementation of the National Health Objectives for the Year 2000. The nutrition objectives for the year 2000 are summarized in Appendix 7B. See also the Dietary Guidelines for Americans in Appendix 7A.

Heartburn. Also called pyrosis. Epigastric pain just below the sternum, often accompanied by regurgitation of acid and the presence of gas in the stomach. It may occur 10 to 15 minutes after eating a heavy meal, especially when the person is in a recumbent position. It is a common symptom of many disorders, such as hiatal hernia and esophageal dysfunction. During pregnancy, heartburn results from upward displacement of the esophageal sphincter because of increased intraabdominal pressure. *Nutrition therapy:* avoid spices, gas formers, alcohol, and large meals. Encourage the patient to eat slowly six to eight small meals a day. Avoid large meals before bedtime and lying in a reclining position after eating.

Heart transplantation. Prior to transplantation, provide 1.0 to 1.2 g of protein per kilogram body weight/day and sufficient calories to maintain or obtain desirable weight. Keep fat intake <30% of total kilocalories, cholesterol <300 mg/day, and limit saturated fat to ≤10%. Undernourished individuals may require aggressive nutrition support for repletion before surgery. It may be necessary to restrict sodium (2 to 3 g/day) and fluids (1 to 2 liters/day). After surgery, advance the diet from clear to full liquid within 48 hours, then progress to solid foods as tolerated. For about 2 months after transplantation, provide basal energy expenditure (BEE) × 1.2 kcal, 1.2 to 1.5 g of protein/kg body weight, and 2 to 3 g of sodium per day. To reduce the risk of bacterial infection, follow food safety precautions and restrict the intake of fresh fruits and vegetables for 2 to 3 months. Monitor nutritional side effects of immunosuppressive drugs (cyclosporine, prednisone, azathioprine). Long-term posttransplant dietary recommendations include the following: kilocalories to maintain desirable body weight; 0.8 to 1.5 g/kg/day of protein as indicated by protein status and renal function; 50% to 60% of calories from carbohydrate and < 30% from fat; restrict saturated fat and cholesterol as previously indicated; avoid simple carbohydrates and concentrated sweets in cases of elevated serum triglyceride levels or steroid-induced diabetes; restrict sodium (2 to 3 g/day) to control edema; give 800 to 1200 mg calcium/day; and daily vitamins and minerals to meet RDA. Restrict caffeine if the patient has arrhythmias.

Heat of combustion. Amount of heat produced (usually expressed in calories) when a unit weight of a substance is oxidized.

Helium dilution. A gasometric procedure for determining body density. The animal or human is enclosed in a chamber of known volume, and a measured volume of helium gas is injected into the chamber. After a period of time long enough to allow mixing of the helium and air in the chamber, a sample is removed and analyzed. The degree to which helium gas is diluted is inversely proportional to the volume or space occupied by the animal or human. The smaller the subject, the more diluted the helium gas will be.

Hematin. Ferriprotoporphyrin hydroxide, a neutral compound in which iron is in the ferric state.

Hematocrit (HCT). The packed cell volume of erythrocytes, expressed as a percentage of the total blood volume. It is useful in clinical analysis of blood and can detect decreases (hemoconcentration) or increases (hemodilution) in plasma volume. The hematocrit is increased in dehydration, shock, and polycythemia vera; it is decreased in anemia, massive blood loss, hyperthyroidism, cirrhosis, leukemia, and water overload. Normal values vary with age and sex. See Appendix 30.

Hematopoiesis. See *Hemopoiesis*.

Heme. Ferroprotoporphyrin or ferroheme, a nonprotein, iron-containing portion of hemoglobin with no net charge. Contains divalent iron and combines with various nitrogenous bases, forming hemochromogens.

Hemeralopia. Glare blindness or day blindness; defective vision in bright light. A term used to describe the condition of reduced *dark adaptation* resulting from vitamin A deficiency, although there may be other causes such as diseases of the retina.

Hemicellulose. Indigestible polysaccharide found in plant cell walls, particularly in woody fibers and leaves. Hydrolysis by alkalis and acids yields xylose, a pentose sugar, other monosaccharides, and uronic acid. It may be digested to some extent by microbial enzymes. Hemicellulose differs from cellulose in that it has fewer glucose units. See also *Dietary fiber* under *Fiber*.

Hemin. Ferriprotoporphyrin or ferriheme, a nonprotein, iron-containing portion of hemoglobin with a net positive charge. Contains trivalent iron.

Hemochromatosis. Disorder of iron metabolism characterized by abnormal deposits of *hemosiderin* in the liver, spleen, and other tissues, causing cellular damage and degeneration. Two forms are known: *genetic* hemochromatosis, resulting in inappropriately high absorption and storage of dietary iron, and *acquired* hemochromatosis secondary to blood transfusion, excessive long-term intake of iron, and iron overload associated with some forms of liver disease. Features of the disorder include high serum ferritin level (above 1000 ng/mL), bronzing of the skin, jaundice, cirrhosis of the liver, abdominal pain, and sclerosis of the pancreas. Treatment is generally by phlebotomy or by chelation of iron. *Nutrition therapy:* reduce the dietary intake of iron to required need and avoid other sources, as from mineral supplements with iron. Individuals with hemochromatosis should avoid high intakes of vitamin C.

Hemodialysis. Process of removing waste products and excess fluid from the blood with the use of an artificial kidney, called a hemodialyzer. The average dialysis treatment usually lasts 3 to 4 hours, depending on the method of hemodialysis chosen and individual patient requirements. Hemodialysis is usually required three times a week and can be performed at home or in a dialysis center, which is either staff-assisted or self-care. There is a loss of nutrients during hemodialysis. About 10 to 15 g of amino acids and 10 to 25 g of glucose are lost per dialysis treatment using glucose-free dialysate; 30% to 40% of the amino acids lost are essential. There is also a loss of minerals and water-soluble vitamins. Catabolism and energy expenditure are increased and constipation is a frequent problem. *Nutrition therapy:* the diet should be individualized and periodically reevaluated as the patient's condition changes. Provide 1.1 to 1.4 g of protein/kg dry weight (actual or desirable body weight at normal hydration) and, per kilogram dry body weight, 30 to 35 kcal for weight maintenance, 35 to 50 kcal for weight gain, or 20 to 30 kcal for weight loss. About 30% to 40% of total calories should come from fat, with a 1:1 ratio of polyunsaturated to saturated fatty acids. The remainder of the nonprotein calories are provided by carbohydrates, with emphasis on complex carbohydrates and about 20 to 25 g of fiber. Suggested daily intakes for minerals and vitamins are: 2 to 3 g sodium, depending on urine output; 1.5 to 3.0 g potassium, 1400 to 1600 mg calcium, 12 to 17 mg/kg phosphorus, about 100 mg elemental iron if the patient is receiving erythropoietin, 0.8 to 1.0 mg folic acid, 10 mg pyridoxine, and 3 to 6 μg vitamin B_{12}. The other water-soluble vitamins (niacin, thiamin, riboflavin, biotin, pantothenic acid, and vitamin C) should be supplemented at the RDA level. The fat-soluble vitamins A, E, and K usually do not need supplementation. Vitamin D may be supplemented, if needed. The recommended fluid intake is 700 to 1000 mL plus urine output in 24 hours. Indications that the patient needs to reduce sodium and fluid intake are increased thirst, edema, and excessive interdialytic weight gains. A fluid weight gain of 1 to 2 lb a day between treatments should be the goal. Monitor the weight, serum albumin, and electrolytes. Some patients continue to lose weight and/or their albumin levels decrease even with aggressive oral nutritional supplementation. An *intradialytic parenteral nutrition* may be considered for these patients for nutritional repletion. Dietary guidelines for infants, children, and adolescents on hemodialysis are the following, based on dry weight, age, and height: calories per kilogram dry body weight: infants, 105 to 115 kcal; 1 to 3 years, 100 kcal; 4 to 10 years, 85 kcal; 11 to 14 years, 60 kcal for boys and 48 kcal for girls; 15 to 18 years, 42 kcal for boys and 38 kcal for girls. Grams protein per kilogram dry weight/day: birth to 1 year, 2 to 6 g; 1 to 2 years, 2 g; and 2 years to adolescence, 1.5 g. For fluids, allow 30 to 35 mL/100 kcal/day plus urine volume. Restrict sodium to 1 to 3 mEq/kg/day (23 to 70 mg/kg) in infants, and 2 to 3 g/day in children and adolescents. Avoid foods high in phosphorus and potassium; if needed, restrict potassium to 1 to 3 mEq/kg/day (40 to 120 mg/kg) in infants, 40 to 60 mEq/day (1500 to 2300 mg) in children <20 kg body weight, and 60 to 70 mEq/day (2300 to 2700 mg) if >20 kg. Supplement calcium to meet RDA levels and multivitamins plus 1 mg folic acid for infants and children up to 11 years.

Hemoglobin (Hb). The oxygen-carrying pigment of red blood cells; a conjugated protein with the prosthetic group, *heme,* attached to the protein moiety, *globin.* Normal Hb values for adults are about 14 to 16 g/100 mL (140 to 160 g/L) of blood. The main function of Hb is to carry oxygen from the lungs to the tissues and transport carbon dioxide back to the lungs. A low Hb content of the blood is a useful indicator of nutritional anemia, which may be caused by a deficiency in iron, copper, folic acid, vitamin B_{12}, and protein. A low level may also indicate protein-energy malnutrition, excessive fluid intake, hyperthyroidism, and severe hemorrhage. Elevated levels may indicate hemoconcentration, dehydration, congestive heart failure, and polycythemia. See also *Anemia and Hemopoiesis.*

Hemoglobin A_{Ic}. See *Glycosylated hemoglobin.*

Hemolysis. Destruction of red blood cells with the liberation of hemoglobin. May be brought about by various hemolytic agents, such as bacterial toxins, bile salts, and the venoms of certain poisonous snakes, or by the production in the body of specific *hemolysins,* a class of immune substances or antibodies elicited as a result of injection of incompatible blood.

Hemopoiesis (hematopoiesis). More appropriately termed "erythropoiesis" or the formation of red blood cells (RBCs). The average life span of an RBC is 4 months or about 127 days. RBCs are continually destroyed and replaced. The process takes place mainly in the bone marrow, and to some extent in the liver and spleen. The primitive RBC matures by passing through several stages in the following order: *proerythroblast* (large cell with a granular nucleus); *basophilic normoblast* (the nucleus is more compact and less granular); *eosinophilic normoblast* (the cell acquires more hemoglobin and becomes acidophilic); *reticulocyte* (an immature or young RBC devoid of a nucleus but rich in ribosomes and nucleic acids); and finally, the mature RBC or erythrocyte. The hemoglobin (Hb) remains in the RBC throughout its life span until the worn-out cell is removed from the circulation by cells of the *reticuloendothelial system* (bone marrow, liver, and spleen). On destruction, the heme portion of Hb is split into *bilirubin* and other pigments that are carried to the liver and secreted in the bile. The released iron can be reutilized to form Hb. Factors needed for normal hemopoiesis include a hormonal factor (erythropoietin, a hematopoietic hormone found in the plasma); a maturation factor (folic acid); and other nutritional factors (good-quality protein, folic acid, ascorbic acid, iron, copper, etc.). In general, all nutrients, plus sufficient calories, are involved directly or indirectly in hemopoiesis.

Hemosiderin. A dark yellow pigment of an iron–protein complex; a storage form of iron found in the liver, spleen, and bone marrow. Unlike *ferritin,* which is a water-soluble complex of iron and protein, hemosiderin is insoluble and granular.

Hemosiderosis. Condition of increased *hemosiderin* accumulation in the liver and other tissues; it occurs when excess iron can no longer be stored as *ferritin* and hemosiderin storage has reached its normal limits. It has been observed with excess iron intake (such as from prolonged iron therapy in non-iron-deficient individuals) and in chronic alcoholism, chronic liver disease, pernicious anemia, and certain types of refractory anemia. Hemosiderosis is increased iron storage without tissue damage, whereas *hemochromatosis* refers to increased storage with associated tissue damage.

HEP. Abbreviation for *high-energy phosphate*.

Heparin. A mucopolysaccharide that acts as an anticoagulant by preventing the conversion of prothrombin to thrombin.

HepatAmine®. Brand name of an 8% amino acid solution with high branched-chain amino acids (36% BCAAs) for parenteral nutrition support in patients with liver disease and hepatic encephalopathy; it is high in arginine and low in methionine. See Appendix 40.

Hepatic Aid II®. Brand name of a powdered nutritional supplement containing essential and nonessential amino acids. It is high in branched-chain amino acids (46% BCAAs) and arginine, and low in aromatic amino acid and methionine; for the dietary management of patients with chronic liver disease. See Appendix 39.

Hepatic coma. A neurologic disorder indicating extensive liver damage, characterized by varying degrees of consciousness, stupor, and lethargy. Other symptoms include personality change, trembling of the hands, loss of memory, hyperventilation, convulsions, and respiratory alkalosis. Endogenous or exogenous products toxic to the brain are not neutralized in the liver, and death may occur. *Nutrition therapy:* the level of protein restriction is very low—about 0.2 to 0.3 g/kg body weight. Protein restriction should be used for as short a time as possible because protein is important in healing the liver tissue. As the patient's condition improves, increase the protein intake by 10 g/day until a normal allowance is consumed. Provide sufficient calories from carbohydrate and fat (2000 kcal/day or more) to keep body tissue breakdown to a minimum. Supplementation with branched-chain amino acids (BCAAs) has been reported to improve nitrogen balance but has not be shown to improve neurologic symptoms. See minimal protein diet under *Diet, protein-modified*. See also *Hepatic encephalopathy*.

Hepatic encephalopathy (HE). A type of brain damage caused by ammonia intoxication due to failure of the liver to convert ammonia into urea, which accumulates in the blood. Clinical signs include personality changes, disorientation, "flapping" tremor of the hands, and spasticity. HE may be mild but could lead to hepatic coma if blood ammonia level is very high. Ammonia is a by-product of amino acid metabolism; it is also produced by the action of intestinal bacteria on dietary protein and gastrointestinal bleeding. Blood ammonia is reduced by neomycin (destroys gut flora) and lactulose (reduces colonic absorption of ammonia by lowering the luminal pH). *Nutrition therapy:* the basic principle of the diet is to reduce ammonia production and avoid tissue protein catabolism. It may not be necessary to reduce protein intake; provide 0.8 to 1.0 g/kg body weight. Restrict protein to 0.5 to 0.7 g/kg if a higher level is not tolerated and antibiotic therapy is not sufficient to control ammonia production. If necessary, reduce protein further to 0.3 to 0.5 g/kg in impending coma. Enteral supplements with branched-chain amino acids (BCAAs) may improve nitrogen balance. Some patients may tolerate a higher protein level supplied from vegetable sources, starch, and dairy products; limit protein from meat sources to 20 to 40 g/day. Protein restriction should be used for as short a time as possible. As the patient's condition improves, increase the protein intake by 10 g each week until a level of 1 g/kg is consumed. Provide sufficient calories [30 kcal/kg or basal energy expenditure (BEE) × 1.2] to maintain a positive nitrogen balance for regeneration of hepatic tissue. Restrict sodium to 2 g or less if ascites and edema are present. Restrict fluids to 1500 mL/day or less, depending on urinary output, serum electrolyte values, and fluid retention.

Hepatic failure (insufficiency). Disorder that occurs in severe hepatic disease. Hepatic failure occurs when liver function diminishes to 30% or less. Characterized by a tendency to hemorrhage; prothrombin and fibrinogen levels are decreased, and blood clotting is delayed. Accompanying psychomotor abnormalities due to accumulation of ammonia in blood leads to *hepatic encephalopathy, hepatic coma,* and death. *Nutrition therapy:* estimated protein needs are 1.0 to 1.2 g/kg desirable body weight (DBW) to achieve a positive nitrogen balance. Patients with clinical signs of encephalophathy may not tolerate this level; restrict protein to 0.7 to 0.8 g/kg DBW but not more than 50 g per day. If there is no improvement, reduce protein by 10 g every 2 to 3 days to a level of 0.5 g/kg. Increase or decrease the protein level as tolerated while ensuring sufficient calorie intake to prevent tissue catabolism. See also *Hepatic coma* and *Hepatic encephalopathy*.

Hepatitis. Inflammation of the liver. May be caused by infectious agents such as viruses and bacteria, alcohol, transfusion of incompatible blood, toxic drugs such as arsenicals, or toxic solvents such as carbon tetrachloride. Characterized by abnormal liver function, marked anorexia, fever, headache, rapid and marked weight loss, jaundice, and abdominal discomfort. It may be mild and brief or severe and life-threatening. The liver is usually able to regenerate its tissue, but severe hepatitis may lead to cirrhosis and chronic liver dysfunction. *Nutrition therapy:* a high-calorie [basal energy expenditure (BEE) × 1.5], high-protein (1.5 to 2.0 g/kg), high carbohydrate (300 to 400 g/day) diet, and moderate to liberal fat as tolerated (50 to 100 g/day). Reduce protein to tolerance level and increase carbohydrates to spare protein. Monitor for signs of liver failure. The objectives of the diet are to aid the regeneration of the liver tissue, to assist in maintaining nitrogen balance, and to ensure glycogen storage. During the acute phase of the disease, a full liquid diet is given, which progresses to a soft diet and eventually to a regular diet as the patient's condition improves. Six to eight small feedings spaced throughout the day are beneficial. Encourage fluids but avoid alcoholic drinks. Supplementation with minerals and vitamins, especially zinc, vitamin C, the B complex vitamins, and vitamin K may be necessary.

Hepatoflavin. Name given to *riboflavin* isolated from the liver. See *Vitamin B_2*.

Hepatomegaly. Enlargement of the liver. It is seen in certain infections; in diseases of the liver, blood, and heart; and in some nutritional deficiency states, such as kwashiorkor.

Herpes. Inflammatory skin disease characterized by the formation of small vesicles or blisters on the skin or mucous membranes. There are many types, but the most common is an acute viral type called *herpes simplex*. Type I is likely to cause oral infections and Type II affects the genitalia and anus. *Nutrition therapy:* provide adequate calories and liberal amounts of protein. Monitor the side effects of drugs used for reducing infection and fever.

Hexose monophosphate shunt (HMS). Also called the "pentose phosphate pathway," "Warburg–Dickens–Lipmann pathway," "phosphogluconate shunt," or "oxidative" shunt; one of two major pathways of glucose metabolism; the other is anaerobic *glycolysis* or the Embden–Meyerhof (EM) pathway. HMS is an aerobic process, whereas the EM pathway is anaerobic; the latter occurs almost exclusively in muscles, whereas HMS is the major pathway in the mammary glands, testes, adipose tissues, leukocytes, and adrenal cortex. Both pathways occur in the liver simultaneously; about 50% of glucose is degraded by the EM pathway and the rest by the HMS. The HMS provides the pentoses, particularly ribose, needed for DNA and RNA synthesis. It yields NADPH (TPNH), which is needed for fat, steroid, and cholesterol synthesis and is important in photosynthesis. Many reactions and enzyme systems (e.g., transketolase, aldolase, dehydrogenase, and isomerase) involved in the HMS are identical to those observed in dark-reaction photosynthesis.

Hexuronic acid. 1. Acid derived from hexose sugar by the oxidation of the group on carbon atom 6. The hexuronic acid derived from glucose is glucuronic acid. 2. Name originally given to a substance isolated from lemon juice, later identified as vitamin C.

Hg. Chemical symbol for mercury.

HHANES. Abbreviation for Hispanic Health and Nutrition Examination Survey.

5-HIAA. Abbreviation for 5-hydroxyindoleacetic acid.

Hiatal hernia. An outpouching of a portion of the stomach upward through the esophageal hiatus of the diaphragm. Usual symptoms include difficulty in swallowing, reflux, heartburn, esophagitis, and discomfort after a heavy meal when lying down or bending over. *Nutri-*

tion therapy: same as for *Gastroesophageal reflux disease.*

High-density lipoprotein (HDL). A plasma lipoprotein that has a density of 1.063 to 1.210 g/dL; classified as the alpha fraction on the basis of electrophoresis. HDL contains about 33% protein, 29% phospholipid, 30% cholesterol, and 8% triglyceride. HDL is involved in the turnover of tissue cholesterol and in the transport of excess cholesterol to the liver, where it is metabolized to bile acids and eventually excreted. High-density lipoprotein (HDL)-cholesterol levels have an inverse relationship with coronary heart disease (CHD). The normal range in serum is 35 to 80 mg/dL (0.9 to 2.1 mmol/liter). Values above 60 mg provide extra protection against CHD; values below 35 mg imply significant risk for CHD. Common causes of reduced serum HDL-cholesterol are cigarette smoking, obesity, lack of exercise, poorly controlled diabetes mellitus, chronic renal failure, hypothyroidism, liver disease, starvation, hypertriglyceridemia, genetic factors, and certain drugs. The National Cholesterol and Education Program (NCEP) recommends that the HDL-cholesterol be checked in all persons with high blood cholesterol ($>$240 mg/dL or $>$6.2 mmol/L), and those with borderline high cholesterol (200 to 239 mg/dL or 5.2 to 6.2 mmol/L) if two other major risk factors for CHD or if CHD itself is present. The optimal ratio of total cholesterol to HDL-cholesterol is $\leq$3.5.

High-energy phosphate (HEP). Compound that contains a labile phosphate bond that yields free energy, varying from 5 to 12 kcal/mol, when the bond is dissociated. Examples are adenoside triphosphate, creatine phosphate, and acetyl phosphate. See also *Phosphate bond energy*.

Hippuric acid. Conjugation product of benzoic acid and glycine; a normal constituent of the urine. Formed largely, if not solely, in the liver as a *detoxication* product of benzoic acid.

Hispanic Health and Nutrition Examination Survey (HHANES). Survey conducted in 1982 to 1984 on more than 4,000 Hispanics, ages 6 to 74 years. Similar methodologies for the NHANES surveys were used to assess nutritional status. See *National Health and Nutrition Examination Survey (NHANES)*.

Histamine. Amine formed by the decarboxylation of histidine; occurs as a decomposition product of histidine and is prepared synthetically. It is a powerful vasodilator and can lower blood pressure. It can be useful in treating various allergies and as a stimulant for gastric pancreatic secretion and visceral muscles. It is used as a diagnostic agent in testing gastric secretion (histamine test). It promotes the contraction of smooth muscles, increases nasal secretions, and relaxes blood vessels and respiratory airways.

Histidine (His). Beta-imidazole alanine, an *essential amino acid*. Although its essentiality for infants and children was established years ago, it was only recently that studies confirmed that adults also need a dietary source of histidine. It is a component of carnosine, anserine, and hemoglobin and a precursor of histamine. See Appendix 9.

Histidinemia. Elevated blood level of histidine and its consequent excretion in the urine (histidinuria). It is a genetic disorder due to lack of the enzyme histidase. The condition is harmless for the majority of affected individuals, although some may develop speech and hearing defects and mental retardation. Avoidance of a high protein intake to restrict dietary histidine may be helpful, although this has not been proven effective.

HIV. Abbreviation for *human immunodeficiency virus*.

Hodgkin's disease. Malignant, enlarged lymph nodes causing fatigue, weight loss, slight fever, cough, dyspnea, and chest pain. *Nutrition therapy:* a diet high in calories, protein, and fluids is recommended. Monitor electrolyte balance and correct weight loss. See *Lymphoma*.

Home-delivered meals. Also called "meals-on-wheels." Administered by the U.S. Department of Agriculture, meals are prepared by approved hospitals or institutions and delivered to persons who are homebound due to old age, illness, or disability. The spouse of an older person, regardless of age or condition, may also receive a home-delivered meal if it is for the best interest of the homebound person. Generally, a hot noon meal and a packaged evening meal are provided 5 days a week.

Homeostasis. Constancy of the internal environment. The ability of the body to maintain a balance among its physiochemical processes; these are dependent on *dynamic states* of metabolism. Homeostatic mechanisms in the body include fluid and pH balance; regulation of body temperature, blood sugar level, heart rate, and pulse rate; and hormonal control.

Hominex™. Brand name of a methionine-free powdered product for use in infants and toddlers (Hominex™-1) and children and adults (Hominex™-2) with vitamin B_6-nonresponsive homocystinuria or hypermethioninemia.

Homocysteine. A demethylated product of methionine; a homolog of cysteine. Present in cells as an intermediate metabolite; capable of conversion to methionine by direct transfer of a methyl group from compounds such as choline and betaine.

Homocystinuria. Inborn error of metabolism due to lack of the enzyme cystathionine synthetase, which is essential for the conversion of homocysteine to cystathionine, both of which are intermediate products formed in the metabolism of methionine. Plasma levels of methionine and homocysteine are elevated, and large amounts of homocystine are excreted in the urine. A similar condition can result from vitamin B_6 insufficiency, since cystathionine synthetase requires the vitamin as a cofactor. The characteristic symptoms of homocystinuria include mental retardation, dislocated lenses, glaucoma, osteoporosis, skeletal deformities, mild mental retardation, thromboembolism, and early atherosclerosis. *Nutrition therapy:* a marked biochemical response to high doses of pyridoxine (vitamin B_6) is seen in about half of the individuals affected. Some individuals also respond to supplemental folate intake of 400 mcg/day. Those who do not respond adequately to pyridoxine require a low-methionine, high-cystine diet and monitoring of plasma methionine and homocysteine levels. Supplementary betaine or choline may help promote remethylation of homocystine. See *Diet, methionine-restricted*.

Homogentisic acid. Intermediate product in the metabolism of phenylalanine and tyrosine. It is excreted in the urine in alkaptonuria and becomes oxidized to a blackish pigment on exposure of urine to air. See *Alkaptonuria*.

Hormone. Organic substance produced by groups of cells or an organ and discharged directly into the bloodstream for specific regulatory action on other organs or tissues remote from its original source. With a few exceptions, hormones are generally manufactured by the endocrine glands. Certain hormones are protein in nature (e.g., parathormone and insulin); some are amino acid derivatives (e.g., thyroxine and epinephrine); and others are steroids (e.g., estrogens and androgens). See Appendix 11B for a summary of hormones.

Hospital Prognostic Index (HPI). Probability of survival in medical and surgical patients based on serum albumin, presence or absence of sepsis and/or cancer, and delayed hypersensitivity.

$$\text{HPI} = 0.9\ (\text{ALB}) - 1.0\ (\text{HST}) - 1.44\ (\text{SEP}) + 0.98\ (\text{DIA}) - 1.09$$

where ALB is serum albumin in g/dL; HST is hypersensitivity skin test where anergy = 2 and reactive = 1; SEP is sepsis where no sepsis = 1 and septic = 2; and DIA is diagnosis where cancer = 1 and noncancer = 2.

Household Food Consumption Surveys. Administered by the U.S. Department of Agriculture, these periodic surveys were conducted from 1909 to 1978 in sample populations primarily to collect data about the production and marketing of food. The plan was to repeat the survey every 10 years. The 1965 survey included dietary intakes, but no physical and biochemical data. The survey was repeated in 1977 to 78 as the *Nationwide Food Consumption Surveys (NFCS)*.

Hp. Abbreviation for *haptoglobulin*.

Human immunodeficiency virus (HIV). A type of retrovirus that causes AIDS. The 1993 CDC classification system for HIV-infected adolescents and adults categorizes persons on the basis of clinical conditions associated with HIV infection and category of CD4+ T-lymphocyte count, as follows: category 1, ≥ 500 cells/μL; category 2, 200 to 499 cells/μL; and category 3, <200 cells/μL. HIV infects T-helper cells of the immune system and results in infection with a longer incubation period, averaging 10 years. The late stage of HIV infection attacks organ systems of the body and develops into AIDS and opportunistic infections such as Kaposi's sarcoma, pneumonia, candidiasis, and tuberculosis. *Nutrition therapy:* the objectives are to optimize nutrient stores and delay the onset of malnutrition, prevent tissue wasting, minimize malabsorption, and initiate nutrition intervention for problems associated with anorexia, diarrhea, gastrointestinal pain, and vomiting. Encourage small, frequent feedings of nutrient-dense foods that are easy to chew and swallow. For other details, see nutrition therapy for *Acquired immunodeficiency syndrome (AIDS)*.

Hunger. Craving for food more pronounced than appetite. A feeling of intermittent, brief cramping sensations of pressure and tension in the epigastric region, later accompanied by weakness and irritability. See *Nutrition, hunger*.

Huntington's disease. An inherited disorder characterized by chronic, progressive chorea and mental deterioration terminating in dementia. It is usually evident in the fourth decade and becomes fatal after 15 years. *Nutrition therapy:* give a high-protein, high-calorie diet modified in consistency from pureed to soft to prevent choking. Semisolid foods are easier to swallow than liquid foods. If necessary, the patient may be tube fed. Because there is loss of control of voluntary movements, most patients need assistance in feeding.

Hyaluronic acid. A viscous, high-molecular-weight mucopolysaccharide found in connective tissue and acts as an intercellular cement that holds the cells together. It also binds water in the interstitial spaces and acts as a shock absorber in the joints.

Hydralazine. An antihypertensive agent used in the management of moderate to severe hypertension and congestive heart failure. Hydralazine is a pyridoxine antagonist and may cause pyridoxine depletion and peripheral neuropathy. Long-term use of the drug may also cause anorexia and retention of sodium and fluid. It is excreted in breast milk. Brand names include Apresazide®, Apresoline®, Serapes®, and Serpasil®.

Hydrochloric acid (HCl). A normal constituent of the gastric juice in humans and other mammals. Its functions in digestion are to denature protein, activate pepsinogen to pepsin, provide an acid medium for absorption of iron, and stimulate the opening of the pylorus.

Hydrochlorothiazide. A thiazide diuretic used in the management of edema and hypertension. The drug enhances the urinary excretion of riboflavin, potassium, magnesium, zinc, and sodium, and thus causes deficiency states if the dietary intake is not adequate. It may also elevate blood glucose, lipid, and uric acid levels and cause dry mouth, increased thirst, loss of appetite, stomach cramping, and constipation or diarrhea. Hydrochlorothiazide is excreted in breast milk. Brand names include Aldoril®, Aquazide®, Esidrix®, Hydrodiuril®, Moduretic®, and Oretic®.

Hydrocortisone. Also called cortisol. A corticosteroid hormone occurring naturally in the body and produced synthetically for use as an anti-inflammatory agent. See *Corticosteroid* for its adverse effects. See also *Cortisol*.

Hydrogenation. A process by which molecular hydrogen is added to the double bonds in the unsaturated fatty acids of triglycerides. Oils are changed to solid fats; the process reduces the biologic value of essential fatty acids when these polyunsaturated fatty acids become saturated.

Hydrostatic pressure. Pressure exerted by a liquid on the surfaces of the walls containing the liquid. In the body, it refers to the blood pressure, which maintains the fluid volume and circulation in the blood vessels.

Hydroxocobalamin. Form of vitamin B_{12} in which the cyanide group is replaced by a hydroxyl group.

Hydroxyapatite. A naturally occurring mineral crystal of the general formula $3Ca_3(PO_4)_2.Ca(OH)_2$. The minerals in the bone are deposited in the organic matrix in a crystal formation similar to that of hydroxyapatite, except that the hydroxyl groups are partially substituted by other elements and radicals such as fluoride and carbonate.

25-Hydroxycholecalciferol (25-HCC). Also called "25-hydroxyvitamin D_3"; now called "calcidiol." See *Vitamin D*.

5-Hydroxyindoleacetic acid. Abbreviated as 5-HIAA; product formed from the breakdown of *serotonin*.

Hydroxylysine. Lysine to which a hydroxyl group has been added. One of the nonessential amino acids, it is found in the structural protein collagen.

Hydroxyproline. Proline to which a hydroxyl group has been added. One of the nonessential amino acids, it is found in the structural protein collagen.

Hydroxyprolinemia. Metabolic disorder due to lack of the enzyme hydroxyproline oxidase. Blood and urine accumulate large amounts of free hydroxyproline, and the condition may lead to mental retardation. At present, no therapy is available.

Hyperaldosteronism. A disorder characterized by increased production of aldosterone by the adrenal cortex. Among the signs and symptoms are muscle spasms of the extremities, hypertension, headache, cardiomegaly, retinopathy, hypokalemia, and paresthesia. *Nutrition therapy:* a diet high in potassium may be required. Restrict sodium intake and hydrate adequately.

Hyperalimentation. Also called "total parenteral nutrition (TPN)" or "total parenteral alimentation (TPA)"; the parenteral administration of all nutrients for patients with gastrointestinal dysfunctions. Although the term "hyperalimentation" is commonly used to designate total or supplemental nutrition by intravenous feedings, it is not technically correct because the procedure does not always involve an abnormally increased or excessive amount of feeding. See *Parenteral feeding (nutrition).*

Hyperammonemia. An inherited metabolic disorder characterized by an elevated blood ammonia level. It is due to a deficiency of the enzyme ornithine transcarbamylase, which catalyzes the reaction between carbamyl phosphate and ornithine to form citrulline. The symptoms include ammonia intoxication, vomiting, lethargy, hypotonia or spasticity, and cerebral and cortical atrophy. *Nutrition therapy:* restriction of dietary protein intake (0.5 g/kg body weight/day) is the mainstay of long-term treatment and results in a reduction of blood ammonia to near-normal levels. Life-threatening hyperammonemic crises are apt to recur unpredictably, especially during infectious illness; they may require a further reduction of protein to a minimum (0.2 to 0.3 g/kg body weight/day). If protein intake from common food sources is to be kept very low, supplementary essential amino acids may be needed.

Hypercalcemia. Abnormally high level of calcium in the blood. It occurs in various clinical disorders, such as hyperparathyroidism; solid tumors of the breast, ovary, and lungs; hyperthyroidism; and toxicity from vitamins A and D. It may also be drug related, involving, for example, the use of thiazide diuretics, chlorthalidone, lithium, and large amounts of calcium-containing antacids. Hypercalcemia may result in vomiting, nausea, muscular weakness, high blood pressure, and renal calculi. Normal serum calcium is 9 to 11 mg/dL (2.25 to 2.75 mmol/L). *Nutrition therapy:* encourage plenty of fluids, especially with diuretic therapy. Avoid excess vitamin D. A calcium-restricted diet is recommended.

Hypercalciuria. Presence of abnormally large amounts of calcium in the urine. High calcium concentration in the urinary tract may form kidney stones. *Nutrition therapy:* encourage plenty of fluids. In Type II absorptive hypercalciuria, restrict both dietary calcium and oxalate. In Type I absorptive hypercalciuria or renal hypercalciuria, calcium intake is not restricted but should not be excessive ($\leq$1000 mg/day). Keep sodium intake below 3 g/day. For more details, see *Urolithiasis.*

Hypercarotenosis. Condition characterized by high levels of carotene in the blood and skin and manifested by a yellow jaundice-like coloration of the skin that is particularly evident in the nasolabial folds, forehead, palms, and soles. But unlike jaundice, in which bile pigments accumulate in the body, the eyes do not become yellow. *Nutrition therapy:* the condition is benign and slowly disappears on reduction of the intake of carotenoid-rich foods from the diet.

Hypercatabolism. Excessive breakdown of reserve tissue or cellular materials to the extent that nutrients are depleted at an abnormally fast rate. The end result of an untreated hypercatabolic state could be fatal. Hypercatabolism is seen in advanced cases of AIDS, burns, severe injuries, cancer, and malnutrition. Immediate nutrition intervention is needed.

Hyperchloremia. Elevated blood chloride level; may be caused by dehydration, excess solute loading, diabetes insipidus, brain stem injury, or excessive administration of solutions containing chloride. Normal range in blood is 95 to 105 mEq/L (95–105 μmol/L).

Hyperchlorhydria. Excessive hydrochloric acid (HCl) in the stomach due primarily to increased secretion of the gastric juice.

Hypercholesterolemia. Condition in which blood cholesterol is above the normal limits (about 200 mg% or more). It is associated with atherosclerosis and other cardiovascular diseases, obstructive jaundice, hyperlipidemia, and excess adrenocorticotropic hormone. The etiologic factors implicated in high serum cholesterol levels are many and varied, but the exact mechanisms are not well understood. Hormonal, genetic, nutritional, and environmental factors have been implicated. Serum cholesterol levels vary with age. In adults, a level between 200 and 239 mg/dL (5.2 to 6.2 mmol/L) is considered borderline high, and ≥240 mg/dL (≥6.2 mmol/L) is considered high blood cholesterol. Abnormalities in serum cholesterol include: elevated total cholesterol, elevated low-density lipoprotein (LDL)-cholesterol, low high-density lipoprotein (HDL)-cholesterol, and a high ratio of total cholesterol or LDL-cholesterol to HDL-cholesterol. *Nutrition therapy:* follow the dietary guidelines recommended by NCEP (National Cholesterol Education Program) and AHA (American Heart Association). Epidemiologic studies have shown that there is a 2% decrease in coronary heart disease (CHD) for each 1% decrease in serum cholesterol. The overall goal is to lower total cholesterol to less than 200 mg/dL (5.2 mmol/L) and LDL-cholesterol below 130 mg/dL (3.4 mmol/L) if without CHD or ≥100 mg/dL (≥2.6 mmol/L) if CHD is present. See *Diet, cholesterol-restricted, fat-controlled*.

Hyperemesis. Severe vomiting which may lead to nutritional inadequacy. *Hyperemesis gravidarum* is abnormal, protracted vomiting seen in pregnancy, causing weight loss and dehydration. *Nutrition therapy:* serve dry meals in frequent small feedings. Avoid forced feedings. Provide high-carbohydrate, dry foods such as soda crackers and melba toast. Beverages should be drunk between meals, not with food. Vitamin and mineral supplements are recommended. Fluid and electrolyte imbalances should be corrected. Parenteral feeding may be necessary if the patient is unable to retain fluids by mouth.

Hyperglycemia. Increased glucose concentration in the blood above normal limits. Glucose levels are above 140 mg/100 mL of blood. This may occur in the following conditions: *diabetes mellitus* due to lack of insulin; increased *epinephrine* secretion; following ingestion of a very high carbohydrate intake (called "alimentary" hyperglycemia); *hyperthyroidism* due to increased hepatic glycogenolysis; increased intracranial pressure (as a result of skull fracture, cerebral hemorrhage, or brain tumor); administration of anesthetics such as ether, chloroform, and morphine; and *hyperpituitarism*. For dietary management of hyperglycemia in diabetes mellitus, see *Diabetes mellitus*.

Hyperglycemic, hyperosmolar, nonketotic syndrome (HHNK). A metabolic disorder in which blood glucose is highly elevated (greater than 600 mg/dL) without ketosis. It leads to high serum osmolarity. Symptoms include polyuria, severe dehydration, polydipsia, and tachycardia. It commonly occurs with Type II diabetes mellitus and is considered one of its acute complications. *Nutrition therapy:* restore electrolyte balance promptly. Rehydration is the mainstay of therapy. Monitor the levels of potassium and blood glucose and adjust the diet accordingly.

Hyperinsulinism. Condition of excessive insulin in the body. Caused either by an overdose of insulin (as in insulin shock) or by overproduction of insulin by the pancreas. The latter is commonly known as *reactive* or *stimulative hypoglycemia* or *functional hyperinsulinism*. See *Hypoglycemia. Nutrition therapy:* dietary modifications vary, depending on the cause. Give glucose solution or fruit juice containing natural sugars in case of insulin overdose. In functional hyperinsulinism, restrict carbohydrate intake, especially sugars and concentrated sweets, to minimize insulin production. A high protein intake (1.5 g/kg body weight/day or more) is recommended; fat furnishes the remaining calories. Provide six small meals, with equal distribution of carbohydrate, protein, and fat. Nonnutritive sweeteners may be used. Emphasize foods high in soluble fiber. Avoid alcohol.

Hyperkalemia. Also called "hyperpotassemia"; abnormally high potassium level in the blood. The normal range in serum is 3.5 to 5.0 mEq/L (3.5 to 5.0 mmol/L). A toxic elevation of serum potassium is observed in cases of renal failure, acute dehydration, Addison's disease, and excessive administration of potassium in the presence of renal insufficiency; it may also be due to massive release of potassium from cells, such as in crash injury, major surgical operations, and gastrointestinal hemorrhage. Symptoms, involving chiefly the cardiac and central nervous systems, include numbness, mental confusion, bradycardia, paralysis of the extremities, and cardiac arrest. *Nutrition therapy:* restrict potassium to 2 to 2.5 g/day (60 mEq). Also avoid potassium-containing salt substitutes. See *Diet, potassium-restricted.*

Hyperkinesis. Also called "hyperactivity" or "attention deficit disorder (ADD)," in children. Condition observed with some children who are excessively restless, inattentive, and disruptive at home or in school due to their abnormally high level of energy. Believed to be due to eating foods with artificial colors and flavors. Studies done to recommend a dietary regimen are not conclusive. See *Diet, Feingold.*

Hyperlipidemia. Nonspecific term that refers to an elevation of one or more lipid constituents of the blood, including glycolipids, lipoproteins, and phospholipids. The preferred term is *hyperlipoproteinemia.*

Hyperlipoproteinemia. One of several types of inherited or acquired disorders of lipoprotein metabolism characterized by elevated cholesterol, triglyceride, and other protein-bound lipids in the blood. Serum cholesterol and triglycerides are dependent on age and sex. (See Appendix 28). The 75th percentile for serum cholesterol is considered significant for hypercholesterolemia, and the 95th percentile is significant for hypertriglyceridemia. *Nutrition therapy:* the overall objectives are to attain and maintain a healthy weight and to reduce elevated cholesterol and/or triglycerides. For details, see *Hypercholesterolemia* and *Hypertriglyceridemia.* See also *Diet, cholesterol-restricted, fat-controlled.* Specific treatment and diet may vary according to the type of disorder. The five major types of hyperlipoproteinemia according to Fredrickson's classification are:

Type I. Characterized by hyperchylomicronemia or an extremely high triglyceride level, with normal or elevated cholesterol levels. It results in recurrent bouts of pancreatitis. It is extremely rare. *Nutrition therapy:* restrict fat intake to 20% or less of calories (30 g/day for adults). Medium-chain triglycerides (MCTs) may be used as source of calories. The cholesterol intake is normal, and the caloric level is adjusted to attain a healthy weight. Alcohol is not allowed.

Types IIa and IIb. In both types, there are increased serum cholesterol levels because the low-density lipoproteins are elevated. Cholesterol levels are elevated at birth and increase with age. At 50 years of age, affected individuals have three to 10 times greater risk of ischemic heart disease. In Type IIa, very-low-density lipoproteins and triglycerides are normal; in Type IIb, both are increased. *Nutrition therapy:* Type IIa requires a restricted cholesterol intake (150 to 200 mg/day) and adequate calories, of which 30% or less is supplied by fats, with a 1:1 ratio of polyunsaturated to saturated fatty acids. Alcohol is permitted, but it should be used in moderation.

Type III. Identified by the presence of elevated prebetalipoproteins, elevated plasma cholesterol, and elevated triglycerides. It is relatively uncommon and is referred to as "broad beta disease." Individuals with this disorder are at risk of developing early coronary disease. *Nutrition therapy:* if weight reduction is indicated, reduce caloric intake; a maximum of 30% is provided by fats with low saturated fatty acids or a polyunsaturated/saturated ratio of 1:1. Cholesterol is restricted to about 200 mg/day. Alcohol is limited to 25 g/day. Complex carbohydrates with dietary fiber are recommended. Nonnutritive sweeteners may be used for calorie-restricted diets.

Type IV. Characterized by elevated prebetalipoproteins, elevated triglycerides, and normal or slightly elevated cholesterol levels. It is also called "carbohydrate-induced hyperlipidemia" or "essential familial hyperlipoproteinemia." It occurs often and is associated with diabetes mellitus, obesity, and artherosclerosis. *Nutrition therapy:* weight loss by caloric restriction usually lowers the triglyceride level of the blood and normalizes glucose tolerance. Limit cholesterol, saturated fats, and sugars. Avoid alcoholic beverages.

Type V. A plasma lipoprotein pattern of hyperchylomicronemia and elevated prebetalipoproteins indicates intolerance to both endogenous and exogenous fat sources. Glucose tolerance and uric acid levels are also abnormal, as seen in cases of diabetic acidosis, obesity, nephrosis, and alcoholism. *Nutrition therapy:* weight reduction is needed for the obese person. Restriction of calories, 25% of which are provided by fats and oils of any kind, is recommended. Cholesterol is limited to 300 mg/day. Alcohol and concentrated sweets are not permitted.

Hypermagnesemia. Presence of excessive amounts of magnesium in the blood; almost always the result of renal insufficiency and the inability to excrete excess magnesium in foods and drugs, especially antacids. Symptoms include muscle weakness, confusion, and a fall in blood pressure. *Nutrition therapy:* avoid magnesium-rich foods.

Hypernatremia. Abnormally high blood sodium level; may be caused by dehydration due to inadequate fluid intake, respiratory loss with fever, hyperventilation of dyspnea, skin losses with burns, or metabolic acidosis; may also be due to excessive solute loading, as with concentrated feedings high in protein and salts without adequate supplemental water intake. Symptoms include thirst, flushed loose skin, tachycardia, hypotension, and hyperosmolarity. Normal range in serum is 135 to 147 mEq/L (135 to 147 mmol/L). *Nutrition therapy:* encourage fluid intake and avoid nutrient-dense or concentrated formula feedings without adequate fluid intake. See also *Dehydration.*

Hyperoxaluria. A rare metabolic disease characterized by increased excretion of oxalate in the urine. The basic difficulty is the inability to metabolize glyoxylic acid. As a result, excess oxalic acid is produced and is precipitated as calcium oxalate in the kidney. The individual usually dies of renal failure in infancy. Calcium oxalate deposits may be found in other tissues. *Nutrition therapy:* encourage plenty of fluids and restrict foods high in oxalate. Also limit calcium intake per day to 800 mg for men and 1000 mg for women to prevent formation of calcium oxalate stones. Avoid high vitamin C intake from foods and supplements (no more than 1000 mg/day). An alkaline-ash diet may be beneficial. See *Diet, oxalate-restricted.* See also alkaline-ash diet under *Diet, ash.*

Hyperparathyroidism. Abnormally increased secretion of parathyroid hormone leading to withdrawal of calcium from the bones. Features of this endocrine disorder include tenderness of the bones, muscular weakness and pain, abdominal cramps, and spontaneous fractures.

Hyperphosphatemia. High blood phosphate level; may be due to acute renal failure, chronic renal insufficiency, hypoparathyroidism, and hypervitaminosis D. Normal range in serum is 3.0 to 4.5 mg/dL (1.0 to 1.4 mmol/L). *Nutrition therapy:* See *Diet, phosphorus-restricted.*

Hyperpituitarism. Pathologic condition due to increased activity of the *hypophysis* (pituitary gland). Symptoms vary, depending on the pituitary cells affected and the type of hormone secreted in excessive amounts. *True hyperpituitarism* is overactivity of the eosinophilic cells (excess growth hormone) resulting in gigantism in children and *acromegaly* in adults.

Hypersensitivity skin test. Intradermal injection of three to five antigens to which most individuals have received prior sensitization (*Candida albicans,* mumps, tuberculin, streptokinase-streptodornase, *Trichophyton*). The diameter of the induration, measured after 24 to 48 hours, is graded 0 if nonreactive, 1 if 5 mm reactive, and 2 if more than 5 mm reactive. A normal positive response in a healthy individual is demonstrated by an area of inflammatory induration more than 5 mm in diameter; relative anergy is a response between 1 and 4 mm, and anergy is a negative response to all antigens. See *Anergy.*

Hypertension (HTN). Also called "high blood pressure"; persistent elevation of *blood pressure* above normal. Hypertension is a major risk factor for development of atherosclerotic, cardiovascular, and kidney disorders. Blood pressure varies considerably among individuals, depending on many factors, such as age, physical constitution, occupation, and health. For adults the average systolic/diastolic pressure is about 120/80 mm mercury (mm Hg). Hypertension may occur at any age, but more frequently in persons over 40 years old, with overweight or obesity as a predisposing factor. About 85% to 90% of the cases are *essential hypertension* (hypertension of unknown cause), which can be influenced by dietary factors. The association between sodium and hypertension has been firmly established. About 50% to 60% of hypertensives are sodium-sensitive; blood pressure goes down when dietary sodium intake is reduced. Public health officials recommend that all people use salt and sodium in moderation. The American Heart Association recommends an intake not exceeding 3 g/day. Low intakes of calcium, magnesium, and potassium have also been implicated in the development of hypertension. *Nutri-*

tion therapy: the current focus is on weight management, sodium restriction, and cholesterol control if elevated. Reduce weight and maintain it to within 15% of desirable weight; restrict sodium intake to 2 to 3 g/day in mild hypertension and restrict further in severe hypertension; lower cholesterol and saturated fat intakes if blood lipids are elevated; maintain adequate intakes of potassium, calcium, and magnesium; limit caffeine intake and restrict alcohol to 1 oz or less per day. A regular exercise program and avoidance of tobacco are also recommended. Mineral and vitamin supplementation, such as potassium and folate, may be required when diuretics and beta-blocker agents are used. For details, see *Diet, calorie-modified, Diet, cholesterol-restricted, fat-controlled,* and *Diet, sodium-restricted.* See also alcoholic beverages in Appendix 33.

Hyperthyroidism. Endocrine disorder caused by excessive secretion of the thyroid hormone as a result of overmedication with potent thyroid drugs, hyperactivity of the thyroid gland, or tumor (toxic adenoma of the thyroid). The clinical syndrome is generally called *exophthalmic goiter* because two thirds of the patients show exophthalmos (i.e., protruding eyes with wide-open lids). Other symptoms include thyroid enlargement, increased basal metabolic rate and pulse rate, nervousness and muscle tremors, and loss of weight. *Nutrition therapy:* a high-calorie, liberal-protein, liberal-carbohydrate diet with calcium, phosphorus, and vitamins D and B complex supplementation is recommended. The basic aim of the diet is to compensate for the increase in basal metabolic rate (3500 to 4000 calories) and nitrogen metabolism (90 to 120 g of protein). A high carbohydrate intake will replenish depleted liver glycogen stores. Vitamin D is essential for the utilization of calcium and phosphorus, and the B complex vitamins are needed because of the increased caloric intake and high basal metabolic rate.

Hypertriglyceridemia. Increased blood levels of triglycerides. Current guidelines for fasting plasma triglycerides are as follows: borderline high (≥200 to 400 mg/dL or ≥2.3 to 4.5 mmol/L); high (400 to 1000 mg/dL or ≥4.5 to 11.3 mmol/L), and very high (>1000 mg/dL or >11.3 mmol/L). Levels greater than 500 mg/dL or >5.6 mmol/L may indicate pancreatitis. Hypertriglyceridemia is generally associated with low levels of high-density lipoprotein-cholesterol. Levels greater than 250 mg/dL appear to be a risk factor for peripheral vascular disease if other risk factors are present. A major cause of elevated triglycerides is obesity. Dietary fat may increase triglyceride levels, and in some individuals, simple sugars and/or excessive alcohol may also increase triglycerides. Other causes of elevated triglycerides include uncontrolled diabetes mellitus, hypothyroidism, chronic renal disease, genetic factors, liver disease, and certain drugs. *Nutrition therapy:* reduce weight and maintain desirable body weight. Decrease simple sugars and sugar-containing foods and use complex carbohydrates high in fiber. Restrict total fat to 30% of kilocalories, saturated fat to 10% of kilocalories, and cholesterol to less than 300 mg/day. Regular exercise is recommended. For persons with serum triglyceride greater than 250 mg/dL, alcohol intake should be substantially reduced or eliminated; abstinence from alcohol is advised if greater than 500 mg/dL. Patients with severe hypertriglyceridemia and chylomicronemia may require a very low-fat diet (10% to 20% of kilocalories) to prevent pancreatitis.

Hypervitaminosis. Vitamin toxicity; a condition in which the level of a vitamin in the blood or tissue is high enough to cause undesirable symptoms. Hypervitaminosis has long been associated with excessive intake of the fat-soluble vitamins, especially vitamins A and D, which are not generally excreted from the body. Toxic effects have also been observed with some of the water-soluble vitamins when these are taken in excessively high therapeutic doses. For details, see under each vitamin.

Hypoalbuminemia. Abnormally low serum albumin concentration. The normal range is 3.5 to 5.0 g/dL (35 to 50 g/L). Albumin levels of 2.8 to 3.4 g/dL (28 to 34 g/L) are considered mild depletion; 2.1 to 2.7 g/dL (21 to 27 g/L), moderate depletion; and below 2.1 g/dL (21 g/L), severe depletion. Depressed serum albumin concentrations have been associated with both decreased albumin synthesis and increased albumin degradation. Conditions associated with decreased albumin synthesis include malnutrition, cirrhosis, carcinoma, hypothyroidism, and acute stress due to surgery, trauma, burns, and infection. Hypoalbuminemia may also follow exposure to various hepatic toxins, including alcohol and carbon tetrachloride. Conditions associated with albumin degradation are those leading to marked catabolism. Hypoalbuminemia results in impaired healing of soft and bony tissues, decreased resistance to infection, depressed gastric and intestinal motility, impaired intestinal absorption of water and electrolytes, and dependent edema and ascites. *Nutrition therapy:* daily allowance for protein per kilogram body weight is 1.2 to 1.3 g for mild depletion, 1.4 to 1.7 g for moderate depletion, and 1.8 to 2.0 g for severe depletion. If there is albuminuria, an amount of protein equal to that lost in the urine should be added to the calculated daily protein allowance. Use protein of high biologic value and ensure sufficient calories to spare protein as energy source. See *Proteinuria*.

Hypocalcemia. Abnormally low blood calcium level; may be due to hypoparathyroidism, chronic renal failure, chronic use of anticonvulsants, vitamin D-deficient rickets, and malabsorption syndromes. Hypocalcemia may result in cardiac cramps, seizures, choreiform movements, increased neuromuscular irritability, and paresthesias of the extremities. *Nutrition therapy:* if intravenous feeding is initially required, give Ca^{2+} gluconate. Provide adequate vitamin D_3. Select foods enriched with calcium.

Hypochloremia. Abnormally low plasma chloride level; may be dilutional with hyponatremia, as in expanded extracellular fluid following trauma and water retention, or due to chloride loss from the gastrointestinal tract, as in vomiting and gastric suctioning, adrenal steroid administration with sodium retention and potassium and chloride loss in urine, and diuretic use with loss of chloride in excess of sodium.

Hypochlorhydria. Abnormally low amount of hydrochloric acid in the stomach; observed in pernicious anemia, sprue, chronic gastritis, and pellagra. Some cases occur in nephritis, diabetes, cholecystitis, and cancer.

Hypoglycemia. Condition characterized by abnormally low blood glucose level. *Spontaneous hypoglycemia,* which occurs without the administration of exogenous insulin, is brought about by any of the following etiologic factors; *hyperinsulinism* (e.g., tumor or hypertrophy of the pancreas), *hepatic disease* (toxic hepatitis and von Gierke's disease), *adrenal hypofunction* (e.g., Addison's disease), *pituitary hypofunction* (e.g., Simmonds' disease), and certain *inborn errors of metabolism* (e.g., sugar malabsorption and leucine-induced hypoglycemia). The symptoms are characteristic of those seen in insulin reaction and include extreme hunger, nervousness, flushing of the skin with profuse sweating, dizziness, palpitations, and apathy. On the basis of *dietary management,* hypoglycemias are grouped into fasting and stimulative types:

Fasting h. Blood sugar level is below 60 mg% before breakfast or after fasting. This may also occur in adrenal or pituitary hypofunction, liver diseases, and other conditions. In contrast to the stimulative type, fasting hypoglycemia becomes more severe if carbohydrate intake is restricted. *Nutrition therapy:* simple sugars will rapidly correct symptoms. Provide a constant and regular supply of glucose, with small, frequent feedings high in carbohydrates and protein.

Reactive (stimulative) h. Also called "spontaneous hypoglycemia" or "functional hyperinsulinism"; hypoglycemia in the absence of an organic lesion. Carbohydrate intake stimulates the pancreas to secrete higher than normal levels of insulin. As a consequence, hypoglycemia occurs 2 to 4 hours after meals; there is no hypoglycemia following fasting and omission of meals. *Nutrition therapy:* provide adequate calories based on individual needs. Carbohydrate restriction is not necessary but refined carbohydrates should be avoided, especially sugars and concentrated sweets. Protein and fat should be taken whenever carbohydrate is consumed to delay gastric emptying and to blunt the postprandial insulin response to carbohydrate. Rather than three large meals, it is better to divide the daily food allowance into six small protein-containing meals. A diet high in soluble fiber is beneficial in controlling blood glucose levels. Restrict alcohol and caffeine according to tolerance.

Hypoglycemia, diabetes mellitus. The most common acute complication among diabetics which may be a result of too much medication, too little food, too much exercise without adequate caloric supply, or alcohol consumption in the absence of food. When blood glucose falls within 50 to 60 mg/dL, the early signs of hypoglycemia are dizziness, pallor, sweating, nervousness, hunger, weakness, and tachycardia. When blood glucose falls below 40 mg/dL, impaired central nervous system functions are manifested, such as confusion, lethargy, slurred speech, lack of motor coordination, and mood changes. Eventually seizures and unconsciousness develop, unless treatment is given immediately. Insulin-dependent diabetics are more prone to hypoglycemia, although it may also occur in non-insulin-dependent persons. *Nutrition therapy:* the initial treatment for the alert person is to give 10 to 20 g of simple carbohydrate, such as oral glucose, orange juice, or apple juice. Wait 15 minutes and retest the blood glucose; if still less than 70 mg/dL, give another 15 g of carbohydrate. Food sources providing 15 g of carbohydrate include ½ cup of apple or orange juice; ¾ cup of gingerale; ½ cup of cola drink; 1 tablespoon of sugar or honey; ½ cup of regular gelatin; 2 tablespoons of raisins; 6 jelly beans; 5 hard candies (Life Savers); and 7 Junior Mints®. When blood glucose has improved to ≥70 mg/dL, give 8 oz of low-fat milk and half a sandwich. In severe hypoglycemia when the patient is comatose or unable to swallow, the administration of glucagon or intravenous glucose is necessary.

Hypokalemia. Abnormally low plasma potassium level; may be due to decreased potassium intake, gastrointestinal tract losses (diarrhea, prolonged vomiting or gastric suction, small bowel fistulas), renal losses with potassium-depleting diuretics, and metabolic alterations with secondary potassium loss (surgical trauma, sepsis, burns). The undesirable effects of hypokalemia include impaired glucose tolerance with impaired insulin secretion, muscle weakness, metabolic alkalosis, and heart failure. *Nutrition therapy:* The diet should include potassium-rich foods such as orange juice, bananas, milk, and potatoes. Potassium supplements or use of potassium chloride salt substitutes may be necessary. Severe hypokalemia may require parenteral administration of potassium.

Hypokinesis. "Deconditioning" of the body due to lack of exercise or physical activity. Prolonged physical inactivity results in stiffness, fatigue, weakness, sensitivity, incoordination, instability, muscular atrophy, ataxia, myocardial ischemia, urolithiasis, and osteoporosis. See also *Nutrition, motor performance*.

Hypomagnesemia. Abnormally low blood magnesium level; may be due to malabsorption (inflammatory bowel disease, gluten enteropathy, radiation enteritis), increased loss (renal magnesium wasting, chronic diarrhea, laxative abuse), inadequate intake, or endocrine disorders. Primary hypomagnesemia is a rare genetic defect due to the inability of the intestinal mucosa to absorb magnesium. Symptoms include muscle cramps, athetoid movements, jerking, tetany with facial twitching, and disorientation. *Nutrition therapy:* encourage foods high in magnesium such as whole grains, legumes, beans, potatoes, green leafy vegetables, and seafoods. Magnesium salts are usually given orally or intravenously.

Hyponatremia. Abnormally low blood sodium level; may be due to chronic wasting illnesses (cancer, liver disease, ulcerative colitis), abnormal loss of sodium without adequate replacement (excessive sweating, adrenal insufficiency, diarrhea), or prolonged, strict sodium restriction with drugs (chlorothiazide, mercurial diuretics, ethacrynic acid, or furosemide). Symptoms include loss of appetite, nausea, vomiting, weakness, irritability, confusion, and muscle weakness. *Nutrition therapy:* treatment with intravenous infusion of a balanced solution aims to restore fluid and electrolyte balance. Give salty broths if hyponatremia is due to salt depletion. No specific diet is required in dilutional hyponatremia, although diuretics or fluid restriction may be beneficial. The most common cause of hyponatremia is fluid overload. Restrict fluids in water intoxication.

Hypophosphatasia. A genetic metabolic disorder resulting from serum and bone alkaline phosphatase deficiency leading to hypercalcemia. Clinical manifestations include severe skeletal defects resembling vitamin D-resistant rickets, dyspnea, cyanosis, failure to thrive, beading of the costochondral junction, and rachitic bone changes. *Nutrition therapy:* none recommended. Vitamin D and phosphorus supplements may be beneficial.

Hypophosphatemia. Abnormally low blood phosphate level; may be due to diminished intake and absorption (starvation, malabsorption syndrome, small bowel bypass) and increased loss (hyperparathyroidism, renal tubular defects, uncontrolled diabetes mellitus). Complications ascribed to hypophosphatemia include osteomalacia, congestive heart failure, respiratory failure, and kidney stones. *Nutrition therapy:* large doses of vitamin D and phosphorus salts given orally. Dairy products, whole grain cereals, legumes, and beans are good sources of dietary phosphorus.

Hypophysis. Preferred name for *pituitary gland*. An endocrine gland about the size of a lima bean located beneath the brain and protected by a saddlelike depression called the *sella turcica*. It is composed of three portions: the anterior, intermediate (*pars intermedia*), and posterior lobes. Each lobe secretes important hormones that regulate vital processes in the body. The anterior lobe secretes somatotropin, thyrotropin, gonadotropins (luteinizing hormone, follicle-stimulating hormone, luteotropic hormone), and adrenocorticotropic hormone. The intermediate lobe secretes the melanocyte-stimulating hormone, and the posterior lobe secretes oxytocin and vasopressin. The hypophysis is called the "master gland" or the "king of all glands" because it regulates the action of many of the other endocrine glands through the different hormones elaborated by its three lobes. See Appendix 11B.

Hypopituitarism. Decreased activity of the *hypophysis* (pituitary gland) caused by a tumor, infarct, hemorrhage, or atrophy. Forms include pituitary myxedema, which is due to lack of the thyroid-stimulating hormone and is similar to myxedema of hypothyroidism; panhypopituitarism, which involves all the hormonal functions of the hypophysis (as in Simmonds' disease); and pituitary dwarfism, which is characterized by cessation of growth and diminished metabolic activities. *Nutrition therapy:* a high-calorie, high-protein diet is recommended, supplemented with vitamins and minerals. Provide frequent, small feedings and avoid dehydration.

Hypotension. Reduced arterial systolic blood pressure below normal. May result from injection of drugs that lower blood pressure, hemorrhage or shock, and suppression of renal blood flow. Primary hypotension is not a disease but is common among young asthenic women. Secondary hypotension is associated with diseases such as myocardial infarction, vascular accidents, cachexia, and fever. Postural hypotension may occur in some debilitated or elderly persons when they assume the upright position and there is exaggerated venous pooling.

Hypothalamus. Area lying at the base of the brain just below the thalamus. It is responsible for the maintenance of body temperature, blood pressure, water regulation, control of satiety and appetite, and other basic functions necessary to life. Because of its close anatomic connection to the pituitary gland, it has been implicated in the control of endocrine gland function.

Hypothyroidism. Endocrine disorder resulting from the decreased activity of the *thyroid gland*. The effects of a decreased thyroid hormone supply in the body are *myxedema* in adults, particularly in women, and *cretinism* in children. Clinical signs are reduced basal metabolic rate as low as 40% below normal; puffy face, hands, and eyelids; easy fatigability; apathy and dullness; and reduced gastrointestinal motility. Blood lipids are often increased. Children have retarded growth and development. Drug therapy is the treatment of choice. *Nutrition therapy:* restrict calories for obese persons. Limit intake of fats and cholesterol-rich foods. Liberal fluid and fiber intake prevents constipation.

Hypovitaminosis. See *Avitaminosis*.

I

I. Chemical symbol for iodine.

Iatrogenic. Term meaning "caused by medical treatment or diagnostic procedures." *Iatrogenic malnutrition* is an induced nutritional deficiency due to drug therapy or certain medical procedures, unless carefully monitored to prevent any nutritional complication. The condition is observed with prolonged use of oral contraceptives, antibiotics, and anticonvulsive drugs; total parenteral nutrition lacking certain elements, particularly trace minerals; complications from ostomies; and gastric stapling in morbid obese patients.

IBC. Abbreviation for *iron-binding capacity*.

IBS. Abbreviation for *irritable bowel syndrome*.

Ibuprofen. A nonsteroidal anti-inflammatory agent; also an antipyretic and analgesic. Used for treatment of rheumatoid arthritis and osteoarthritis and for relief of aches and pains. Its most frequent adverse effects involve the gastrointestinal tract and it may cause anemia due to gastrointestinal bleeding and peptic ulceration; it may also cause stomatitis, epigastric pain, cramping, decreased appetite, and weight gain due to fluid retention. It should be taken with milk or food. Alcohol should be avoided. Brand names include Advil®, Medipren®, Midol®, Motrin®, Nuprin®, and Pamprin®.

IBW. Abbreviation for *ideal body weight*. See *Weight*.

ICDA. Abbreviation for *International Committee of Dietetic Associations*.

Icteric. Pertaining to icterus or jaundice. See *Jaundice*.

Icteric index. Measure of the yellow color in blood plasma; expressed as a comparison with the color of the serum with the color of a 1:10,000 potassium dichromate solution. The normal range is from 4 to 6. Values higher than 6 are indicative of jaundice or icterus.

IDDM. Abbreviation for insulin-dependent diabetes mellitus. See under *Diabetes mellitus*.

Idiopathic. 1. Without any known origin; self-originated. 2. Pertaining to disease of unknown cause (e.g., idiopathic celiac disease).

Idiopathic thrombocytopenic purpura (ITP). Bleeding into the skin, gums, nose, and other organs caused by platelet destruction. Acute ITP in children may be due to a viral infection that lasts for periods ranging from several weeks to a few months. Chronic ITP is more common in adults and adolescents and lasts for a longer period. *Nutrition therapy:* since treatment is usually by drugs (corticosteroids) and splenectomy, nutritional guidelines are directed to the side effects of corticosteroids and to surgical care.

IDL. Abbreviation for *intermediate-density lipoprotein*.

IDPN. Abbreviation for *intradialytic parenteral nutrition*.

IEM. Abbreviation for *inborn errors of metabolism*.

Ifosfamide. An antineoplastic agent. Side effects include nausea and vomiting. Excreted in breast milk. Breast feeding is not recommended during chemotherapy because of risk of serious side effects. The brand name is IFEX®.

Ileitis. Acute or chronic inflammation of the lower ileum, although other parts of the intestine may also be affected by edema, fibrosis, and ulceration. Clinical symptoms include abdominal cramps, loss of weight, bloody diarrhea, and progressive anemia. *Nutrition therapy:* correct the diarrhea and lessen the abdominal pain and irritation with an elemental diet for acute cases. Correct anemia and weight loss with a high-calorie, high-protein diet, with vitamin and mineral supplementation. For chronic cases, maintain a normal diet but monitor any fat and lactose intolerance. Supplementary iron and B complex vitamins may be needed.

Ileostomy. A surgical opening is created from the ileum to empty fecal matter outside the abdomen via an intestinal tube. May be needed in patients suffering from cancer of the intestines, Crohn's disease, or ulcerative colitis. Decreased transit time and undigested food excretion are common in ileostomates. The output from the stoma is more liquid than in a *colostomy* and there is greater loss of water, electrolytes, and other minerals. *Nutrition therapy:* a low-residue diet is given several days before the operation, and only liquids 24 hours before surgery to decrease intestinal residue. After surgery, the diet progresses as follows: clear to full liquids, soft low residue, and regular as tolerated. Progress to a regular diet after the bowel edema has subsided. Begin with foods low in fiber and gradually increase fiber intake as tolerated. Avoid foods that are gaseous or that may cause stoma obstruction. Encourage fluid intake (at least 2 liters/day). An average ileostomate drains about 0.5 to 1 liter of fluid per day from the intestinal contents, carrying with it sodium and potassium salts. Electrolyte and fluid losses must be replaced. If the patient is dehydrated or urine output is low, encourage an additional 1 to 2 liters/day of fluids, preferably electrolyte-supplemented beverages. Bile salt deficiency may require fat restriction to prevent steatorrhea and gallstone formation. If weight loss occurs due to reduced fat intake, give MCT oil or nutritional supplements with medium-chain triglycerides to provide additional calories. Supplemental fat-soluble vitamins may be needed.

Ileum. Third and last portion of the small intestine, extending from the jejunum to the large intestine. It is the site for absorption of bile salts, vitamin B_{12} and intrinsic factor, disaccharides, and mineral salts. See also *Absorption, nutrients*.

Ileus. An obstruction of the intestines; may be a mechanical blockage or caused by immobility of the bowel. In most patients, ileus of the gastrointestinal tract is localized to the stomach and colon. The ileus associated with severe closed head injury produces gastric atony that resolves after approximately 4 days. Absorption can occur without significant distension if nutrients can be delivered directly to the small intestinal mucosa.

IM. Abbreviation for "intramuscular."

Imipramine. A tricyclic antidepressant. Has a peculiar taste and may cause anorexia, stomatitis, weight loss, and riboflavin depletion. It is excreted in milk in small quantities. Alcohol should be avoided completely. The drug can increase the intoxicating effects of alcohol. Brand names include Impril®, Norfranil®, and Tofranil®.

Immobilization. Patients who need prolonged bed rest, and those who have paralysis, fractures, multiple trauma, and the like, excrete large amounts of calcium and go into negative nitrogen balance. Because of the demineralization during immobilization, the serum calcium level is elevated. This does not require increased dietary intake, but can be reversed by quiet standing for a few hours or by changes in weight-bearing activities.

Immun-Aid®. Brand name of a specialized formula for metabolically stressed and immunocompromised patients, including those with trauma, burns, AIDS, sepsis, and cancer. Contains arginine, glutamine, and branched-chain amino acids. See Appendix 39.

Immune system. The organs of the immune system include the bone marrow, thymus, lymph nodes, lymphatic vessels, tonsils, adenoids, and spleen. These organs are also known as the "lymphoid organs" because of their role in lymphocyte maturation and activation. The primary function of the immune system is to protect the body against foreign invaders called antigens. The immune response leads either to secretion of specific antibodies involving B cells or B lymphocytes derived from bone marrow (humoral immunity), or to cell-mediated killing of the pathogen involving T cells or T lymphocytes that mature in the thymus (cellular immunity). The immune system also protects the body from invasion by creating local barriers and inflammation. The local barriers provide chemical and mechanical defenses through the skin, the mucous membranes, and the conjunctiva. Inflammation draws leucocytes to the site of injury where these phagocytes engulf the invading organisms. The humoral response and the cell-mediated response develop if these first-line defenses fail or are inadequate to protect the body.

Immunocompetence. The ability of an immune system to mobilize antibodies and other responses to protect the body from invading organisms or foreign stimuli. Individuals who are immunocompromised include those with cancer and on chemotherapy, human immunodeficiency virus (HIV) infection, thermal injury, diabetes, severe trauma, and those receiving immunosuppressive therapy. Immunocompetence is lowered in protein-energy malnutrition and other nutrient deficiencies. Immunologic indexes associated with nutritional status are the total lymphocyte count and delayed cutaneous hypersensitivity. See *Delayed cutaneous hypersensitivity* and *Total lymphocyte count*. See also *Immunonutrition*.

Immunoglobulin (Ig). An antibody produced by the lymphoid tissue in response to bacteria, viruses, and other antigens. There are five kinds of immunoglobulin: IgA, which is the major antibody in the mucous membrane lining the intestines and in the bronchi, saliva, and tears; IgD, which increases during allergic reactions to milk and various toxins; IgE, which reacts with antigens that cause hypersensitivity reactions characterized by wheal and flare; IgG, which reacts to bacteria, fungi, and viruses,; and IgM which is found in circulating fluids and is the dominant antibody in ABO incompatibilities.

Immunonutrition. Or nutritional immunology. Term used to describe the effect of nutrients on the immune system. Nutrients influence the immune response in the following manner: *protein* deficiency results in decreased production of immunoglobulins and depressed cellular immunity, including cutaneous hypersensitivity and phagocyte function; *arginine* stimulates lymphocyte mitosis and suppresses tumor growth; *glutamine* is a substrate for macrophages and T lymphocytes; *ribonucleotides* play a vital role in T lymphocyte-mediated immunity; *omega-3 fatty acids* form prostaglandins and leukotrienes and also lessen immunosuppression associated with burns; *omega-6 fatty acid* deficiency results in depressed antibody responses and lymphoid atrophy; excessive *polyunsaturated fatty acid* (PUFA) induces atrophy of lymphoid tissue and delays T cell immune response; *vitamin A* deficiency reduces lymphocyte response to antigens and mitogens; *vitamin B_6 deficiency is associated with impaired cellular immunity and decreased antibody response; vitamin C* deficiency impairs phagocyte function and cellular immunity; *vitamin D* deficiency causes anergy in the delayed hypersensitivity skin test; *vitamin E* deficiency decreases antibody response to T cell-dependent antigens; *copper* deficiency is associated with increased rate of infections, impaired antibody response, and depressed reticuloendothelial system function; *iron* is necessary for optimum neutrophil and lymphocyte function; *magnesium* deficiency causes depressed immunoglobulin level, thymic hyperplasia, and impaired humoral and cell-mediated response; *manganese* is required for normal antibody synthesis and secretion; *selenium* deficiency reduces antibody responses; *zinc* deficiency is associated with susceptibility to infection, depressed circulating thymic hormone, and abnormal cell-mediated immunity; and *calorie* deprivation is associated with decreased circulating immune complex levels. *Protein-energy* malnutrition affects all levels of the immune response.

Immunosuppressive therapy. Generally consists of a combination of pharmacologic agents, including prednisone, cyclosporine, and azathioprine. Managing the nutritional side effects of immunosuppressive therapy requires a diet high in protein and restricted in simple carbohydrate to counteract the catabolic effects of high-dose steroids and to minimize glucose intolerance. Carbohydrates should supply no more than 50% of calories; emphasize complex carbohydrates and fiber. Because corticosteroid and cyclosporine therapy may also contribute to atherosclerotic disease, the diet should also be low in cholesterol and saturated fats. Fat should supply no more than 35% of kilocalories, with a polyunsaturated/saturated ratio of >1 and cholesterol <400 mg/day. Avoid alcohol. See also *Diet, low-microbial*.

Impact®. Brand name of a tube-feeding formula for critically ill patients with depressed immune function. Has added arginine, omega-3 fatty acids, and RNA nucleotides. Impact® with Fiber contains 10 g of fiber per liter. See also Appendix 39.

Inanition. Wasting of the body due to complete lack of food; a state of starvation.

Inborn error of metabolism (IEM). A large group of inherited disorders due to a deficiency or absence of a protein involved in the metabolic pathway either as an enzyme, carrier, receptor, or other functional role. Examples are galactosemia, phenylketonuria, and tyrosinemia. Treatment for many of these disorders may require dietary intervention. The benefits obtained from dietary modification vary from marginal relief of symptoms to effective palliation and control. See *Nutrition, genetics* and Appendix 44.

Incomplete protein. See *Protein classification* and *Protein quality*.

Index of nutrient quality (INQ). A ratio indicating the *nutrient density* of a food, calculated as follows:

$$INQ = \frac{\%\ \text{RDA of a nutrient for an individual}}{\%\ \text{energy requirement for an individual}}$$

Thus, INQ varies among different individuals. In general, if a food has an INQ of 1 and if a serving portion provides at least 2% of the recommended dietary allowance (RDA) for a nutrient, that food is said to be a good source of the nutrient. Food with an INQ of 1.5 and that furnishes 10% of the U.S. RDA per serving portion is considered an excellent source of the nutrient.

Indian Health Service. Nutrition and dietetic services directly managed by the Public Health Service (PHS) or contracted with the tribal councils. The food distribution program in Indian reservation operates like the Food Stamp Program and is funded by the U.S. Department of Agriculture.

Indigestion. Also called *dyspepsia*. Faulty or incomplete digestion of food. See *Dyspepsia*.

Indomethacin. A nonsteroidal anti-inflammatory agent for the relief of joint pains and inflammation associated with arthritis. Adverse nutritional effects include decreased serum ascorbic acid level, decreased absorption of amino acids and xylose, iron deficiency anemia secondary to gastrointestinal blood loss, and anorexia. The drug may also cause sodium and fluid retention, abdominal distress, and bloating. Excreted in breast milk. Brand names are Indocin® and Indocin®-SR.

Infant feeding. Breast feeding is generally accepted as the most desirable method of feeding an infant. Human milk is considered an ideal food. Except for iron and vitamins C and D, it contains adequate amounts of all nutritional factors needed by the newborn infant. The other advantages of breast feeding are as follows: babies are less likely to be overfed; breast milk is always fresh, bacteriologically safe, and nonallergenic; dental development is promoted; and breast milk contains immune bodies that make the infant more resistant to infection. Compared to cow's milk, breast milk proteins are more easily digested and its higher cholesterol content helps develop the infant's central nervous system. The mother's uterus also contracts to normal size more rapidly when she breast feeds. Other maternal benefits are convenience, lower cost, and suppression of ovulation. Contraindications to breast feeding include a history of tuberculosis, severe chronic illness, mastitis, insufficient milk production, certain medications, drug abuse, poor maternal health, acute infections, emotional and mental stress, alcoholism, and another pregnancy. In such cases, artificial feeding or bottle feeding with cow's milk or a commercially prepared infant milk formula is satisfactory for infants. Sometimes mixed feeding or a combination of breast feeding and artificial feeding is used, especially when breast milk is not sufficient or when the mother works outside the home. The introduction of solid foods to supplement milk is usually started at 4 months of age with iron-fortified rice cereal, followed by strained vegetables and fruit juices at 5 months; protein foods such as cheese, lean meat, and egg yolk at 6 to 8 months; and whole egg at 10 to 12 months of age. The kind, amount, and consistency of foods introduced may vary among infants. See also *Nutrition, infancy*. For the composition of milk and selected formulas for infant feeding, see Appendix 37. See also *Nutrition, infancy*.

Infantile colic. A disorder among infants characterized by unexplained paroxysms of crying and even agonized screaming, irritability, and distended stomach. About 25% of infants who develop colic in the first weeks of life may outgrow it after 4 months of age. The exact etiology is not known, although hypersensitivity to cow's milk and excessive intestinal gas are possible causes. It has been observed that the mother's smoking habit and maternal exposure to drugs may cause infantile colic. *Nutrition therapy:* if there are no symptoms of food intolerance, dietary changes are not necessary. The baby should be fed in the upright position and trained for slow sucking to minimize the amount of swallowed air by limiting the periods of sucking to 10 minutes.

Infantile eczema. Also called "atopic dermatitis." A disorder of unknown etiology seen in infants. Associated with physical, emotional, and hypersensitivity factors. Some symptoms are fragility of the skin with lesions, accompanied by severe itchiness, dryness, and crusting; capillary dilatation; and edema and erythema. *Nutrition therapy:* if hypersensitivity to milk is involved, give a hypoallergenic formula. If the infant is ready for solid foods, determine which items are allergens and follow the approach used for food allergy. See *Food allergy*.

Infectious mononucleosis. Also called "glandular fever." May be caused by the Epstein–Barr herpes virus. Characterized by chills, headache, sore throat, fever, stomach aches, chest pains, breathing difficulty, and visible swelling of the neck and other glands. *Nutrition therapy:* maintain fluid, protein, and electrolyte balance. Provide liberal calories with an N/C ratio of 1:150 (N = nitrogen in g/day and C = kcal/day).

Inflammatory bowel disease (IBD). General term for chronic ulcerative colitis and Crohn's disease, which are two distinct disorders, although both involve mucosal inflammation of the intestines. Crohn's disease (regional enteritis) usually affects the terminal ileum and right colon, and ulcerative colitis usually affects the rectum and descending colon. Protein-energy malnutrition and other nutrient deficiencies are common in patients with IBD. For further details. see *Crohn's disease* and *Ulcerative colitis*.

INH. Abbreviation for isonicotyl hydrazide. See *Isonicotinic acid hydrazide*.

Injury. Wound or damage, as in tearing or rupture of tissue. See *Wound*.

Inositol. Water-soluble, cyclic, 6-carbon compound closely related to glucose. It exists in nine forms, but only myoinositol demonstrates any biologic activity. Myoinositol is present in relatively large amounts in the cells of practically all animals and plants. In animal cells, it occurs as a component of phospholipids; in plant cells, it is found as *phytic acid,* an organic acid that binds calcium, iron, and zinc in an insoluble complex and interferes with their absorption. In addition to occurring in foods, inositol is synthesized in the cells. It is stored largely in the brain, muscles, liver, and kidneys. Like choline, inositol has a lipotropic effect and exists in cells as a phosphatide (phosphatidylinositol). It is not classified as a vitamin, because it is present in practically all plants and animal tissues in concentrations higher than those normally associated with vitamins. There is no demonstrable requirement for inositol in humans, although there is growing evidence of altered inositol metabolism in certain clinical situations likely to induce deficiency, such as diabetes mellitus and multiple sclerosis. Experimental animals fed semipurified diets lacking in inositol develop alopecia, dermatitis, retarded growth, and fatty liver. To prevent inositol deficiency, the American Academy of Pediatrics has recommended the addition of inositol to formulas based on non-cow's milk protein.

Insensible fluid losses. Fluid excreted from the body not readily seen or felt, such as losses from *insensible perspiration,* vaporization from the lungs, and losses from fecal matter. In monitoring daily fluid intake, add 500 mL as insensible fluid losses to the previous day's urinary output.

Insensible perspiration. Water lost through the skin that is not noticeable because evaporation takes place immediately. It is important for the maintenance of body temperature. Water lost daily through insensible perspiration has been estimated to be 15 mL/kg for adults and 30 mL/kg for infants. About 400 mL of water is lost by diffusion in insensible perspiration, in contrast to about 300 mL in sweating. This loss of body weight through vaporization of water can be measured and may be used as an indirect method for estimating the basal heat production of a person.

Insulin. Hormone secreted by the beta cells of the islets of Langerhans of the pancreas. It is a protein with a molecular weight of 6000 and is composed of two polypeptide chains, A and B, with 21 and 30 amino acids, respectively. Insulin secretion is stimulated by carbohydrates, amino acids, and pancreozymin. Before active insulin is released, it exists as proinsulin. The latter has a connecting peptide called "C-peptide" which is cleaved off, leaving the active insulin molecule. The C-peptide has no hormonal function but is useful as a diagnostic test for *diabetes mellitus*. The only hormone that lowers blood sugar, insulin inhibits the breakdown of glycogen into glucose; promotes glycogenesis, or the conversion of glucose to glycogen in the liver; fosters lipogenesis, or the formation of fat from glucose; and increases cell permeability to glucose. In the muscles, insulin stimulates protein synthesis by inhibiting protein breakdown to glucose. Lack of insulin leads to *diabetes mellitus*.

Insulin-dependent diabetes mellitus (IDDM). A form of diabetes mellitus that requires insulin administration to maintain normoglycemia. Persons with IDDM are prone to ketosis. See *Diabetes mellitus*.

Insulin preparations. Insulin is commercially prepared in either amorphous or crystalline form from animal pancreas (beef, pork, sheep) or by biosynthesis with *Escherichia coli,* using recombinant DNA technology. The latter is called human insulin and has the advantage of eliminating allergic reactions that may occur in using animal insulins. The preparations vary in solubility, onset, peak, and duration of action. They are sold either unmodified (regular insulin), in combination with basic proteins (globin insulin and protamine insulin), or with crystalline zinc salts (protamine zinc insulin) to make the compound less soluble and less absorbable, thus effecting a longer duration of action. Fever, infection, pregnancy, surgery, and hyperthyroidism significantly increase insulin requirement. Vomiting, kidney and liver diseases, and hypothyroidism may decrease required insulin dosage.

Insulin pump. A means of providing insulin by inserting a syringe subcutaneously; this is a portable unit that can be operated manually or attached to a microcomputer that regulates insulin flow. The main purpose of this infusion device is to maintain euglycemia for 24 hours without interfering with the person's daily activities. Only regular or rapid-acting insulin is used for pump therapy, which can be given as a continuous infusion, called the "basal rate," or as an intermittent bolus given prior to eating to control the postprandial rise in blood glucose.

Insulin shock. The preferred term is "insulin reaction" or "hypoglycemic episode"; a reaction of the body due to a very low blood sugar level because of overdosage of insulin. It is characterized by a feeling of hunger, weakness, nervousness, double vision, shallow breathing, sweating, pallor, headache, and dizziness. If the blood glucose level falls below 40 mg/dL, the patient develops mental confusion, slurred speech, muscular twitching, convulsions, loss of consciousness, and eventually coma. *Nutrition therapy:* see under *Hypoglycemia.*

Insulin unit. The physiologic activity of insulin is expressed in *international units (IU).* One IU is equivalent to 0.125 mg of the international standard preparation or $\frac{1}{23}$ mg of a standard preparation of crystalline zinc insulin.

Intensive insulin therapy (IIT). A method of diabetes management using multiple injections of insulin each day or using a pump for continuous subcutaneous insulin infusion. Proven to be effective for normoglycemia and in preventing or delaying chronic complications of insulin-dependent diabetes mellitus. It requires more frequent monitoring of blood glucose and balancing of food, activity, and insulin requirements. See also *Diabetes Control and Complications Trial (DCCT).*

Interferon. A family of glycoproteins that interferes with the replication of various viruses (hence the name) and also affects cell growth and immunologic processes by activating or suppressing selected components of the immune system. Interferons are released by cells in response to a variety of agents, including viruses, microorganisms, and endotoxins.

INTERHEALTH. An international collaborative project initiated by the World Health Organization's Division of Noncommunicable Diseases. Participating nations work toward prevention and control of common risk factors for noncommunicable diseases (NCD), such as cardiovascular disease, cancer, diabetes, and osteoporosis. The INTERHEALTH Nutrition Initiative aims to evaluate global trends in food and nutrient intake related to chronic disease prevention, assess common international nutrition policies relevant to NCD risk reduction, and explore the characteristics of lifestyle interventions for NCD prevention.

Intermediary metabolism. Synthesis (anabolism) and degradation (catabolism) of the cell constituents of living organisms. In the intact cell, both processes go on simultaneously, and energy released from the degradation of some compounds may be utilized in the synthesis of other cellular components. In a broad sense, intermediary metabolism refers to all the chemical reactions taking place inside the body, ranging from the ingestion of foodstuffs to the discharge of ultimate chemical products and excretion of metabolic end products. See Appendices 12, 13, 14, and 15.

Intermediate-density lipoprotein (IDL). A subclass of lipoprotein with a density between that of very-low-density lipoprotein and low-density lipoprotein; classified as the broad beta fraction on the basis of electrophoresis. It is formed in plasma from the action of lipoprotein lipase on chylomicrons and very-low-density lipoprotein. Cholesterol and phospholipids are the lipids present in the greatest amount.

Intermittent peritoneal dialysis (IPD). Procedure used for short-term peritoneal dialysis requiring 36 to 40 hours of treatment; usually given for 8 to 10 hours at night, four times a week. Up to 3 L of peritoneal dialysate are infused, with two cycles performed each hour. This procedure can be performed manually or automatically with a machine. The use of a cycler for IPD is called automated peritoneal dialysis or APD.

International Committee of Dietetic Associations (ICDA). Formed in 1952 with the American Dietetic Association as a founding member. It aims to raise the level of the dietetic profession in its member countries. There are about 30 member countries that sponsor the International Congress of Dietetics (ICD) every 4 years. Starting in 1994, ICDA will publish a new international letter linking dietetic professionals worldwide with nutrition news exchanges of global interest.

International Life Sciences Institute (ILSI). A nonprofit, worldwide foundation established in 1978 to advance the understanding of scientific issues relating to nutrition, food safety, toxicology, and environmental safety. It is affiliated with the World Health Organization and with the Food and Agriculture Organization for specialized consultation services. See also *Nutrition Foundation, Inc.*

International unit (IU). Figure that represents the biologic activity of a nutrient or substance. It is a specific reference standard of known potency that produces specific effects over a specified period of time in a laboratory animal.

Interstitial. Situated in the interspaces of tissue or between parts. See interstitial fluid under *Water compartment, body.*

Intestinal failure. A functioning gut mass below the minimal amount necessary for the adequate digestion and absorption of nutrients. It is the only absolute indication for total parenteral nutrition. See also *Nutrition, gut integrity.*

Intestinal juice. *Succus entericus.* A straw-colored alkaline fluid secreted by the intestinal mucosa; contains enzymes that complete the hydrolysis of carbohydrate, protein, and fat. Mixed with it are the pancreatic secretions containing enzymes and the hormone enterokinase. See Appendix 11.

Intestinal lipodystrophy. Also called "Whipple's disease." A rare disease with insidious onset, more common in males than in females; characterized by infiltration of the small intestines with macrophages containing glycoprotein and bacilli. Clinical signs include malabsorption, anemia, endocarditis, hypoproteinemia, lymphadenopathy, edema, and abnormal skin pigmentation. Central nervous system involvement may be serious. This systemic disease is usually fatal. *Nutrition therapy:* correct weight loss, anemia, malabsorption, and hypoproteinemia with a high-calorie (about 1.2 to 1.5 times the basal energy expenditure), high-protein (1.5 g/kg/day), and moderately low-fat diet supplying 25% of caloric needs. Supplemental vitamins and minerals, especially iron, calcium, and fat-soluble vitamins, promote recovery. Edema, dehydration, and electrolyte losses should be corrected.

Intestinal lymphangiectasia. Disorder characterized by increased intestinal pressure dilating the lymphatics and discharging fluids into the bowel. Protein loss is lessened when the fluid is digested and reabsorbed. In addition, the absorption of fats and fat-soluble vitamins is reduced. *Nutrition therapy:* provide adequate calories and protein and monitor fat absorption. A diet using medium-chain triglycerides is beneficial. Supplemental fat-soluble vitamins prevent deficiencies.

Intestine. Part of the digestive tract that extends from the stomach (pylorus) to the anus; divided into the small and large intestines.

Large i. The last portion of the gastrointestinal tract, extending from the ileum to the anus. It is about 5 feet long and is divided into three parts: the *cecum, colon,* and *rectum*. Although no digestion takes place in the large intestine, it serves as the site for absorption of water and unabsorbed products of digestion and for temporary storage of feces until they are eliminated. The microflora inhabiting the large intestine can synthesize some vitamins (especially vitamin K) and can hydrolyze crude fiber to some extent. See also *Fiber*.

Small i. The portion that extends from the pylorus to the large intestine at the cecum. It is about 20 feet long and divided into three parts: the *duodenum, jejunum,* and *ileum*. The small intestine is the main site for the digestion and absorption of food. See also *Absorption, nutrients*.

Intradialytic parenteral nutrition (IDPN). A method of infusing glucose and amino acids into the venous drip chamber of the hemodialysis machine. IDPN is considered when oral and enteral suplementation have been unsuccessful in correcting protein-energy malnutrition. The composition of the IDPN solution is based on the extent of depletion and metabolic status of the patient. Approximately 1.0 to 1.5 liters of solution is infused during the dialysis procedure.

Intralipid®. Brand name of a 10% or 20% fat emulsion for intravenous administration; made of soybean oil, egg yolk, phospholipid, and glycerin. Provides 1.1 kcal (at 10%) and 2 kcal (at 20%) per milliliter. See also Appendix 41.

Intraperitoneal parenteral nutrition (IPPN). Infusion of amino acid solutions in place of dextrose to improve the nutritional status of patients undergoing peritoneal dialysis. IPPN is considered in patients undergoing peritoneal dialysis who are not able to tolerate adequate enteral feedings. Substituting 2 liters of amino acids in place of dextrose in one or two dialysis exchanges per day will help maintain positive nitrogen balance and body weight.

Intrinsic factor. A glycoprotein in the gastric juice that combines with vitamin B_{12} and aids in its absorption from the small intestine. A deficiency in this factor results in *pernicious anemia*.

Introlan™. Brand name of half-strength formula for initiating enteral feeding or for making a transition from total parenteral nutrition (TPN). See Appendix 39.

Introlite™. Brand name of half-strength tube feeding formula for the transitional or intolerant feeder; provides 0.5 kcal/mL. See also Appendix 39.

Inulin. Polysaccharide composed of *fructose* units; found in Jerusalem artichokes and dahlia tuber. It is used as a test for kidney function, since it is completely filtered by the glomerulus and is not reabsorbed or excreted by the kidney tubules. See also *Kidney*.

Inversion. Conversion of sucrose in solution into equal amounts of glucose and fructose by the action of acid or enzyme (invertase). The mixture is called *invert sugar*.

Iodine (I). A trace mineral that is a dietary essential; an important constituent of the thyroid hormones *thyroxine* and *triiodothyronine,* which are necessary for several metabolic functions, including lipid, carbohydrate, and nitrogen metabolism; growth development and reproduction; oxygen consumption; and regulation of basal metabolic rate. Iodine also plays an important role in fetal brain development independent of its action via the thyroid hormones. A deficiency in iodine leads to a wide range of diseases that vary in severity with the degree of iodine deficiency, from *simple goiter* with a barely visible to grossly enlarged thyroid gland to *cretinism* with mental retardation. Endemic goiter and the more severe forms of iodine deficiency continue to be a worldwide problem. The introduction of iodized salt in the United States sharply reduced the incidence of endemic goiter, although isolated cases are still seen in certain areas. Various *goitrogens,* such as those found in the cabbage family and cassava, can also prevent the thyroid from accumulating iodine and converting it into active thyroid hormones. The iodine level in foods varies greatly, depending upon the environment in which they are grown and produced. Seafoods are the richest natural source; iodized salt is another reliable source, providing 76 μg of iodine per gram of salt. Iodine is rapidly and almost completely absorbed and transported to the thyroid gland, where it is found in greatest concentration, although all body tissues and secretions contain trace amounts; excretion is mainly in the urine. Excess iodine ingestion is regulated, within limits, by decreased iodine uptake by the thyroid and increased excretion in the urine. Chronic toxicity with grossly excessive intake can cause *thyrotoxicosis,* an enlarged, hyperactive thyroid, seen among some inhabitants of Japan, who consume as much as 25,000 μg of iodine from seaweed, and in Tasmania, among the elderly population accustomed to low iodine intakes, when iodine was substantially increased by iodization of bread. See Appendix 1 for the recommended dietary allowances.

Iodine-131. Radioactive isotope of iodine; it has a half-life of 8 days. Useful in the diagnosis and treatment of thyroid gland disorders, determination of blood plasma volume and cardiac output, and as a diagnostic aid prior to surgery for the location of brain tumors.

Iodine-deficiency disorders (IDDs). Inclusive term for the various effects of iodine deficiency on growth and development. These effects include goiter, cretinism, impaired mental functioning associated with reduced thyroxine level, and increased frequency of stillbirths and infant mortality. Iodine deficiency during pregnancy causes low blood levels of thyroxine in the mother and the fetus, resulting in irreversible impairment of brian development.

Iodine number. Also called "iodine value"; the number of grams of iodine taken up by 100 g of fat. It is a quantitative value that reflects the amount of fatty acids and the degree of unsaturation of a fat or an oil. The value ranges from 10 for coconut oil to 200 for safflower oil.

Iodized salt. Table salt that contains 1 part sodium or potassium iodide per 5000 to 10,000 parts (or 0.01%) of sodium chloride.

Iodopsin. Visual violet; a light-sensitive violet pigment of the cones in the retina that is important for vision. It contains vitamin A.

Iodothyroglobulin. Globulin–iodine complex found in the thyroid gland; serves as the prosthetic group of *thyroxine*.

Iron (Fe). A trace mineral essential to the body; a constituent of hemoglobin, myoglobin, and various oxidative enzymes. Iron is necessary for the prevention of nutritional anemia and plays an important role in respiration and tissue oxidations. The hemoglobin of the red blood cells and the myoglobin of the tissue cells is vital for oxygen transport to the cells and storage within the cells, whereas the iron-containing enzymes within the cells are associated with metabolic oxidation. Good food sources are liver and other glandular organs, meats, eggs, seafoods, whole-grain or enriched cereals, and dried fruits. Cooking in steel woks, the composition of which is about 98% iron, increases the iron content of the food. This should be considered in calculating iron intake. There are two forms of food iron: heme (organic) and nonheme (inorganic). Heme iron is absorbed more efficiently than nonheme iron and is independent of vitamin C and iron-binding chelating agents. The absorption of nonheme iron can be enhanced by ascorbic acid when the two nutrients are ingested together. It can be inhibited by several factors, such as calcium phosphate, bran, phytates, polyphenols in tea, and antacids. Only about 10% of dietary iron is absorbed in the upper part of the small intestine; absorbed iron is transported by blood as *transferrin* to the bone marrow for hemoglobin synthesis, or is removed from blood by cells for use by respiratory enzymes and as cell constituents. Iron is stored in the liver and other tissues as *ferritin* and *hemosiderin*. Absorbed iron is lost only by desquamation from the alimentary, urinary, and respiratory tracts and by skin and hair losses; iron released from the breakdown of hemoglobin is reutilized. In adult men, the total iron loss that needs replacement is about 1 mg/day; in adult women, there is an additional 5 to 32 mg/month menstrual loss. A deficiency of iron results in *anemia* of the hypochromic, microcytic type. Some people are genetically at risk from iron overload, or *hemochromatosis*. About 2000 cases of iron poisoning occur each year in the United States, mainly in young children who ingest the medicinal iron supplements formulated for adults. Other than long-term ingestion of home brews made in iron vessels, there are no reports of iron toxicity from foods in people without genetic defects that increase iron absorption. See Appendix 1 for the recommended dietary allowances.

Iron-binding capacity (IBC). The relative saturation of the iron-binding protein transferrin. This protein is usually 25% to 30% saturated, representing the serum iron content. In iron deficiency there is decreased saturation to less than 18%. The amount of transferrin bound to iron in relation to the amount remaining free to combine with iron determines the iron-binding capacity (IBC), and the amount of additional iron that can be bound to transferrin is called the "unsaturated iron-binding capacity (UIBC)." Both values together represent the *total iron binding capacity (TIBC)*.

Iron overload. Excessive accumulation of iron in the body due to inadequate excretion of iron and the storage capacity is saturated. It is a dangerous condition that may result in damage to the liver, heart, pancreas, and other organs. A serum ferritin level greater than 1000 μg/mL suggests iron overload. See *Hemochromatosis*.

Irradiation. Exposure to or treatment with ultraviolet rays from sunlight, roentgen rays, gamma rays, or radiation from radioactive materials such as cobalt-60. Has various uses, such as conversion of provitamin D to its active form, destruction of microorganisms in food, and therapeutic treatment of malignancies. See *Radiotherapy*.

Irritable bowel syndrome (IBS). Also called "mucous colitis" and "spastic colon." A common gastrointestinal disorder characterized by chronic and recurrent pain or cramping in the abdominal region, bowel dysfunction varying from diarrhea to constipation, and flatulence accompanied by belching and bloating. The exact etiology of IBD is unknown but it is often associated with emotional upsets and long periods of stress. *Nutrition therapy:* avoid gas-forming foods such as beans and the cruciferous vegetables such as broccoli, brussels sprouts, and cabbage; increase dietary fiber intake to 25 to 30 g/day from whole grains, fruits, and vegetables; avoid large meals; eliminate food intolerances, especially lactose-containing foods; and reduce total fat intake. Limit caffeine and alcohol intake. Fatty foods have been shown to exacerbate diarrhea. Multivitamin and mineral supplementation may be necessary.

Islets of Langerhans. Also called "islands of Langerhans." Cellular masses in the pancreas that produce *glucagon* by its alpha cells, *insulin* by its beta cells, and pancreatic polypeptide. Impairment or destruction of these cells is associated with hyperglycemia and diabetes mellitus.

Isoalloxazine. A heterocyclic yellow flavin that is a constituent of riboflavin together with ribose.

Isoascorbic acid. A geometric isomer of ascorbic acid with only slight vitamin C activity. It is a strong reducing agent, and is used in food as an antioxidant and in cured meats to speed up color fixing.

Isocal®. Brand name of a lactose-free, isotonic, low-residue nutritional product for oral or tube feeding; contains calcium and sodium caseinates, soy protein isolate, maltodextrin, soy oil (80%), MCT oil (20%), and lecithin. Also available with a higher nitrogen content (Isocal® HN) for mildly stressed patients; as a high-calorie, high-protein formula (Isocal® HCN) for hypermetabolic patients; and in a flexible pouch (Isocal® II Entri-Pak) for closed enteral feeding systems. See Appendix 39.

Isolan™. Brand name of a standard isotonic and lactose-free formula for oral consumption or tube administration. See Appendix 39.

Isoleucine (Ile). Alpha-amino-beta-methylvaleric acid. An essential branched-chain amino acid rarely limiting in foods. Manifestations of deficiency include weight loss, tremors, skin desquamation, decreased plasma cholesterol, and elevated plasma lysine, phenylalanine, serine, tyrosine, and valine. See also *Maple syrup urine disease*.

Isomil®. Brand name of a milk-free formula for infants with milk protein hypersensitivity and intolerance to lactose or galactose; made of soy protein isolate. Also made without added sucrose (Isomil® SF) for infants and children with sucrose intolerance and with added soy fiber (Isomil® DF) for the dietary management of diarrhea in infants and toddlers. See Appendix 37.

Isoniazid. Generic name for *isonicotinic acid hydrazide*.

Isonicotinic acid hydrazide (INH). An antituberculosis drug chemically related to pyridoxine that acts as an antagonist to vitamin B_6. Its prolonged administration induces pyridoxine deficiency; it may also interfere with vitamin D metabolism and cause niacin and folate depletion. It has a mild monoamine oxidase inhibitor effect and can cause a tyramine-type reaction with certain foods. Other adverse effects include nausea and vomiting, loss of appetite, epigastric distress, dry mouth, and cheilosis. Vitamin B_6 supplementation may be indicated with long-term use. Brand names are Panazid®, Laniazid®, Nydrazid®, and Turbizid®.

Isoriboflavin. An isomer of riboflavin; acts as a metabolic antagonist and competes with the vitamin.

IsoSource®. Brand name of a lactose-free, isotonic formula for oral and tube feeding. Contains 50% of the fat calories as canola oil and 50% as medium-chain triglycerides. Also available with high nitrogen (IsoSource® HN) and very high nitrogen (IsoSource® VHN) for patients with elevated protein needs. IsoSource VHN has 12.5 g/liter dietary fiber from hydrolyzed guar and soy. See Appendix 39.

Isotein HN®. Brand name of a powdered, lactose-free, high-protein product for oral and tube feeding. Low in sodium and contains medium-chain triglycerides. See also Appendix 39.

Isotonic. Or isoosmotic. A solution that has the same solute concentration as that found in the intracellular and extracellular fluid. The osmolarity of an isotonic enteral product is about 300 mOsm/kg.

Isotretinoin. Synthetic analog of vitamin A used in the treatment of severe acne when other treatments are ineffective. Adverse effects resemble those associated with vitamin A toxicity and it may cause liver damage, increased risk of coronary heart disease, and peripheral vascular disease. It is contraindicated during pregnancy and may damage a developing fetus. The brand name is Accutane®.

Isovaleric acidemia. A genetic disorder of leucine metabolism due to a deficiency of the enzyme isovaleryl CoA dehydrogenase. This leads to increased levels of isovaleric acid in the blood and urine. Affected infants have vomiting, metabolic acidosis, mental retardation, and a characteristic odor similar to that of sweaty feet. *Nutrition therapy:* The condition may be treated by restricting the intake of leucine. This is done by lowering the intake of protein until the amount of leucine ingested equals that required for growth. Supplemental glycine and carnitine may be beneficial.

ITP. See *Idiopathic thrombocytopenic purpura*.

ITT. Abbreviation for *intensive insulin therapy*.

IU. Abbreviation for *international unit*.

I-Valex®. Brand name of a leucine-free formula for infants and toddlers (I-Valex®-1) and children and adults (I-Valex®-2) with isovaleric acidemia and other disorders of leucine catabolism. See Appendix 37.

J

J. Symbol for *joule*.

Jaundice. Also called "icterus." The yellowish discoloration of the skin, mucous membranes, and certain body fluids due to the accumulation of bile pigments in the blood. It may be due to increased production of bile pigments from hemoglobin or may result from the failure of the liver to excrete bilirubin because of an injury to the liver cells or obstruction to the flow of bile. Jaundice is generally classified into three types: *hemolytic jaundice* due to abnormally large destruction of red blood cells, as in pernicious anemia, malaria, yellow fever, and other infections; *obstructive jaundice* due to complete or partial interference with the flow of bile anywhere along its course from the hepatic lobules to the duodenum; and *toxic jaundice* due to damage or injury of the liver cells by toxic substances such as various poisons, drugs, and viral infections. In all types of jaundice, treatment should be directed to the cause and not to the symptoms. *Nutrition therapy:* the anorexia, nausea, and vomiting that accompany jaundice require small feedings of foods that are tolerated. In chronic obstructive jaundice, steatorrhea can be controlled by restricting the intake of long-chain fatty acids. The use of medium-chain triglycerides (2 tablespoons/1000 kcal) should be considered in severe cases of steatorrhea, especially when a high caloric intake is needed. Supplementation with fat-soluble vitamins is recommended. See also *Hepatitis*.

Jaw wiring. Or maxillomandibular fixation. Wiring the jaws restricts eating to liquids or liquefied foods that can be taken through the straw. *Nutrition therapy:* cooked soft foods can be blenderized to a liquid consistency. Careful attention should be made to the combination of liquids and supplements that will provide adequate nutrients.

JCAHO. Abbreviation for *Joint Commission on the Accreditation of Healthcare Organizations*.

Jejunal hyperosmotic syndrome. Preferred term for *dumping syndrome*.

Jejunostomy. Creation of an artificial opening through the abdominal wall into the jejunum for feeding. Jejunostomy permits enteral feeding in patients with an upper gastrointestinal tract obstruction or fistula, esophageal reflux, inability to protect the airway from aspiration, ulcerative or neoplastic disease of the stomach, and impaired gastric emptying. A feeding jejunostomy can be placed by surgical, endoscopic, or laparoscopic procedure. *Nutrition therapy:* small, frequent feedings, with extra care to avoid diarrhea and gas formation. Because jejunostomy bypasses the stomach, there is likelihood of the dumping syndrome. Start with dilute or isotonic formula. Gradually increase the feeding to the desired volume and then increase the strength. A formula that is lactose-free and low in osmolarity may be tolerated better. Consider enteral products with added fiber if jejunostomy feeding is for long-term use.

Jejunum. The second portion of the small intestine between the duodenum and the ileum; it is about 8 feet long. For its role in digestion and absorption of nutrients, see Appendix 11 and *Absorption, nutrients*. *Jejunitis* is the inflammation of the jejunum. *Jejunectomy* is the excision of all or part of the jejunum. See also *Jejunostomy*.

Jevity®. Brand name of a lactose-free, isotonic tube feeding formula containing 14 g of fiber per liter; made of calcium and sodium caseinates, hydrolyzed cornstarch, soy polysaccharide, MCT oil, and corn oil. See Appendix 39.

JODM. Abbreviation for *juvenile-onset diabetes mellitus*.

Joint Commission on the Accreditation of Healthcare Organizations (JCAHO). A private, nonprofit agency that sets guidelines for the operations of hospitals and other health care facilities to ensure high standards of patient care. A staff of medical inspectors is drawn from members of the American College of Physicans, American College of Surgeons, American Hospital Association, and American Medical Association. The team of inspectors (who examine the operation of the hospital by invitation) submits a written report and recognizes compliance with standards by granting a certificate of accreditation for 1 to 3 years.

Joule (J). The international unit (SI) of energy. It is the amount of energy expended when 1 kg of a substance is moved 1 meter by a force of 1 newton. In the metric system, energy is measured in joules. One joule (kJ) equals 4.184 kcal.

Junk food. Layman's term for food with low nutrient value except calories. Usually rich in fats, simple sugars, and/or starches. Similar to foods with "empty" calories.

Juvenile-onset diabetes mellitus (JODM). Diabetes mellitus among the young, in some cases starting in childhood. Classified as insulin-dependent *diabetes mellitus*. See *Diabetes mellitus*.

Juvenile-onset obesity (JOO). Obesity that develops during childhood; characterized by large amounts of fat stored. The child also has an excessive number of adipose cells.

K

K. 1. Chemical symbol for potassium. 2. Abbreviation for "kilo," as in kilogram (kg) and kilocalorie (kcal); generally, lower case k is used. 3. Abbreviation for *katal*.

K-40. Potassium-40, a radioactive isotope naturally present in the body. Potassium 40 measurement is used to determine body composition. It is based on the assumption that body potassium is found in constant concentration in the muscles and lean portions of the body but is not present in fat. Body potassium therefore becomes an index of lean body mass.

Kaposi's sarcoma (KS). A cancer of the reticuloendothelial cells usually associated with AIDS and lymphomas. Characterized by brownish or purplish papules on the skin that progressively spread to the viscera and lymph nodes. Lesions in the mouth and esophagus may result in dysphagia; lesions in the bowel may cause ulceration and gastrointestinal distress leading to poor nutritional intake. *Nutrition therapy:* if the patient is receiving chemotherapy and radiotherapy, management is similar to that of cancer. See *Chemotherapy* and *Radiotherapy*. See also *Dysphagia*.

Katal. An enzyme unit in moles per second defined by the *SI system*. Abbreviated K or kat.

Kerasin. A cerebroside occurring in the brain. On hydrolysis, it yields a fatty acid, galactose, and sphingosine.

Keratin. Insoluble protein of hair, hooves, nails, and feathers. Contains a large amount of sulfur and is a commercial source of the amino acid cystine. Not hydrolyzed by digestive enzymes of humans.

Keratinization. Process of becoming horny due to the development of *keratin*. In vitamin A deficiency, keratinization of epithelial tissue occurs throughout the body, first in the salivary glands and then in the respiratory tract, eyes, and skin. Secondary infections from cracks in this dry layer in the respiratory tract lead to pneumonia and, in the eyes, to *xerophthalmia*.

Keratitis. Inflammation of the cornea of the eye. May be due to infection, vitamin A deficiency, allergy, or injury to the eye.

Keratomalacia. Softening and death of cells of the cornea of the eye. The earliest sign is dryness of the conjunctiva, which may lead to ulceration and infection. Effective treatment at this stage is followed by corneal scarring and opacity. If the process is not stopped by treatment, it leads to perforation of the cornea, prolapse of the iris, and infection of the whole eyeball. Healing results in scarring of the whole eye and frequently in total blindness. Keratomalacia is seen in severe vitamin A deficiency, although it may also be due to other diseases causing corneal lesions, such as trauma, bacterial infection, measles, and repeated exposure to excessive dust.

Keratosis. Any skin disease characterized by horny growth. In *follicular keratosis,* the skin becomes rough, dry, and scaly, and the hair follicles are blocked with plugs of keratin, which appear as prominent projections along the upper forearm and thighs, as well as along the shoulders, back, abdomen, and buttocks. Because of its appearance, the condition is commonly called "gooseflesh" or "toad skin." It is seen in vitamin A deficiency.

Keshan disease. Syndrome named after the region in China where it is endemic. It is considered to be due primarily to a deficiency in *selenium.* Symptoms include low blood and hair selenium levels and cardiomyopathy with high mortality rates in children and women of childbearing age.

Keto-. Prefix denoting the presence of the ketone or carbonyl group (C=O). A *keto acid* is a compound that contains both the carbonyl group (C=O) and the carboxyl group (—COOH).

Ketoacidosis. Accumulation of ketones in the body resulting in an abnormal increase in hydrogen ion concentration (acidosis). Occurs in faulty carbohydrate metabolism and is a complication of *diabetes mellitus;* characterized by nausea, vomiting, dyspnea, Kussmaul breathing with a fruity odor of the breath, and mental confusion. It leads to coma if untreated.

Ketogenesis. Formation of ketone or acetone bodies in the liver. Ketone bodies formed in the liver are oxidized in other tissues, especially in the muscles. When ketone formation exceeds ketone oxidation, ketone bodies accumulate and cause *ketosis.*

Ketogenic. Capable of being converted into ketone bodies. The ketogenic substances in metabolism are the fatty acids and certain amino acids. See also *Diet, ketogenic,* and *Ketogenic amino acid* under *Amino acid.*

Ketogenic/antiketogenic ratio. Ratio between substances in the diet that give rise to ketone bodies (ketogenic factors) and those that favor ketone oxidation (antiketogenic factors). The *ketogenic factors* are precursors of ketone bodies such as fatty acids and the ketogenic amino acids. The *antiketogenic factors* are precursors of glucose; these include carbohydrates, glucogenic amino acids, and the glycerol portion of fat. For clinical purposes, the following simplified ratio may be used:

$$\frac{\text{Ketogenic factors}}{\text{Antiketogenic factors}} = \frac{0.5\,P + 0.9\,F}{0.5\,P + 0.1\,F + 1.0\,C}$$

where P, F, and C represent, respectively, the number of grams of protein, fat, and carbohydrate in the diet. A ketogenic diet with a ketogenic/antiketogenic ratio of 3:1 will produce a state of ketosis. See *Diet, ketogenic.*

Ketone. Compound containing the carbonyl group (C=O); derived from oxidation of a secondary alcohol. The ketone test in the urine is done with a dipstick reagent or a test tablet. See also *Ketonuria.*

Ketone bodies. Also called "acetone bodies"; collective term given to the intermediate products of fatty acid degradation. These include acetoacetic acid, beta-hydroxybutyric acid, and acetone and are present in the blood in very small amounts under ordinary conditions. Ketone bodies formed in the liver are normally oxidized in other tissues. However, ketone bodies tend to accumulate when the rate of production becomes so great that the organism cannot burn them at a sufficiently rapid rate. See *Ketosis.*

Ketonemia. Presence of ketone (acetone) bodies in the blood above normal levels; characterized by a fruity breath odor.

Ketonex®. Brand of branched-chain amino-acid-free dietary product for infants and toddlers (Ketonex®-1) and children and adults (Ketonex®-2) with maple syrup urine disease (branched-chain ketoaciduria).

Ketonuria. Presence of ketone (acetone) bodies in the urine. Abnormal amounts in the urine indicate rapid catabolism of fats, as in uncontrolled diabetes mellitus, starvation, or other disorders associated with increased fat metabolism.

Ketose. Carbohydrate containing the ketone group. Fructose is a ketose.

Ketosis. Clinical condition in which *ketone bodies* accumulate in the blood and appear in the urine; characterized by a sweetish acetone odor of the breath. Ketosis may be caused by a disturbance in carbohydrate metabolism (as in uncontrolled diabetes mellitus), by a dietary intake quite low in carbohydrate but very high in fat (as in a ketogenic diet), or by a diminution in carbohydrate catabolism with consequent high mobilization of body fats (as in starvation). Uncontrolled ketosis leads to *acidosis*. It causes a decrease in the alkali reserve in plasma, hyperventilation, low CO_2 tension in alveolar air, and an increase in urinary ammonia; in advanced states, it causes low blood pH.

Ketoxylose. Also called "xylulose." One of the few L-sugars found in nature. It is excreted in the urine of humans with a hereditary abnormality in pentose metabolism.

Kidney. One of a pair of bean-shaped organs located in back of the abdominal cavity on either side of the spine just below the spleen on the left and the liver on the right. The functioning unit is the *nephron,* which consists of a *glomerulus,* or tuft of capillaries, that is surrounded by *Bowman's capsule*. This capsule is attached to a long, winding *tubule* through which passes the fluid from the blood contained in the glomerulus. Urine is produced in the nephron and emptied into the *pelvis* of the kidney. From here, urine flows into the *ureter,* which is a muscular tube extending from the kidney to the *urinary bladder*. The chief functions of the kidney are to maintain constant blood composition and volume by its unique filtering system; to maintain normal pH of body fluids by excreting an acid urine and by synthesizing ammonia whenever necessary; and to excrete body wastes or metabolic by-products. There are at least 2 million nephrons, which filter more than 2500 pints of blood daily. The pituitary hormone (antidiuretic hormone) helps the kidneys to regulate *water balance*. See also *Urine*.

Kidney (renal) clearance test. Test of *kidney function* by measuring the ability to excrete waste products such as creatinine, urea, inulin, or dye in the urine. The quantity excreted per minute divided by the amount present in 1 mL of plasma is the urinary clearance.

Kidney transplantation. The surgical implantation of a kidney from a living related donor or a cadaver; a widely accepted and successful method of treating end-stage renal failure. As in other organ transplantation, the recipient is given immunosuppressive drugs before and after transplantation to minimize infection and rejection of the foreign organ. Nutrition support to normalize the patient's nutritional status is an important part of the therapy. *Nutrition therapy:* in the acute post-transplant period (up to 8 weeks after transplantation), protein intake is at 1.3 to 2.0 g/kg of normalized body weight (NBW), i.e., dry weight or adjusted body weight if the patient is obese; calories at 1.3 to 1.5 × basal energy expenditure (BEE) or 30 to 35 kcal/kg NBW; 2 to 4 g sodium; 2 to 4 g potassium if hyperkalemic; 800 to 1500 mg calcium; vitamins and minerals at RDA levels; and no fluid restriction unless fluid retention worsens. Carbohydrates should provide no more than 50% of calories; emphasize complex carbohydrates and fiber and limit simple carbohydrates. Fat should provide no more than 35% of calories, with a polyunsaturated/saturated ratio of > 1 and cholesterol < 400 mg/day. Avoid alcohol. In the chronic post-transplant period, protein intake is at 1 g or higher, if prednisone is used, per kilogram of body weight; calories at 1.2 to 1.3 × BEE or to maintain desirable body weight based on activity level; 2 to 4 g of sodium if hypertension is present; unrestricted potassium; 800 to 1500 mg of calcium; vitamins and minerals at the RDA level; and unrestricted fluids. If there is renal insufficiency or chronic rejection, restrict protein to 0.6 to 0.8 g/kg NBW. Monitor fluid and electrolytes, especially sodium and potassium. Dialysis may be resumed, depending on the severity of rejection. See also *Transplantation*. When a patient receives a normal kidney from a donor, he or she is usually given high doses of immunosuppressive drugs (such as prednisone azathioprine and cyclosporine A) to prevent immune rejections. Mineral balance, particularly for calcium and phosphorus, may be affected by these drugs; thus, their intake should be increased 1.5 times the normal recommended dietary allowance. Hyperglycemia and edema are other side effects that require carbohydrate and sodium restriction. Following successful kidney transplantation, protein restriction is no longer needed, as nitrogen retention is not a problem. Maintain desirable body weight and monitor any hyperlipidemia, hypercholesterolemia, and hypertension. With prednisone therapy, however, increase protein intake to 1.5 to 2 g/kg/day for adults. In cases of transplant rejection, resume monitoring of protein, fluid, and electrolytes, especially sodium and potassium. Dialysis may be resumed, depending on the severity of rejection.

Kilocalorie (kcal). See *Calorie*.

Kindercal™. Brand name for a nutritionally complete liquid formula for children 1 to 10 years of age. Used as tube feeding or oral supplement for children with trauma, chronic illness, or failure to thrive. See Appendix 37.

Kinky hair disease. See *Menke's disease*.

Kjeldahl method. Method of determining the amount of nitrogen in an organic compound by digestion with sulfuric acid and conversion of the amine group in the amino acid to ammonia. Nitrogen is calculated by measuring the amount of ammonia formed. In foods, most of the nitrogen comes from protein (about 16% nitrogen). Thus the *crude protein* content in foods may be determined by multiplying the total "Kjeldahl nitrogen" by the factor 6.25.

Knee height. An anthropometric measurement used to estimate the stature of a person who is bedfast or chairbound or who has spinal curvature. To measure knee height, bend the left knee and ankle at 90 degree angles and place the fixed blade of a caliper under the heel of the foot and the sliding blade on top of the thigh about 2 inches from the kneecap. Read the measurement (in cm) and calculate the stature (also in cm) from the formulas below:

Stature for males (in cm)

6 to 16 yrs: Stature = (knee height × 2.22) + 40.54

19 to 59 yrs: Stature = (knee height × 1.88) + 71.85

Stature for females (in cm)

6 to 18 yrs: Stature = (knee height × 2.15) + 43.21

19 to 59 yrs: Stature = (knee height × 1.86) − (age × 0.05) + 70.25

Age is rounded to the nearest whole year. Divide the stature measurement (cm) by 2.54 to get the height in inches. See also *Weight, computed.*

Koagulation vitamin. See *Vitamin K.*

Kofranyi–Michaelis respirometer. Lightweight portable *respirometer* that makes possible the systematic measurement of energy expenditure during work. The instrument weighs only about 5 lb and can be worn on the back.

Koilonychia. Spoon-shaped nails; a nail deformity in which the outer surface becomes concave. Occurs in severe iron deficiency anemia.

Koladex®. Brand name of a cola-flavored beverage containing 100 g of dextrose per 10-oz bottle. It is used for the standard glucose tolerance test.

Korsakoff's syndrome. Set of symptoms characterized by confusion, loss of memory, and irresponsibility. Often, degenerative changes occur in the thalamus as a result of thiamin and vitamin B_{12} deficiencies. Seen in chronic alcoholism, usually together with *Wernicke's encephalopathy.*

Kosher. Term for the Jewish Kashruth dietary law which means "fit to eat." Kosher foods are classified into three groups: those that are inherently kosher (pareve) and may be eaten in their natural state, such as grains, fruit, vegetables, tea, and coffee; those that require some form of processing to be kosher, such as meat, poultry, and cheese; and those that are inherently not kosher, such as pork products, shellfish, and fish without scales and fins. Kosher meat may come only from cloven-hooved creatures, such as cows, sheep, and goats, and animals that graze and chew their cud. Products must be certified by a rabbi in order to be considered kosher. Meat and dairy products may not be eaten together. Generally after eating meat, an interval of 3 to 6 hours must elapse before eating dairy products. It is also necessary to maintain separate sets of utensils, one for the preparation and service of meat and poultry and the other for dairy foods.

Krebs cycle. Tricarboxylic acid cycle or citric acid cycle. It is a cycle of reactions in which acetylcoenzyme A combines with oxalacetate, forming seven intermediary products—*citrate, cis-aconitate, isocitrate, alpha-ketoglutarate, succinate, fumarate,* and *malate*—and eventually reforming *oxalacetate,* which is set free to unite with another molecule of acetylcoenzyme A, thus repeating the cycle. The Krebs cycle is considered the final common pathway in the oxidation of carbohydrate, protein, and fat to carbon dioxide and water, with the release of energy. The acetylcoenzyme A that unites with oxalacetate may come from the anaerobic phase of carbohydrate metabolism, from fatty acid breakdown, and indirectly from ketogenic amino acids. Some of the intermediate products in the cycle may also be formed from glucogenic amino acids and during fatty acid breakdown. For details, see Appendix 15.

Krebs–Henseleit cycle. Also called "ornithine cycle" in urea formation. See *Urea.*

KS. Abbreviation for *Kaposi's sarcoma.*

Kwashiorkor. A form of protein-energy malnutrition associated with extreme protein deficiency; seen principally in children shortly after weaning to a diet high in starch and low in protein. Unlike *ma-*

rasmus, kwashiorkor occurs even with adequate caloric intake. It is characterized by retarded growth, anemia, edema, fatty infiltration of the liver, pigmentary changes of the skin and hair, gastrointestinal disorders (especially diarrhea), muscle wasting, delayed wound healing, and psychomotor changes. Water and electrolyte imbalance is brought about by hypoalbuminemia, diarrhea, and decreased cell enzyme and endocrine functions. Severe dehydration may cause heart failure. Kwashiorkor is a public health problem in many countries where the quality and/or quantity of protein intake are below minimal requirements. Although this severe protein deficiency syndrome is especially prevalent in malnourished children, it may also occur in hospitalized patients suffering from long-term illness who are in hypercatabolic states with insufficient protein intake. The onset may be rapid; visceral protein levels are depressed, although body weight and anthropometric measurements are preserved. The condition is called *marasmic kwashiorkor* in individuals and children whose weight for height is less than 70% of standard for age and sex. *Nutrition therapy:* immediate measures should be taken to correct dehydration and electrolyte imbalance. As the patient improves, a high-protein diet (2.5 to 4 g/kg ideal body weight/day for young children) is initially provided by nonfat milk solids or skimmed milk powder added to a mixed diet. Refeeding should be gradual. Provide sufficient calories from carbohydrate to spare protein. If necessary, use tube feeding. Currently, many commercial nutrient formulas of high protein-calorie density are available. Mineral and vitamin supplementation is recommended, because magnesium and potassium depletion and vitamin A deficiency may be serious. Compare kwashiorkor with *Marasmus*.

Kynurenine. An intermediate product in tryptophan metabolism. Kynurenine in mammalian liver may be hydrolyzed to anthranilic acid and alanine, requiring pyridoxal phosphate as a cofactor. Large amounts of kynurenine are excreted in the urine in pyridoxine deficiency. It is found in normal urine in trace amounts but may be temporarily increased after tryptophan administration.

Kyphosis. Also called "humpback." Abnormal curvature of the spine. Often the result of bad posture, vitamin D deficiency, certain types of arthritis, osteoporosis, or tuberculosis of the spine.

L

L. Abbreviation for liter and also lung.

Labeling. Accompanying printed information about the product (food, drug, cosmetics, tool, etc.), usually following established regulations. See Appendices 5 and 6.

Lactaid®. Commercial preparation of liquid *lactase* enzyme, which can hydrolyze up to 99% of the *lactose* in milk when added 24 hours before use; also available in caplet form for oral ingestion with foods containing lactose.

Lactalbumin. One of the milk proteins; the others are casein and lactoglobulin. It is identical to serum albumin and is not easily precipitated by acids. The soft, flocculent curds are easy to digest, in contrast to the large, hard curds of casein. Thus the higher ratio of lactalbumin to casein in human milk (one-half of human milk protein is lactalbumin) is an advantage for infant feeding. Although the total amount of protein in cow's milk is about two times that of human milk protein, the ratio of lactalbumin to casein is only 1:5. Also, the higher lactose/protein ratio in human milk is favorable. See *Infant feeding* and *Lactose*.

Lactase. Enzyme that splits *lactose* into glucose and galactose. It is present in the intestines of all young mammals but may become deficient with age and in certain conditions affecting the small intestine. Lactase deficiency causes intolerance to lactose, which is characterized by bloating, cramping, flatulence, and diarrhea. See *Lactose intolerance*.

Lactase deficiency. See *Lactose intolerance*.

Lactate dehydrogenase (LDH). Enzyme that reversibly catalyzes the reduction of pyruvic acid to lactic acid or the oxidation of lactic acid to pyruvic acid. It is present in the tissues and released into the blood when there is tissue necrosis, as in liver damage, myocardial infarction, and renal tubular necrosis. Normal range in serum is 100 to 190 mU/mL (100 to 190 U/liter).

Lactation. The period of milk secretion; the secretion of milk by the mammary gland. The amount of milk produced is affected by several factors, including the nutritional status of the mother, frequency of sucking by the young, ingestion of medicinal or food galactogens, and hormonal control. See Appendix 1 for the recommended dietary allowances for lactation. See also *Nutrition, lactation*.

Lacteals. Tiny vessels in the villi of the wall of the small intestine through which chylomicrons are absorbed. These ducts empty into the lymphatic system.

Lactic acid. An acid produced by fermenting lactose (milk sugar); also formed from sucrose, glucose, or maltose by the action of lactic acid bacteria. In mammalian tissues, lactic acid is formed by the reduction of pyruvate when oxygen is lacking and pyruvate cannot be oxidized and is channeled into the *Krebs cycle*.

Lactobacillus bifidus factor. A protective factor in human milk that encourages growth of beneficial bacteria in the infant's intestines; one of the reasons for encouraging breast feeding.

Lactobacillus bulgaricus factor. A growth factor for certain microorganisms identical to *pantothenic acid,* a water-soluble vitamin.

Lactobacillus casei factor. Former name for *folic acid,* a water-soluble vitamin.

Lactoflavin. Riboflavin of milk; the name given to the greenish-yellow fluorescent flavin pigment of whey. A former name is "lactochrome." See *Vitamin* B_2.

Lactofree®. Brand name for a milk-based, lactose-free formula for infants and children with feeding problems related to lactose intolerance.

Lactose. Also called "milk sugar"; a disaccharide that occurs naturally only in the milk of mammals. It is hydrolyzed by acid or the enzyme lactase into the monosaccharides glucose and galactose. Lactose is less readily digested than other disaccharides. The undigested lactose passes into the large intestine, where it is fermented by normal colonic bacteria to form lactic acid and hydrogen. These fermentation products can osmotically draw water into the lumen, causing cramping and diarrhea.

Lactose intolerance (malabsorption). Condition characterized by intestinal symptoms such as bloating, cramping, flatulence, and diarrhea that follow the ingestion of lactose. Lactose intolerance varies in degree among individuals and may be primary or secondary. *Primary* lactose intolerance is due to lack of the enzyme lactase. Congenital lactase deficiency is rare, but the condition may develop after weaning and with maturity. Adult lactase deficiency is quite common in many nonwhite and ethnic populations, including blacks, Orientals, Jews, Mexicans, and American Indians. *Secondary* or acquired lactose intolerance is due to diseases affecting the intestinal mucosa (as in acute enteritis, celiac sprue, inflammatory bowel disease), and can occur after small bowel or gastric surgery and after periods of disuse of the intestinal tract (as in starvation and prolonged parenteral nutrition), causing lactase deficiency with atrophy of the small intestine. Secondary lactose intolerance is transient and disappears when the disease resolves. Primary lactose intolerance seems to be permanent, although the extent of intolerance varies and the severity of symptoms is related to the amount of lactose ingested. Most lactose-intolerant individuals usually have no difficulty handling small amounts of lactose. See *Diet, lactose-restricted.*

Lactose tolerance test. Measurement of the blood glucose level following the ingestion of a lactose dose. In the adult, an increase less than 20 mg/dL after a 50-g lactose load is strongly suggestive of lactase deficiency.

Lactulose. A synthetic derivative of lactose used as a laxative and as an adjunct to protein restriction in the treatment of fulminant hepatic failure. Because lactulose cannot be digested, colonic bacteria metabolize it to lactic acid, which removes ammonia from the blood and reduces the degree of hepatic encephalopathy. Lactulose also increases the water content of the stools. Brand names include Cephulac®, Cholac®, Chronulac®, Constilac®, and Constulose®.

Laënnec's cirrhosis. Also called "portal" or "alcoholic cirrhosis"; the end result of prolonged dietary inadequacy, particularly of protein, coupled with the toxic effects of alcohol on the liver. See *Cirrhosis.*

Laetrile. The compound 1-mandelonitrile-β-glucuronic acid, found in the seeds of certain fruits and nuts. It has been claimed, but not supported experimentally, that laetrile has therapeutic value in the treatment of cancer due to the cyanide it contains, which acts specifically to destroy the cancer cells. Normal animal cells, however, lack the enzymes required to release cyanide from mandelonitrile. Laetrile has also been erroneously promoted as vitamin B_{17}; there is no evidence of physiologic or biochemical abnormality when it is not included in the diet.

Large-for-gestational-age infant (LGAI). An infant with accelerated fetal growth and birth weight well above normal, usually beyond the 90th percentile of appropriate weight-for-gestational-age infants. Often influenced by maternal diabetes mellitus and Beckwith's syndrome. Signs include hypoglycemia and respiratory distress.

Lauric acid. A saturated fatty acid containing 12 carbon atoms; found in coconut oil, laurel oil, and spermaceti.

Laxative. An agent that promotes evacuation of the bowel by increasing the fecal bulk, by softening the stool, or by lubricating the intestinal wall. Laxative abuse may cause malabsorption, steatorrhea, dehydration, weight loss, hypoalbuminemia, vitamin D deficiency, and osteomalacia. See *Bisacodyl, Mineral oil, Phenolphthalein,* and *Psyllium.*

LBM. Abbreviation for *lean body mass.*

LBWI. Abbreviation for *low-birth-weight infant.*

LCFA. Abbreviation for long-chain fatty acid. See under *Fatty acid.*

LCT. Abbreviation for *long-chain triglycerides.*

LDH. Abbreviation for *lactate dehydrogenase.* See also *Serum enzymes.*

LDL. Abbreviation for *low-density lipoprotein.*

Lead (Pb). Trace element required in very small amounts for normal growth and health. Experimentally induced lead deficiency is associated with growth retardation; hypochromic anemia with decreased iron stores in the blood, liver, and spleen; and decreased hepatic concentrations of glucose, triglycerides, and phospholipids. The average intake of lead is about 300 μg/day, although this is highly variable. The lead content of foods varies with soil and environmental conditions; in addition, smaller amounts of lead are ingested with drinking water and with foods from lead-soldered cans, and are inhaled from the atmosphere and from cigarette smoke. Absorption is age dependent; infants and children absorb nearly 40%, whereas adults absorb only 5% to 10%. In the body, most of the lead is deposited in the bone (about 90%) and the rest in soft tissues, notably the liver and kidney. The observation that humans accumulate lead slowly and gradually with age implies that the capacity for excretion is not adequate to maintain overall homeostasis. Lead intake is of concern not for its beneficial properties but rather for its toxic effects, which can be especially damaging to children. The most common consequences of lead intoxication are anemia, stunted intellectual development, hypertension, and renal damage.

Lean body mass (LBM). Active tissue mass; the part of the body weight concerned with energy metabolism. It is a measure of body composition, taken as the difference between body weight and the total mass of the adipose tissue, the extracellular fluid, and the skeleton. The minimum essential fat content is only about 2%.

Lecithin. Phosphatidyl choline, a phosphatide consisting of glycerol, two molecules of fatty acids, phosphoric acid, and choline. It is widely distributed in animal cells, especially nerves. Lecithin exists in the cell as a dipolar ion; choline is a strong base, and phosphoric acid is a moderately strong acid. This is significant in fat transport. See *Lipotropic agent (factor)* and *Phospholipid*.

Let-down reflex. The release (let-down) of milk from breast tissues to the nipple area triggered by the action of the hypothalamus on the pituitary gland. The latter releases oxytocin, which "lets down" breast milk from storage sites. If the let-down reflex does not function, little milk is available to the infant.

Leucine (Leu). Alpha-aminoisocaproic acid; an *essential amino acid* that is strongly ketogenic; acetoacetic acid is its chief catabolic product in the liver. Leucine deficiency results in loss of appetite; apathy; irritability; weight loss or poor weight gain; decreased plasma leucine; and increased plasma isoleucine, methionine, serine, threonine, and valine.

Leucine-induced hypoglycemia. A rare inborn error of metabolism due to a deficiency of a lyase enzyme needed in the degradation of leucine. Infants with this disorder fail to thrive and may have convulsions, delayed mental development, severe acidemia, and life-threatening, profound hypoglycemic attacks. Acute management is directed at correcting the profound hypoglycemia. Longer-term management is directed at reduction of leucine intake by moderate protein restriction in small, frequent meals and with nocturnal feedings if required to avoid hypoglycemia. See *Diet, leucine-restricted*.

Leucovorin. Generic drug name for the calcium salts of folinic acid, an active metabolite of folic acid. Folinic acid circumvents the metabolic block produced by folic acid antagonists (such as methotrexate) and is used as an antidote to these drugs. It is also used to treat megaloblastic anemia due to folate deficiency.

Leukemia. Type of blood cancer due to proliferation of leukocytes and decreased production of red blood cells. The etiology is unknown, although ionizing radiation, the effects of certain chemicals, heredity, and hormonal abnormalities may be causative factors. There are two kinds: acute and chronic. Generally, leukemia is characterized by diffuse replacement of bone marrow with proliferating leukocyte precursors, abnormal numbers and forms of immature white blood cells in the circulation, and infiltration of lymph nodes, spleen, liver, and other organs. Usually occurs with sudden onset; visible signs are fatigue, pallor, and extreme weakness. Patients have frequent hemorrhaging and mouth ulcers. Some patients have been successfully treated with chemotherapy and with bone marrow transplantation. *Nutrition therapy:* serve attractive meals in small frequent feedings, providing 30 to 35 kcal and 1.5 g of protein per kilogram of body weight. Increase calories and protein in febrile conditions and for repletion. Alter the diet, as necessary, to alleviate the side effects of chemotherapy and stabilize the patient's nutritional status prior to bone marrow transplantation. For further dietary recommendations, see *Bone marrow transplantation, Cancer,* and *Chemotherapy*.

Leukocyte (leucocyte). Also called "white blood cell (WBC)"; the nonpigmented, nucleated cell of the blood. WBCs play an important role in the body's defense mechanisms. They can destroy disease-causing organisms. The normal WBC count is about 5000 to 10,000/mm^3. This is increased in acute infections and leukemia and decreased in *leukopenia*.

Leukopenia. An abnormal reduction in the number of white blood cells as a result of decreased production by the bone marrow or increased destruction of the cells, usually in the spleen. The condition may be congenital or brought about by malignancy, folic acid deficiency, or some unknown causative factor (pernicious type).

Levan. Also called "fructosan"; a *homopolysaccharide* composed of fructose units. It is derived chiefly from Jerusalem artichokes and certain grasses. *Inulin* is a levan.

Levodopa. The levorotatory isomer of dihydroxyphenylalanine and the metabolic precursor of dopamine. It is used in the treatment of parkinsonian syndrome. The drug's effectiveness is decreased by high intakes of protein and pyridoxine; vitamin supplements containing pyridoxine should be avoided. Levodopa increases the need for ascorbic acid and vitamin B_6; decreases the absorption of tryptophan, phenylalanine, and tyrosine; and increases the urinary excretion of sodium and potassium. It may also cause anorexia, nausea, vomiting, and abdominal distress. It may be present in breast milk. Long-term use and high doses may cause vitamin B_6 deficiency and hypokalemia, especially if laxatives and potassium-losing diuretics are taken concurrently. Brand names are Dopar®, Laradopa®, and Sinemet®.

Levulose. Another name for *fructose* or fruit sugar to indicate that it is levorotatory.

LGAI. Abbreviation for *large-for-gestational-age infant*.

Light adaptation. Changes occurring in the eye opposite in nature to those observed in *dark adaptation,* i.e., constriction of the pupil, diminished sensitivity of the retina, bleaching of the visual purple, and a change in pH from an alkaline to an acid reaction.

Lignin. An indigestible compound occurring in the cell walls of plants. It is a noncarbohydrate, insoluble *dietary fiber* and consists of a multiringed alcohol network. It is resistant to hydrolysis by digestive enzymes, strong acids, and alkalis, and is not attacked to any extent by intestinal microorganisms.

Linoleic acid. An *essential fatty acid;* a polyunsaturated fatty acid with 18 carbon atoms in its chain and two double bonds, one of which is in the omega-6 position. It is the major omega-6 fatty acid in foods; it is found in linseed, safflower, cottonseed, and soybean oils; fish oils; and animal tissues. Human milk contains about four times more linoleic acid than cow's milk. Linoleic acid is a dietary essential (i.e., the body cannot synthesize it). It is essential for growth and prevention of dermatitis, serves as a precursor of arachidonic acid, and helps form hormone-like compounds called *eicosanoids*. The omission of linoleic acid from the diet results in failure of growth, scaly skin, hair loss, poor muscle tone, fatty liver with hepatomegaly, decreased levels of prostaglandins, increased susceptibility to infection, and delayed wound healing. Linoleic acid deficiency was reported in the early 1970s among hospitalized patients fed exclusively with intravenous fluids without fat. Patients with malabsorption, as in cystic fibrosis, may also be deficient in linoleic acid. Infants receiving milk formula lacking in essential fatty acids develop an infantile eczema characterized by leathery skin with desquamation and oozing. Addition of trilinolein to the diet (5% to 7% of total calories) results in the disappearance of symptoms. The need for *essential fatty acids* will be met if linoleic acid is present in the diet at a level of at least 1% of total caloric requirements. See also *Fat intake, recommended.*

Linolenic acid. The alpha form is an essential fatty acid. It is a polyunsaturated fatty acid with 18 carbons and three double bonds, one of which is in the omega-3 position. It is found chiefly in linseed oil and fish oils. Alpha linolenic acid is the major omega-3 fatty acid in foods. It is considered a nutrient because of its growth-promoting effect; it plays a vital role in the immune system of the body. It produces hormone-like compounds called *eicosanoids.*

Lipemia. An increased level of blood lipids. *Temporary absorptive lipemia* occurs after a fatty meal as chylomicrons discharged into the blood plasma cause a rapid rise in lipid, chiefly as triglycerides, along with a small amount of protein. *Idiopathic lipemia* (unknown etiology) is frequently observed in diabetic acidosis and glycogen storage disease.

Lipid. Member of a large group of organic compounds insoluble in water and soluble in fat solvents, e.g., chloroform, ether, benzene, petroleum, and carbon disulfide. Lipids are divided into two groups: nonpolar lipids, which occur mainly as esters of fatty acids, which are water-insoluble and have to be hydrolyzed before entering the body; and polar or amphipathic lipids, which have fatty acids plus a polar radical such as a carbohydrate, lecithin, sphingomyelin, or phosphate-containing amino alcohol. Lipids of nutritional importance are *fatty acids,* particularly essential fatty acids; triglycerides or neutral *fats;* phosphatides, especially *lecithins;* terpenes, especially *carotene;* and steroids such as *cholesterol* and the *adrenocortical steroids.* Lipids are essential components of the cell and the body's response to injury and sepsis, in addition to serving as the major form of energy storage. See Appendix 10.

Lipid malabsorption. Interference with lipid absorption brought about by bile deficiency, lack of the pancreatic enzyme *lipase,* defective intramucosal metabolism (as in nontropical sprue and adrenal hormone insufficiency), lymphatic obstruction, and impaired lipoprotein synthesis.

Lipisorb®. Brand name of a nutritionally complete supplement with 85% of fat as MCT oil; useful in the management of patients with fat malabsorption. See Appendix 39.

Lipochromes. Plant pigments that are soluble in fats and organic solvents. Examples are chlorophyll and carotenoids.

Lipofuscin. Ceroid pigments that occur as brown spots on the skin when lipid breakdown products accumulate. Studies indicate that lipofuscin contributes to the aging process of cells.

Lipogenesis. 1. Synthesis of lipids or formation of body fat. 2. Specifically, the formation of fatty acids in the liver. Any dietary constituent that supplies acetate (e.g., carbohydrate and protein metabolites) may contribute to lipogenesis. In obese persons, the rate of lipogenesis may be five times greater than normal.

Lipoic acid. 6,8-Dithio-*n*-octanoic acid, a cyclic disulfide. It is also called "thioctic acid," "protogen," "pyruvate oxidation factor (POF)," and "*L. casei* acetate factor." It is needed for oxidative decarboxylation of alpha keto acids such as pyruvic acid and alpha-ketoglutaric acid. Lipoic acid functions as a hydrogen and an acyl acceptor. It is an essential growth factor for various organisms. Although its function is closely associated with that of thiamin, lipoic acid is not considered a vitamin because it appears to be synthesized in adequate amounts in the mammalian cell. See also *Lipothiamide pyrophosphate.*

Lipoid. A substance resembling fat in appearance and solubility but containing groups other than glycerol and fatty acids, which make up the true fats.

Lipolysis. 1. Breakdown or degradation of lipids. 2. Specifically, the splitting of fat into glycerol and fatty acids in the liver. Glycerol enters the pathway of carbohydrate metabolism as glycerol-3-phosphate. The fatty acids are degraded by a series of reactions leading to the ultimate product, *acetyl CoA*.

Lipoprotein. Compound protein formed when a simple protein unites with a lipid. It has the solubility characteristics of protein; hence it is involved in lipid transport from the intestinal tract and liver to a variety of tissue sites. Five types circulate in the blood: *chylomicrons;* alpha lipoprotein or *high-density lipoprotein (HDL);* prebeta lipoprotein or *very-low-density lipoprotein (VLDL);* beta lipoprotein or *low-density lipoprotein (LDL);* and broad beta or *intermediate-density lipoprotein (IDL)*. Chylomicrons primarily contain triglycerides; serum cholesterol is a component of HDL, LDL, and VLDL. A high HDL-cholesterol level correlates with decreased risk for coronary heart disease (CHD), whereas a high LDL-cholesterol level correlates with increased risk. The role of VLDL, which is 50% triglycerides, is not as clearly related to CHD risk as that of LDL. Levels of lipoproteins in the blood are influenced by diet, age, weight change, emotions and stress, drugs, illness, and a number of hormones (insulin, thyroxine, adrenal hormones, and anterior pituitary hormones). See *Hyperlipoproteinemia*.

Lipoprotein lipase. Also called *clearing factor;* this lipase catalyzes the hydrolysis of fats present in chylomicrons and lipoproteins. It is found in various tissues and is important in the mobilization of fatty acids from the blood into the depot fats.

Liposuction. Also called "suction lipectomy." A method for removing adipose tissue with a suction pump device; used primarily to reduce fat deposits around the abdomen, breasts, legs, face, and upper arms.

Liposyn®. Brand name of a 10% and a 20% fat emulsion for intravenous administration; made with safflower oil, soybean oil, egg phosphatides, and glycerin. Each milliliter provides 1.1 kcal (10% emulsion) and 2.0 kcal (20% emulsion). See also Appendix 41.

Lipothiamide pyrophosphate (LTPP). A conjugate of thiamin pyrophosphate and *lipoic acid* believed to be the active catalyst in the oxidative decarboxylation of alpha keto acids such as pyruvic acid.

Lipotropic agent (factor). Any substance capable of transporting or mobilizing fat and preventing or correcting the fatty liver of choline deficiency. The lipotropic agents are choline, betaine, methionine, inositol, serine, and lecithin. The exact mechanism by which these substances prevent fatty liver is not known. However, it has been suggested that neutral fats are converted to choline-containing *phospholipids* to mobilize the fat and thus prevent deposition in the liver.

Lithiasis. Formation of stones, or calculi. See *Cholelithiasis* and *Urolithiasis*.

Lithium (Li). A trace element that may be essential; plays a role in maintaining normal health and function in animals. Diets very low in this element have resulted in fewer offspring in rats and goats. In pharmacologic amounts, lithium is effective in the treatment of bipolar disorder (manic–depressive psychosis). Lithium therapy is not recommended for children under 12 years. Its side effects include polydipsia, polyuria, and renal damage. Retention of fluid and sodium may occur, which may require a sodium-restricted diet.

Liver. Also called "hepatic gland"; largest gland of the body, comprising about 3% of body weight, and located in the upper right quadrant of the abdomen. It is said that no other organ is concerned with so many varied functions as the liver, so that it is fittingly called the "warehouse and chemical manufacturer" of the body. More than 500 functions have been identified which include manufacture of vital substances (bile, prothrombin, fibrinogen, heparin, and urea); regulation of bodily processes (detoxification, reticuloendothelial activity, blood volume, and blood sugar level); metabolism of carbohydrate, protein, lipid, vitamins, and minerals; and storage of nutrients and other substances (protein reserves; glycogen; iron; copper; and vitamins A, D, K, and some B complex vitamins). The liver detoxifies many substances, including poisons, nicotine, and alcohol.

Liver disease. Any of the disorders that causes damage to the hepatic cells. Etiology may be unknown or congenital, but most liver diseases are caused by infection, biliary obstruction, heart disease, or by the toxic effects of alcohol and poisonous substances. Common symptoms of liver disease are anorexia, jaundice, fatigue, weakness, and an enlarged liver. For specific clinical signs, see under each liver disorder. *Nutrition therapy:* by understanding the numerous functions of the liver, one appreciates the rationale of nutrition therapy for each liver disease. The general guidelines are as follows: high calorie intake, allowing 35 to 45 kcal/kg of actual body weight, with carbohydrates as the main source; high protein level up to 1.5 times the RDA, except in hepatic coma when protein is restricted; and moderately low fat to make up the rest of the energy needs. If there is steatorrhea, use MCT oil. Sodium is restricted up to 2 g/day and fluids up to 1500 mL/day in case of edema or ascites. Avoid alcohol. Vitamin and mineral supplements are almost always needed, considering the role of the liver in the manufacture, metabolism, and/or storage of these nutrients. In advanced liver disease, the need for water-soluble vitamins may be as much as fives times the RDA. Of the fat-soluble vitamins, vitamin K supplement is emphasized. Monitoring serum levels of the nutrients helps determine the kind and amount of supplements needed for a particular liver disease. For specific nutrition therapy in liver disease, see *Ascites, Cholestatic liver disease, Cirrhosis, Hepatic coma, Hepatic encephalopathy, Hepatic failure,* and *Hepatitis.*

Liver failure. See *Hepatic failure.*

Liver function test (LFT). A diagnostic aid to evaluate various functions of the liver and the stage of liver disease. LFT includes alkaline phosphatase, aspartate transaminase, prothrombin time, serum ammonia, serum bilirubin, and serum glutamic pyruvic transaminase. See Appendix 30 for normal blood values.

Liver transplantation. A major surgery transferring liver from a donor with healthy organs but who is brain-dead to treat end-stage liver disease. Liver transplantation has become an established treatment in hepatic failure. *Nutrition therapy:* the goals are to correct malnutrition, particularly muscle wasting, edema, electrolyte imbalance, and abnormal blood glucose and lipids. Provide a pretransplantation diet that will correct nutritional deficiencies. Recommended intakes are 30 to 35 kcal/kg dry weight or 1.2 × basal energy expenditure (BEE), 1.0 to 1.5 g/kg/day protein as tolerated, carbohydrate and fat as needed, and vitamins and minerals at RDA levels. Adjust sodium and fluids if appropriate. Restrict protein to tolerance in hepatic encephalophathy and impending coma. Immediately after and for 2 months post-transplant, provide calories at 1.5 to 1.7 × BEE, protein at 1.5 to 1.75 g/kg/day, 30% of total calories as fat and 50% to 60% as carbohydrate, 2 to 3 g/day sodium, vitamins and minerals to meet RDA, and fluids as needed. TPN may be necessary if oral or enteral feedings can not be initiated within 5 days after surgery. Monitor potassium, phosphorus, and magnesium for refeeding syndrome. Feeding by mouth begins when postoperative ileus resolves. Start with liquids and progress to solid foods. Reduce TPN as enteral nutrition is increased. Supplementation with enteral formulas may be necessary, either orally or by tube feeding, to meet energy and protein needs. Exceptions from high protein intake are cases of increased serum ammonia and the amino acids that are precursors for neural transmitters (dopamine, norepinephrine, and serotonin), when branched-chain amino acids are increased and aromatic amino acids are restricted. Specialized formulas are available for modified amino acid levels. (See Appendix 39.) Monitor for side effects of immunosuppressive drugs and other medications, and adjust the diet accordingly. The long-term post-transplant diet aims to prevent high blood sugar, hyperlipidemia, and excessive weight gain. Provide calories at BEE × 1.1 to 1.3, protein at 1 g/kg/day, fat at 30% of total calories (low in saturated fats), carbohydrate as the remainder of calories (high in fiber and reduced in simple sugars), and vitamins and minerals to meet RDA. Restrict sodium if there is fluid retention. Use of a low-microbial diet is essential pre- and postoperatively, in conjunction with antibiotics, to reduce the risk of infection. See also *Transplantation* for general guidelines.

Load test. Also called "saturation test"; a method of assessing the nutritional status of a particular nutrient by measuring its urinary excretion after administration of a test dose to a person on a controlled intake. It is assumed that an individual whose tissues are saturated with the nutrient will retain little and excrete most of the dose, whereas one with low nutrient reserves will retain a large amount in order to saturate tissue levels.

Lofenalac®. Brand name of an infant formula that is low in phenylalanine; used in the dietary treatment of *phenylketonuria*. It is prepared by enzymatic digestion of casein followed by chemical treatment to remove 95% of the phenylalanine. See also Appendix 37.

Lomustine. An alkylating agent used in the treatment of malignant neoplastic diseases. It interferes with the function of DNA and RNA. Side effects include loss of appetite which may persist for 2 to 3 days after a dose; stomatitis; nausea and vomiting; and anemia. Excreted in breast milk. Breast feeding is not recommended during chemotherapy. Brand name is CeeNu®.

Lonalac®. Brand name of a low-sodium, high-protein product in powder form prepared from casein, lactose, and coconut oil; used as a dietary source of protein when sodium restriction is severe, as in congestive heart failure, nephrosis, and hepatic cirrhosis with ascites. See also Appendix 39.

Long-chain triclycerides (LCTs). Fats containing fatty acids longer than lauric acid (C12). Naturally occurring fats are composed predominantly of LCT containing palmitic (C16), stearic (C18), oleic (C18 with one double bond), and linoleic (C18 with two double bonds) acids. LCT should be restricted in the dietary treatment of *malabsorption syn-*

dromes and replaced with *medium-chain triglycerides (MCT)*.

Lou Gehrig's disease. See *Amyotrophic lateral sclerosis*.

Lovastatin. A lipid-lowering drug used for the treatment of Type IIa and Type IIb hyperlipoproteinemia; it reduces serum low-density lipoprotein (LDL) cholesterol and increases high-density lipoprotein (HDL) cholesterol. The brand name is Mevacor®.

Low-birth-weight infant (LBWI). The infant is considered LBW if the birth weight is less than 2500 g (5 ½ lb), regardless of gestational age. Infants born before 25 weeks rarely survive. If the weight is less than 1500 g (3 ½ lb), the infant is considered very-low-birth-weight. Tiny premature infants encounter problems such as feeding intolerances, respiratory distress, fluid and electrolyte imbalances, intracranial hemorrhaging, anemia, immunulogic abnormalities, cardiovascular difficulties, and hypoglycemia. *Nutrition therapy:* LBW infants usually need intensive care. Intravenous glucose solution and electrolytes are immediately given. Later, parenteral nutrition is introduced gradually, especially for infants with gastrointestinal problems. Care is taken to prevent protein-energy malnutrition with a proper choice of commercial formulas. As the infant improves, he or she can be gradually weaned from hyperalimentation to enteral feeding. Frequent oral stimulation will promote the development of the suckling and swallowing reflexes. The energy needs of LBW infants who receive parenteral nutrition are usually less (about 70 to 90 kcal/kg/day) than those who receive enteral feeding (120 to 130 kcal/kg/day) because the latter allows for nutrient losses due to poor digestion and absorption capacity. Protein needs range from 2 to 3 g/kg/day. The initial protein source is in the form of crystalline amino acid solutions. Fluid requirements are variable because of environmental and physiologic factors. Initially, 40 to 80 mL/kg/day of fluids may be given. After the second week of life, most LBW infants receive 120 to 180 mL/kg/day. Glucose is the typical carbohydrate source, calculated at 3.4 kcal/g. However, it is important to monitor glucose intolerance (hyperglycemia), which is especially common among VLBW infants.

Low-density lipoprotein (LDL). A plasma lipoprotein that has a density of 1.019 to 1.063 g/mL; classified as the beta fraction on the basis of electrophoresis. LDL contains 20% to 25% protein, 15% to 22% phospholipid, 45% to 50% cholesterol, and 10% to 13% triglycerides. LDL cholesterol is the most atherogenic lipoprotein. Individuals with elevated LDL have Type II hyperlipidemia. Because LDL is difficult to measure, the level is calculated by the following formula (all values in mg/dL):

$$\text{LDL} = \text{Total cholesterol} - \left(\frac{\text{triglycerides}}{5} + \text{HDL}\right)$$

A general optimal goal is to lower LDL-cholesterol to less than 130 mg/dL (3.4 mmol/L). Individuals with high LDL-cholesterol should follow the Step-One diet recommended by NCEP (National Cholesterol and Education Program), based on the following high-risk levels: $\geq$ 160 mg/dL ($\geq$ 4.1 mmol/L) for individuals without coronary heart disease (CHD) and with fewer than two risk factors, 130 to 159 mg/dL (3.4 to 4.1 mmol/L) for individuals without CHD, but with two or more risk factors, and $\leq$100 mg/dL ($\leq$2.6 mmol/L) for individuals with CHD. See *Coronary heart disease* and *Hyperlipoproteinemia*.

Low-sodium syndrome. Disturbance characterized by the following set of symptoms: weakness, lethargy, loss of appetite, nausea and vomiting, confusion, acid–base disturbance, and abdominal pain with general muscular cramps, renal damage, oliguria and later uremia, and possibly convulsions and shock. It is caused by prolonged periods of very low sodium intake, adrenal cortical insufficiency, and marked losses of body fluids and electrolytes, as in very hot weather or excessive perspiration, severe burns, marked diarrhea, and vomiting. Prompt provision of salt will correct the condition.

Luteotropic hormone (LTH). Also called "luteotropin," "prolactin," and "lactogenic hormone"; a hormone secreted by the anterior lobe of the hypophysis (pituitary gland). It is helpful in the development of the mammary glands and also initiates milk secretion. It stimulates the corpus luteum to synthesize progesterone and estrogens.

Lycopene. The principal red pigment present in tomatoes, watermelon, and pink grapefruit. It is an isomer of carotene but has no vitamin A activity.

Lymph. Fluid obtained from lymphatic ducts; one of the circulating fluids of the body. When fat globules are present, lymph turns milky. It resembles blood plasma in appearance and composition, except that lymph contains colorless cells (lymphocytes) and has a lower protein content than plasma. Its main function is to nourish and bathe cells by circulating substances from the blood into the tissues. It is important in fat absorption.

Lymphocyte. Type of white blood cell that arises in the reticular tissue of the lymph glands. There is a single nucleus and the cell is surrounded by a nongranular protoplasm. Lymphocytes are important for antibody formation. There are two major classes of lymphocytes: the B cells which mature in bone marrow, and the T cells which mature and multiply in the thymus. See *Total lymphocyte count.*

Lymphoma. Abnormal growth of the lymphoid tissue; usually malignant but may be benign. Characterized by enlarged lymph nodes, weakness, fever, anemia, anorexia, and weight loss. The liver and spleen may be enlarged, and bone lesions and gastrointestinal disturbances may also occur. Hodgkin's disease is one of several kinds of lymphomas. Treatment usually includes radiotherapy and chemotherapy. *Nutrition therapy:* encourage a high-calorie, high-protein intake to correct weight loss and to increase resistance to infections. Limit fat intake to 30% of total calories/day because a high fat (saturated and polyunsaturated) intake may be immunosuppressive. Monitor for side effects of chemotherapy or radiotherapy and adjust the diet accordingly as tolerated. See *Chemotherapy* and *Radiotherapy.* See also *Cancer cachexia.*

Lysin. An *antibody* with destructive action on cells and tissues, causing dissolution or breakdown.

Lysine (Lys). An *essential amino acid;* the limiting amino acid in many cereal proteins, especially gliadin. Deficiency in humans may cause nausea, vomiting, dizziness, and anemia, in addition to growth failure in the young. In the tissues, this basic amino acid readily converts its epsilon carbon to carbon dioxide and helps to form glutamic acid. However, it does not exchange its nitrogen with other circulating amino acids, a property unique to lysine. See also *Amino acid.*

Lysine intolerance. Also called "hyperlysinemia." A congenital metabolic disorder caused by the inability to hydrolyze the amino acid lysine due to insufficient levels of the enzyme lysine ketoglutarate reductase. Symptoms include nausea, vomiting, mental retardation, episodes of coma, and high blood levels of lysine, arginine, and sometimes ammonia. Whether or not dietary treatment is effective in hyperlysinemia is controversial. It may be prudent to restrict protein intake to 1 g/kg body weight.

Lysolecithin. A substance obtained by partial hydrolysis of lecithin, with only one fatty acid liberated. It aids in the emulsification of dietary lipids.

Lysophosphatides. Substances that are destructive to red blood cells. They are produced from lecithin by the action of the enzyme lecithinase, which is present in the venom of poisonous snakes.

Lysozyme. Enzyme that digests certain high-molecular-weight carbohydrates and some gram-positive bacteria. It is present in tears, saliva, mucus and nasal secretions, and other body fluids. Its activity is reduced by generalized malnutrition, particularly by vitamin A deficiency.

M

M. 1. Symbol for mega-. 2. Symbol for molar concentration. 3. Abbreviation for metastasis, as in the malignant stage of cancer.

MAC. Abbreviation for *mid-arm circumference*.

Macrocyte. A giant red blood cell. *Macrocythemia* is an abnormal number of macrocytes in the blood. *Macrocytic anemia* is a type of anemia in which the red blood cells are unusually large, as in folic acid or vitamin B_{12} deficiency.

Macronutrients. Nutrients that are present in relatively high amounts in the body, constituting about 0.005% of body weight (50 ppm) or above. Protein, fat, water, and major minerals are macronutrients. See also *Nutrient classification*.

Macrophages. Cells of the reticuloendothelial system; they ingest foreign particles such as bacteria and cellular debris by *phagocytosis*.

Magnacal®. Brand name of a high-calorie liquid formula that is low in residue and lactose-free; provides a concentrated source of calories (2 kcal/mL) for patients with restricted fluid allowance or increased energy needs. Suitable as an oral supplement or for tube feeding. See also Appendix 39.

Magnesium (Mg). A major mineral essential to plants and animals; a component of chlorophyll in green plants. In humans, it is needed in relatively large amounts. In adults, it comprises about 0.05% of body weight, 60% of which is in bones and teeth, about 40% in the muscle and soft tissues, and the remainder in extracellular fluids. As a cofactor in reactions involving adenosine triphosphate, magnesium is important in carbohydrate, protein, and fat metabolism. More than 300 enzymes are known to be activated by magnesium. It also plays an important role in neuromuscular transmission and activity. Dietary magnesium deficiency is uncommon because it is widely distributed in nature, particularly in legumes, nuts, and unrefined grains; other good sources are cocoa, soybeans, and green leafy vegetables. Magnesium deficiency, or hypomagnesemia, occurs when the dietary intake is poor and associated with disease states involving intestinal malabsorption and/or decreased renal function, as in protein-calorie malnutrition, chronic alcoholism with malnutrition, and renal disease involving tubular dysfunction. Magnesium deficiency also occurs in long-term use of magnesium-free parenteral feeding; when there is excessive loss from the body, as in burns; and in increased urinary magnesium excretion with drugs such as the cardiac glycosides and the loop diuretics, furosemide and ethacrynic acid. Deficiency states affect the neuromuscular, cardiovascular, and renal systems; characteristic symptoms are muscular twitching and muscle weakness, convulsions, tachycardia, nausea, and vomiting. Recent studies have linked a low intake of magnesium to cardiovascular disease, diabetes, and hypertension. Magnesium toxicity and hypermagnesemia occur primarily when there is severe renal insufficiency and when magnesium salts or magnesium-containing antacids and cathartics are administered in large doses. Signs of toxicity include nausea and vomiting, hypotension, paralysis of voluntary muscles, and somnolence. See Appendix 1 for the recommended dietary allowances. See also *Hypermagnesemia* and *Hypomagnesemia*.

Malabsorption syndrome. Set of symptoms indicating defective absorption of nutrients, which may include carbohydrates, protein, fat, vitamins, and minerals, as well as total calories. It may occur as a result of structural changes in the alimentary tract or its adjacent organs, failure of food to reach the absorptive surfaces, maldigestion, diversion of foodstuffs to intestinal organisms, failure of the absorptive mechanism, or intestinal resection. Symptoms include abdominal distention and pain, steatorrhea and diarrhea, anorexia and muscle wasting with loss of weight, and anemia. Examples of disorders that show the malabsorption syndrome are *celiac disease, cystic fibrosis, carbohydrate intolerance,* and *pancreatic insufficiency*.

Malacia. Morbid softening or softness of a tissue or part, as in *osteomalacia,* or softening of bones.

Malaria. An infectious disease caused by one of four species of *Plasmodium,* a protozoan transmitted by *Anopheles* mosquito bites. It may also be spread by blood transfusion or by the use of an infected hypodermic needle. *Nutrition therapy:* see *Fever*. Frequently, the liver is enlarged and liver functions are impaired; in this case, dietary modifications for liver disease are followed.

Malnutrition. 1. Simply stated, any disorder of nutrition; bad or undesirable health status due to either *lack* or *excess* of a nutrient supply. 2. State of impaired biologic activity or development due to a discrepancy between the nutrient supply and the nutrient demand of cells. Malnutrition may be classified into three categories: malnutrition associated with poverty or an inadequate food supply; malnutrition associated with ignorance and indifference; and malnutrition secondary to such factors as diseases, alcoholism, drug abuse, and mental illness. Based on percentage of expected weight for age, malnutrition may be mild or first degree if weight for age is 75% to 89% of standard, moderate or second degree if 60% to 75% of standard weight, and severe or third degree if less than 60% of standard. All cases where edema is present, irrespective of weight of the individual, are included in the third degree. Undesirable medical practices that contribute to malnutrition in hospitalized patients include: withholding meals for diagnostic tests; maintaining NPO status for several days; prolonged use of glucose and saline intravenous fluids; failing to observe patient food intake; inadequate monitoring of nutritional status; delaying nutritional support until depletion is advanced; and failing to recognize increased metabolic needs. If allowed to develop or if left untreated, malnutrition may lead to increased incidence of infection, decreased immune competence, decreased ventilatory response, delayed wound healing, and fluid and electrolyte imbalance. See also *Protein-energy malnutrition*. See also *Deficiency disease* and *Nutritional deficiency (inadequacy)*.

Maltase. Also called "alpha-glucosidase"; an enzyme found in yeast and intestinal juice. It hydrolyzes *maltose* to two glucose units.

Maltose. Also called "malt sugar"; a disaccharide made up of two molecules of glucose. It is formed as an intermediate product of starch hydrolysis. During digestion, maltose is hydrolyzed by the enzyme maltase to two glucose units.

MAMC. Abbreviation for *mid-arm muscle circumference*.

Manganese (Mn). An essential trace mineral widely distributed in plant and animal cells. The human body contains 12 to 20 mg of manganese, which is concentrated in the liver, pancreas, kidney, skin, muscles, and bones. Manganese is a component of two metalloenzymes and is a catalyst for a number of enzymes involved in glucose and fatty acid metabolism and urea formation. It is also needed for bone development, skin integrity, and utilization of thiamin. Main food sources are nuts, legumes, and unrefined grains; tea, leafy vegetables, and fruits contain moderate amounts. Dietary manganese deficiency in humans has not been observed. Experimentally induced deficiency results in weight loss, transient dermatitis, retarded growth of hair and nails, changes in hair color, and decreased serum cholesterol and triglycerides. Cases of suboptimal manganese status have been found in selected populations, such as in children with inborn errors of metabolism (phenylketonuria, maple syrup urine disease, and galactosemia), in children and adults with epilepsy, and in persons with exocrine pancreatic insufficiency and active rheumatoid arthritis. Individuals with these conditions may have a special need for manganese. Toxicity from dietary sources appears to be highly unlikely. The few cases of manganese toxicity have resulted from ingesting large doses of the mineral supplement for 4 to 5 years, drinking well water contaminated with manganese from buried batteries, and prolonged exposure and inhalation of ore dust. Toxicity results in dementia and a psychiatric disorder resembling schizophrenia, followed by a crippling neurologic disorder that is similar to Parkinson's disease. See Appendix 2 for the estimated safe and adequate daily dietary intake.

Mannitol. A sugar alcohol obtained from the hydrogenation of *mannose*. Commercially it is extracted from certain seaweeds. It has a sweetening power, as does glucose, but yields only half as many calories because it is only partially absorbed.

Mannose. A monosaccharide containing 6 carbon atoms (a hexose); does not occur free in nature. It is found in legumes in the form of *mannosan,* a partially digestible polysaccharide.

MAO. Abbreviation for *monoamine oxidase*.

Maple syrup urine disease (MSUD). Also called "branched-chain ketoaciduria." An inborn error of metabolism due to a defect in the oxidative decarboxylation of the branched-chain amino acids leucine, isoleucine, and valine. This leads to the accumulation of these keto acids in the blood and cerebrospinal fluid and consequent excretion in the urine, to which they impart an odor of maple syrup or burnt sugar. Symptoms usually appear within the first week of life; these include difficulty in feeding with jerky aspirations, vomiting, periods of rigidity and flaccidity, seizures of the grand mal type, hypoglycemia, and possibly death. Survivors generally have severe brain damage. *Nutrition therapy:* treatment consists of restricting the dietary intake of leucine, isoleucine, and valine. Formulas specifically designed for MSUD (Analog MSUD, Ketonex®, Maxamaid, Maxamum, MSUD Diet Powder) are supplemented with a small amount of cow's milk or infant formula to provide the BCAA needed to support growth and development; small amounts of low-protein foods are used for the older child. Monitor and adjust BCAA intake to maintain plasma leucine levels between 2 and 5 mg/dL. Levels above 10 mL/dL are associated with ketoacidemia and neurologic symptoms. During illness, decrease BCAA and increase carbohydrates. A rare form of MSUD responds to 10 mg/day of thiamin hydrochloride for 3 weeks. See *MSUD Diet Powder*.

Marasmic-kwashiorkor. Protein-energy malnutrition characterized by clinical findings of both *marasmus* and *kwashiorkor* due to inadequate intake of energy and protein, plus superimposed infection or catabolic stress. See *Protein-energy malnutrition*.

Marasmus. 1. Form of extreme undernutrition due primarily to a lack of calories and protein. It was formerly called "protein-calorie malnutrition" (PCM). The preferred designation now is "energy-protein malnutrition (EPM)" or "protein-energy malnutrition (PEM)." 2. Infantile atrophy that occurs almost wholly as a sequel to acute disease, especially diarrheal diseases. Marasmus is characterized by loss of weight, retarded growth and development in infants and children, loss of subcutaneous fat, and wasting of muscle tissue. The word "marasmus" comes from a Greek word meaning "wasting." Other clinical symptoms due to protein deficiency are edema, skin changes, anemia, enlarged liver, and increased susceptibility to infections, accompanied by high fever. Marasmus is a major public health problem affecting as many as 50% of children in the developing countries. *Nutrition therapy:* depends on the presence of complications such as dehydration, electrolyte imbalance, vitamin deficiencies, and infections. Fluid and electrolyte imbalances should be corrected promptly. Initially, start with small feedings and gradually increase the amounts of protein and calories. Do not overfeed. Monitor electrolytes and blood sugar for signs of *refeeding syndrome*. The initial dietary intake recommended for children is about 50 kcal/kg and 1 g/kg of protein; the ultimate goal for optimal recovery is an intake of 175 kcal and 3 to 4 g/kg/day of protein by the second week. More aggressive nutritional support may be required if weight loss is more than 30%. For adults, the usual amounts for protein are 1.2 to 1.3 g/kg for mild protein depletion, 1.4 to 1.5 g/kg for moderate depletion, and 1.6 to 1.7 for severe depletion. Provide 15 to 20 kcal/kg and gradually increase to 35 to 40 kcal/kg of desirable body weight, or an amount necessary to achieve a gradual weight gain of 1 to 2 lb/week.

Maxamaid XP. An orange-flavored powder free of phenylalanine and formulated for the 2- to 8-year-old child with phenylketonuria. Contains L-amino acids, carbohydrates, minerals, trace elements, and vitamins.

Maxamum XP. A phenylalanine-free powdered product designed for children 8 years of age and older and for pregnant women with phenylketonuria.

MCFA. Abbreviation for medium-chain fatty acid. See under *Fatty acid*.

MCT. Abbreviation for *medium-chain triglycerides*.

MCT oil. A special dietary supplement for use as a substitute for or as a supplemental source of fat calories for patients in whom conventional food fats are poorly digested, absorbed, or utilized. It is made from fractionated coconut oil and contains 67% octanoic (C-8) and 23% decanoic (C-10) fatty acids; provides 115 kcal/tablespoon. See *Medium-chain triglycerides*.

MCV. Abbreviation for *mean corpuscular volume*.

ME. Abbreviation for *metabolizable energy*.

Meals-on-Wheels. See *Home-delivered meals*.

Mean corpuscular volume (MCV). Determined from the following formula:

$$MCV = \frac{HCT}{RBC} \times 10$$

where HCT and RBC represent values for hematocrit and red blood cell count, respectively. MCV represents the average size of red blood cells, which may be abnormally small (microcytic), abnormally large (macrocytic), or normal (normocytic). The normal MCV is between 82 and 98 μm^3. Elevated levels may indicate macrocytosis due to a deficiency of folate or vitamin B_{12} or excess alcohol consumption. Depressed levels indicate microcytosis due to iron deficiency or malabsorption, increased iron requirement, and blood loss.

Medical foods. Products specifically formulated to be consumed or administered enterally under the supervision of a physician and that are intended for the specific dietary management of a disease or condition for which distinctive nutritional requirements, based on recognized scientific principles, are established by medical evaluation. Medical foods are used for the primary treatment of metabolic and genetic disorders.

Medical history. Record of pertinent information about an individual or patient that includes past illnesses, familial tendencies toward certain diseases, general health status since birth, and present complaints. Routinely taken when a patient is admitted to a hospital or seeks a medical consultation. It is a diagnostic aid in the management of a patient and a technique used in *nutrition surveys* and *nutritional assessment* to detect secondary or conditioning factors in nutritional inadequacy. Patients are at high nutrition risk if their medical history indicates the following: increased metabolic needs (infection, fever, trauma, burns, cancer, hyperthyroidism, pregnancy); increased nutrient losses (open wound, draining fistula or abscess, chronic blood loss, chronic renal dialysis, exudative enteropathies); recent major surgery or illness; diseases of the gastrointestinal tract; chronic diseases (diabetes mellitus, hypertension, hyperlipidemia, chronic lung, liver, or renal disease); and prolonged comatose state.

Medical nutrition therapy. The use of specific nutrition services to treat an illness, injury, or condition. It consists of two phases: nutrition assessment and nutrition therapy. The latter includes diet therapy, counseling, and/or specialized nutrition therapies. The services of qualified health professionals, e.g., registered dietitians and dietetic technicians, are provided in a variety of settings, including inpatient and outpatient services, ambulatory settings, long term care facilities, rehabilitation centers, schools, offices, and homes.

Mediterranean diet. Refers to the typical food intake in Greece, Italy, Portugal, and Spain. Although fat consumption is higher than in the United States, the use of olive oil, which is rich in monounsaturated fatty acids, is said to be beneficial to health. Additionally, the Mediterranean diet includes more fresh fruits, vegetables, whole grains and breads, and moderate consumption of red wine with meals.

Medium-chain triglycerides (MCT). Fats composed of fatty acids shorter than lauric acid, predominantly saturated fatty acids with 6 to 10 carbon atoms. Compared with long-chain triglycerides, MCT have the following advantages: lower melting point, faster rate of hydrolysis, less need for bile acids, easier dispersion in water, smaller quantity incorporated into lipid esters, and less tendency to storage in the liver. MCT preparations are useful in the dietary management of *malabsorption syndromes* such as pancreatitis, postgastrectomy, sprue, cystic fibrosis, and chyluria. It is widely used in pediatric and enteral formulas and is believed to be an ideal substrate during stress and sepsis. Ingestion of large amounts may produce nausea, abdominal distention, cramping, and diarrhea. MCTs do not provide essential fatty acids.

Megacolon. An enlarged bowel due to an abnormality in the dilatation of the colon. Usually affects elderly persons who have constipation problems or use excessive amounts of laxatives. Tumors and strictures may also cause obstruction of the elimination process. Signs are distention, flatulence or incontinence, nausea, fatigue, and headache. *Nutrition therapy:* provide adequate fluids and fiber, particularly prune juice, fresh fruits, leafy vegetables, legumes with skin, and whole-grain cereals.

Megaloblastic anemia. Type of anemia characterized by an increased level of *megaloblasts,* which are primitive nucleated red blood cells much larger than the mature normal erythrocytes. The megaloblastic anemias of pregnancy and infancy respond readily to folic acid therapy and an adequate balanced diet. See also *Anemia.*

Megavitamin therapy. Use of massive doses of vitamins to cure hyperactivity and other behavioral abnormalities. Advocates of megavitamin therapy or orthomolecular psychiatry propose that optimum molecular concentrations of vitamins are essential for proper mental functioning. However, this theory was based on studies that failed to meet the requirements of a proper scientific protocol. Currently, the American Academy of Pediatrics and the American Psychiatric Association feel that megavitamin therapy is unproved in terms of safety and efficacy.

Menadione. Synthetic compound with vitamin K activity; used in the prevention and treatment of hypoprothrombinemia secondary to factors that limit absorption or synthesis of vitamin K. It is two to three times more potent than the naturally occurring vitamin K. The brand name is Synkayvite®.

Menaphthone. Vitamin K_3 or *menadione.*

Menaquinone. Vitamin K_2; a homolog of vitamin K synthesized in animals. It is found in meats, especially liver, and in eggs and cheese, but the major source is bacterial synthesis in the intestines. Formerly called "farnoquinone."

Ménétrier's disease. Also called "giant hypertrophic gastritis." A rare disease characterized by massive enlargement of the gastric mucosa and hypersecretion of gastric acid. Hyperplastic cells that cover the walls of the stomach cause anorexia, nausea, vomiting, gastritis, and abdominal disturbances. There may be iron depletion from chronic blood loss and loss of plasma protein from the enlarged gastric folds, which may result in edema. *Nutrition therapy:* dietary modifications are aimed at correcting the clinical signs and symptoms. Provide a high-calorie, high-protein diet and reduce edema with sodium restriction (2 to 3 g/day). Monitor nutrient interactions of drugs used. See also *Gastritis.*

Meniere's syndrome. A chronic disorder of the inner ear with signs of vertigo, nausea and vomiting, blurred vision, and nerve deafness. In some cases, allergy, trauma, or infection accompanies the disorder or may be the cause. *Nutrition therapy:* restrict fluid and salt intake to reduce the pressure on the labyrinth. Eliminate any allergenic food.

Meningitis. Inflammation of the *meninges,* the membranes covering the brain and spinal cord. Often caused by bacterial and viral infections. Symptoms include fever, severe headaches, nausea and vomiting, stiff neck, and tachycardia. Onset is sudden, and if untreated it could lead to septic shock and respiratory failure. *Nutrition therapy:* intravenous fluids and nasogastric tube feeding may be required until dehydration is corrected. When tolerated, gradually progress to oral feeding of a high-calorie, high-protein diet, with at least 2 liters of fluids/day. Monitor caloric intake to avoid weight loss and to offset the catabolic effects of fever.

Menke's disease. Also called "kinky hair disease." An inborn error of *copper* metabolism characterized by brittle, kinky hair texture resembling steel wool, accompanied by all the overt symptoms of gross copper deficiency except anemia. It is due to the intestinal failure to absorb copper, which accumulates in the intestinal mucosa. The liver becomes refractory to copper intake. *Nutrition therapy:* some improvement is achieved with parenteral copper administration (not by the oral route), which restores the *ceruloplasmin* level to normal, although the neurologic abnormalities do not improve. The disease is usually fatal.

Menstruation. Also called "menses"; the periodic cycle, usually 28 to 30 days, characterized by uterine bleeding or menstrual flow. It normally lasts for 3 to 7 days, with a bloody discharge of about 125 mL. Iron lost during menstruation can be as much as 1.5 mg/day, with an average of 0.7 mg/day, which is taken into consideration in establishing the recommended dietary allowance for women who are still menstruating. See also *Premenstrual syndrome*.

Mental illness. Also called "psychiatric disorder." Any disturbance in the adaptive and emotional balance of an individual. *Nutrition therapy:* nutritional care must be individualized. Among the feeding problems that may be encountered are the following: a depressed patient loses interest or appetite; an overactive patient may not sit long enough to eat; a delusional patient may develop fears and suspicions about food; an emotionally insecure patient may indulge in overeating for personal satisfaction; a patient with anorexia nervosa is difficult to feed; and a patient undergoing shock therapy may need a high caloric intake. See *Nutrition, psychiatry*.

Mercaptopurine. A chemical analog of the physiologic purines adenine and hypoxanthine; a component of various chemotherapeutic drugs used in the treatment of acute leukemia. It may antagonize pantothenic acid and produce a sprue-like syndrome. It may also cause weight loss, anemia, nausea, vomiting, and diarrhea. The brand name is Purinethol®.

Mercury (Hg). Heavy liquid metal used in thermometers and other scientific instruments; also used as a dental amalgam. Its salts are used as antiseptics, diuretics, fungicides, and parasiticides. The environmental concern regarding mercury in the food supply focuses on its toxic effects on the tissues, particularly those of the brain. Toxicity is due to the binding of tissue proteins and interference with cellular metabolism. The dangerous forms of mercury are the alkyl derivatives methylmercury and ethylmercury. Population outbreaks of poisoning have been reported from the ingestion of fish and seafood exposed to methylmercury contamination and of seed grains previously treated with mercurial fungicides. Characteristic symptoms of mercury poisoning are visual abnormalities, tremors, proteinuria, apathy, and mental deterioration. The U.S. Department of Agriculture has set a safe level guideline of 0.5 ppm mercury in fish.

Meritene®. Brand name of a milk-based nutritional product, in liquid or powder form, for oral supplementation or tube feeding. Contains 55 g of lactose/1000 mL. See Appendix 39.

Metabolic body size. Also called "physiologic size"; the active tissue mass of an individual. It is determined by raising the body weight in kilograms to the three-fourths power.

Metabolic cart (MC). A portable cart used for indirect *calorimetry*. It measures the amount of oxygen and carbon dioxide in a respiratory gas sample. Data are used to calculate the resting energy expenditure (REE) and respiratory quotient (RQ) as follows:

$$REE = 3.9\ VO_2 + 1.1\ VCO_2 \times 1.44$$

where VO_2 is oxygen consumed (mL/min) and VCO_2 is carbon dioxide produced (mL/min).

$$RQ = VCO_2 \div VO_2.$$

Metabolic chamber. A room-sized chamber that permits continuous and long-term analysis of exhaled gases; used in indirect *calorimetry*.

Metabolic formulas. A line of medical foods and formulas designed to meet the special nutrient needs of infants, children, and adults with inherited metabolic disorders. Examples are Cyclinex®, Glutarex®, Ketonex®, MSUD Diet Powder, and Tyromex®.

Metabolic pool. 1. Phrase descriptive of the manner in which a nutrient can change, combine with, or participate in metabolic reactions. 2. Components, indistinguishable as to origin, that may be employed for either synthetic or degradative processes. When the end products of digestion (e.g., amino acids, fatty acids, glycerol, and glucose) are absorbed, they enter the metabolic pool and intermingle with other substances, or they are metabolized for various bodily functions.

Metabolic water. See under *Water*.

Metabolism. The sum of all the chemical changes in the body. There are two phases: *anabolism,* or constructive metabolism, which is concerned with the building up of materials and tissues; and *catabolism,* or destructive metabolism, which is the breaking down of materials and tissues. See also *Intermediary metabolism*.

Metabolizable energy (ME). The portion of gross food energy capable of transformation in the body for useful work (or net energy) and for basal metabolism. It does not include heat losses from urinary and fecal excretions.

Metalloprotein. A compound protein; a protein combined with a metal-containing prosthetic group. Examples are ferritin (iron-containing protein), carbonic anhydrase (zinc-containing protein), and ceruloplasmin (copper-containing protein). See also Appendix 9.

Metallothionein. The most abundant nonenzymatic zinc-containing protein; also contains cysteine, cadmium, copper, and iron. It may have a role in the detoxification of metals and the metabolism of sulfur-containing amino acids.

Methemoglobin. Also called "ferrihemoglobin." A hemoglobin molecule in which the iron component is oxidized to the ferric state and can no longer carry oxygen. It is present in the blood as a product of normal metabolic activity in trace amounts only. See *Hemoglobin*.

Methionine (Met). An *essential amino acid;* contains sulfur and the labile methyl group. It is one of the *lipotropic agents* and participates in methylation reactions; hence it is important in protein and fat metabolism. Methionine deficiency results in poor weight gain; decreased plasma cholesterol; increased plasma phenylalanine, proline, serine, threonine, and tyrosine; and decreased plasma methionine.

Methotrexate. A folic acid antagonist; a potent anticancer drug. It inhibits dihydrofolate reductase and thus decreases the formation of tetrahydrofolic acid. It also causes malabsorption of folate, vitamin B_{12}, fat, and carotene. Other adverse effects include anorexia, weight loss, stomatitis, gingivitis, megaloblastic anemia, nausea, vomiting, and diarrhea. It is excreted in breast milk. Brand names are Folex® and Rheumatrex®.

Methylcellulose. Preparation of indigestible polysaccharide that provides bulk and satiety value. It is prescribed for constipation and in the dietary management of obesity.

Methyldopa. An antihypertensive agent that is structurally related to *catecholamines* and their precursors. It increases the need for folic acid and vitamin B_{12}, and may also cause glossitis, dry mouth, constipation, and weight gain due to salt and fluid retention. It is excreted in breast milk. High protein meals and iron supplements may reduce absorption and drug effect. Brand names are Aldoclor®, Aldomet®, and Aldoril®.

3-Methylhistidine. Compound released during the breakdown of actin and myosin. The 24-hour urinary excretion of 3-methylhistidine is an indicator of change in skeletal muscle mass and in the rate of muscle protein breakdown. Its excretion is increased in sepsis, trauma, and other catabolic conditions.

Methylmalonic acidemia. A metabolic disorder characterized by high levels of methylmalonic acid in blood and urine due to defective reduction or transport of cobalamin. Symptoms include acidosis, hypoglycemia, ketonuria, and elevated plasma ammonia and lactate. *Nutrition therapy:* maintain adequate energy intake to prevent tissue catabolism; encourage fluids to prevent dehydration, normalize blood ammonia, and excrete abnormal metabolites in the urine; and correct electrolyte imbalances. Protein intake may range from 1.0 to 1.5 g/kg/day. Response to protein varies. Some people require little or no protein restriction, whereas others may require severe protein restrictions. The vitamin B_{12}-responsive patient may respond to pharmacologic doses of 1 to 2 mg/day.

Metoprolol. A beta-adrenergic blocking agent used in the management of hypertension, myocardial infarction, and angina. It may cause dry mouth, nausea, vomiting, abdominal cramping, and constipation or diarrhea; it may also increase blood urea nitrogen and serum triglycerides. The brand name is Lopressor®.

Metronidazole. An antibacterial and antiprotozoal agent used for the treatment of trichomonal infections and intestinal amebiasis. It has a sharp, unpleasant, metallic taste. May cause loss of appetite, altered taste, dry mouth, glossitis, nausea, vomiting, epigastric distress, and abdominal cramping. It is excreted in breast milk. It should not be taken with alcohol due to the risk of a disulfiram-like reaction. Brand names are Flagyl®, Metro I.V.®, and Protostat®.

Mg. Chemical symbol for magnesium.

MI. Abbreviation for *myocardial infarction.*

Micelle. 1. Dispersed particles in a colloidal system that are held in a particulate form because of their special physicochemical properties. 2. One of the submicroscopic structural units of protoplasm. 3. It is an essential step in the digestion and absorption of fats; an emulsified compound formed by positioning the hydrophobic radicals of the molecule toward the center and the hydrophilic part toward the outside. Thus, the compound becomes more water-miscible.

Microbiologic assay. Means of analyzing nutrients, especially vitamins and amino acids, by the use of microorganisms. A suitable microorganism is inoculated into a medium containing all the needed growth factors except the one nutrient under examination. The rate of growth is proportional to the amount of the particular nutritional factor added to the medium. The commonly used test microorganisms and the vitamins determined are as follows: *Lactobacillus fermentum 36* (thiamin), *Lactobacillus casei* (riboflavin and folic acid), *Saccharomyces carlsbergensis* (vitamin B_6 and inositol), and *Lactobacillus leichmannii 313* (vitamin B_{12}).

Microcyte. Red blood cell that is smaller than normal. *Microcytic anemia* is a type of anemia in which the red blood cells are smaller than normal. This is seen in iron-deficiency anemia.

Microlipid®. Brand name of a safflower oil supplement to oral or tube feeding formulas; used as a concentrated source of calories when protein and carbohydrate need to be restricted, as in renal failure and chronic obstructive pulmonary disease. Provides 4.5 kcal/mL.

Micronutrients. Nutrients present in the body in amounts of less than 0.005% of body weight (50 ppm). Examples are trace minerals, vitamin B_{12}, and pantothenic acid.

Midarm circumference (MAC). Also called "upper-arm circumference." Measurement (in centimeters) at the midpoint between the tip of the acromial process of the scapula and the olecranon process of the ulna. MAC is an indicator of fat stores. Measurements greater than the 50th percentile are acceptable; the 40th to 50th percentile indicates mild fat depletion; the 25th to 39th percentile indicates moderate fat depletion; and measurements below the 25th percentile indicate severe fat depletion.

Midarm muscle circumference (MAMC). Measurement of midarm muscle circumference provides an indirect assessment of skeletal muscle protein reserves. This is obtained by measuring the midarm circumference (MAC) and the triceps skinfold (TSF). The midarm muscle circumference (in centimeters) is then determined from the following formula:

$$\text{MAMC} = \text{MAC} - (0.314 \times \text{TSF})$$

where MAC and TSF, respectively, represent values for midarm circumference (in centimeters) and triceps skinfold (in millimeters) measurements. Values greater than 85% are considered acceptable; 76% to 85% indicates mild depletion; 65% to 75% moderate depletion; and less than 65%, severe depletion. See Appendix 23.

Migraine. A recurring vascular headache characterized by severe pain, nausea and vomiting, visual disturbances, and vertigo. Dietary factors may trigger a migraine attack. Food allergy, alcohol, tyramine, histamine, phenylethylamine, nitrites, monosodium glutamate, and aspartame appear to have some link with migraine. *Nutrition therapy:* the diet should be individualized. Eliminate or restrict foods known to trigger a migraine attack. Foods commonly associated with migraine are red wine, beer, chocolate, coffee, tea, nuts, fermented foods, hot dogs and luncheon meats, tuna, mackerel, and cola drinks. Establish a tolerance threshold for the offending substance.

Milk-alkali syndrome. Occurs in persons with peptic ulcer who consume large amounts of milk and alkalis for a long period of time. The symptoms are hypercalcemia, calcium deposition in soft tissues, vomiting, gastrointestinal bleeding, and high blood pressure.

Mineral. An inorganic element that remains as ash when food is burned. Analysis of mineral ash may show as many as 40 kinds, but only 17 are essential to human nutrition, and 9 are probably essential. The criteria that determine the essentiality of a mineral are as follows: a deficiency state occurs with a diet considered adequate in all respects except for the mineral under study; there is a significant response (growth or alleviation of signs of deficiency) when a supplement of the mineral is given; the response is repeatedly demonstrable; and the deficiency state correlates with a low level of the mineral in the blood or tissues. The essential major minerals (macrominerals) or those required in amounts of 100 mg/day or more are *calcium, chloride, magnesium, phosphorus, potassium, sodium,* and *sulfur*. The essential trace minerals (microminerals) or those required in amounts of less than 100 mg/day are *chromium, cobalt, copper, fluoride, iodine, iron, manganese, molybdenum, selenium,* and *zinc*. Probably essential trace minerals are *arsenic, boron, cadmium, lead, lithium, nickel, silicon, tin,* and *vanadium*.

Mineral functions. Minerals make up about 4% of body weight. In general, the role of minerals in the body is classified as either *structural* or *regulatory*. The function of a mineral is structural when it is an integral part of a cell, tissue, or substance, e.g., calcium, phosphorus, and magnesium in bones and teeth; sulfur in hair, insulin, and thiamin; iron in hemoglobin; and chloride in hydrochloric acid of gastric juice. Regulatory functions include maintenance of water and acid–base balance, muscle contractility, nerve irritability, and actions as cofactors of enzyme systems.

Mineral oil. Liquid petrolatum preparation used as a lubricant laxative. It has no caloric value, but it interferes with the absorption of carotene and the fat-soluble vitamins A, D, E, and K; it also decreases the absorption of calcium and phosphorus. Brand names are Agoral®, Kondremul®, and Milkinol®.

Minoxidil. A vasodilator used in the treatment of refractory hypertension. Sodium and fluid retention occur frequently and may result in edema and weight gain. Restrict sodium intake to 2 g/day to minimize fluid retention. Trade names are Loniten®, and Rogaine®.

Mn. Chemical symbol for *manganese*.

Mo. Chemical symbol for *molybdenum*.

Moducal®. Brand name of a refined, readily digestible carbohydrate source for use in patients with increased caloric requirements or limited intake. It is bland and nearly tasteless, and can be mixed with foods and beverages without appreciably altering their taste. Consists of glucose polymers and small amounts of glucose, maltose, and isomaltose. Provides 3.8 kcal/g of powder.

Molality. Number of moles of solute per 1000 g of solvent; usually designated by a small "m." A *1-molal solution* contains 1 mole (gram molecular weight) of solute in 1000 g of solvent. See *Osmolality*.

Molarity. Number of moles of solute per liter of solution; usually designated by a capital "M." A *1-molar solution* contains 1 mole (gram molecular weight) of solute in 1 liter of solution. See *Osmolarity*.

Molybdenum (Mo). An essential trace mineral; it functions as a constituent of three enzymes involved in the metabolism of sulfur and purines and the transfer of electrons for the oxidation/reduction process. Molybdenum deficiency in goats fed purified rations results in decreased food intake, retarded weight gain, impaired reproduction, and shortened life expectancy. Molybdenum deficiency has been reported in one individual on long-term total parenteral nutrition who developed a variety of symptoms, including headache, night blindness, lethargy, disorientation, hypermethioninemia, and increased urinary excretion of xanthine and sulfite. Treatment with molybdenum resulted in clinical improvement and normalization of sulfur metabolism. The congenital deficiency of molybdenum-pterin cofactor leads to a lack of sulfite oxidase and xanthine dehydrogenase, causing mental retardation, bilateral dislocation of the lens, and severe neurologic dysfunction. Little is known about the chemical form or bioavailability of molybdenum in foods. The estimated average intake from food and water is about 0.2 mg/day. Good food sources are milk and milk products, whole grains, legumes, and cereals. Molybdenum is rapidly absorbed from the gastrointestinal tract and is excreted mainly in the kidneys. Because molybdenum is antagonistic to copper, the adverse effects of molybdenum toxicity are similar to those of copper deficiency. Toxicity in animals results in severe diarrhea, anemia, retarded growth, weakness, stiffness, and loss of hair color. See Appendix 2 for the estimated safe and adequate daily dietary intake.

Monoamine oxidase (MAO). An enzyme, found mainly in nerve tissue and in the liver and lungs, that catalyzes the oxidative deamination of various amines, including epinephrine, norepinephrine, dopamine, and serotonin. Inhibition of MAO results in an increase in the concentration of these amines throughout the body. The increase in free serotonin and norepinephrine and/or alterations in the concentration of other amine neurotransmitters in the central nervous system are believed to have an antidepressant effect. Thus the *MAO inhibitor* drug is used in patients with neurotic or atypical depression. The drug may cause a hypertensive reaction, severe headache, tachycardia, and intracranial hemorrhage when taken with foods high in *tyramine* and caffeine. Brand names are Marplan®, Nardil®, and Parnate®. See also *Diet, tyramine-restricted.*

Monosaccharides. Group of carbohydrates that are simple sugars; composed of one sugar unit that cannot be hydrolyzed into smaller units. They are classified according to the number of carbon atoms they contain as triose, tetrose, pentose, and hexose (3, 4, 5, and 6 carbon atoms, respectively). Monosaccharides of nutritional importance are glucose, ribose, fructose, and galactose. See also Appendix 8.

Monosodium glutamate (MSG). Chemical used to enhance flavor in foods. It is used extensively in Asian cookery and is thought to cause the *Chinese restaurant syndrome*. This food additive is avoided in a restricted sodium diet below 2 g/day.

Monounsaturated fatty acid (MUFA). A fatty acid that has one unsaturated linkage or double bond in its carbon chain. See under *Fatty acid*.

Morning sickness. A common condition during the early months of pregnancy characterized by nausea and vomiting, often in the morning. Cooking odors may be a problem for some women. The condition usually disappears by the 12th week of pregnancy, although prolonged, persistent vomiting may develop in about 2% of pregnant women. *Nutrition therapy:* small, frequent dry meals of easily digested carbohydrate foods are usually better tolerated. Liquids are best taken between meals. Restrict fat if not tolerated. Vitamin B_6 supplement (50 to 100 mg three times a week) may relieve the nausea. Weight should be closely monitored. Other feeding methods should be considered if vomiting continues beyond 18 weeks of pregnancy. See also *Hyperemesis*.

MOSF. Abbreviation for *multiple organ systems failure*.

Motor neuron disease (Lou Gehrig's disease). See *Amyotrophic lateral sclerosis*.

Mottled teeth. Also called "dental fluorosis"; condition of the teeth in which the enamel appears dull, rough, and chalky, with white patches separated by yellow or brown staining, giving the characteristic mottled appearance. In severe cases, pits and depressions may be present on the surface. All the teeth may be affected, but mottling is usually seen on the incisors of the upper jaw. The condition does not impair health. It is common in many parts of the world where the fluoride content of the water is high (3 to 5 ppm). However, it occurs only when high fluoride ingestion takes place during tooth development and ceases afterward. Other conditions (genetic and other nonnutritional factors) may also cause mottling.

MRFIT. Abbreviation for *Multiple Risk Factor Intervention Trial*.

MSUD. Abbreviation for *maple syrup urine disease*.

MSUD Diet Powder. A product formulated with an amino acid mixture free of the branched-chain amino acids (BCAAs) leucine, isoleucine, and valine, and containing corn syrup solids, modified tapioca starch, and corn oil. It is designed for the dietary management of maple syrup urine disease in infants and children, and may be helpful in other disorders of BCAA metabolism such as hypervalinemia, leucine-induced hypoglycemia, and isovaleric acidemia.

Mucositis. Inflammation of a mucous membrane, such as the lining of the mouth and throat. It can affect any part of the alimentary tract and may lead to malabsorption and ulceration. Mucositis is a common side effect of drugs, especially antineoplastic agents.

MUFA. Abbreviation for *monounsaturated fatty acid*.

Multiple organ systems failure (MOSF). A condition that needs intensive care due to the failure of two or more organ systems at the same time, such as *a* combination of failures in the cardiac, respiratory, renal, or hepatic systems. *Nutrition therapy:* provide close monitoring of blood levels of nutrients to ensure appropriate nutrition, particularly amino acids, glucose, and vitamins. Energy and protein needs generally parallel increases in metabolism and skeletal muscle breakdown. Evaluate the necessity for parenteral or tube feeding if oral intake is inadequate.

Multiple peripheral neuritis. See *Guillain-Barré syndrome*.

Multiple Risk Factor Intervention Trial (MRFIT). A longitudinal epidemiologic study dealing with the effect of diet and other risk factors related to cardiovascular diseases. Sponsored by the National Heart, Lung, and Blood Institute (NHLBI). Results from this study served as basis for dietary guidelines on sodium, cholesterol, and fat (Appendix 7A) and the National Cholesterol Education Program (NCEP).

Multiple sclerosis. Central nervous system disorder of unknown etiology. It develops as an acute disease and runs intermittently, with exacerbations and remissions. There is destruction of the myelin sheaths of the brain and spinal cord. The symptoms are weakness, incoordination, strong jerky movements of the limbs, and slurred speech. Loss of hand coordination, involuntary movement, visual impairment, and dysphagia may cause self-feeding problems and loss of interest in food. Muscle tone can be either spastic or flaccid. Weight gain is common as a consequence of decreased activity levels. Neurogenic bowel can cause either constipation or diarrhea. *Nutrition therapy:* dietary modification is highly individualized, depending on the severity of the neural damage. Modify food texture and give small meals and snacks. Avoid highly salty foods if steroid therapy is used. Monitor body weight to avoid obesity. The role of fat as a causative agent has been implicated; it may be advisable to reduce fat intake. Daily intake of 20 g of saturated fat or less enables patients with multiple sclerosis to remain ambulant and less fatigued. Diets high in polyunsaturated fatty acids and gluten-free have been proposed but have not been proved effective. Tube feeding is given if the person cannot chew or swallow. Use oral feeding as long as the gastrointestinal tract is functional. Otherwise, parenteral nutrition may be needed. For further details on feeding, see *Dysphagia* and *Self-feeding problems*.

Muscular dystrophy. Disorder of striated or skeletal muscles characterized by progressive atrophy, increased urinary creatinine excretion, increased oxygen consumption, and necrosis of muscle fibers leading to paralysis. It is rare in humans, and if it does occur, its onset is during childhood. Usually the child cannot close the lips, chew, or swallow easily. Lack of physical activity may lead to overweight. *Nutrition therapy:* modify the consistency of foods if there is difficulty in chewing and swallowing. Encourage self-feeding with special devices. In serious cases of dysphagia, tube feeding of a prepared formula is recommended. Monitor body weight to avoid obesity. See also *Diet, dysphagia*.

Myasthenia gravis. A neuromuscular disorder characterized by fatigue and exhaustion of the muscular system without sensory disturbance or atrophy. It may affect any muscle of the body, but especially those of the face, lips, tongue, throat, and neck. It may be due to lack of acetylcholine. Other symptoms are fatigue, dysphagia, faint voice, and dysarthria. *Nutrition therapy:* encourage feeding by mouth unless the dysphagia is severe. Allow plenty of time for feeding, with rest periods during the meal. Modify the consistency of the diet and use foods high in nutrient density. Parenteral fluids may be necessary in acute exacerbation of muscular weakness. As the patient improves, progress from clear liquids to soft, easy to chew foods given in small, frequent meals. Assistance in eating may be required. Encourage intake of at least 2 liters/day of fluids. Monitor nutritional side effects of drugs and modify the diet accordingly. It may be necessary to restrict sodium to 2 g/day, increase protein (1.2 to 1.5 g/kg), and reduce simple carbohydrates. Lecithin and choline have been proven successful in certain patients, although more research is needed.

Myelin sheath. Whitish cylindrical covering of nerve fibers rich in lipids.

Myeloma. Cancer of the plasma cells which have invaded the bone marrow, resulting in abnormal immunoglobulin. The bones usually affected are the vertebrae, ribs, pelvic bone, and flat bones of the skull. Symptoms include nausea and vomiting, anorexia, weight loss, severe pain of the bones, susceptibility to infections, easy fatigability, fragility of bones, and a tendency to bleed. *Nutrition therapy:* adjust the calorie level to correct weight loss. The main feeding challenge is to increase the patient's appetite, which is further decreased if the patient receives radiotherapy and steroid therapy. For further dietary guidelines, see *Cancer*. See also *Chemotherapy* and *Radiotherapy*.

Myelomeningocele. A congenital condition due to the failure of the neural tube to close during the development of the embryo. Symptoms usually include varying degrees of paralysis of the lower extremities, musculoskeletal defects, joint deformities, and hip dysplasia, which are all detectable at birth. *Nutrition therapy:* prevent excessive weight gain due to limited physical activity, decreased muscle mass, and lower basal calorie needs. Assess energy needs based on desirable weight for height, or provide approximately 50% of calories for school-age children. It may be necessary to reduce intake to 7 kcal/cm height to decrease weight. Supplementary vitamins and minerals may be required with restricted caloric intake. Provide liberal fluids to prevent urinary tract infections. Folate supplements (400 mg/day) during pregnancy may reduce the incidence of myelomeningocele in subsequent pregnancies.

Myocardial infarction (MI). Also called "coronary thrombosis" and commonly called "heart attack." It results in the death of part of the heart muscles. Acute MI causes permanent damage of a heart muscle, as in a thrombotic occlusion of a branch of an atherosclerotic coronary artery. It is accompanied by severe pain, shock, and cardiac dysfunction and may result in sudden death. *Nutrition therapy:* on the first day, give caffeine-free and clear liquids; during the next few days, provide full liquid to semisolid foods in small, frequent feedings to avoid heart strain. Give foods that are soft in texture; avoid gas formers and extremes in food temperatures; avoid caffeine or limit caffeine-containing beverages to no more than 3 cups/day; and restrict sodium to 2 g/day if there is hypertension or congestive heart failure. On discharge, the usual dietary recommendation is to reduce weight and maintain desirable body weight, if the person is overweight, and follow a cholesterol-lowering diet if blood lipids are high. See *Diet, cholesterol-restricted and fat-controlled*.

Myogen. Muscle protein present in the sarcoplasm and not within the muscle fibrils. It comprises about 20% of the total muscle protein.

Myoglobin. Hemoglobin of muscle. It is an iron–protein complex responsible for the color of muscle meat. Oxygen is stored temporarily as oxymyoglobin; that is, oxygen is loosely bound to the ferrous iron, giving a bright red color. *Metmyoglobin* is a brown pigment formed when ferrous iron is oxidized to the ferric state.

Myoinositol. The biologically active form of *inositol*. It is present in relatively large amounts in the cells of practically all animals and plants. See *Inositol*.

Myosin. A muscle protein found in the fibrils that constitutes about 65% of the total muscle proteins. It combines with actin to form *actomyosin,* which is responsible for the contractile and elastic properties of muscle.

Myristic acid. Saturated fatty acid containing 14 carbon atoms. It is found in nutmeg, butter, coconut oil, and spermaceti.

Myxedema. *Hypothyroidism* in adults. It is more prevalent in women than in men; usually insidious, with gradual retardation of physical and mental functions. The disorder is characterized by decreased basal metabolic rate; dry, thick skin; puffy face and eyelids; enlargement of the tongue; sparse dry hair; husky voice; and slurred speech. There is general mental deterioration and decreased reproductive activity. The symptoms may be completely reversed by suitable therapy with thyroid preparations. *Nutrition therapy:* control and monitor weight to prevent weight gain due to decreased basal metabolism.

N

Na. Chemical symbol for *sodium*.

NAD. Abbreviation for *nicotinamide adenine dinucleotide*.

NADP. Abbreviation for *nicotinamide adenine dinucleotide phosphate*.

Nasogastric (NG) tube. Tube that is inserted into the nose and passed through the esophagus and then the stomach. See *Tube feeding*.

National Academy of Sciences (NAS). A private, nonprofit organization of distinguished scholars engaged in scientific and engineering research for the furtherance of science and technology. It advises the federal government of the United States on scientific and technical matters. In 1916, NAS organized the *National Research Council (NRC)*.

National Center for Nutrition and Dietetics (NCND). The public education initiative of the American Dietetic Association (ADA) and its foundation. The purposes of the center include identifying and providing food and nutrition information to meet the needs of dietitians and consumers; working closely with librarians and at least 300 databases to locate and obtain the needed information; and providing a hotline staffed by dietitians to answer consumers' calls about food and nutrition.

National Child Nutrition Project. Nonprofit, voluntary agency that works with individuals, state, and local groups to improve delivery of nutrition services. Initiates *Hunger Task Forces* to extend food assistance benefits to those in need, and supports food stamp outreach campaigns. It publishes *Food Action*.

National Cholesterol Education Program (NCEP). This program conducts nationwide education efforts to modify risk factors in adults for elevated blood cholesterol (total cholesterol and low-density lipoprotein (LDL)-cholesterol); releases results of studies through its Expert Panel on Detection, Evaluation, and Treatment of High Blood Cholesterol, also known as the Adult Treatment Panel (ATP); and recommends the two-step diet for hypercholesterolemia. The main goal for adults is to keep total serum cholesterol under 200 mg/dL (<5.18 mmol/L) and LDL-cholesterol under 160 mg/dL (<4.14 mmol/L) in individuals without coronary heart disease (CHD) and with fewer than two risk factors; under 130 mg/dL (<3.37 mmol/L) in individuals without CHD and with two or more risk factors; and ≤100 mg/dL (≤2.59 mmol/L) in individuals with coronary heart disease. See *Hypercholesterolemia*. See also *Diet, cholesterol-restricted, fat-controlled*.

National Dairy Council (NDC). Nonprofit research and educational organization supported by the dairy industry. Serves as a national resource agency in nutrition education, maintaining cooperative relations with government, professional, educational, and consumer groups. Publishes the *Dairy Council Digest* bimonthly.

National 5-a-Day Program. The first national nutrition program to approach

Americans with a simple, positive message to eat five servings or more of fruits and vegetables every day (at least three fruits and two vegetable servings). It is sponsored jointly by the National Cancer Institute (NCI) and the Public Health and Human Services (PHHS), and more than 800 licensed food industries and organizations and 49 state and territorial health departments are participating in the program.

National Health and Nutrition Examination Survey (NHANES). A continuing national program to obtain information on the health and nutritional status of the American people. NHANES I, conducted from 1971 to 1974, included 28,043 persons sampled statistically from 48 states ranging in age from 1 to 74 years. Dietary data, biochemical tests, clinical examinations, and anthropometric measurements were evaluated. NHANES II was conducted from 1976 to 1980 in 50 states with a total of 27,801 persons surveyed, including 6-month-old infants. The methods used were the same as those for NHANES I, except that a special study was made on iron, zinc, and copper, and more information was collected about vitamin and mineral supplements. The Hispanic Health and Nutrition Examination Survey (HHANES), conducted from 1982 to 1984, included about 16,000 Hispanics in the United States, using the same methods of study as NHANES II. NHANES III is in progress. This ongoing survey was planned to follow people throughout their lives using the same methodology as NHANES I and II. Starting with the years 1988 to 1994, the same people will be studied at regular intervals thereafter. The main results from NHANES I revealed the prevalence of anemia among children and older people; low vitamin A intake among blacks, and low serum transferrin among children. In NHANES II, anemia continued to be a problem, especially among women of child-bearing age, children, and adolescent boys. Calcium intake was inadequate among females. NHANES III preliminary results revealed the prevalence of obesity, especially among adolescent females.

National Institutes of Health (NIH). Research unit of the Public Health Service in the U.S. Department of Health and Human Services engaged in clinical research on diseases of public health importance. The NIH supports research and training programs in nutrition related to health maintenance, disease prevention and treatment, and human development throughout the life cycle.

National Nutrient Data Bank. The primary source of information in the United States on the nutrient content of foods. The data come from academic and government laboratories as well as private industry. More than 800,000 records are stored in the data bank and at least 6000 are added monthly. The information is used to develop the nutritional composition of foods from different parts of the country.

National Nutrition Consortium, Inc. An organization composed of professional societies in the fields of nutrition, food, and dietetics. Provides leadership in the development and coordination of food and nutrition policies at national and local levels. Publications include materials on nutrition labeling and interpretation of complicated nutrition information for public understanding.

National Nutrition Monitoring System (NNMS). Set of activities to provide the scientific foundation for the maintenance and improvement of the nutritional status of the population and the nutritional quality and healthfulness of the national food supply in the United States. The objectives of the NNMS are to ensure the adequacy of nutrients for all Americans, ensure the safety and quality of the food supply, develop a better database for decisions on national nutrition policies, and guide appropriations for nutrition and health. The NNMS coordinates the data of the *National Health and Nutrition Examination Survey (NHANES)* and the *National Food Consumption Surveys (NFCS),* especially on the following components: nutritional surveillance, food production and marketing, food consumption and nutritional status surveys, dissemination of nutrition information, and development of methods for these components or activities.

National Nutrition Surveillance System. Conducted by the Centers for Disease Control (CDC), this project was started in 1973 for a continuous monitoring of the nutritional status of specific high-risk population groups in at least 32 states. The purpose of the nutrition surveillance system is to identify nutritional problems and determine the kind of intervention needed. Data collected include height, weight, hematocrit, hemoglobin, height for age, and weight for height.

National Research Council (NRC). Organized in 1916 by the National Academy of Sciences (NAS), it engages in research and advises the federal government on scientific and technical matters. The NRC functions in accordance with general policies determined by the NAS and is currently the main operating agency for the NAS in providing services to the government, the public, and the professional communities. The 10th edition of the Recommended Dietary Allowances (RDA: see Appendix 1) was a project approved by the NRC's governing board, whose members were drawn from the councils of the National Academy of Sciences, the National Academy of Engineering, and the Institute of Medicine. The 10th edition of RDA was mainly the work of the *Food and Nutrition Board (FNB).*

National School Lunch Program (NSLP). Administered by the U.S. Department of Agriculture, state departments, and local school districts, it provides full price, low-cost, or free nutritious lunches to school children coming from families with incomes ranging between 130% and 185% of poverty level. NSLP has full participation in public schools and voluntary participation in private schools.

National Science Foundation (NSF). This foundation was established in 1950 to improve scientific research and education in the United States. Grants are awarded to universities and other nonprofit institutions to support research. The foundation also maintains a register of scientific personnel.

Nationwide Food Consumption Surveys (NFCS). The U.S. Department of Agriculture conducts nationwide food consumption surveys to evaluate the kinds and amounts of foods people are eating and to gather other related information on food consumption to guide in the development of food plans and policies. To cite one survey, the sixth NFCS, conducted in 1977 to 1978, included 8661 individuals from a statistical sample of households from 48 states. Dietary data from 24-hour recall and 2-day food records were analyzed for nutrient density for 14 nutrients, and these were related to factors such as income level, race, employment status, educational level of the head of the household, and geographic region. Alcohol consumption was also studied.

Nationwide Nutrition Network. A national referral service provided by the American Dietetic Association for registered dietitians. It links together dietitians and physicians, industry groups, companies, and other agencies who need the services of a dietitian. A consumer hot line is also available.

Natural killer cell (NKC). Non-T, non-B large granular lymphoid cells that play a role in immune surveillance against virus-infected and cancer cells. The stress of malnutrition depresses the activity of the natural killer cells.

NE. Abbreviation for *niacin equivalent.*

Necrotizing enterocolitis (NEC). Infection of the intestinal tract mucosa by enteric pathogens, characterized by respiratory distress syndrome, vomiting, distended abdomen, diarrhea, and sepsis. It occurs primarily in premature and low-birth-weight infants. Surgical resection of the affected area may be necessary. *Nutrition therapy:* initially, give nothing by mouth; use parenteral feeding only. As the patient's condition improves, progress to oral feeding of a predigested formula such as Pregestimil® and then resume standard formula feeding, preferably lactose-free.

NEFA. Abbreviation for *nonesterified fatty acid.* See under *Fatty acid.*

Neopham®. Brand name of a 6.4% solution of essential and nonessential amino acids for intravenous infusion in total parenteral nutrition.

NephrAmine®. Brand name of a 5.4% solution of essential amino acids plus histidine for parenteral nutrition of patients with renal failure. It is designed to be infused with a concentrated source of calories. See also Appendix 40.

Nephritis. Any acute or chronic inflammation of the kidney. The disease is called *glomerulonephritis* if the inflammation is primarily of the glomeruli. It is called *pyelonephritis* if it is caused by bacterial invasion from the urinary tract. Nephritis may follow such diseases as scarlet fever, tonsillitis, and influenza. For further details of dietary modifications, see *Glomerulonephritis* and *Pyelonephritis.*

Nephrolithiasis. Formation of stones in the kidney or the disease condition that leads to their formation. Kidney stones are of various types, shapes, and sizes. Causes include chronic infection of the kidney, stagnation of the urine, prolonged confinement in bed, dietary factors (e.g., excessive oxalates, urates, calcium or magnesium trisilicate), and certain congenital biochemical abnormalities. A hot climate may contribute to the formation of kidney stones. *Nutrition therapy:* Encourage intake of plenty of fluids (1.5 to 2 liters/day) to keep the urine dilute. Dietary manipulation may be beneficial in supplementing the effect of medication and influencing the urine pH. Consider an alkaline-ash diet in uric acid and cystine stones that precipitate in an acidic urine; an acid-ash diet in stones consisting of calcium and magnesium phosphates, carbonates, and oxalates that precipitate in an alkaline urine; a low-oxalate diet (40 to 50 mg/day) with adequate calcium intake to bind oxalate in hyperoxaluria; a low-calcium diet in hypercalciuria associated with an extremely high calcium intake; and limit protein intake to the level of the RDA in hyperuricosuria. The formation of struvite stones containing ammonium, magnesium, and phosphate is not influenced by diet. See *acid-ash diet, alkaline-ash diet, low-calcium diet,* and *oxalate-restricted diet* under *Diet.*

Nephrosclerosis. Hardening of the renal arteries seen in renal hypertension and often associated with arteriosclerosis. As a rule, it occurs in adults after 35 years of age and may be benign for many years. Characterized by albuminuria and nitrogen retention, retinal changes, fibrosis of the glomeruli, degeneration of the renal tubules, and, ultimately, renal insufficiency. Death usually results from circulatory failure. *Nutrition therapy:* protein intake of 1 g/kg/day should consist of good-quality proteins to replace losses, especially albumin. Sodium is restricted to 2 g/day to reduce edema and help control hypertension. Fluid restriction is recommended in renal failure. Cholesterol and fat should be monitored as necessary. Weight reduction is recommended for the obese to lessen the work of the circulatory system. Vitamin and mineral losses should be replaced. See also *Diet, cholesterol-restricted, fat-controlled.*

Nephrotic syndrome (NS). Also called nephrosis. A group of symptoms characterized by marked proteinuria, hypoalbuminemia, edema, and hyperlipidemia. NS is due to an alteration in the capillary basement membrane of the glomerulus, causing loss of large amounts of protein into the urine. It may also be caused by chronic glomerulonephritis and other diseases, infections, allergic reactions, and drugs. Urinary protein losses in adults may range from 3 to 16 g/day, and children can lose up to 50 mg of protein/kg body weight daily. Albumin synthesis by the liver cannot compensate for the excessive urinary losses, resulting in hypoalbuminemia. The syndrome is usually reversed by corticosteroid or immunosuppressive drugs, but it may progress to chronic renal failure. *Nutrition therapy:* the aim of the diet is to minimize edema, control hypertension, decrease urinary albumin loss, retard progression of renal disease, and prevent protein malnutrition. Control of hyperlipidemia is also desirable. Provide 0.6 to 0.8 g of protein/kg per day plus 1.0 g/day of high-biologic protein for each gram of urinary protein lost daily; 35 kcal/kg/day, with 30% to 40% of total calories from fat and the remainder of nonprotein calories from carbohydrate; and 1.5 to 2 g of sodium/day to manage hypertension and edema. Increase protein to 1.0 g/day if renal function is good. Fluid intake is generally not restricted, unless urine output is less than 1 liter per day. Fluid intake should equal urine output. Weight reduction is recommended if the person is overweight or obese. Reduce elevated blood lipids by restricting intake of saturated fat to less than 10% of total calories and cholesterol to 300 mg or less per day. Supplemental B vitamins, calcium, and zinc may be needed if daily protein intake is 60 g or less.

Nepro®. Brand name of a specialized high-calorie, low-electrolyte and low-fluid formula for dialyzed patients with chronic or acute renal failure. It is lactose-free and provides 2 kcal/mL. May be given orally or by tube. See Appendix 39.

Net protein utilization (NPU). Proportion of the nitrogen intake that is retained in the body. It is calculated by multiplying the biologic value of a protein by its digestibility factor.

Neuritis. Inflammation of a nerve or nerves; usually associated with pain, anesthesia or paresthesia, paralysis, muscle degeneration, and loss of reflexes. Symptoms vary according to the cause, the location, and the nerves involved. See also *Polyneuritis*.

NHANES. Abbreviation for *National Health and Nutrition Examination Survey*.

Ni. Chemical symbol for nickel.

Niacin. Generic name for pyridine-3-carboxylic acid (nicotinic acid) and its corresponding amide (nicotinamide or niacinamide); a member of the B complex water-soluble vitamins. The term "niacin" was originally proposed to avoid association of the vitamin with nicotine of tobacco. Nicotinamide is a constituent of two coenzymes, nicotinamide adenine dinucleotide (NAD) and nicotinamide adenine dinucleotide phosphate (NADP), which act as hydrogen and electron acceptors and donors, respectively, and that function in the metabolism of carbohydrate, fat, and protein; rhodopsin synthesis; and cellular respiration. Niacin also has two pharmacologic actions at high dosage: peripheral vasodilation (mainly nicotinic acid) and serum cholesterol reduction. Niacin deficiency may be one of the features of a general nutritional deficiency. A multiple deficiency syndrome, called *pellagra,* is seen in maize-eating areas in association with diets providing low levels of niacin equivalents. The characteristic symptoms are dermatitis, diarrhea, inflammation of the mucous membranes, and dementia in severe cases. Niacin is found in most animal and plant foods, but often in a form that is unavailable. Good food sources are meats, fish, poultry, liver, and enriched cereals. Another source of niacin is biosynthesis from the amino acid tryptophan; 1 mg of niacin is formed from 60 mg of tryptophan. This conversion requires *pyridoxal phosphate*-dependent enzymes and hence adequate dietary intake of pyridoxine, or *vitamin* B_6. Ingestion of large doses of nicotinic acid, but not of the amide, may produce a transient sensation of burning, flushing, and tingling or stinging of the skin; long-term use may cause xerostomia, activation of peptic ulcer, blurred vision, hyperglycemia, jaundice, and liver damage. See Appendix 1 for the recommended dietary allowances.

Niacinamide. Also called "nicotinamide"; the amide form of nicotinic acid. See *Niacin.*

Niacin equivalent (NE). Sum of the two forms in which niacin is made available to the body, as the preformed niacin and that derived from tryptophan (60 mg of tryptophan = 1 mg of niacin).

Nickel (Ni). A trace element essential to several animals that probably is also required by humans. Nickel is present in RNA and DNA. Its precise biologic role has not been clearly defined, although recent findings indicate that nickel functions as a cofactor or structural component in specific enzymes involved in intermediary metabolism. Nickel deficiency in animals results in depressed growth and hematopoiesis, and in changes in the levels of iron, copper, and zinc in the liver. Nickel interacts directly or indirectly with at least 13 essential minerals. Of the dietary interactions, the one with iron is perhaps the most significant; nickel enhances the absorption and metabolism of iron. In humans, about 10 mg of nickel is present in the body, with the largest proportions in the skin and bone marrow. Liver and muscle concentrations appear to be most responsive to the level of dietary intake, which is about 0.3 to 0.6 mg/day, mainly from plant foods. Rich food sources are nuts, dried beans and peas, chocolate, and grains. Average absorption is about 3% and is enhanced during pregnancy and with high intake. Nickel is a cause of allergic contact dermatitis. Toxicity through oral intake is unlikely, although chronic exposure and inhalation of substantial quantities in animal studies resulted in degeneration of heart muscle, brain, lung, liver, and kidney.

Nicotinic acid. Pyridine-3-carboxylic acid, a member of the B complex vitamins. See *Niacin.*

NIDDM. Abbreviation for non-insulin-dependent diabetes mellitus. See *Diabetes mellitus.*

Night blindness. Also called "nyctalopia"; a condition of defective or reduced vision in the dark, especially after coming from bright light. When temporary, it may be due to vitamin A deficiency, and it responds to suitable vitamin supplementation. Night blindness occurs when there is insufficient vitamin A to bring about prompt and complete regeneration of the *visual purple*. Night blindness not responsive to vitamin A supplements may be due to diseases of the retina and other factors not related to vitamin A. *Nutrition therapy:* vitamin A (10,000 IU recommended) daily for 1 week in the form of retinol. Continue with an adequate vitamin A intake daily. See also *Dark adaptation*.

NIH. Abbreviation for *National Institutes of Health*.

Nitrofurantoin. An antimicrobial used for prevention and treatment of infections in the urinary tract. It may decrease serum folate and cause megaloblastic anemia; it may also cause peripheral neuritis, anorexia, and abdominal cramps. It may be present in breast milk. Brand names include Furadantin®, Macrobid®, and Macrodantin®.

Nitrogen (N). A chemical element essential to life; found free in the air and in combinations of proteins and other organic compounds. Plants can use nitrogen from the soil, and nitrogen-fixing bacteria can use free nitrogen from the air. Animals and humans, however, can utilize nitrogen only when it is supplied from foods. It is an important constituent of all animal and plant tissues and a unique element in *proteins*. See also *Amino acid*.

Nitrogen balance. Measurement of the state of nitrogen equilibrium in the body. An individual is said to be in nitrogen balance or equilibrium when the nitrogen intake (from food eaten) equals the nitrogen output (in urine, feces, and perspiration). A *positive nitrogen balance* exists when nitrogen intake is above output. This can be brought about by growth, pregnancy, lactation when the mother is storing protein as milk, or recovery from illness or trauma. A *negative nitrogen balance* exists when intake is below output, as in fasting, fever, surgery, burns, or shock following an accident. A negative nitrogen balance is undesirable because body protein is being broken down more rapidly than it is being built up. A method of determining nitrogen balance is as follows:

$$\text{Nitrogen balance} = \left(\frac{\text{protein intake}}{6.25}\right) - (\text{UUN} + 4)$$

where protein intake (in grams) is calculated from a 24-hour food record and urinary urea nitrogen (UUN), also in grams, is analyzed from a 24-hour urine specimen of the same period. The factor 4 is additional nitrogen lost from the feces and skin. To be assured of a positive nitrogen balance, a 3-day average value of more than 0.04 g of nitrogen/kg body weight/day is desirable.

Nitrolan™. Brand name of a high-protein, lactose-free formula for moderately stressed patients. See Appendix 39.

Nonesterified fatty acid (NEFA). Also called "free fatty acid." See under *Fatty acid*.

Non-insulin-dependent diabetes mellitus (NIDDM). Or Type II diabetes mellitus. See *Diabetes mellitus*.

Nonprotein nitrogen (NPN). The total nitrogen of the blood, excluding that from protein. This may come from urea, uric acid, creatine, creatinine, etc.

Norepinephrine. Also called "noradrenaline." A catecholamine hormone secreted by the *adrenal medulla*. It is a demethylated epinephrine synthesized from tyrosine and liberated at the ends of sympathetic nerve fibers after stimulation. It causes an increase in blood pressure by increasing peripheral resistance, with little effect on cardiac output. Unlike epinephrine, it has little effect on carbohydrate metabolism and oxygen consumption. See also *Epinephrine*.

Nortriptyline. A tricyclic antidepressant used for the relief of emotional depression. It has a peculiar taste and may cause dry mouth, irritation of the tongue and mouth, fluctuation of blood sugar levels, and constipation. It may be present in breast milk. Brand names are Aventyl® and Pamelor®.

Novamine®. Brand name of an amino acid solution for intravenous administration in total parenteral nutrition by peripheral vein or central infusion. Contains both essential and nonessential amino acids and is available in 8.5%, 11.4%, and 15% concentrations. See also Appendix 40.

NPN. Abbreviation for *nonprotein nitrogen*.

NPU. Abbreviation for *net protein utilization*.

NRC. Abbreviation for *National Research Council*.

NSF. Abbreviation for *National Science Foundation*.

NSI. Abbreviation for *Nutrition Screening Initiative*.

NSLP. Abbreviation for National School Lunch Program.

Nucleic acid. A highly complex portion of nucleoproteins that yields a mixture of purines and pyrimidines, a ribose or deoxyribose component, and phosphoric acid on complete hydrolysis. The two general types of nucleic acid are *ribonucleic acid (RNA)* and *deoxyribonucleic acid (DNA)*.

Nucleotide. Phosphate ester of the nucleoside. Examples are adenylic acid, guanylic acid, and cytidylic acid. Nucleotides of importance in metabolism are adenosine diphosphate (ADP) and adenosine triphosphate (ATP). See *Adenosine phosphates*.

Nursoy®. Brand name of a hypoallergenic formula for infants sensitive to milk and lactose; contains soy protein isolate, sucrose, tapioca, dextrose, oleo, coconut oil, and soy oil. See also Appendix 37.

Nutraceutical. Term proposed for a new category of compounds that are not included in the current U.S. Food and Drug Administration (FDA) regulations for foods and drugs. It refers to any substance that may be considered a food or part of a food and provides medical or health benefits, including the prevention and treatment of diseases. Nutraceuticals include isolated nutrients, dietary supplements, diets, genetically engineered "designer" foods, and herbal products.

Nutramigen®. Brand name of a hypoallergenic formula for infants sensitive to intact protein, cow's milk, and/or lactose; contains a high percentage of free amino acids. The remainder of the protein is in the peptide form. Made with casein hydrolysate, sucrose, modified cornstarch, and corn oil. See also Appendix 37.

Nutren®. Brand name of a lactose-free nutritional supplement. Available in concentrations providing 1.0, 1.5, and 2.0 kcal/mL. Nutren® with fiber contains 14 g of fiber/liter. See Appendix 39.

Nutrient. Any chemical substance needed by the body for one or more of the following functions: to provide heat or energy, to build and repair tissues, and to regulate life processes. Although nutrients are found chiefly in foods, some can be synthesized in the laboratory (e.g., vitamins) or in the body (biosynthesis). See also *Nutrition*.

CARBOHYDRATE
- Glucose

FAT
- Linoleic acid
- α-Linolenic acid

PROTEIN
- *Essential Amino Acids*
  - Histidine
  - Isoleucine
  - Leucine
  - Lysine
  - Methionine
  - Phenylalanine
  - Threonine
  - Tryptophan
  - Valine
- *Conditionally Essential Amino Acids*
  - Arginine
  - Cysteine
  - Glutamine
  - Taurine
  - Tyrosine
- *Nonessential Amino Acids*
  - Alanine
  - Asparagine
  - Aspartic acid
  - Cystine
  - Glutamic acid
  - Glycine
  - Hydroxylysine
  - Hydroxyproline
  - Proline
  - Serine

MINERALS
- *Major Minerals*
  - Calcium
  - Chloride
  - Magnesium
  - Phosphorus
  - Potassium
  - Sodium
  - Sulfur
- *Trace Minerals*
  - Chromium
  - Cobalt
  - Copper
  - Fluoride
  - Iodine
  - Iron
  - Manganese
  - Molybdenum
  - Selenium
  - Zinc
- *Probably Essential Trace Minerals*
  - Arsenic
  - Boron
  - Cadmium
  - Lead
  - Lithium
  - Nickel
  - Silicon
  - Tin
  - Vanadium

VITAMINS
- *Fat-soluble Vitamins*
  - A
  - D
  - E
  - K
- *Water-soluble Vitamins*
  - Ascorbic acid
  - Biotin
  - Cobalamin
  - Folacin
  - Niacin
  - Panthothenic acid
  - Riboflavin
  - Thiamin
  - Vitamin B_6
  - Vitamin B_{12}
- *Vitamin-like Substances*
  - Carnitine
  - Choline
  - Bioflavonoids
  - Inositol

WATER

Nutrient classifications. Nutrients are classified according to the amount present in the body, chemical composition, essentiality, and function:

Amount present in the body. See *Macronutrients* and *Micronutrients*.

Chemical composition. The two categories are *inorganic* (water and minerals) and *organic* (carbohydrate, protein, fat, and vitamins). These six major groups of nutrients are composed of individual nutrients as listed above.

Essentiality. The term "essential" refers to any nutrient that the body cannot make in sufficient quantity to meet its physiological needs; it must be supplied preformed from foods. A "nonessential" nutrient can be synthesized in the body in sufficient quantity. All nutrients are *physiologically essential*. The term "nonessential nutrient" is therefore misleading and should be more appropriately called "nondietary essential." Some of the "nonessential" nutrients become "essential" during a diseased condition or a developmental state when the body cannot manufacture enough to meet the high metabolic demand. Such nutrients are called "conditionally essential," and include arginine, choline, cysteine, glutamine, inositol, and tyrosine.

Function. Nutrients are grouped according to three general functions: *source of energy* (carbohydrate, protein, and fat); *growth and repair of tissues* (protein, minerals, vitamins, and water); and *regulation of life processes* (protein, minerals, vitamins, and water).

Nutrient interrelationships. Cellular metabolism is an integrated, coordinated chain of reactions, and interference with any reaction affects the whole system, which includes all nutrients. The close interrelationship of the six major groups of nutrients is best summarized as follows: Protein, fat, and carbohydrate metabolites enter a common pathway to yield energy with the help of enzyme systems containing vitamins and minerals as cofactors. Water is the circulating medium for all reactions. Specific interrelationships can exist between two nutrients, such as iron–copper, calcium–phosphorus, cobalamin–cobalt, water–sodium, and folic acid–vitamin C. Specific interrelationships also exist among several nutrients, such as calcium–phosphorus–magnesium–vitamin D, cobalt–iron–zinc, and niacin–tryptophan–pyridoxine. One recent example is the involvement of manganese, zinc, and copper in preventing peroxidation or breaking up of polyunsaturated fatty acids. See also Appendix 15.

Nutrient stores. Nutrient stores or reserves in the body affect the pathogenesis of deficiency diseases. The capacity to store and the duration before nutrient stores are depleted vary with each nutrient. The body cannot store amino acids for more than a few hours, but it can store minerals in the bones for a few years. Calcium reserves can last as long as 7 years. Water loss after 4 days is critical. Iron stores for menstruating women are depleted in 3 months, compared to 2 years for men. Thiamin reserves are depleted faster than those of the other B vitamins, i.e., 2 months and 5 months, respectively. More studies are needed to confirm the length of time needed to deplete nutrient stores in the body.

NutriHep™. Brand name of a specialized enteral formula for hepatic patients who are fluid restricted and protein intolerant; provides high branched chain and low aromatic amino acids. See Appendix 39.

Nutrilan™. Brand name of a flavored nutritional product for partial or total oral supplementation. Available in chocolate, strawberry, and vanilla flavors. See Appendix 39.

Nutrilipid®. Brand name of a fat emulsion for intravenous administration. Provides 1.1 kcal/mL (at 10%) and 2 kcal/mL (at 20%). See Appendix 41.

Nutrition. Simply stated, the study of food in relation to health. As defined by the Food and Nutrition Council (of the American Medical Association), nutrition is the "science of food, the nutrients and other substances therein, their action, interaction and balance in relation to health and disease, and the processes by which the organism ingests, digests, absorbs, transports, utilizes and excretes food substances." Nutrition deals with the physiologic needs of the body in terms of specific nutrients, the means of supplying these nutrients through adequate diets, and the effects of failure to meet nutrient needs. In addition, nutrition is concerned with the social, economic, cultural, and psychological implications of food and eating. The basic concepts in nutrition may be summarized as follows: (1) Adequate nutrition is essential for health. (2) A number of compounds and elements broadly classed as protein, fat, carbohydrate, minerals, vitamins, and water are needed daily in the food of humans. (3) An adequate diet is the foundation of good nutrition, and it should consist of a wide variety of natural foods. (4) Many nutrients should be provided preformed in food, whereas a few may be synthesized within the body. (5) Nutrients are interrelated, and there must be metabolic balance in the body. (6) Body constituents are in a dynamic state of equilibrium. (7) Human requirements for certain nutrients are known quantitatively within certain limits. The search for quantitative determination for the others has been going on for over a century. (8) The effects of nutritional inadequacy are more than physical; behavioral patterns and mental performance are also affected. (9) The nutritional status of populations and individuals can be measured for certain nutrients. However, for other nutrients, techniques of assessment (dietary, clinical, and biochemical) have yet to be refined. (10) Proper education, technical expertise, and the use of all resources in applied nutrition and food technology will help upgrade the nutritional status of people. (11) The biologic meaning of food is attributable to the three functions of nutrients. To an individual or family, food is eaten for more than its physiologic, social, and aesthetic values. (12) The study of nutrition as a subject or course has a broad scope and is interrelated with many allied fields, such as physiology, biochemistry, food technology, dietetics, public health, behavioral sciences (sociology, anthropology, and psychology), and many branches of medicine (anatomy, preventive medicine, pediatrics, etc.). For nutrition objectives for the next decade, see *Healthy People 2000* and Appendix 7B.

Nutrition, adolescence. The adolescent period is characterized by an accelerated growth rate and intense activity with physical, social, emotional, and mental changes. It is a transition period between childhood and adulthood; girls mature earlier than boys. The nutritional needs of adolescents are unique, conditioned primarily by the building of new body tissues, the demands of increased physical activity, and to some extent the emotional changes attending maturation. In general, the growing adolescent requires a high caloric intake, an abundance of good-quality protein, and a liberal intake of minerals and vitamins. Because of the onset of menstruation, adolescent girls have specific nutrient needs, especially for iron, protein, and other nutrients essential for blood formation. Nutrition education is focused on the eating habits of adolescents because of the following problems that affect nutrient intake: (1) irregularity of meals and skipping breakfast; (2) poor choice of snack items, with a tendency to eat foods with empty calories or junk foods such as cakes, cookies, and soft drinks; and (3) anxiety about figure development, which causes some girls not to eat enough. Anorexia nervosa and bulimia may occur in this group. On the other hand, obesity that may have started in early childhood may continue into adolescence. For recommended dietary allowances, see Appendix 1.

Nutrition, adulthood. Adulthood is the period of life when one has attained full growth and maturation. The onset of this stage varies among individuals, and there are no clear-cut age boundaries. However, in relation to dietary needs, adulthood pertains to the years between ages 25 and 50 without stresses such as pregnancy, lactation, and convalescence. Proper nutrition needs to be emphasized in adulthood, since it is the longest period of the life cycle and possibly the years of peak productivity. The population in this age group that needs special mention are the female adults, ages 18 to 40, who use oral contraceptive drugs. Ideally, one should reach adulthood with broad familiarity with and acceptance of different foods, as well as sound food habits. Adults tend to resist changes in their food habits, hence the importance of proper training both in food selection and in regularity of eating as early in life as possible. Another aim of good nutrition throughout adulthood is the maintenance of a healthy body weight. It is recommended that the daily caloric allowances be reduced with increasing age. See Appendix 1 for recommended dietary allowances. See also Appendices 5 and 7 for dietary guidelines.

Nutrition, aging process. Theoretically, aging is a continuous process from conception until death. However, in the young growing organism, the building-up processes exceed the breaking-down processes, so that the net result is a picture of growth and development. Once the body reaches adulthood, the process is reversed. Although the rate of degenerative changes is slow during middle adulthood, it increases as the individual approaches the geriatric age. See *Nutrition, geriatric*.

Nutrition, alcoholism. Chronic alcoholism is a complex condition involving psychological, social, and physiologic factors. Excessive intake of alcohol is a prominent contributor to and causative factor in cirrhosis of the liver, one of the 10 leading causes of death in the United States. An alcoholic usually has faulty eating habits and an inadequate diet, consuming as much as 50% to 60% of the total caloric intake from ethanol. *Cirrhosis of the liver* in chronic alcoholism is caused mainly by dietary deficiencies, particularly thiamin and other B complex vitamins and protein. The metabolic explanation for vitamin B-induced deficiencies is based in part on the involvement of the B complex vitamins in enzyme systems needed to oxidize alcohol (1 g of alcohol yields 7 kcal), which takes place in the liver. Excess alcohol has toxic effects on the central nervous system (e.g., alcoholic dementia and Korsakoff's psychosis), heart, kidneys, and other organs of metabolism. Nutritional inadequacy is sometimes aggravated by chronic gastritis resulting from habitual drinking. See also *Alcohol, Fetal alcohol syndrome*, and *Wernicke–Korsakoff syndrome*.

Nutrition, anemias. Nutritional anemias probably constitute the most common nutritional diseases in humans. This statement is not difficult to believe because of the many etiologic factors that lead to anemia. Nearly all nutrients are involved directly or indirectly. *Protein* is needed for the globin portion of hemoglobin (Hb); *iron* is an integral part of heme; *copper* catalyzes the utilization of iron both in the intestinal tract and at the tissue level; *pyridoxine* helps in the formation of the pyrrole ring of Hb; *riboflavin* is important in protein synthesis; *ascorbic acid* increases the absorption of iron by keeping it in its ferrous state and catalyzes the conversion of folic acid to folinic acid; and *cobalamin* and *folinic acid* are needed for normal maturation of red blood cells and in the synthesis of methyl groups needed for heme structure. Because of their interrelationships, other nutrients are indirectly involved. See also *Anemia* and *Hemopoiesis*.

Nutrition, antibiotics. The increasing use of antibiotics necessitates a review of how nutrition is affected by this special group of drugs. There are several mechanisms or modes of action to explain their nutritional influence. Many antibiotics have direct effects on the gastrointestinal tract (e.g., nausea, anorexia, glossitis, stomatitis, and diarrhea). Certain antibiotics bind some nutrients (e.g., tetracycline binds protein, and penicillin and sulfonamides bind serum albumin). Other antibiotics increase the volume and/or frequency of stools. In general, antibiotics alter microflora and inhibit bacterial synthesis of certain vitamins. For the nutritional side effects of antibiotics, see *Aminoglycoside, Amphotericin*

B, Cephalosporin, Chloramphenicol, Clindamycin, Cycloserine, Daunorubicin, Erythromycin, Gentamicin, Griseofulvin, Moxalactam, Penicillin, and *Tetracycline*.

Nutrition Assessment. This is a comprehensive process of identifying and evaluating the nutritional needs of a person using appropriate, measurable methods. Nutritional assessment consists of gathering information and evaluating the data using four techniques: history taking, nutritional anthropometry, physical examination, and biochemical tests. *History taking* includes medical, social, dietary, and related background information about the client or patient. Risk factors associated with nutrition from history-taking include drastic weight loss or gain; anorexia, chronic illnesses, and recent major surgery; chewing and swallowing difficulties; drug addiction; habitual intake of oral contraceptives, catabolic steroids, antibiotics, and other drugs with significant nutrient–drug interaction; and socioeconomic factors such as poverty, lack of education, and inadequate or poor food habits. The latter may be assessed by a 24-hour food recall and/or by keeping a food record from 3 to 7 days. *Anthropometric measurements* include data on height and weight, frame size, body mass index (BMI), skinfold thickness, midarm muscle circumference (MAMC), upper arm muscle area (UAMA), and knee height. See Appendices 18 to 23 for details. *Physical examination* can reveal possible deficiency signs of malnutrition based on clinical signs and symptoms associated with lack of a nutrient as summarized in Appendix 26. *Biochemical tests* include data from laboratory analysis for various nutrients and related substances from the blood and urine. Common biochemical indices include: serum albumin, serum albumin/globulin ratio, serum transferrin, total iron binding capacity, total lymphocyte count, complete blood cell profile, lipid profile, somatomedin-C, nitrogen balance, creatinine–height index, serum enzyme levels, urinary ketones, urinary nitrogen, and others. For details, see Appendices 27 to 31. A meaningful nutritional evaluation consists of interpreting the results from these four techniques of measuring nutritional assessment. This should be followed by a nutritional intervention/plan of action to correct any deficiency and to aid in therapy. See also *Subjective global assessment*.

Nutrition Assistance Program for Puerto Rico. Administered by the U.S. Department of Agriculture, this program has the same eligibility guidelines as the Food Stamp Program. The Puerto Rico Commonwealth government, operating under the block grant program, screens recipients who receive the cost of food benefits in cash. Half of the Commonwealth administrative costs is set up by their legislature.

Nutrition, bone. Bone is a metabolically active tissue, constantly turning over through resorption or breakdown of bone cells and formation of osteoblasts or bone mineralization. These processes are necessary for the reshaping and repositioning of bones to bear the changing weight shifts of the body. Bone formation starts with fetal development. Rapid bone growth and calcification occur up to adolescence. Peak bone mass is attained in the early twenties, at age 23 on the average. Bone is made up of an organic matrix that contains the protein collagen, which acts as the supporting material for mineral crystals. The process of depositing calcium and phosphorus in the form of hydroxyapatite is called "calcification" or "ossification." The bone shaft, which is the rigid part, also contains sodium, zinc, fluoride, and carbonates. The nutritional implications of bone formation are focused not only on protein and mineral needs, but also on other nutrients, such as vitamins C and D, and on the proper ratio of calcium to phosphorus (about 1:1) for the growing years. See *Nutrition, dental health* and *Nutrition, growth and development*. See also *Osteoporosis* and *Osteodystrophy*.

Nutrition, cancer. To date, cancer is the second leading cause of death in the United States. Present knowledge about the relationship of nutrition and cancer is not conclusive for two main reasons: (1) the causative factors of malignant tumors are numerous and often undetermined, and (2) the carcinogenic stimulus affects the organism in varying degrees, depending on factors such as the amount or dose of the stimulus, the period of exposure, and the susceptibility of the individual. The observation that cancers of the stomach and liver are the prevalent types is of interest, as these organs are directly involved in nutrient utilization. In addition, certain substances found in foods are carcinogenic. Among these are the aflatoxin of moldy peanuts, cycasin in cycad meal, polyphenols in tea, excess selenium in the diet, and a toxic substance developed in overheated fats. Studies on dietary factors in relation to cancer prevention include the following: vitamins A, C, E, and selenium, which are antioxidants, protect the DNA from electron-seeking compounds so that it is not altered; dietary fibers are recommended, because they bind carcinogens in the feces; cruciferous vegetables such as cabbage, cauliflower, and brussels sprouts contain indoles and phenols, which are believed to reduce cell division; a diet excessively high in phosphates may stimulate nucleic acid synthesis, as well as malignant growth of tumors; restriction of essential amino acids (e.g., phenylalanine) may inhibit the cancerous growth of cells; a diet low in pyridoxine may be beneficial, because pyridoxine tends to stimulate malignant growth; and excessive intake of certain food additives and nonnutritive sweeteners may promote or cause cancer. Evidence also suggests that calcium and vitamin D may offer some protection against colorectal cancer. Recommendations from the National Cancer Institute and the American Cancer Society to reduce cancer risk include the following dietary guidelines: maintain desirable body weight; eat a varied diet and include a variety of fruits and vegetables daily; eat more high-fiber foods; cut down on total fat intake to 30% or less of total calories; and limit the consumption of alcoholic beverages and nitrite-preserved and smoked foods.

Nutrition, chemical dependence (drug abuse). The extent of malnutrition seen in chemical dependency depends on the degree of substance abuse, presence of underlying medical conditions, and changes in appetite. Most drug users have poor eating habits. "Street drugs" usually contain adulterants that are appetite depressants or laxatives. Besides the primary factor of malnutrition (i.e., inadequate food ingestion), the conditioning factors of malnutrition observed with chemical dependency include digestive malfunctions, malabsorption, increased nutrient requirements to metabolize the drug, poor utilization due to inactivated enzymes, impaired nutrient storage, and increased losses of nutrients through diuresis and diarrhea. Hormone production may be upset for the pregnant women and birth defects in the fetus are harmful consequences. Vitamins and minerals that are often deficient include thiamin, vitamin B_6, vitamin B_{12}, folate, pyridoxine, vitamin C, calcium, magnesium, and zinc. Chemical dependence impairs the body's physiologic systems. Malnutrition can exacerbate these impairments and increase the risk of infections and diseases due to an already compromised detoxification system. The nutritional care of drug addicts undergoing treatment includes: an evaluation of the addict's nutritional status; supervision of nutritional rehabilitation through delivery of palatable, nourishing meals and supplements; monitoring of body weight; monitoring of gastrointestinal problems (e.g., anorexia, nausea, vomiting, diarrhea), especially during the initial phase of detoxification; and provision of nutrition counseling.

Nutrition, childhood. Childhood is the period of life between infancy and puberty, generally from ages 1 to 12 years. Nutritional needs during this period cannot be generalized because of the following factors: spurts of growth occur, and growth patterns among children vary; the kind and size of food servings should be adjusted with age, development, and appetite; and outside influences such as school activities and play affect eating habits; and so on. Thus nutritional needs for specific ages from 1 to 12 years are considered separately. See *Nutrition, preschool age; Nutrition, schoolchildren;* and *Nutrition, toddler*.

Nutrition counseling. Providing expert advice to help a person with current or potential nutrition problems. The process involves a great deal of knowledge and skill in interviewing and educating people. Trying to make dietary changes is difficult because of a person's lifetime habits, personal tastes, ethnic or cultural background, educational level, attitudes toward food, economic obstacles, social lifestyle, and other factors. Effective nutritional counseling should start with establishing rapport, stating the objectives of the session and the benefits derived from dietary adherence, and identifying possible problems that the client may encounter with the dietary regimen or meal plan. After the initial counseling session, an aftercare plan or follow-up is essential. Long-term adherence and positive results leading to the client's recovery or maintenance of good nutritional status are the most challenging aspects of nutritional counseling. The guidelines of LEARN (i.e., listen, explain, acknowledge, recommend, and negotiate) lead to effective nutrition counseling.

Nutrition, coronary heart disease. Diets low in saturated fat and cholesterol and rich in fruits, vegetables, and grain products that are good sources of fiber (particularly soluble fiber) are beneficial in reducing the risk of coronary heart disease (CHD). Of the dietary factors for increased risk of coronary heart disease, excessive intakes of saturated fat and cholesterol are the major contributors to elevated levels of total cholesterol and low-density-lipoprotein cholesterol (LDL-cholesterol). Water-soluble fibers can lower blood lipid levels by binding with bile acids and increasing fecal cholesterol excretion. Public health authorities therefore recommend that Americans consume a high-fiber diet that derives less than 30% of calories from total fat, less than 10% from saturated fat, and less than 300 mg of cholesterol per day. Evidence suggests that lowering the level of cholesterol in the blood of healthy adults can lower the risk of CHD. For every 1 mg/dL increase in cholesterol, the risk increases as much as 1%. High-density lipoprotein (HDL) cholesterol correlates inversely with CHD risk; HDL below 35 mg/dL indicates risk. Also, if LDL to HDL ratio is greater than 5, it indicates risk. Trans-fatty acids increase risk, but the use of monounsaturated fatty acids (10 to 15% of kcal/day) decreases risk. Walking vigorously 30 minutes three times a week has protective effects against CHD.

Nutrition, dental health. The time intervals involved in the development of *teeth* are important considerations in any discussion of the relationship of nutrition to tooth development. The life history of a tooth may be divided into three main eras: the period during which the crown of the tooth is forming and calcifying in the jaw; the period of maturation when the tooth is erupting into the oral cavity and its root or roots are forming; and the maintenance period when it is in full function in the oral cavity. Prenatal factors affect *deciduous teeth* (baby teeth) far more than factors after birth. At the early stage of the 6th fetal week, the tooth buds start to form; calcification begins in the 16th week. Eruption of baby teeth starts at around the age of 7 months; the last baby molar comes out at the age of 2 years. Calcification of the permanent teeth starts soon after birth, and eruption of the first permanent tooth occurs at about 6 years of age, when the child is starting to lose baby teeth. Hence, an adequate diet is imperative as early as the first trimester of pregnancy. All nutrients, directly or indirectly, play an important role in dental development, as does the maintenance of dental health. The nutrients directly involved are *protein,* for organic matrix formation; *calcium, phosphorus, magnesium,* and *vitamin D* for deposition of the mineral compound *apatite* into the matrix structure; *ascorbic acid,* involved in mineral utilization and for cementum formation to connect the tooth to the bone structure and to the gum tissues; *vitamin A* for the proper functioning of enamel-forming cells to achieve a smooth, even enamel as well as a deposit of sound dentin; and *fluoride* to harden the enamel and prevent dental caries. Although nutrition influences dental health profoundly, nonnutritional factors are also important, such as oral hygiene, regular visits to the dentists, and fluoridation of the community water supply.

Nutrition, dermatology. The *skin* is a tough but resilient covering of the body with protective, regulatory, excretory, and sensory functions. The first layer (epidermis) is the site of keratin and vitamin D synthesis; the second layer (dermis), which is rich in blood vessels, nerves, and sweat and sebaceous glands, is a storehouse for water, blood, and electrolytes. The innermost layer (subcutaneous tissue) is a storehouse for body fats and helps to support the body. Protein deficiency results in dryness, scaliness, pallid appearance, and inelasticity of the skin, often with brownish pigmentation on the face. In kwashiorkor there is extensive hypopigmentation and dryness of the skin and hair, with the characteristic "flag sign" of the hair. Lack of *essential fatty acids* results in eczematous skin lesions that are probably related to the seborrheic dermatitis of *pyridoxine* deficiency. The skin changes caused by lack of *vitamin A* include desquamation of epithelial cells; keratinization; and dry, rough gooseflesh with follicular hyperkeratosis. On the other hand, excessive carotene (provitamin A) results in jaundice-like yellow discoloration of the skin. In *riboflavin* deficiency, dermatitis is of the seborrheic type, with fine oily scales, especially around the nose and lips. Symmetric dermatitis, (i.e., dark red eruptions and desquamation appearing bilaterally) is characteristic of pellagra (*nicotinic acid* deficiency). Dry, "crackled," and scaly dermatitis is observed in *biotin* deficiency. *Ascorbic acid* deficiency results in inelastic skin with a tendency to petechial hemorrhages.

Nutrition, diet therapy. Also called "therapeutic nutrition". Dietary modifications for the treatment or management of a disease or specific illness have certain nutritional effects, depending on the type of therapeutic diet and the length of time it is used. For example, the clear liquid diet is nutritionally inadequate; sodium-restricted diets reduce the palatability of food; fat-restricted diets decrease absorption of fat-soluble vitamins; a high polyunsaturated fatty acid diet tends to lower serum cholesterol; a high protein intake increases the requirement for riboflavin and pyridoxine; a high-carbohydrate diet tends to elevate serum triglycerides and to increase the thiamin requirement; and prolonged sodium restriction, especially below 500 mg, leads to the *low-sodium syndrome*. Throughout this dictionary, each disease or disorder has a section on *nutrition therapy* that discusses the recommended dietary modifications and rationale for the diet. Details about the general characteristics of the diet (foods allowed and avoided) are under the *Diet* series. See *Nutrition therapy*.

Nutrition, disaster feeding. Or emergency feeding. Refers to the provision of meals to persons suddenly deprived of food as a result of either artificial or natural catastrophes such as famines, floods, industrial accidents, fires, and wars. In emergency feeding, the long-range objective is to sustain adequate nourishment until the victims return to their normal pattern of life. The immediate aim, however, is to provide water and warm food to maintain body temperature and to give nourishment as soon as possible. Less emphasis is placed on dietary adequacy, especially when the emergency is brief. Distribution of food should be orderly, and the feeding program should include the rescue workers. Special attention should be given to vulnerable groups such as infants, children, pregnant women, and those under pathologic stress. Health care facilities as well as homes must be prepared for disaster feeding. In stockpiling supplies, problems due to lack of electricity, water supply, cooking and refrigeration facilities, waste disposal, etc., should be considered. Have enough water and beverages for a week and food for at least 3 days. Canned foods, ready-to-eat packaged foods, and nonperishables such as instant dry milk and powdered beverages are preferred. Include nonfood items, such charcoal or wood, battery-operated gadgets, lighter, matches, manual can opener, disposable eating ware, and sanitary garbage disposal. Water may be disinfected by adding 3 drops of bleach per quart of water.

Nutrition, diseases. The nutritional effects of diseases are collectively considered as secondary factors or "conditioning factors" of nutritional inadequacy. In general, a diseased condition leads to anorexia. Bed rest increases calcium and nitrogen excretion; diarrheal diseases increase motility of the intestines, and the frequent bowel movements obviously cause reduced nutrient intake, maldigestion, and malabsorption. Many diseases, particularly gastrointestinal disorders, are characterized by nausea and vomiting, affecting electrolyte and water balance. Renal diseases increase the loss of nitrogen. For other examples of the nutritional effects of diseases, see the discussion under each disease. Medical treatment of diseases usually involves the use of drugs, which affect nutrition in many ways. See *Nutrition, drug–nutrient interactions*.

Nutrition, drug–nutrient interactions. Nutrients and drugs interact in many ways, affecting nutrient utilization and the effectiveness of drug therapy. To cite the most common examples, antacids destroy thiamin; appetite depressants such as amphetamines may irritate the gastric mucosa and cause nausea and vomiting; mineral oil decreases absorption of fat-soluble vitamins; chelating agents bind many minerals and reduce their availability; cation-exchange resins also reduce available sodium, potassium, and calcium; antimetabolites for cancer therapy antagonize the physiologic role of some vitamins; barbiturates have antivitamin action against folic acid; and corticosteroids alter electrolyte, carbohydrate, and fat metabolism. The presence of food enhances absorption of some drugs whereas other drugs are best absorbed in the fasted state. The relative composition of the diet also influences drug absorption, e.g., fat promotes the absorption of some drugs. The side effects of some drugs, such as nausea, loss of appetite, and altered taste sensation may decrease food intake and create nutritional problems. Generally, drug-induced nutrient deficiencies develop slowly and the rate is hastened by high dosage, multiple drug regimens, poor diet, and chronic illness. This dictionary contains the names of 150 drugs, entered alphabetically, and discusses the nutritional implications of each drug.

Nutrition Education and Training Program. The U.S. Department of Agriculture administers this program through the state departments of education. Local schools provide the staff, including a dietitian, to conduct nutrition education about nutritionally balanced meals, the use of the Dietary Guidelines for Americans, weight control, and other nutrition topics for food service personnel, teachers, students, and their parents.

Nutrition, emotional stability. Well-nourished individuals generally have a cheerful disposition and a positive attitude toward changes and can adjust easily to various situations. By contrast, undernourished individuals tend to be nervous, tense, apathetic, dull-looking, and irritable. Examples of the relationship between nutrition and emotion include the following: (1) women who are emotionally distressed require a higher intake of calcium (and possibly other nutrients) than happy, relaxed women; (2) the well-known depression of pellagra starts with irritability, headache, and sleeplessness in the early stages, followed by loss of memory, hallucinations, and severe depression in the advanced stages; (3) persons starved for a long time are prone to excitement and hysteria; (4) fear, anger, and worry stimulate adrenaline secretion, which in turn increases the loss of nitrogen from the body; and (5) thiamin-deficient individuals show moodiness, uneasiness, and disorderly thinking. The allusion to thiamin as the "morale vitamin" originated from its ability to alleviate certain mental and emotional depressions. See also *Nutrition, mental health* and *Nutrition, stress*.

Nutrition, exercise. It is now common knowledge that regular exercise is important to one's physical, mental, and emotional health. Exercising conditions the body for greater stamina, increases energy output for weight reduction, maintains normal serum cholesterol and blood glucose levels, and relieves stress. During exercise, more endorphins, which are natural tranquilizers, are secreted. Regular exercise maintains a healthy cardiovascular system, lowers blood pressure, improves pulmonary function, and builds immunity, thereby prolonging life. Physical fitness lowers the risk for some types of cancer and Type II diabetes mellitus. Weight-bearing exercises build bone strength and prevent osteoporosis. Persons confined to bed lose muscle tone and show mineral imbalance (see *Immobilization*). Normally, it is not necessary to consult a physician for an exercise regimen. Some individuals will need medical clearance before starting an exercise program (e.g., cardiac disease, hypertension, breathing problems, some types of chronic illness). See also *Nutrition, physical health* and *Nutrition, sports*.

Nutrition Foundation, Inc. A public nonprofit institution established in 1941 for "the advancement of nutrition knowledge and its effective application in improving the health and welfare of mankind." It is supported by food and allied industries. The foundation publishes *Nu-*

trition Reviews, monographs, and pamphlets for the layman. It also sponsors conferences on nutrition. In 1985, the Nutrition Foundation, Inc. merged with the International Life Sciences Institute (ILSI) becoming the North American branch for ILSI. In 1990, ILSI-NFI published the sixth edition of the nutrition classic, *Present Knowledge in Nutrition*.

Nutrition, gastronautic. See *Nutrition, space feeding*.

Nutrition, genetics. Every person is genetically unique because of various hereditary factors. The multifactorial inheritance pattern results from the interaction of genes, such as from parents to offspring, or with environmental variables. The relationship between genetics and nutrition is best observed in disorders known as *inborn errors of metabolism (IEM)*. These inherited enzyme malfunctions of absorption, reception action, or utilization may lead to malnutrition and specific damage to a given organ system. See Appendix 44 for a summary of nutrition therapy in inborn errors of metabolism. An area of study is the relationship of genetics and nutrition in the familial tendency or predisposition to some disorders, such as arthritis, cancer, diabetes mellitus, heart disease, hyperlipidemia, hypertension, obesity, and osteoporosis.

Nutrition, geriatric. In the United States life expectancy is 72 years for men and 79 years for women. By the year 2000, it is estimated that one out of four persons will be 65 years old and above. This geriatric group presents problems that contribute to *nutritional inadequacy*. These include poor dentition or loss of teeth; loss of appetite and acuity of taste and smell; reduced secretions of digestive enzymes; lack of neuromuscular coordination; reduced cellular metabolism; reduced circulatory and excretory functions; hormonal changes; and a tendency to develop osteoporosis, pernicious anemia, and many metabolic disorders. Aside from these physiologic changes are socioeconomic and psychological factors that affect nutriture. Most common among the aged are depression, boredom, inactivity, lack of interest in the environment, anxiety, faulty eating habits that resist change, and economic insecurity. Major indicators of poor nutritional status in older Americans include the following: inappropriate or inadequate food intake; 5% weight loss in 1 month or 10% in 6 months; <3.5 g of serum albumin; lymphocyte count <1500 cells; change in functional status; reduction in midarm circumference; increase or decrease in triceps skinfolds; significant obesity; and other nutrition-related disorders. From the public health standpoint, people over 65 years of age represent 14% of the U.S. population, and it is estimated that by the year 2040, about 22% will have reached age 65 years. About 85% of them have nutrition-related medical problems, such as osteoporosis, obesity, hypertension, cardiovascular diseases, and diabetes mellitus. Increasing the RDA for calcium is currently under study due to the high incidence of osteoporosis among the elderly. There is need to revise the recommended dietary allowances (Appendix 1) for persons above 51 years because of the lack of stratification for the geriatric ages, particularly from age 65 to over 90 years old. The creation of the National Institute on Aging is a positive approach by the government to meet the needs of the elderly. Nutrition programs such as Meals-on-Wheels and Congregate Meals provide one-third of their RDA. Senior citizens in nursing homes are assured quality care with the implementation of OBRA guidelines. See also *Nutrition Screening Initiative*.

Nutrition, growth and development. In a broad sense, growth and development refers to the increase in the size and number of cells, as well as cell maturation for functional processes of the body. Thus, the study of nutrition as related to growth and development covers the entire life cycle of an individual. For nutrition principles for each period of life, see *Nutrition, adolescence; Nutrition, adulthood; Nutrition, geriatric; Nutrition, infancy; Nutrition, pregnancy; Nutrition, preschool age; Nutrition, schoolchildren;* and *Nutrition, toddler*.

Nutrition, gut integrity. The gastrointestinal tract has high nutrient needs because of high turnover of epithelial cells. Nutrition depletion is associated with a loss of gut mass and function (i.e., integrity). Gut function begins to deteriorate when nutrients are lacking, as in starvation. This may result in atrophy of the gut, particularly the jejunum, impairment of the immune functions of the GI tract, and a reduced capacity of the gastrointestinal tract to limit bacterial translocation (invasion). See *Bacterial translocation*.

Nutrition, heart disease. The 1990 CDC National Center for Health Statistics reported that the leading cause of death (44% of total deaths in the USA) is due to cardiovascular diseases (CVD). The American Heart Association estimated that in 1993, almost 70 million Americans had one or more forms of CVD, with hypertension as the leading condition affecting 64 million, followed by heart attacks, strokes, and rheumatic heart disease. The major risk factors identified in CHD are a family history of heart disease, hypercholesterolemia (especially elevated low-density-lipoprotein cholesterol), hypertension and other vascular disorders, obesity, hyperlipidemia, diabetes mellitus, reduced physical activity, cigarette smoking, and a stressful lifestyle. The incidence of CHD is higher among males than among females. The National Institutes of Health urges people over 20 years of age to have their total serum cholesterol, serum high-density-lipoprotein and low-density-lipoprotein cholesterol, and triglyceride levels checked for preventive measures against early heart diseases. Many national programs (such as the National Cholesterol Program, HeartGuide, National High Blood Pressure Education Program, and NHLBI Smoking Education Program) aim to reduce the incidence of heart diseases. Advances in medicine and public health nutrition help reduce morbidity and mortality rates from CVD. See Appendices 7A and 7B for the objectives of Healthy People 2000 and the dietary guidelines for healthy Americans.

Nutrition, hunger. Hunger is discomfort, weakness, or pain caused by lack of food. It is the consequence of being unable to obtain food from nonemergency channels due to poverty, environmental isolation, food inaccessibility, and other factors. Food shortages occur predominantly in Asia, Africa, and Latin America. All together, these areas account for 50% of the world's population, but they have only 25% of the world's food supply. World food production has not increased at a rate faster than or equal to the rate of population growth. The world population is expected to double by the year 2026. Unfortunately, at present, many depressed areas of the world are already experiencing famine and hunger. The United Nations reported that over 500 million people in the world are malnourished. Over 20 million people die each year due to hunger-related causes and lowered resistance to infections because of malnutrition. At least one-third of these deaths involve children below 5 years of age. In the United States, 20 million Americans are said to suffer from hunger and inadequate nutrition for several weeks due to domestic interruption, poverty, family dislocation, and other situational problems. A national program to fight domestic hunger should include: strengthening key government food programs such as the Food Stamp Program, the School Breakfast Program the School Lunch Program, the Summer Food Program, and the Home Delivered Meal Program for senior citizens; coordinating private and government services to create both short-term and permanent programs to alleviate hunger, such as the Self-Help and Resource Exchange (SHARE), the Food Policy Councils, and Farmers' Market Coupon Program; increasing funding for nutrition education programs, such as the Expanded Food and Nutrition Education Program (EFNEP) and the Nutrition Education and Training Program (NETP); implementing a comprehensive and responsive national nutrition monitoring system to track trends and evaluate intervention efforts; and providing job training, child care and housing assistance, adequate minimum wages, extended unemployment benefits, equitable public assistance benefits, and charitable tax incentives. The ultimate objective of a national program to prevent hunger is to provide food security, which means that nutritionally adequate and affordable meals are available at all times through conventional food sources.

Nutrition, immune system. The human body has the ability to identify and defend itself against the invasion of foreign substances through its complex system of nonspecific response barriers (e.g., skin, mucous membranes, and phagocytic cells) and specific response mechanisms with antigens and antibodies. The organs of the immune system include the tonsils and adenoids, lymph nodes and lymphatic vessels, thymus (T cells), Peyer's patches, spleen, and the bone marrow, where B cells (B lymphocytes) mature. Good nutrition helps maintain the integrity of the immune system and builds the body's resistance to disease. Protein-calorie malnutrition impairs the body's immune system and is associated with increased incidence of immunodeficiency infections. To cite a few cases, a breakdown of the first barrier against infection (skin) may be due to deficiencies in protein; zinc; and the A, C, and B complex vitamins. Lack of protein reduces biosynthesis of lymphocytes and immunoglobulins or antibodies. Iron is needed for optimal immune function of neutrophils and lymphocytes. Selenium promotes antibody production and protects the phagocytic function of the neutrophils. *Anergy* occurs in protein-energy malnutrition (PEM), which is observed in cancer, bacterial and viral infections, multiple trauma, burns, uremia, liver disorders, and other debilitating diseases. Arginine, glutamine, magnesium, manganese, the omega-3 and omega-6 fatty acids, and vitamin E also have an effect on the immune system. For more details, see *Immunonutrition*.

Nutrition, infancy. For the past years, the leading causes of death among infants in the United States have been low birth weight and prematurity. However, the infant mortality rate has decreased significantly due to improvements in sanitation, modern drug therapy, and preventive public health measures. The first 12 months of life are characterized by the most rapid rate of growth and development of the entire lifetime; thus, an infant's nutritional needs merit special attention. Recommended caloric intake is about 110 to 117 kcal/kg/day, using the higher level for the first months of age. About two-thirds of the caloric needs should come from milk and the rest from added carbohydrates and, later, from supplementary foods. *Protein* needs should be within 2 to 2.2 g/kg/day, which is easily supplied by adequate milk intake. *Fat* is important primarily to provide essential fatty acids comprising at least 1%

to 2% of total caloric needs. This is adequately supplied by human milk or whole cow's milk. If protein intake is inadequate, niacin and choline, which are synthesized from tryptophan and methionine, respectively, may be deficient. *Fluid* requirement is about 4.5 to 5 oz/kg/day; it is the most variable nutrient need because of fluctuations in the infant's activities and the environmental temperature. *Mineral* needs are generally met during the first 3 months of age by human milk or by the prescribed milk formula. However, after 3 months, the infant's iron stores are depleted, and there is need for other foods and supplementation. *Vitamins* needed by the infant are ample in the milk supply except for vitamins A, D, and C, which must be supplemented (see Appendix 1). See also Appendix 37 for the composition of milk and selected formulas for infant feeding. The following are the main criteria of normal growth and development during infancy: *steady weight gain* of 5 to 8 oz/week, which slows down toward the end of the first year to about 4 oz/week (this means that an average infant should have doubled and tripled his or her birth weight at the end of 5 and 12 months, respectively): *normal increase in body length* is about 10 inches (50% more) at the end of the first year; firm, well-formed muscles with moderate subcutaneous fat; normal sleeping habits and happy disposition; and normal dental and motor development. See *Nutrition, dental health; Nutrition, mental health;* and *Nutrition, motor performance*.

Nutrition, infection. Blood levels of vitamin A are sufficiently reduced in acute infections that xerophthalmia and keratomalacia frequently develop in children receiving diets deficient in this vitamin. Any intestinal infection producing malabsorption interferes with the absorption of fats and fat-soluble vitamins. Clinical manifestations of thiamin, folic acid, vitamin B_{12}, and ascorbic acid deficiencies are all related to a preceding infection of a vulnerable host. Increased losses of calcium and phosphorus are seen in tuberculosis. Losses of sodium, chloride, potassium, and phosphorous are seen in diarrheal diseases of infectious origin. Hookworm infections can be responsible for enough loss of blood to induce anemia. Nutritional losses due to the infectious process can be significant. These losses may be due to decreased intestinal absorption and loss of nutrients in the gut. The presence of fever increases the rate at which nutrients are utilized. The hypermetabolism and hypercatabolism seen in severe infections significantly increase nutrient requirements, leading to further nutrient depletion. Nutritional deficiencies impair immune function which, in turn, leads to decreased resistance to infection.

Nutrition labeling. The Nutrition Labeling and Education Act (NLEA) of 1990 was signed into law and became effective in May 1994. All packaged foods bear the new format "Nutrition Facts" in bold letters. Information on the specific amounts for the following nutrients per serving of the food is mandatory: total food energy, total energy from fat, total fat, saturated fats, cholesterol, sodium, total carbohydrates, sugars, dietary fiber, and protein. The only vitamins and minerals that must be listed on the label are vitamin A, vitamin C, calcium, and iron. Also, health claims are permitted only if valid links between diet and prevention of the disease or disorder have been scientifically established. To date, the FDA has approved the following relationships: calcium and osteoporosis; sodium and hypertension; dietary fat and cancer; fruits, vegetables, and fiber-containing products and cancer or coronary heart disease; and dietary saturated fat and cholesterol and risk of coronary heart disease. Moreover, foods that contain more than 20% of the Daily Value (DV) for fat, saturated fat, cholesterol, and sodium are disqualified from inclusion in health claims. The Food and Drug Administration also replaced the U.S. RDA values with two sets of standard values. See *Reference Values (RV)* for details. See also Appendices 5 and 6.

Nutrition, lactation. The human milk of an adequately nourished lactating mother offers nutritional, immunological, and psychological benefits not paralleled by bottle feeding with milk formulas. On the average, a lactating mother produces 25 oz of milk daily. Nutrient requirements of the mother during *lactation* are increased to provide for normal secretion of milk and for recovery from pregnancy and delivery. Thus, the *caloric requirement* is increased by 500 to 750 calories over that of the nonpregnant woman. This is approximately 90 kcal/100 mL of milk secreted to provide for energy expenditure of the secretory function and for the caloric content of milk. Similarly, intakes for *protein, minerals* (particularly *calcium*), and *vitamins* are increased (see Appendix 1). To be assured of adequate fluid intake, it has been suggested that a lactating mother drinks a glass of fluid at each meal and each time the baby is breast-fed. Certain drugs are excreted in breast milk; breast feeding may be contraindicated or the infant must be observed for adverse side effects when these medications are taken. Examples are chloramphenicol, chlorpromazine, cyclophosphamide, glutethimide, hydralazine, phenobarbital, phenytoin, prochlorperazine, sulfasalazine, and valproic acid. Caffeine and alcohol also pass into the milk; limit the intake while lactating. Obese mothers who are breastfeeding are not placed on a low-calorie diet. Caloric intake of <1500 kcal/day can result in reduced milk production and inadequate nutrients for the infant. See also *Infant feeding* and *Lactation*.

Nutrition, mental health. The relationship of nutrition and mental health is well established; proper nourishment is conducive to mental efficiency, the ability to concentrate, and the ease of adjustment to environmental changes. By contrast, nutrient inadequacy may result in permanent changes in the size or chemical composition of the brain, thus affecting mental capacity; inborn errors of metabolism retard mental development (e.g., phenylketonuria and galactosemia); and specific nervous lesions occur as a result of nutrient deficiencies, as in *pellagra, kwashiorkor, marasmus,* and *beriberi*. For further details, see also *Nutrition, emotional stability* and *Nutrition, psychiatry*.

Nutrition monitoring. Process of providing information on a regular basis about the role and status of nutritional factors that relate to health. Nutrition monitoring, therefore, extends over a period of time, with repeated nutritional assessment to measure changes. See also *National Nutrition Monitoring System*.

Nutrition, motor performance. The ability of the body to coordinate its movements and to perform other physical activities has its roots in the motor development during infancy and childhood. Evidence of the importance of adequate nutrition for motor development comes from studies of kwashiorkor, beriberi, and other deficiency diseases; phenylketonuria and other inborn errors of metabolism; and *hypokinesis*. Lack of exercise and prolonged immobilization, as in bedridden patients, lead to negative nitrogen and mineral balances. The physiologic effects of moderate exercise on nutriture and good health in general are as follows: it promotes pulmonary ventilation, circulation, and oxidation processes; it stimulates appetite, flow of digestive juices, and peristalsis; it improves muscle tonus and sleep habits; and it increases strength, mental acuity, and resistance to infection. See also *Nutrition, exercise,* and *Nutrition sports*.

Nutrition, neonatology. The study of specialized care of newborn infants. The first 4 weeks after birth is called the "neonatal period." Newborn care is categorized into Level I: full-term neonates; Level II: normal (full term) neonates, but considered high-risk for various medical reasons; and Level III: critically ill neonates requiring intensive care. The last group is classified as either low-birth-weight (LBW) infants weighing less than 2500 g (5½ lb) at birth or very-low-birth-weight (VLBW) infants weighing less than 1500 g (3½ lb). Level III neonates have a smaller metabolic reserve than those in Levels I and II and are unable to store fat and glycogen during the last period of gestation. Their glycogen reserve may be only 110 kcal/kg, and their fat reserve may be only 1% of their total body weight (compared to 16% in the full-term baby). Because the basal metabolic need of LBW infants is 50 kcal/kg, careful monitoring and adequate provision of energy intake are very important. The American Academy of Pediatrics recommends the following nutrients per 100 kcal: 2.7 to 3.1 g protein; 4.3 to 5.4 g fat, of which 300 mg are essential fatty acids; 2.3 to 2.7 mEq sodium, 2.0 to 2.4 mEq chloride; 1.8 to 1.9 mEq potassium; 140 to 160 mg calcium; 95 to 108 mg phosphorus; 6.5 to 7.5 mg magnesium; 1.0 to 1.8 mg iron; 0.5 mg zinc; 90 μg copper; 5 μg manganese; and 5 μg iodine. For vitamins, the total amount per day is 1400 IU vitamin A; 500 IU of vitamin D; 5 to 25 IU vitamin E; 35 mg vitamin C; 300 μg thiamin; 400 μg riboflavin; 6 mg niacin; 15 μg vitamin B_6 per gram protein; 50 μg folic acid; 0.3 μg vitamin B_{12}; 2 mg pantothenic acid; and 35 μg biotin. High-risk neonates can be given parenteral feedings of dextrose, amino acids, electrolytes, vitamins, minerals, and trace elements. See also *Nutrition, prematurity*.

Nutrition, nervous system. Proper nutrition is important for the integrity of nerve cells and the normal functioning of the nervous system. An obvious example is the feeling of restlessness or irritability when one is hungry. All nutrients are involved in maintaining a healthy nervous system, notably the following: *carbohydrate* (specifically, glucose) is the chief energy source utilized by the brain; *protein* is needed to synthesize the enzyme systems to oxidize glucose and provide the aromatic and acidic amino acids needed for brain function; and *B complex vitamins* are involved as cofactors for these enzymes. Certain vitamin deficiencies lead to neurologic changes varying from mild symptoms to gross anatomic nerve lesions; for example, lack of *thiamin* results in neuromuscular incoordination (nystagmus, hyperesthesia, ataxia, cramps, and loss of tendon reflexes), brain lesions and degeneration of nerve fibers and myelin sheaths, with death of the parent nerve cells; *riboflavin* deficiency is characterized in the early stages by photophobia and affects nicotinic acid metabolism, in which riboflavin is indirectly involved; lack of *nicotinic acid* is associated with the psychosis of pellagra; and *pyridoxine* deficiency leads to sensory neuritis, convulsions, hyperesthesia, and loss of positional sense. Neurologic disorders that affect nutritional status because of self-feeding problems and metabolic changes include Alzheimer's disease, cerebral palsy, Down syndrome, epilepsy, Guillain-Barré syndrome, Lou Gehrig's disease, multiple sclerosis, and Parkinson's disease. Many of these illnesses require individualized nutrition support. See also *Nutrition, emotional stability; Nutrition, mental health;* and *Nutrition, psychiatry*.

Nutrition, ophthalmology. It has long been established that the muscles, nerves, and mucous membranes of the eyes, as well as the visual process, are affected by nutrition. The ability to see or adapt in dim light depends in part on adequate *vitamin A* intake. Ocular changes associated with a lack of vitamin A are xerosis, keratomalacia, Bitot's spots, night blindness, and xerophthalmia. *Riboflavin* deficiency leads to corneal vascularization, conjunctivitis, photophobia, and burning and itching of eyes. Lack of *pyridoxine* results in angular blepharitis and conjunctivitis. The *essential amino acids, vitamin C,* and *nicotinic acid* maintain the integrity of the eye lens and prevent cataracts.

Nutrition, parasitism. Infestation of the human host with *parasites,* particularly in the digestive tract, is a conditioning factor of nutritional inadequacy. The common intestinal parasites are tapeworms, hookworms, pinworms, whipworms, ascarides, flukes, *Trichinella,* and *Trichuris*. The extent of their harmful effects depends on the tissues invaded, the types of secretion they produce, their rate of growth and multiplication, and the ability of the body to protect itself from these effects. The principal ways in which parasites harm the host are as follows: mechanical injury (tissue lesions and hemorrhage), inflammation, and pain; obstruction (e.g., blocking ducts); toxicity of their secretions; and robbing the host of its nutrient supply, especially proteins and vitamins. The nutritional implications of these parasitic effects do not need further elaboration. Preventive measures against parasitism include *nutrition education* (a well-nourished body can build antibodies that neutralize the toxic effects of parasites); *environmental sanitation* (e.g., proper sewage disposal, safe water and food supplies, educated food handlers and consumers); and *parasitic control* (medical attention and use of chemicals or other means of interfering with the life cycle of the parasite). See also *Nutrition, resistance to disease*.

Nutrition, physical health. The state of nutrition is easily reflected in a person's appearance. The physical signs of good nutrition are normal weight for height, body frame, and age; firm, moderately padded muscles; clear and slightly moist skin with good color; well-formed jaw and teeth; soft, glossy hair and clear, bright eyes; a well-formed trunk; good appetite and abundant energy; endurance for physical work and resistance to disease; and, in general, a happy personality. See Appendix 26 for the physical assessment of nutritional status. See also *Nutrition, sports*.

Nutrition, preadolescence. See *Nutrition, puberty*.

Nutrition, pregnancy. Recent data indicated that of the 5 million pregnancies annually in the United States, only 75% resulted in live births. Statistics of live birth could have been improved if pregnant women who did not seek prenatal care were counselled to seek help before the second trimester of pregnancy. Good nourishment during pregnancy is important for meeting not only the mother's needs but also those of the growing fetus (prenatal nutrition). Good maternal nutrition results in a lower incidence of abortion and miscarriage; fewer stillborn and premature infants and infants with congenital malformations; fewer complications during pregnancy (e.g., toxemias and anemias) and delivery (e.g., prolonged labor, premature separation of the placenta, and hemorrhaging); healthier full-term babies; and reduced infant mortality and morbidity rates. Pregnancy imposes a physiologic stress on the mother, and the most important changes with nutritional implications are an increase in basal metabolic rate (about 25% in the latter half of pregnancy); a tendency to retain water; decreased gastric acidity and intestinal motility, with frequent impairment of digestion and absorption in the early stage and constipation in the last trimester; simple glycosuria; hormonal changes (increased activities of progesterone, gonadotropin, estrogen, and adrenal steroid hormones); and a positive nitrogen balance and an increase in plasma volume, with corresponding decrease in hemoglobin concentration. With these significant changes in mind, dietary intakes try to provide for the increased maternal metabolic activities; the nutrient needs of the growing fetus; the development of reproductive tissues (uterus, placenta, etc.) and the mammary glands; and the nutrient reserves to allow for losses during delivery. In general, all nutrient requirements are increased from the nonpregnant to the pregnant state. For further details, see Appendix 1. Noteworthy is the need to regulate weight gain during pregnancy. The desirable weight gain is about 20 to 25 lb throughout the gestation period, distributed as follows: normal weight of the infant at birth = 7 to 7½ lb; weight of the uterus, placenta, and membrane = 3 to 3½ lb; weight of amniotic fluids = 2 lb; weight of the mammary glands and tissues = 1 to 1½ lb; and the remaining weight is in the form of maternal body water and increased blood volume. The total weight gain of 23 lb (average) should be about 5, 8, and 10 lb for the first, second, and third trimesters, respectively. In view of the rapid fetal growth and development that occur as early as the second month of pregnancy, it is recommended that *protein, mineral,* and *vitamin* intakes be increased as early as possible. For a healthy pregnant woman, adequate vitamin and mineral intakes will be met by a variety of foods in recommended amounts, according to the food guide pyramid. For strict vegetarians, supplementation with iron, zinc, folate, and vitamin B_{12} is necessary. Caffeine should be limited to the equivalent of 2 to 3 cups of regular coffee per day. Alcohol should be avoided because its consumption is associated with *fetal alcohol syndrome*. See also *Nutrition, teenage pregnancy*.

Nutrition, prematurity. A premature or preterm infant born before the 38th week of pregnancy is less developed than a full-term baby born between the 38th and 42nd week of gestation. Premature infants have special nutritional needs because their nutrient requirements are high relative to their body size and weight. Their digestive capacity is usually small; the sucking and swallowing reflexes are not well developed; respiration is irregular, with a tendency to regurgitate and aspirate foods; achlorhydria is common; gastric emptying is delayed; and gastrointestinal motility is sluggish. Fat absorption is poor, resulting in a loss of fat-soluble vitamins and calcium along with fecal fat. The dietary management of a premature infant has to be individualized, depending on the anatomic facilities and physiologic condition. A major nutrition goal is to provide nutrients for the development of new lung tissues to improve ventilation. Daily nourishment should provide at least 125 kcal/kg and 4.5 g of protein/kg body weight with vitamin and mineral supplementations. A recommended feeding regimen for a premature infant is as follows: for the *first 12 hours,* nothing is given orally; for the *next 24 hours,* a 5% glucose solution is given at 2- to 3-hour intervals, starting first with a half teaspoonful and gradually increasing by half teaspoonfuls; for the *fourth 12 hours,* the mixed formula is given at 2- to 3-hour intervals, gradually increasing the amount; for the *third day* and thereafter, a full concentration of the prescribed formula is given, starting with small amounts and gradually increasing as tolerated by the infant. Vitamin and mineral supplements are given by the second week. Preterm infants less than 36 weeks of gestation need sodium supplementation because they excrete more sodium and are at greater risk for dehydration than full-term babies. See also *Nutrition, infancy* and *Nutrition, neonatology*.

Nutrition, preschool age. The group composed of 1- to 5-year-old children constitutes a far more nutritionally vulnerable group than infants. Recognizing that ages 1 to 5 years are the most formative years of child development in all aspects of personality (physical, mental, and social), the importance of nutrition cannot be overemphasized. For a quantitative evaluation of dietary adequacy for ages 1 to 3 and 4 to 6 years, see Appendix 1. See also *Nutrition, toddler*.

Nutrition, psychiatry. Humans associate varying emotional experiences with food, such as social acceptance or rejection, feelings of security or anxiety, pleasure or pain, satisfaction or frustration, and many other feelings. In the mentally ill, these associations are often exaggerated, and feeding problems are more difficult to manage. Well-known conditions with underlying psychological disorders are obesity with uncontrollable overeating, anorexia nervosa, chronic alcoholism, and drug addiction. Other psychiatric disorders that affect food intake include mood disorders (major depression and delusions) and schizophrenia. Anorexia and weight loss are common in depression, although some individuals may have increased appetite and weight gain. Psychotic behavior and delusional thinking in schizophrenia may lead to avoidance of certain foods and abnormal meal patterns. An area that needs evaluation are the effects of psychotropic drugs on nutritional status. Many of these drugs cause gastrointestinal distress, dry mouth, increased carbohydrate cravings, and weight changes. Antidepressants can cause either anorexia or overeating.

Nutrition, puberty. Puberty is the period during which the reproductive organs become functionally operative and secondary sex characteristics start to develop, with accompanying increased rates of growth and metabolism. Its onset varies from 10 to 15 years of age, with 12 to 13 years of age being the average for boys and 2 years earlier for girls. For nutritional considerations, see also *Nutrition, adolescence*.

Nutrition, public health. The theory and practice of nutrition as a science through organized community efforts, with the family as the smallest unit under study. The overall aim of public health nutrition is to improve or maintain good health through proper nutrition. Of the 10 leading causes of death, five are associated with diet (atherosclerosis, diabetes, heart disease, stroke, and some kinds of cancer), and three with excessive alcohol consumption (accidents, cirrhosis, and suicide). Food components that constitute current public health issues are energy; sodium; alcohol; iron; calcium; and total fat, saturated fat, and cholesterol. Other food components considered to be potential public health issues include dietary fiber; fluoride; folacin; zinc; and vitamins A, B_6, and C. Various agencies, governmental and nongovernmental, are concerned with nutrition work at local, national, or international levels. See Appendix 49.

Nutrition, resistance to disease. See *Nutrition, immune system* and *Nutrition, infection*.

Nutrition, respiratory function. Nutrition affects pulmonary function in three areas: ventilatory drive, respiratory muscles, and gas exchange (i.e., oxygen consumption and carbon dioxide production). If the person is malnourished, the ventilatory drive is insufficient to respond to hypoxia and hypercapnia. The respiratory muscles become weak and lack endurance. As body weight is lost, there is a proportional loss of strength of the diaphragm. This is evident by reduced vital capacity and inspiratory/expiratory pressures. All these events lead to respiratory failure. It has been observed that 50% of patients with acute respiratory failure are malnourished and malnutrition leads to ventilatory failure. In addition, there is impairment of the immune system which makes the tracheobronchial mucosa susceptible to bacterial infection. See also *Respiratory failure*.

Nutrition risk. The patient at high risk for malnutrition has one or more of the following: severe underweight (less than 80% of standard for height); severe overweight (more than 120% of standard for height); recent unintentional 5% weight loss in 1 month or 10% loss in 6 months; serum albumin <3.5 g/dL; total lymphocyte count <1500 cells; on NPO or nothing by mouth for more than 5 days while getting simple intravenous solutions such as dextrose or saline; protracted nutrient losses such as in malabsorption or short bowel syndrome, fistula, draining abscesses or wounds, or renal dialysis; increased metabolic demands due to, e.g., trauma, burns, or sepsis; taking drugs such as steroids, immunosuppressants, and antitumor agents that have antinutrient or catabolic properties; problems with chewing or swallowing; and substance abuse, especially alcohol abuse.

Nutrition, schoolchildren. The current focus on child nutrition programs is based on studies that good nutrition improves not only overall health status, but also the learning capabilities and attention span of school children. Feeding this age group is said to be the best and most cost-effective health preventive measure today. Cognizant of these facts, the school lunch and school breakfast programs have received more financial support and a general improvement in the nutriture of this age group is expected in the year 2000. Dietary principles in feeding schoolchildren are the same as those for preschoolers, but food intake differs in quantity and variety. Due to school activities, children from 6 to 12 years of age are more independent, and this includes their choice of food outside the home. Unless supervised by nutrition educators, as in the case of *school lunch programs,* feeding time and food selection may be erratic. Hurried meals, skipping breakfast, poorly selected snacks, and lack of appetite due to social and school interests are the common feeding problems faced by parents and teachers, which can be corrected by child nutrition programs in schools.

Nutrition screening. Process of identifying clinical characteristics known to be associated with malnutrition in order to identify persons at risk and to plan appropriate nutrition therapy. Information gathered in nutrition screening includes: age; height; usual and present weight; change in appetite; difficulty with chewing or swallowing; presence of nausea, vomiting, or diarrhea; and serum albumin, hemoglobin, hematocrit, and total lymphocyte counts.

Nutrition Screening Initiative (NSI). A consortium composed of the American Academy of Family Physicians, the American Dietetic Association, the National Council on the Aging, and a blue-ribbon Advisory Board of 30 health care-related organizations. It emphasizes the vital role of nutrition screening as a first step in identifying nutritional risk factors, especially among the elderly. A public awareness checklist and two screening levels have been developed to recognize these risk factors. The checklist, Determine Your Nutritional Health, assists the elderly or those who interact with them to recognize food and lifestyle habits, conditions, and diseases that affect their nutritional health. Those with scores that indicate nutritional risk are referred to individuals or organizations that provide nutrition services. The Level I Screen evaluates weight, height, body mass index, eating habits, and socioeconomic and functional status. In addition to the Level I components, the Level II screen includes anthropometric measurements; laboratory tests; cognitive, emotional and functional status; living environment; clinical signs of nutrient deficiency; and chronic medication use.

Nutrition, skin condition. See *Nutrition, dermatology*.

Nutrition, space feeding. Also called "gastronautics"; the preferred name is "astrophysiologic dietetics," signifying that feeding an organism beyond the earth's environment is both an art and a science. Space begins about 120 miles above the earth and is devoid of air, friction, gravity, and oxygen. Thus, the very nature of a trip into space causes nutritional problems. Although the water requirement remains the same, caloric needs are reduced because of weightlessness and confinement, resulting in sedentary activity. About 50% of the total caloric intake should come from fat (this is an advantage, as fat catabolism requires less oxygen); 15% of calories is provided by protein; and the rest is carbohydrates. Fecal losses are minimized by the use of low-residue diets. Mineral levels for calcium, sodium, potassium, and magnesium are increased by 10% above normal needs. Foods used should be specially prepared to tolerate weightlessness; sudden fluctuations of pressure, temperature, light, radiation, and humidity; and other conditions during the flight. The overall aim of the feeding system is to promote maximum efficiency of the crew, whose movements are restricted even in mastication. Evidently, there are changes in ingestion and digestion, taste acuity, nitrogen and calcium needs, and metabolism of food. Besides providing nutrient adequacy, foods should be palatable, easy to manage, and not subject to deterioration. Thus, the types of foods chosen are bite-sized finger foods, precooked foods, and freeze-dried foods, each packaged to serve three purposes—storage container, food utensil, and eating device. For example, the food supply of the Gemini and Apollo missions consisted of 50% rehydratable foods requiring addition of water before ingestion and 50% bite-size foods that rehydrate in the mouth. The 4-day cycle menu supplied at least 2500 kcal/day/person. Fruit-flavored drinks were fortified with calcium lactate. Fresh fruits were provided for consumption within 3 days. Fruit bars (170 kcal/bar) were carried in suit pockets. For food safety, the microbiologic criteria of space foods exceed those for commercial foods on earth. Additionally, the psychological effects of food become more significant as the duration of the mission is prolonged.

Nutrition, sports. Recommended intakes of calories and the major nutrients for an athlete vary according to age, sex, body size, type of sports activity, and duration and intensity of the athletic event or training. For example, a large-framed teenage male athlete on endurance training needs more than an adult female tennis player. The caloric needs of an athlete should, therefore, be individualized and calculated accordingly. Half (50%) of total calories should come from complex carbohydrates, 5% from simple carbohydrates, 15% from protein, and 30% from fats. Athletes who compete in prolonged endurance events and those who train exhaustively on successive days should consume 60% to 70% of calories from complex carbohydrates. The RDA for protein (0.8 to 1.0 g/kg body weight) is adequate if the athlete is not engaged in intensive physical activity. The protein allowance is increased to 1.1 to 1.2 g/kg body weight during vigorous exercise and greater activity, and 1.3 to 1.5 g/kg for endurance athletes and during periods of intensive training or competition. Body builders, weight lifters, and those involved in intensive strength training may need up to 2.0 g/kg of body weight. Contrary to popular myth, a very-high-protein diet is not recommended. With an intake of at least 3000 kcal/day, vitamin and mineral requirements should be adequately met, using the dietary guidelines and recommended food groups. However, amenorrheic athletes may require supplementary calcium to make up for decreased absorption in prolonged training and lowered estrogen levels. Athletes should avoid alcohol and caffeine. Five to six meals and snacks, properly spaced throughout the day, are better than three big meals. A pre-game meal is best eaten 3 hours before the event to allow for digestion. Avoid gas-forming foods and replace fluid losses. An athlete may lose more than 2 liters of sweat per hour in a hot environment while performing or during prolonged exercise. Electrolytes lost by sweating can be replaced easily from foods eaten and plain cool water or low-calorie beverages. Sugar-containing beverages with more than 10% carbohydrate can delay gastric emptying and may stimulate the release of insulin, resulting in low blood sugar levels. The following are recommended fluid intakes for an athletic event: 16 oz, 2 hours before the event; 20 oz, 10 to 15 minutes before the event; 4 to 6 oz every 15 minutes during the event; and 16 oz for every pound lost after the event. The intake of caffeine-containing beverages should not exceed more than 300 mg/day or approximately 2 to 3 cups of coffee. (See Appendix 35.) There is no need for special tablets when natural foods rich in sodium, potassium, and magnesium will supply these minerals. See also *Carbohydrate loading*.

Nutrition, starvation. A state of malnutrition arises from prolonged starvation caused by any of the following circumstances: lack of food, as in famine, war, or poverty; the presence of a conditioning factor, especially disorders of the digestive tract and malabsorption syndrome; and the effect of certain toxemias that may be metabolic or infectious in origin. The outstanding changes in starved individuals are generalized atrophy, or wasting of tissues, with loss of body fat and increased extracellular fluids; shrinkage of lean tissues, with muscular weakness and loss of skin elasticity; disturbance of water balance as a result of loss of plasma proteins and electrolyte changes, with consequent nutritional edema; digestive and renal disorders, especially nocturia and "starvation diarrhea"; lack of hormone and enzyme production, leading to decreased metabolic activities; loss of sexual function; amenorrhea in women; lowered blood pressure; mental restlessness and physical apathy; cancrum oris; and many other clinical and anatomic lesions, depending on the severity of starvation (partial or complete deprivation of food) and the length of time an individual is starved. Death is likely to occur when lean tissues are depleted over 40% and the body mass index falls below 13. In developing countries with population groups suffering from severe starvation, immediate causes of deaths are pneumonia, dehydration from diarrhea, and cardiac arrhythmias. The rehabilitation of the starved person should be gradual. Immediately correct any existing electrolyte imbalances. Thereafter, dietary needs are met by calories in the form of warm, easily digested, bland foods given in small, frequent feedings and then gradually increased as tolerated. Recommended first foods are skim milk and simple carbohydrates. Medical supervision and parenteral feedings are needed for severely starved persons. See also *Nutrition, hunger* and *Starvation*.

Nutrition, stress. Stress is any stimulus that interferes with the normal equilibrium of the body with varying rate of wear and tear on vital activities. Stress factors may be internal or "built-in" ones such as genetic disposition, age, or sex, or they may be external or manageable ones such as poor diet, alcoholism, and drug abuse. Therefore, stress management has become a necessary consideration in nutrition assessment, care, and planning based on human needs and an essential component in current promotional programs. Factors that disturb the normal equilibrium of the body are called "physiologic stresses." They include physical; mental; emotional stimuli, such as pain, anxiety, fear, time pressure, and worry; and metabolic factors. Illness and surgery also create stress. Physiologic stresses stimulate the sympathetic nervous system. The initial responses include: increased catecholamines and hormones leading to hyperglycemia, glucogenesis, gluconeogenesis, and increased secretions of glucagon and glucocorticoid and antidiuretic hormones. Respiratory rate and heart rate are elevated. Peristalsis is decreased and nutrient absorption is reduced. A person under stress may experience nausea, vomiting, anorexia, and oliguria. Energy requirement may increase as much as 100%, depending on the severity of stress factors (see Appendix 24). The body's basal protein needs may increase as much as 300% in severe trauma. Undoubtedly, other dietary factors (e.g., fats, vitamins, and minerals) play important roles in the overall process of nutrition and stress interactions.
Another group of stress factors are environmental in nature and are called "psychosocial" factors. They include poverty, feelings of isolation, insecurity, or powerlessness, and others. These are perpetuated by society's values and attitudes. Emotional tension from multiple causes is the most common agent of human stress. It can contribute to cardiac and gastrointestinal diseases, especially if the body is conditioned by malnutrition, faulty diet, or poor housing, as in the case of high-risk families in the grip of poverty. Both physiologic and psychosocial stresses can depress the immune function of the body and decrease its resistance to disease.

Nutrition surveillance. Recommendations from the World Food Conference of 1974 included a global nutrition surveillance system to monitor the food and nutrition conditions of populations at risk, and to provide a method of rapid and permanent assessment of all factors that influence food consumption patterns and nutritional status. In contrast to nutrition screening which identifies individuals at risk, nutrition surveillance determines status and necessary intervention at the community, regional, national, or international level. Nutrition surveillance activities include monitoring national and regional indicators of nutrition status, evaluating specific nutrition programs, and predicting food consumption inadequacy for early intervention. See also *National Nutrition Monitoring System*.

Nutrition survey. Study of the nutritional status of a population group in a given area of operation. The population may be homogeneous (e.g., teenage girls or diabetics) or heterogeneous (e.g., hospital patients). The survey may be focused on various factors, such as age, sex, race, or socioeconomic, geographic, physiologic, or pathologic condition, depending on the aims of the study. In general, the main objectives of the nutrition survey are to determine the extent of malnutrition and ascertain feeding problems, to provide ways and means of correcting or preventing nutritional problems, and to help in nutrition education, economic planning, and other programs for the improvement of the health status of the population or group. The main techniques used in nutrition surveys are: *dietary surveys,* which are food consumption data to detect inadequate diets and faulty food habits (e.g., food accounts, food frequency questionnaires, food recall, and food records); *medical history* to know the past and present illnesses and identify secondary factors of nutritional deficiencies; *clinical tests* or *physical examinations* to detect signs and symptoms of malnutrition; *biophysical tests* that reveal anatomical or *tissue* changes related to malnutrition; and *biochemical analyses* (e.g., analysis of blood and urine constituents) to detect abnormal nutrient levels. Examples of nutrition surveys conducted in the United States include the Ten-State Nutrition Survey; the National Health and Nutrition Examination Survey (NHANES) I, II, and III; and the United States Department of Agriculture Nationwide Food Consumption Surveys (NFCS).

Nutrition, teenage pregnancy. Adolescent or teenage pregnancy is still a major public health problem in the United States. Of the industrialized countries in the world, the United States has the highest frequency of teenage pregnancy. Pregnant adolescents are nutritionally at risk and require nutritional care throughout the gestation period. For optimal reproductive performance, a biologically mature female is the first requisite. Many pregnant adolescents have not reached this stage, which is at least 5 years postmenarche. The growth demands of a teenager plus the anabolic demands of pregnancy per se and the growing fetus require adequate health care, including proper nutrition. Because of the accompanying complex effects of socioeconomic, emotional, and physical factors, teenage pregnancy presents the greatest challenge for nutrition counseling. Assessment and management of nutrition services for pregnant adolescents should be done by a dietitian of an established adolescent program, with prenatal health clinic visits and parent sessions. The latter involve educating parents to continue supportive counseling at home. Significant nutrient-related risk factors for pregnant teenagers include: low pregnancy weight gain; low pregnancy weight for height and other evidence of malnutrition; excessive prepregnancy weight for height; low gynecological age, i.e., age of onset of pregnancy minus age of menarche; unhealthy lifestyle such as the use of drugs, alcohol, or cigarettes; a history of eating disorders; and the presence of anemia, toxemia, and other chronic diseases. To ensure optimal fetal development and growth, and at the same time provide for her own needs as a growing adolescent, a pregnant teenager requires increased intakes of calories and nutrients that exceed the dietary allowances for a mature pregnant woman. An appropriate weight gain for a pregnant adolescent is 14 to 15 kg (about 30 lb) throughout the gestation period, or a rate of 500 g/week. See Appendix 1.

Nutrition, therapeutic. See *Nutrition, diet therapy*.

Nutrition therapy. Also called "therapeutic nutrition." Based on the assessment of the nutritional status and needs, the plan for nutrition therapy (intervention) in the treatment of a disorder or illness includes diet therapy, nutrition counseling, and/or use of specialized nutrition therapies such as supplementation with nutritional or medical foods for those unable to obtain adequate nutrients through food intake only, enteral nutrition delivered via tube feeding into the gastrointestinal tract for those unable to ingest or digest food, and parenteral nutrition delivered via intravenous infusion for those unable to absorb nutrients. For further details, see *Diet therapy, Nutrition counseling, Parenteral feeding, Total parenteral nutrition,* and *Tube feeding*. See also *Medical nutrition therapy, Nutrition assessment, Nutrition screening,* and the series on *Diet* for the various dietary modifications.

Nutrition, toddler. The nutritional problem of the child from 1 to 3 years of age is only an extension of his or her needs from infancy. The primary concern is to increase gradually the kind and amount of food and to lessen the number of feedings to three meals with in-between snacks. Significant in this period is the establishment of proper food habits at home, hence the need for nutrition education for mothers. For quantitative consideration of nutrient needs for children ages 1 to 3 years, see Appendix 1.

Nutrition, tongue and mouth conditions. Certain oral changes are associated with specific nutrient deficiencies. Two striking examples are the tongue and mouth lesions caused by a lack of riboflavin and nicotinic acid. In *ariboflavinosis,* angular stomatitis, cheilosis, and a purplish or magenta tongue are the characteristic lesions. In *pellagra,* the tongue is swollen and has a beefy red color; its papillae are smooth or denuded, and the tongue assumes a mushroom appearance. Mucous membranes and lips are also red and often fissured. The tongue and mouth lesions associated with lack of other B complex vitamins are as follows: a clean, pale, and smooth tongue (chronic atrophic glossitis) in pernicious anemia due to *vitamin* B_{12} deficiency; seborrheic angular dematitis in *vitamin* B_6 deficiency; and endematous glossitis resulting from a lack of *folic acid.*

Nutrition, vegetarianism. Researchers have observed that in general, people who do not eat meat have relatively low risks for chronic degenerative diseases, such as coronary heart disease, hypertension, cancer, diabetes mellitus, obesity, and rheumatoid arthritis. Meatless vegetarian meals tend to be low in fat, but high in fiber and complex carbohydrates. Most vegetarian diets can be nutritionally adequate with careful planning. In general, the more variety of food products used, especially the addition of some animal protein foods, the easier it is to provide all nutrient needs. Hence, persons who add milk, cheese, and/or eggs to their vegetarian diet are less likely to encounter nutritional deficiencies than the pure vegetarians. The vegans, fruitarians, and persons who observe the Zen macrobiotic diet may have difficulties in obtaining good quality protein, riboflavin, vitamin B_{12}, vitamin D, iron, calcium, and zinc, unless carefully planned by using a wide variety of plant foods and by adding supplements. See *Protein, supplementary value* and *Vegetarian.*

Nutrition, weight control. Weight control is concerned with bringing a person's body weight to its healthy level. An increase of body weight of 20% or more than the so-called ideal weight in the form of adipose tissues is an established health hazard (see *Obesity*). Body weight as a parameter is the sum of fat, water, protein, and minerals in the skeletal system. The first two components are particularly variable. The currently used height–weight tables according to body frame should be considered guidelines only, because there are other varying factors among individuals that affect body weight, such as age, body composition, nutritional status, and state of health. Other methods to estimate body lean mass or subcutaneous fat should be used, such as skinfold and circumference measurements, body density, and bioelectric impedance analysis (BIA). Of the weight control problems, obesity is the most studied, because of its prevalence and its physiological and psychological consequences. It is estimated that 25% of the American population weighs more than 20% over the ideal body weight. Preliminary results from Phase I of the National Health and Nutrition Survey (NHANES III) on adolescents revealed that 21% were obese. Over 20 million adult Americans who are obese need a weight control program that is effective for long-range maintenance of healthy body weight. There is a high dropout rate among weight reducers which is not surprising, given the complexity of obesity from the standpoint of etiology or origin; genetic make-up; physiological factors; dietary regimen used; physical activities; and psychological, emotional, and social factors. Metabolic variabilities that exist are due in part to the many hormones involved in the regulation of hunger, appetite, satiety, and energy utilization. At least 20,000 weight-loss methods have been surveyed by the Health, Weight and Stress Program at Johns Hopkins University; fewer than 6% have been found effective or safe. The recommended weight control programs use a multidisciplinary approach including dietary means, behavioral changes, regular exercise, and monitoring of gradual weight loss with the help of a support group and professionals. Professional supervision is especially needed in treating morbid obesity using a very-low-calorie diet (500 to 800 kcal/day). Other weight-loss methods that require medical supervision are the use of surgical procedures and drug therapy; these have side effects that should be carefully considered. Two main problems of treating morbid obesity are recidivism or the tendency to regain lost weight, and the high cost of some weight loss programs. In addition to obesity, extreme underweight conditions also need medical attention. See *Anorexia nervosa* and *Bulimia*.

Nutrition, wound healing. Wound healing involves complex metabolic reactions that require the following nutrients to speed up tissue repair and skin closure: *glucose* for immediate energy; *amino acids,* particularly arginine and methionine, for protein synthesis of collagen and other structural tissues; *glutamine* for the synthesis of purines and pyrimidines; *fatty acids* for normal cell membranes; *thiamin* as a coenzyme factor for lysyl oxidase needed in collagen formation; *vitamin A* for normal epithelial tissues and collagen cross-linking; *vitamin C* for the hydroxylation of proline and lysine needed to synthesize collagen; *vitamin K* for normal blood coagulation; *calcium* as a cofactor for blood clotting and to hasten the action of collagenases; *iron* and *copper* for red blood cells and as cofactors for collagen polymerization; *magnesium* as a cofactor to hasten synthesis of polypeptide chains; and *zinc* to help cope with stress factors and as enzyme cofactor in the synthesis of nucleic acids and in many enzymatic processes. Maintaining good nutriture to resist infections while wounds are healing cannot be overemphasized.

Nutritional adaptation. Maintenance of a fairly constant blood or tissue composition despite deficiencies or excesses of nutrients in the food supply. Any change in the diet that would create a condition of nutritional stress is checked by the body's ability to adapt itself to maintain constancy in nutrient composition. An increased supply of a nutrient leads to the accentuation of systems associated with excretion of this excess nutrient, increased metabolism, or the preferential use of the nutrient as a source of energy. A reduced supply of a nutrient leads to decreased activity of systems associated with its metabolism, e.g., the unusual thrift in the use of a scant supply seen in fasting. Eventually, an adaptive increase or decrease in the utilization of a nutrient takes place.

Nutritional anthropometry. Measurement of the physical dimensions and gross composition of the body at different age levels and degrees of nutrition. These measurements include height; weight; circumference of the arm, chest, and hip (ACH index); skinfold thickness, waist/hip ratio, elbow breadth, growth charts, and many others. Nutritional anthropometry is a useful aid in the assessment of the nutritional status of an individual or a population group. See Appendices 18 to 23.

Nutritional care. A set of activities to provide a diet adequate in nutrients needed by an individual and to help with proper eating habits. The nutritional care process starts with *nutrition screening*. This is followed by the four-step process of assessment, planning, implementation, and evaluation or review of the care plan. In healthcare facilities, the patient's medical record is the basic means of communication among health professionals. An organized system of charting that is commonly used is the problem-oriented medical record (POMR) system. It starts with the collection of a database and identification of any problem concerned with providing quality healthcare to the patient. The problem could be medical, nursing, nutritional, behavioral, economic, psychosocial, environmental, or other. Examples of nutritional problems include anorexia, food allergies, dysphagia, missing teeth, inability to self-feed, and drug–nutrient interactions. Each problem is clearly defined, followed by specific treatment plans, implementation, patient education, and follow-up. A suggested format in charting is to use the acronym SOAP as an outline, which stands for subjective (S), objective (O), assessment or analysis (A), and plan (P). The nutritional plan includes intervention, monitoring, and follow-up, written as progress notes.

Nutritional deficiency (inadequacy). Condition of the body that may arise as a result of a lack of one or more nutrients in the diet (primary factor of nutritional inadequacy) and/or a breakdown of one or more of the bodily processes concerned with nutrient utilization (secondary factor of nutritional inadequacy). The underlying reasons for a dietary lack, either in quantity or in quality, may in turn be due to poverty, ignorance of proper nutrition practices, faulty selection of food, lack of facilities for preservation and storage, and overpopulation. The secondary factors are sometimes referred to as "conditioning factors of nutritional inadequacy" (i.e., nutrient deficiencies occur even if the diet is adequate) because they interfere with bodily processes, as listed below:

1. *Factors that interfere with ingestion:* loss of teeth, anorexia, self-feeding problems, neuropsychiatric disorders, and therapy that reduces taste acuity and appetite.
2. *Factors that interfere with digestion and absorption:* diarrhea, achlorhydria, biliary disease, gastrointestinal surgery, and therapy with mineral oil.
3. *Factors that interfere with utilization:* liver disease, diabetes mellitus, hypothyroidism, malignancy, alcoholism, and antimetabolite and sulfa drug therapy.
4. *Factors that increase nutritive requirements:* strenuous physical activ-

ity, fever, delirium, hyperthyroidism, growth, pregnancy, and lactation.

5. *Factors that increase excretion:* polyuria, lactation, excessive perspiration, and therapy that causes diuresis.
6. *Factors that increase nutrient destruction:* achlorhydria, lead poisoning, and alkali and sulfonamide therapy.

See also *Deficiency disease, Nutrition,* and *Nutritional status.*

Nutritional status. Also called "nutriture." State of the body resulting from the consumption and utilization of nutrients. Clinical observations, biochemical analyses, anthropometric measurements, and dietary studies are used to determine this state. Poor nutritional status includes undernutrition, nutritional deficiencies and imbalances, and dehydration; it also includes obesity and other nutrient excesses. The following are risk factors as well as major and minor indicators of poor nutritional status: *Risk factors* are inappropriate food intake, poverty, social isolation, dependency or disability, acute or chronic diseases and conditions, chronic medication use, and advanced age. *Major indicators* include weight loss, under- or overweight, low serum albumin, change in functional status, inappropriate food intake, $<$10th percentile midarm muscle circumference, $<$10th percentile or $>$95th percentile triceps skinfold, and nutrition-related disorders. *Minor indicators* include alcoholism, cognitive impairment, chronic renal insufficiency, malabsorption syndromes, anorexia, dysphagia, fatigue, change in bowel habits, dehydration, poor oral or dental status, poorly healing wounds, fluid retention, loss of subcutaneous fat and/or muscle mass, and reduced iron, ascorbic acid, and zinc. See *Nutritional assessment.*

Nutritional tool. Any material, device, instrument or technique used to carry out the nutritional activities and services of a nutritionist; includes a wide array or variety of resources, such as the basic food guides, the recommended dietary allowances (RDA), medical and food records, anthropometric measurements, growth charts, food exchange lists, food and nutrient labels, food composition tables, meal plans and recipes, food models, printed materials, software and other audiovisual aids, and many others. See Appendices.

Nutritionist. A professional who teaches and/or applies the science of nutrition for the improvement of health and control of disease. He or she may organize and conduct training programs for paraprofessionals and members of allied professions on food and nutrition, coordinate the nutritional activities of public and private health and related agencies, plan and conduct meetings on nutrition, and participate in nutritional surveys.

Nutrition Today Society. A nonprofit professional organization that aims to increase public awareness of new developments in nutrition. It publishes *Nutrition Today* and produces audiovisual learning materials and other nutrition resources.

Nutrivent™. Brand name of a specialized enteral liquid formula designed to reduce CO_2 production in pulmonary patients; high in protein and provides 1.5 kcal/mL. See Appendix 39.

Nyctalopia. See *Night blindness.*

Nystagmus. Also called "nystaxis." Involuntary rapid and rhythmic eye movements. Seen in thiamin deficiency, certain eye diseases, and a number of neuromuscular disorders.

O

Oatrim. A fat substitute derived from oat fiber; designed to replace saturated fat and to provide the cholesterol-lowering benefits of beta-glucan fiber in oats. Can be used in baked products, dips, sauces, dressings, cheeses, margarine, and frozen desserts.

Obesity. Also called "adiposity"; state of malnutrition in which the accumulation of depot fat is so excessive that functions of the body are disturbed. An individual is considered *obese* when the body weight is 20% or more above the desirable weight because of adiposity. There are several causative factors (genetic, traumatic, environmental, etc.), but ultimately, the overall picture is the result of caloric intake in excess of caloric output. *Simple obesity* is due to overeating (sometimes called *exogenous obesity*), reduced physical activity, decreased basal metabolism, or a combination of these factors. Some cases are due to a metabolic disturbance that favors adiposity, as in hypothyroidism, lesions of the hypothalamus, hypofunction of the gonads, hyperfunction of the adrenal cortex, and pituitary obesities. Based on the percentage of weight above the standard or the *body mass index* (BMI), the degree of obesity may be classified as follows: mild (weight 20% to 40% above standard or >25 to 30 BMI); moderate (40% to 60% above or >30 to 35 BMI); medically significant (60% to 80% above or BMI >35 to 40); morbid (100% to 120% above or >40 to 45 BMI); and gross obesity (>120% above or >50 BMI). Weight loss is crucial for individuals whose BMI is greater than 35 or who weigh more than 60% above the standard, especially those with diabetes mellitus, hypertension, and other health problems. The pattern of fat distribution in the body, as determined by the *waist-to-hip measurement,* is another way of assessing obesity. Individuals with fat deposit around the waist and upper abdomen (called android or "apple shape" obesity) are at greater risk for cardiovascular disease, hypertension, and non-insulin-dependent diabetes mellitus than individuals with fat deposit around the thighs and buttocks (called gynoid or "pear shape" obesity). *Nutrition therapy:* a weight loss program combined with behavior modification and increased physical activity or exercise. Consider the individual's weight history, medical history, usual diet, behavioral eating patterns, resources, motivation, weight goal, and support group intervention. A reduction of 500 kcal/day from the usual intake, while keeping activity constant, should bring about a weight loss of 1 lb/week. Weight loss is greater if physical activity is increased. A weekly loss greater than 2 lb/week is not advisable unless under the supervision of a physician. Calorie levels below 1200 kcal/day are marginal in nutrient content and may require vitamin and mineral supplementation. Distribute total calories into 55% to 60% carbohydrate, less than 30% fat, and a protein allowance of 1.0 g/kg with a minimum of 60 g/day. A more liberal approach to weight reduction is to reduce the fat intake without restricting calories, coupled with an exercise program to burn calories. Plan three moderate meals per day with snacks, if desired. Emphasize food selection from a variety of fruits, vegetables, grain products, complex carbohydrates, and fiber. For further details, see *Nutrition, weight control*. See *Low-calorie diet* and *Very-low-calorie diet* under *Diet, calorie-modified*. See also *Body mass index* and *Waist/hip ratio*.

OBRA. Abbreviation for *Omnibus Budget Reconciliation Act* of 1987.

OGTT. Abbreviation for oral glucose tolerance test. See *Glucose tolerance test*.

OHA. Abbreviation for *oral hypoglycemic agent*.

Oleic acid. Monounsaturated fatty acid containing 18 carbon atoms. The double bond is in the *cis* configuration. It is one of the most abundant fatty acids in nature, occurring widely in animal and vegetable fats and oils.

Olestra. A fat substitute that is a sucrose polyester. Not absorbed by the gastrointestinal tract and has no caloric value. It reduces the absorption of cholesterol and fat-soluble vitamins. Undergoing evaluation by the Food and Drug Administration (FDA); not currently available.

Oligosaccharides. Group of *carbohydrates* that yield 2 to 10 simple sugars or monosaccharide units on hydrolysis. See Appendix 8.

Omega fatty acid (OFA). Fatty acid designated by the position of the double bond closest to the omega or the methyl (CH_3—) end of the carbon chain. The three classes of OFA of dietary significance are omega-3 FA (O3FA), such as α-linolenic acid, eicosapentaenoic acid, and docosahexaenoic acid; omega-6 FA (O6FA), such as linoleic acid and arachidonic acid; and omega-9 FA (O9FA) such as oleic acid. Omega-9 is a monounsaturated fatty acid. The polyunsaturated omega-3 (α-linolenic) and omega-6 (linoleic) fatty acids are essential in the diet. Diets rich in omega-3 fatty acid are associated with a low incidence of cardiovascular disease and improvement of autoimmune diseases such as rheumatoid arthritis, lupus erythematosus, and multiple sclerosis. Evidence also suggests that diets high in O3FA may lower serum triglycerides and total cholesterol, increase high-density lipoproteins (HDL), prolong bleeding times, decrease platelet aggregation and plasma fibrinogen, and reduce tumor growth. The potential for health benefits from O3FA exists, but a great deal more research is necessary. O3FA is found in fish oils (salmon, mackerel, tuna, and anchovy). See also under *Fatty acid*.

Omnibus Budget Reconciliation Act (OBRA). Federal legislation of 1987 affecting nursing home care. It emphasizes comprehensive assessment, quality of care, quality of life, and residents' rights issues. The new Medicaid provisions now provide funds for nutrition assessment, counseling, and education. Dietitians and dietetic technicians are members of the interdisciplinary health team to carry out the goals of OBRA.

Oncotic pressure. Pressure exerted by proteins that helps to regulate the distribution of fluids on both sides of a semipermeable membrane. In this manner, plasma proteins regulate water balance in the body. The oncotic pressure gradient is the pressure difference between the osmotic pressure of the blood and that of the lymph or tissue fluids. It maintains the water balance of these two compartments. If it is disrupted, as in hypoalbuminemia, fluid accumulates in the interstitial tissues, resulting in edema. See also *Osmotic pressure*.

Open heart surgery. An open heart surgical procedure performed with the use of a cardiopulmonary machine, as in a coronary artery bypass surgery. *Nutrition therapy:* before and after surgery, monitor fluid and electrolyte balance and modify the diet as needed. Provide adequate protein and calories for wound healing. Supplement with vitamins and minerals, especially vitamins A, C, K, iron, and zinc. Other dietary changes are made, depending on medical problems such as edema, hypercholesterolemia, and obesity. Promote weight control, exercise, and a healthy eating habit. See *Diet, prudent*.

Opportunistic infection. An infection caused by an organism that does not ordinarily cause disease but can become pathogenic in individuals with impaired immunity, such as in AIDS, or under certain disorders such as cancer and diabetes mellitus. In addition to fever, opportunistic infections often cause diarrhea, weight loss, malabsorption, and other problems that result in malnutrition.

Opsin. Protein occurring in both the rods and the cones of the retina. It reacts with retinaldehyde in the dark to form the visual pigment rhodopsin.

Opsonin. Form of antibody that can render bacteria and cells more susceptible to *phagocytosis*.

Oral contraceptive. Commercial preparation of an estrogen–progestin combination for prevention of conception. Oral contraceptives may produce a wide variety of metabolic changes and may cause alterations in carbohydrate and lipid metabolism; decreased blood levels of vitamin B_6, vitamin B_{12}, ascorbic acid, folate, riboflavin, magnesium, and zinc; and increased urinary excretion of xanthurenic acid and kynurenine, indicating interference with the normal pathway of tryptophan metabolism. It may also cause nausea, vomiting, fluid retention, and weight gain. Brand names include Brevicon®, Demulen®, Loestrin®, Nordette®, Norlestrin®, Ortho-Novum®, Ovcon®, and Ovral®.

Oral glucose tolerance test (OGTT). See *Glucose tolerance test*.

Oral hypoglycemic agent (OHA). Medication taken by mouth which lowers the blood glucose level. The most commonly used OHAs are the sulfonylureas. At least six compounds are available, classified as first generation (*tolbutamide, acetohexamide, tolazamide, chlorpropamide*) and second generation (*glyburide* and *glipizide*). Their duration of action ranges from 6 to 72 hours. They stimulate pancreatic insulin release, increase the number of insulin receptor sites or decrease insulin resistance, and decrease glucose production in the liver. OHAs are indicated for Type II diabetics who are either obese or of normal weight but in whom hyperglycemia still persist, despite compliance with the prescribed meal plan and exercise regimen. They should not have complications such as acidosis, infections, severe trauma, hepatic or renal dysfunction, or diarrhea. OHAs are *contraindicated* for patients with juvenile diabetes, gestational diabetes, and for those who will undergo surgery. Side effects of OHAs include hypoglycemia, skin rash, itchiness, headache, nausea, diarrhea, and rarely, anemia. The side effects occur less often with second-generation sulfonylureas. See also *Diabetes mellitus* and *Insulin*.

Oral rehydration solution (ORS). Solution developed by the World Health Organization to rehydrate children suffering from severe diarrhea where intravenous fluids would otherwise be necessary. It can be made at home and given as a drink or through an enteral feeding tube. To make ORS, add to 1 liter of water the following: 3.5 g of sodium chloride, 2.5 g of sodium bicarbonate, 1.5 g of potassium chloride, and 20.0 g of glucose.

The solution should be made up fresh every 24 hours. The World Health Organization ORS contains 90 mEq sodium, 20 mEq potassium, 80 mEq chloride, 30 mEq bicarbonate, and 20 g glucose. Commercial ORSs are also available and are gaining popularity for use with adults. There are two types based on sodium content: a *rehydration solution,* used for treating dehydration and containing 75 to 90 mEq/L of sodium, and a *maintenance solution,* used to maintain hydration after dehydration treatment and containing 40 to 60 mEq/L of sodium. See Appendix 38.

Ornithine. Diaminovaleric acid, an amino acid obtained from arginine by the hydrolytic removal of urea. Together with citrulline, arginine, and aspartic acid, it is an intermediate product in the cyclic process of urea formation.

Ornithine transcarbamylase deficiency (OTC). A sex-linked recessive disorder in the conversion of ornithine and carbamyl phosphate to citrulline. It is characterized by hyperammonemia, increased urinary orotic acid, and normal levels of citrulline, argininosuccinic acid, and arginine. *Nutrition therapy:* depending on individual tolerance, restrict protein intake to 1 to 2 g/kg/day. Supplement with L-arginine (1 g/day for infants, and 2 g/day for older children) to prevent arginine deficiency and to assist in waste nitrogen excretion. Ensure adequate intakes of calories, vitamins, and minerals according to the RDA.

Osmolality. The number of osmols or milliosmols of solutes per kilogram of water. Osmolality refers to the number of particles (solutes) in 1 liter of solvent, whereas *osmolarity* refers to the number of particles per liter of the solution (solute plus solvent). The normal osmolality of extracellular fluid is 280 to 300 mOsm/kg (280 to 300 mmol/kg). An osmolality below this range indicates overhydration and hyponatremia. Above this range is a condition of hyperosmolality that is associated with certain clinical signs and symptoms, such as hypernatremia and dehydration. Hyperosmolality may be brought about by feeding formulas of high osmolality or osmolarity. Serum osmolality is one of the most reliable parameters of fluid balance. It can be calculated indirectly using the following formula:

$$\text{Serum osmolality} = 2 \times \text{Na (mEq/liter)} + \frac{\text{BUN}}{3} + \frac{\text{glucose}}{18}$$

where BUN = blood urea nitrogen in mg/dL and glucose is in mg/dL. A measured osmolality as reported in a chemistry profile is usually slightly higher than a calculated osmolality. See also *Osmolarity* and *Osmotic pressure*.

Osmolality, formula. The osmolality of an enteral formula is determined by the number and size of particles. Smaller particles such as glucose, free amino acids, and electrolytes contribute more to osmolality than bigger particles such as proteins and carbohydrates. Fats do not increase osmolality because of their insolubility in water. Isotonic formulas generally have an osmolality of 300 mOsm/kg. Gastric emptying is slowed by solutions with higher osmolality and they may cause gastric retention. Many tube feedings are hyperosmolar with respect to body osmolality. If given too rapidly, a hyperosmolar solution will cause water to be drawn into the gastrointestinal tract, causing nausea, vomiting, and diarrhea. Diluting the formula or slow initial administration with gradual increases in strength or amount will enhance tolerance. Formulas with osmolalities of ≤500 mOsm/kg can be given full strength at the start. Osmolality also affects the solute load and water requirement. Formulas that cause a large renal solute load may cause clinical dehydration. The American Academy of Pediatrics suggests that formulas for infant feeding should have osmolalities of less than 460 mOsm/kg. Enteral products for adults that exceed this limit should be diluted in strength if used for infants. See *Renal solute load*.

Osmolarity. The number of osmols or milliosmols per liter of solution. Osmolarity therefore considers both solute and solvent, and is affected by the temperature of the solution and the number of particles present in it. Thus, osmolarity is more appropriate in describing the characteristics of an enteral or parenteral nutritional solution. The terms *hypotonic, isotonic,* and *hypertonic* are used to compare the mOsm/liter of a particular solution with that of plasma. For example, a hypertonic enteral formula given by jejunostomy feeding may result in diarrhea. Serum osmolarity can be estimated by the following formula:

$$\text{Serum osmolarity (mOsm/liter)} = 2(\text{Na} + \text{K}) + \frac{\text{Glucose}}{18} + \frac{\text{BUN}}{3}$$

where Na + K are in mEq/liter and glucose and BUN (blood urea nitrogen) are in mg/dL. Normal range is 275 to 295 mOsm/liter.

Osmolite®. Brand name of an isotonic, lactose-free liquid formula for tube feeding or oral supplementation. Also available in higher nitrogen concentration as Osmolite® HN. See Appendix 39.

Osmosis. Passage of solvent through a membrane from a dilute solution to a more concentrated one to equalize the osmolality of the fluids on both sides of the membrane.

Osmotic pressure. "Attractive" or drawing force that many substances in solution exert on water molecules (or solvents). It is the force exerted by a solute that causes the solvent to pass through the semipermeable membrane until the concentration on both sides is approximately equal. The osmotic pressure of body fluids is due to the presence of various electrolytes and crystalloids. It is the number of particles, and not their size, that influences osmotic pressure. This pressure is the fundamental force underlying physiologic processes such as the interchange of materials between the blood and tissue cells, the excretion of urine, and the regulation of blood volume. However, osmotic pressure in the body is not constant. It varies as a result of metabolic processes and the concentrations of various constituents of the intra- and extracellular fluids. See also *Oncotic pressure*.

Ossification. Process of bone formation. The *organic matrix* becomes strong and rigid as a result of calcification or deposition of minerals, chiefly calcium and phosphorus salts plus magnesium, carbonate, citrate, fluoride, and others in trace amounts. The structure resembles that of *hydroxyapatite* crystals, and the process is catalyzed by *vitamin D*.

Osteoarthritis. Also called "degenerative" or "hypertrophic arthritis"; a joint disorder characterized by degeneration of the articular cartilage and bony outgrowths around the joints. It is a painful disease associated with advancing age and is often seen in overweight or obese persons. Extra body weight puts a strain on the weight-bearing joints such as the ankles and knees. *Nutrition therapy:* a weight reduction program is recommended for the obese. Assure adequate intakes of calcium and vitamin D at levels specified in the RDA. Monitor nutritional side effects of drugs. Self-feeding devices may assist in eating if joint pain limits mobility and use of the hands. See *Self-feeding problems*.

Osteodystrophy. A bone disorder, usually associated with calcium–phosphorus imbalance and renal insufficiency. See *Renal osteodystrophy*.

Osteomalacia. Also called "adult rickets"; softening of the bones resulting from a lack of vitamin D, calcium, and/or phosphorus. There is inadequate mineralization of the organic matrix of bones, resulting in skeletal deformities, pain of the rheumatic type in bones of the legs and lower part of the back, and spontaneous multiple fractures. It is common among women who have become depleted of calcium because of repeated pregnancies and those who have little exposure to sunlight. Osteomalacia may occur when there is interference with fat absorption or defective metabolism of vitamin D, such as seen in severe liver disease, end-stage renal failure, and prolonged use of anticonvulsant drugs. *Nutrition therapy:* prevention through adequate intakes of vitamin D, calcium, and phosphorus at levels specified in the RDA. Monitor drugs that deplete vitamin D and calcium. Assure vitamin D (5 to 10 μg or 400 to 800 IU daily) from natural or fortified food sources, or sun exposure. Calcium supplement may be necessary. If osteomalacia is already present, a daily dose of 25 to 125 μg (1000 to 5000 IU) vitamin D_3 is recommended; if there is malabsorption, the dose is 1250 μg (50,000 IU) daily.

Osteomyelitis. Local or generalized infection of the long bones and bone marrow due to certain bacteria, such as *Escherichia coli, Staphylococcus aureus,* and *Streptococcus pyogenes*. Fever, chills, nausea and vomiting, acute joint pain, regional muscle spasm, and bone pain are the usual symptoms. The patient may be on parenteral antibiotics and bed rest for several weeks. Surgical removal of necrotic bone and tissue may be necessary. *Nutrition therapy:* maintain fluid balance and ensure a liberal calorie (basal energy expenditure × 1.3) and protein intake (1.2 to 1.3 g/kg body weight).

Osteopenia. A condition of subnormal bone density; the rate of bone matrix synthesis is not great enough to equal the rate of bone lysis. May be caused by cancer or hyperthyroidism. Associated with deficiencies of copper, manganese, and zinc, with overall reduction of bone calcification. Osteopenia has been observed to increase the incidence of fractures among the elderly. *Nutrition therapy:* similar to that of osteoporosis, with emphasis on intake of trace minerals.

Osteoporosis. Bone disorder characterized by a reduction in total bone mass without a change in chemical composition. It occurs when the rate of bone resorption exceeds the rate of formation, resulting in bone loss and increased porosity. Bone formation remains the same with age, whereas bone resorption increases with age. The result is a gradual loss of bone. Osteoporosis is seen in persons after age 50, especially in women after menopause. Practically all people begin to lose bone at age 55. Osteoporosis is a public health problem and has been recognized by WHO as a disease of both sexes, although it is 8 times more prevalent in women. It affects more than 25 million Americans and is implicated in more than 1.3 million fractures annually. Clinical manifestations are low back pain that is sometimes severe, kyphosis of the dorsal spine, and skeletal fractures. The etiology of osteoporosis is multifactorial and includes the following: decreased estrogen production associated with menopause, long-time low intake of calcium and decreased absorption of calcium, reduced circulating level of calcitriol, increased urinary loss of calcium, cigarette smoking, excessive alcohol consumption, and reduced physical activity and immobilization. Medications such as phenytoin, thyroid hormone, and corticosteroids also contribute to osteoporosis either by promoting calcium loss from the bone or by interfering with calcium absorption. *Postmenopausal osteoporosis* is seen in women within 15 to 20 years after menopause; it primarily involves the trabecular bone and is characterized by fractures in the pelvis and proximal femur, and painful or deforming "crush" fractures of the lumbar vertebrae. *Age-associated osteoporosis* is seen after age 70 and affects both men and women; it involves both the trabecular and cortical bones and is characterized by wedge fractures of the thoracic vertebrae, back pain, loss of height, and kyphosis or "dowager's hump." The loss in height may be as much as 4 to 8 inches. Prevention of osteoporosis focuses on increasing *peak bone mass* and reducing age-related bone loss. Calcium absorption from the intestines decreases with age, as does the usual dietary intake of calcium. Although the role of dietary factors is not clear, calcium deficiency must be considered a possible risk. The severity of bone loss can be ameliorated if there is a liberal intake of calcium throughout life, especially among women. Those who chronically have low intakes of calcium and are inactive appear to have a higher risk of osteoporosis than those who are physically active and have developed maximum bone mass. However, not all persons with low calcium intake develop osteoporosis. Treatment is aimed at reducing bone resorption which includes exercise, estrogen replacement therapy (ERT), calcitonin, biophosphonate, and dietary measures, including calcium and vitamin D (alone or in combination), and fluoride. *Nutrition therapy:* adequate calcium intake throughout life, especially during the teens and early adulthood to optimize peak bone mass and reduce the severity of bone loss. The Consensus Development Conference on Osteoporosis has recommended a daily calcium intake of 1000 mg for premenopausal and 1500 mg for postmenopausal women, and 1200 mg for adolescents. There is still no good information on requirements for men. Maintain an adequate source of vitamin D (5 to 10 μg or 400 to 800 IU), especially if sun exposure is limited, as with institutionalized elderly. Avoid excessive intake of nutrients or dietary components that alter calcium needs, either by increasing urinary calcium loss (e.g., protein, sodium, caffeine) or by decreasing calcium absorption or availability (e.g., dietary fiber, oxalate, phytate, and phosphorus). Avoid smoking and restrict alcoholic beverages to no more than 1 to 2 drinks/day. See *Calcium, Bone mass,* and *Peak bone mass*. See also *Nutrition, bone* and *Vitamin D*.

Osteosarcoma. A malignant bone tumor of unknown etiology. It occurs primarily in adolescents and elderly people. Clinical signs and symptoms include weight loss, fatigue, fever, and pain in the affected area. *Nutrition therapy:* guidelines are directed to the use of anticancer drugs, weight loss, and fever. Maintain a well-balanced diet consisting of foods tolerated by the patient, given in six feedings per day.

Ostomy. The surgical procedure of creating an opening, or stoma, in the wall of the abdomen to allow the passage of intestinal contents from the bowel. Disorders that may require ostomy procedures include colon cancer, Crohn's disease, severe diverticulitis, and ulcerative colitis. The most common sites for ostomies are the ileum (ileostomy) and the colon (colostomy). *Nutrition therapy:* immediately after the surgery, IV feeding is given for 2 to 3 days until bowel sounds return. The diet progresses from clear liquids to soft, low-residue foods. Gradually introduce fiber as tolerated. Restrict only those foods that are not tolerated and produce excessive gas or loose stools. Avoid tough skins from fruits and vegetables and other foods that may cause stoma obstruction. Take plenty of fluids (at least 8 to 10 cups per day), especially if the ostomy output is excessive. When steatorrhea occurs, restrict fat and/or use MCT oil. For further details see *Colostomy* and *Ileostomy*.

Ovalbumin. The albumin of egg white. Although it is considered a simple *protein* because of its solubility and other physical properties, some authorities classify it with glycoproteins, as it contains small amounts of carbohydrate (less than 4% hexose).

Overeaters Anonymous. A self-help support group of volunteer members who meet regularly to discuss problems and help each other in coping with compulsive eating, binging, purging, and other concerns about weight control.

Overweight. Term applied to a person whose weight is about 10% to 20% above the standard weight for individuals of the same age, sex, and height. An athlete with well-developed muscles may be overweight but not obese. See also *Obesity* and *Weight control*.

Ovoflavin. Name give to *riboflavin* in egg. See also *Vitamin* B_2.

Oxalate. Salt of oxalic acid; the end product of both glyoxylic and ascorbic acid metabolism. It occurs primarily in foods of plant origin; rich food sources are spinach, beets, rhubarb, berries, plums, and tea. Oxalates form insoluble calcium salts, rendering calcium unavailable for absorption. Urinary excretion of oxalate (oxaluria) is normally less than 60 mg/24 hours; small increases in urinary oxalate concentration enhance the potential for crystal formation. See also *Hyperoxaluria*.

Oxidation–reduction (oxido–reduction). Oxidation involves a loss of electrons, and reduction involves a gain of electrons. Thus the electron donor or acceptor is itself oxidized or reduced during the process. Oxidation and reduction reactions occur simultaneously in cellular respiration.

Oxycalorimeter. Apparatus that determines the caloric value of food by measuring the amount of oxygen consumed. It is used in *indirect calorimetry*.

Oxygen debt. Deficit of oxygen that arises during continuous or intense exercise when energy cannot be adequately supplied by oxidative means and lactic acid accumulates faster than it can be oxidized. When exercise is finished, the depth of respiration is increased to provide more oxygen to oxidize the excess lactic acid that has accumulated. See *Lactic acid*.

Oxythiamin. Analog of thiamin in which a hydroxyl group is substituted for the free amino group on the pyrimidine ring. It is an antimetabolite and displaces thiamin from the tissues.

P

P. Chemical symbol for *phosphorus.*

PAA. Abbreviation for *plasma amino acids.*

PABA. Abbreviation for *para-aminobenzoic acid.*

Pancreas. Glandular organ extending across the upper abdomen close to the liver. It secretes into the intestinal tract the *pancreatic juice,* which contains enzymes that act on protein, fat, and carbohydrate. It also produces the hormones *insulin, glucagon,* and *somatostatin* secreted by its islets of Langerhans. For more information on pancreatic enzymes, see Appendix 11.

Pancreas transplantation. Alternate treatment for persons with Type I diabetes mellitus to improve quality of life and to prevent the progression of chronic complications, such as nephropathy, neuropathy, and retinopathy. *Nutrition therapy:* in the *pretransplantation* phase, energy requirement is estimated at 1.2 × basal energy expenditure; protein is 1.0 g/kg body weight per day, but restricted if there is renal failure; fat provides 30% of total calories and the balance of energy needs comes from carbohydrates. Limit saturated fats. Fiber is given as tolerated. Sodium is restricted to 2.0 g/day (90 mEq), but in the absence of hypertension and edema, a "no added salt" diet is allowed. Potassium is maintained at about 2.5 g (60 mEq) per day, adjusted according to blood values, and calcium is from 800 to 1200 mg/day. A low-microbial diet is not necessary until after surgery. In the *immediate post-transplantation* phase, TPN may be needed if oral or enteral route is not possible within a week after surgery. Once bowel function returns, start enteral feeding and wean gradually from TPN. Energy requirement, based on desirable body weight, is 30% more than basal energy expenditure and protein intake is increased to 1.2 to 1.5 g/kg/day. Fat provides up to 30% of total calories/day. Sodium, potassium, and calcium needs are the same as in pretransplantation. To be emphasized is a low-microbial diet for at least 3 weeks after surgery. During the *long-term post-transplantation* period, patients tend to gain weight and/or have hyperlipidemia. Evaluate for signs of diabetic gastroparesis and continue glucose monitoring. Otherwise, nutrient requirements prior to surgery are the same for long-term care, except for calcium which is increased to 1500 mg/day. Emphasize complex carbohydrates and fiber. A low-microbial diet is not necessary but food safety precautions should be continued for several months. For general guidelines, see *Transplantation.*

Pancreatectomy. Partial or complete removal of the pancreas. Decreased glucose tolerance and a brittle-type diabetes mellitus develop; other problems associated with pancreatic insufficiency interfere with utilization of food, causing malabsorption and weight loss. *Nutrition therapy:* after surgery, follow the routine progression from clear to full liquids, then soft diet low in fat and simple sugars. Short-term parenteral nutrition may be necessary if enteral feeding is not tolerated. Hyperglycemia and brittle diabetes may require a more precise dietary adjustment to insulin dose. Emphasize complex carbohydrates and use of glucose polymers (oligosaccharides). Eventually, a calorie-controlled diabetic diet is needed. See also dietary modifications for *Pancreatic insufficiency*.

Pancreatic insufficiency. Deficiency of pancreatic secretion, especially of the digestive enzymes (see Appendix 11). It may be congenital, as in cystic fibrosis, or caused by a number of diseases, such as carcinoma of the pancreas and pancreatitis. It may also result from pancreatectomy or from the destruction of exocrine function by ligation of the pancreatic duct. Symptoms include recurrent attacks of abdominal pain, alteration of bowel habits, and steatorrhea. Weight loss and malnutrition are due to malabsorption of proteins, fats, and fat-soluble vitamins. *Nutrition therapy:* weight loss is corrected with a high caloric intake, usually between 2500 and 3500 kcal/day. Protein intake is 1.5 to 2.0 g/kg/day, and carbohydrate is the main source of energy (at least 60% of total calories) in the form of simple sugars and easy-to-digest starches. Fat, which is poorly tolerated, is provided as medium-chain triglycerides, about 4 tablespoons/day. Generally, pancreatic enzymes given with meals relieve the condition. Provide vitamin and mineral supplementation, especially fat-soluble vitamins in water-miscible forms. In severe cases of acute pancreatic insufficiency, enteral feeding of an elemental formula may be necessary.

Pancreatic juice. Digestive juice produced by the pancreas and secreted into the duodenum. It is alkaline (pH 7.5 to 8.0) and contains some protein and electrolytes, mainly sodium, potassium, bicarbonate, and chloride ions. It contains a number of enzymes involved in the digestion of protein, fat, and carbohydrate. For further details, see Appendix 11.

Pancreatin. Commercial preparation made from the pancreas of animals, usually the ox or hog. It contains the enzymes of pancreatic juice and is used to correct pancreatic insufficiency.

Pancreatitis. Inflammation of the pancreas. *Acute pancreatitis* is characterized by severe epigastric pain, nausea, vomiting, fever, and decreased peristalsis. The serum amylase level is elevated. However, the serum calcium level is decreased in some patients. If chylomicronemia occurs, withdrawal of long-chain fatty acids from the diet effectively relieves the clinical sign. *Chronic pancreatitis* is characterized by a disturbance in the functioning of the pancreas, leading to inadequate production of digestive enzymes. About 25% of patients develop pancreatic insufficiency. As a result, stools become bulky and foul-smelling, with increased excretion of protein, carbohydrate, and fat. Pancreatitis is associated with alcoholism, biliary tract disease, abdominal trauma, and hyperlipidemia. *Nutrition therapy:* the basic aim is to rest the pancreas by restricting foods that stimulate its action. In the acute stage, nothing is given by mouth, and total parenteral nutrition may be necessary initially. Tube feeding is used when the patient is not vomiting and the abdomen is not tender and painful. Some clinicians recommend the nasojejunal route, rather than nasogastric feeding, to lessen pancreatic secretions. Oral feeding is given within a week, starting with clear liquids. An elemental formula with medium-chain triglyceride supplementation minimizes digestion. Progress the diet to full liquid and soft diet with pancreatic enzyme taken with meals and snacks. A diabetic diet may be needed by persons with insulin insufficiency. Caloric intake is individualized; allow for losses from malabsorption. Additional calories may be supplied with medium-chain triglyceride oil. Protein is moderately high (1.2 to 1.3 g/kg/day) to compensate for losses. Fat is limited to 20% of total calories, but this may be increased to 30% in the absence of steatorrhea and with the administration of pancreatin. Vitamin and mineral supplements are given, especially thiamin, folate, vitamin B_{12}, iron, and calcium. Alcoholic beverages and gastric stimulants such as caffeine and spices are avoided. Restrict fiber as long as steatorrhea is present. For chronic pancreatitis, pancreatic enzymes with meals and a low-fat diet are effective therapeutic measures. In some cases, supplemental vitamin B_{12} may be needed. Avoid alcoholic beverages.

Pancrelipase. A substance containing the enzymes lipase, amylase, and protease. It is used in the treatment of malabsorption syndrome caused by pancreatic insufficiency, as in cystic fibrosis and chronic pancreatitis. The dosage is determined by the fat content of the diet. About 2000 USP units of lipase activity are given before or with each meal for each 5 g of dietary fat. Pancrelipase may decrease folate absorption. Brand names are Cotazym®, Ilozyme®, Pancrease®, and Viokase®.

Pangamic acid. D-Gluconodimethylaminoacetic acid, a natural substance present in foods that was first prepared from apricot pits and later from rice, liver, blood, and yeast. Little is known about the role of pangamate at the molecular level, although scientists in Europe and the Soviet Union have recorded enhanced oxygen uptake and better adaptation to hypoxia and strenuous exercise. Some effects on cardiovascular function and lowering of cholesterol have also been reported. Because of these physiologic effects, pangamic acid has been designated "vitamin B_{15}." However, there is no evidence that its lack in the diet results in a deficiency disease.

Pantothenic acid. A B complex vitamin derived from the Greek word meaning "from everywhere." It is present in all living cells and was formerly called "chick antidermatitis factor" and "filtrate factor." As a component of coenzyme A and of the acyl carrier protein, pantothenic acid plays an important role in acetylation and acylation reactions: oxidation of keto acids and fatty acids; synthesis of triglycerides, steroids, phospholipids, and fatty acids; formation of acetylcholine; and synthesis of porphyrin for hemoglobin formation. As *acetyl CoA,* the vitamin functions in the metabolism of carbohydrate, protein, and fat. It is also necessary for the maintenance of normal skin and for the development of the central nervous system. Pantothenic acid is widely distributed among foods; good sources are liver, egg, meat, whole grain cereals, and legumes. It is readily absorbed from the small intestine and stored to some extent in the liver and kidneys. Dietary deficiency in humans has not been observed, although pantothenic acid has been implicated in the *burning feet syndrome* observed among prisoners in World War II, which responded to pantothenic acid administration but not to other members of the vitamin B complex. Experimental deficiency states in humans fed a semisynthetic diet virtually free of the vitamin or given a metabolic antagonist (omega-methylpantothenic acid, pantoyl taurine, or phenyl pantothenate) produced a syndrome characterized by fatigue, dermatitis, muscle cramps, paresthesia in extremities, susceptibility to infection, and loss of antibody production. Large amounts of pantothenic acid appear to help humans withstand stress, perhaps due to its effect on the adrenal gland. Deficiency in animals results in a wide range of defects such as growth retardation, abortion, infertility, graying of hair, severe dermatitis, myelin degeneration, convulsions, gastrointestinal disorders, adrenal cortical failure, and sudden death. Pantothenic acid is relatively nontoxic; however, ingestion of large amounts may cause diarrhea and water retention. See Appendix 2 for the estimated safe and adequate daily dietary intake.

Para-aminobenzoic acid (PABA). Formerly classified with the B complex vitamins. However, present knowledge maintains that it is not a vitamin for humans, although it plays an indirect role as a component of *folic acid*.

Para-aminosalicylic acid (PAS). A synthetic antituberculosis agent, commercially available as the acid and the sodium salt. The most frequent adverse effects are gastrointestinal disturbances including nausea, vomiting, anorexia, abdominal pain, and diarrhea. Malabsorption of vitamin B_{12}, folic acid, iron, and lipids may also occur, possibly as the result of increased peristalsis. Production of PAS in the United States was discontinued temporarily in late 1991.

Parasympathetic nervous system. Part of the autonomic nervous system comprising preganglionic fibers that arise from the midbrain, the medulla, or the sacral region of the spinal cord. It controls the activity below the level of consciousness, e.g., gland secretion, intestinal action, and heart function.

Parathormone (PTH). Parathyroid hormone or parathyrin, a protein consisting of a single polypeptide chain with alanine as its N-terminal amino acid. Secreted by the parathyroid glands, it exerts a profound effect on the metabolism of calcium and phosphorus. Administration of the hormone raises the blood calcium level and lowers the blood phosphorus level, increases the elimination of phosphorus in the urine, causes migration of calcium from the bones if there is an insufficient supply of this element in the food, and increases the *phosphatase* activity of the serum.

Parathyroid glands. Two pairs of small endocrine glands located in the posterior part of the thyroid gland, a pair in each lobe, one below the other. They are reddish or yellowish brown, egg-shaped bodies weighing a total of 0.1 to 0.2 g in humans. Their secretion is concerned chiefly with calcium and phosphorus metabolism. Insufficient secretion results in a decrease of calcium and phosphorus ions in the blood, causing *tetany*. Oversecretion results in increased calcium in the blood and increased excretion of phosphate through the kidneys. As a result, there is muscular weakness, and calcium phosphate stones are formed.

Parenteral feeding (nutrition). A means of providing nutrients by routes other than through the mouth and digestive tract, such as subcutaneous, intramuscular, or intravenous feeding. A common example is the standard intravenous (IV) dextrose (D_5W) with added electrolytes through the peripheral veins. This provides an immediate source of energy (about 170 kcal/liter/day) until the patient can eat normally. Parenteral nutrition is appropriate when the oral or enteral route is inadequate or contraindicated. It can be used in addition to enteral feedings, or used alone. If parenteral feeding is the main source of nutrition, other nutrients have to be given via the small veins, usually in the arm (called peripheral parenteral nutrition or PPN), or centrally into the superior or inferior vena cava or the jugular vein (called central parenteral nutrition or CPN). Central parenteral nutrition is also called total parenteral nutrition (TPN) or intravenous hyperalimentation (IVH) which is a misnomer as parenteral nutrition does not contain an excessive or "hyper" amount of any nutrient. The decision to use PPN or CPN is based on the number of calories needed, duration of parenteral nutritional support necessary, and the osmolarity of the solution. Hypertonic solutions cannot be given in peripheral veins that have low blood flow; the area can become infiltrated and inflamed, or a thrombosis can occur. If administered via CPN, a hypertonic solution can be quickly diluted by the rapid flow of blood. PPN is usually used for brief periods, but central vein parenteral nutrition (CPN) may be used for a longer time. See also *Total parenteral nutrition*.

Parenteral nutrients. Almost any solution needed can be formulated from the nutrient sources available. Dextrose solutions and fat emulsions are the energy substrates commonly used. Synthetic crystalline L-amino acid solutions are used to provide nitrogen for protein synthesis. Vitamins, minerals, and fluid (water) are added to meet the patient's needs. See Appendices 40, 41, and 42.

Carbohydrate (dextrose). Carbohydrate is given as a dextrose monohydrate that provides 3.4 kcal/g. Dextrose solutions are available in 5%, 10%, 20%, 30%, 50%, and 70% solutions. The osmolarity of a dextrose solution increases with its concentration. See Appendix 43. The 5% solution is the only product within the range of normal serum osmolarity (275 to 296 mmol/kg); all the other solutions are hypertonic. The 5% and 10% solutions can be given by PPN; all the other solutions are given by CPN. The most widely used in parenteral nutrition is the 50% solution.

Fat emulsion. To prevent essential fatty acid deficiency, fat emulsion is given several times a week, if it is not used daily as a calorie source. The minimum dose to prevent linoleic acid deficiency is about 2% to 4% of the total calories daily, but the total amount infused should not exceed 2.5 g/kg/day. Fat emulsions are derived from soybean oil and safflower oil; these are available in concentrations of 10% (provides 1.1 kcal/cc) and 20% (provides 2.0 kcal/cc). The emulsions are made isotonic by the addition of glycerol, giving an osmolarity of 270 to 340 mmol/liter. Fat infusion should be introduced slowly, and gradually increased in kind and amount. It is contraindicated in patients with hyperlipidemia, acute pancreatitis, and egg allergy. See Appendix 41.

Fluid (water). Adult patients generally require about 35 mL of water/kg body weight/day. Provide a minimum of 1 mL/kcal or 100 mL of free water per gram of nitrogen. Fluid is restricted immediately after myocardial infarction or in kidney failure, but is increased when fluid losses have to be replenished, as in wound drainage, diarrhea, fever, and other hypercatabolic states. When fluid restriction is necessary, 20% fat emulsion, up to 70% dextrose solution, and up to 10% amino acid solutions can be used to increase the nutrient density. Fluid monitoring is very important, especially with the elderly, infants, and young children.

Minerals. Mineral requirements should be adequately met and adjusted as needed to correct electrolyte imbalances. The six major electrolytes are sodium, potassium, magnesium, calcium, chloride, and phosphate. Electrolytes are added to some of the commercially prepared amino acid solutions (See Appendix 40). Specific electrolytes may be reduced, depending on the disorder (e.g., sodium in cardiovascular and renal conditions, and potassium in scanty urinary output). Serum levels of the trace elements, in addition to those of the macrominerals, must be monitored. Trace elements such as copper, iodine, selenium, and zinc have been found to be deficient in prolonged use of total parenteral nutrition. Supplementation that meets established guidelines for parenteral trace elements is also commercially available.

Protein (amino acids). Several amino acid formulations are available in 3.0%, 3.5%, 5.0%, 7.0%, 8.5%, 10.0%, and 15% solutions, with or without added electrolytes. Introduce protein solutions with care to avoid ammonia intoxication. For patients with renal or hepatic failure, the infusion rate is slower. Special formulations are available commercially for such conditions; those containing branched-chain amino acids have been effective. See Appendix 40. The patient's protein needs and underlying disease condition should be considered when selecting an amino acid solution.

Vitamins. The U.S. recommended dietary allowances for vitamins do not apply to parenteral nutrition, because the absorptive process is by-passed. Parenteral vitamin supplementation is based on recommendations of the American Medical Association Nutrition Advisory Group (AMA/NAG). Several commercial products are available that encompass these guidelines. See Appendix 42.

Paresthesia. An abnormal sensation characterized as burning, pricking, or tingling. It is seen in *beriberi* and other disorders involving the nerves and spinal cord.

Parietal. 1. Of or pertaining to the walls of a cavity. Parietal cells found on the margin of the peptic glands of the stomach that secrete hydrochloric acid. 2. The parietal bone that forms part of the sides and top of the skull.

Parkinson's disease. A neurologic condition marked by muscular rigidity and tremors, abnormal gait and balance, slurred speech, poor mastication, and dysphagia. It is a progressive disabling disease characterized by a low concentration of the neurotransmitter *dopamine* at the basal ganglia of the brain. Of unknown etiology, it responds to therapy with the drug *levodopa* or its combination with carbidopa (Sinemet®). Patients with Parkinson's disease show mild to moderate nutritional depletion with weight loss, loss of subcutaneous fat, marginal albumin level, and varying degrees of swallowing difficulty. Dysphagia becomes a significant feeding problem in the final stages of the disease when rigidity is the prominent feature. *Nutrition therapy:* focus on the dysphagia, difficulty in self-feeding, and the nutrient–drug interactions. If the patient is taking levodopa, a high level of protein interferes with this drug. Restrict protein to 0.8 g/kg/day. Patients responding poorly to levodopa may benefit from the intake of low-protein meals for breakfast and lunch, followed by an evening meal that provides the balance of 0.8 g/kg/day allowance for protein. Give proteins of high biologic value, modifying the consistency according to the chewing and swallowing abilities of the patient. Serve the protein-rich foods mainly in the evening meal. This may reduce tremors and allow some normal functioning during the day. However, consumption of a larger protein meal in the evening can result in suboptimal levodopa effect and increased rigidity, trembling, and slowed movements overnight. Tyrosine (a precursor of dopamine) may be beneficial. Limit the intake of pyridoxine to less than 5 mg/day to make levodopa more effective. Avoid constipation and encourage the intake of fluids. Promote independence with self-feeding. Get the patient in the best possible position for feeding and use adaptive feeding devices such as cups with double handles, utensils with built up or loop handles, and rimmed dishes. Serve small, frequent meals and use semisolid foods rather than fluids if swallowing is a problem. For other details, see *Dysphagia*. See also *Diet, protein redistribution* and *Diet, 7C:1P*.

Parorexia. Abnormal craving for special foods, as opposed to *anorexia*.

Patch test. Skin test for allergy administered by applying the suspected antigen to a filter paper and placing it on a certain patch of skin. The area is uncovered after 2 to 4 days and compared with an unpatched skin area.

Pb. Chemical symbol for lead.

PBI. Abbreviation for *protein-bound iodine*.

PCM. Abbreviation for *protein-calorie malnutrition*.

PCR. Abbreviation for *protein catabolic rate*.

Peak bone mass. Bone growth is characterized by an increase in longitudinal growth and in bone mass. Peak bone mass is attained about 5 to 10 years after adolescence or in early adulthood. Men have greater peak bone mass than women because of their larger frame size. Peak bone mass is related to dietary calcium intakes, the sex hormones, and the extent of weight-bearing activity. Adequate dietary calcium (about 1200 mg/day for adults in their mid-twenties) is important for achieving peak bone mass. This is a preventive measure against *osteoporosis* in later life. See also *Bone mass*.

Pectic substances. Also called "pectins." Group designation for complex colloidal carbohydrate derivatives that occur in or are prepared from plants. They contain a large proportion of anhydrogalacturonic acids and are considered soluble fibers. Pectins have considerable water-holding capacity and can bind ions and organic materials such as bile acids. Apples, citrus fruits, and strawberries are rich in pectins. See *Dietary fibers* under *Fiber*.

Pedialyte®. Brand name of a ready-to-use oral electrolyte solution for maintenance of water and electrolytes during mild to moderate diarrhea and vomiting in infants and children. See Appendix 38.

PediaSure®. Brand name of a lactose-free, isotonic, enteral formula for children aged 1 to 6 years old who are undernourished because of illness, poor appetite, or inability to eat. PediaSure® with Fiber contains 1.2 g of soy fiber per 8 fl oz. See Appendix 39.

PEG. Abbreviation for percutaneous endoscopic gastrostomy. See *Gastrostomy*.

Pellagra. From *pelle,* meaning "skin," and *agra,* meaning "rough." A deficiency disease due to lack of niacin (nicotinic acid); characterized by dermatitis, diarrhea, dementia, and eventually death if unremedied. These pellagrous symptoms are commonly referred to as the classic "four Ds." The skin changes occur in several stages: there is thickening and pigmentation with temporary redness similar to that of sunburn, then atrophic thinning with dark red eruption, and finally desquamation. In the severe stage the involved parts erupt and swell, with ulceration and infection. Typical pellagrous dermatosis is symmetric, clearly demarcated, and hyerpigmented. Parts of the body exposed to sunlight are commonly affected, e.g., the cheeks, neck, hands, and forearms. The digestive disturbances include sore mouth with angular stomatitis, bright red or scarlet tongue with glossitis, achlorhydria, and nausea and vomiting, followed by severe diarrhea. The neurologic disturbances include insomnia, irritability, poor memory, confusion, delusions of persecution, hallucinations, and dementia. *Nutrition therapy:* oral dose of 50 to 100 mg of niacin (about 10 times the RDA), followed by 5 to 20 mg daily as a dietary supplement. Maintain a normal diet with emphasis on good sources of niacin (lean meats, fish, poultry, and whole grains). See also *Casal's necklace*.

Pelvic inflammatory disease (PID). Inflammation of the pelvic cavity that affects the fallopian tubes and ovaries. Symptoms include abdominal, pelvic, and low back pains accompanied by fever and vaginal discharge. *Nutrition therapy:* maintain a well-balanced diet and control body weight. Restore fluid and electrolyte balance. Increase caloric intake if fever is a serious problem. Frequent feedings are suggested, with supplementary vitamins and minerals.

PEM. Abbreviation for *protein-energy malnutrition*.

Penicillamine. A degradation product of all penicillins; used in the treatment of Wilson's disease, cystinuria, and lead intoxication. Penicillamine can chelate copper, cystine, mercury, lead, and other heavy metals by forming soluble complexes that are excreted by the kidneys. However, the drug can also chelate iron, zinc, and pyridoxine and may cause depletion of these nutrients. It may also cause partial or total loss of taste perception, particularly for saltiness and sweetness, as well as anorexia and weight loss. Adverse gastrointestinal effects include diarrhea, nausea, vomiting, dyspepsia, epigastric pain, colitis, and reactivation of peptic ulcer. Brand names are Cuprimine® and Depen®.

Penicillin. Natural or semisynthetic antibiotic produced by or derived from certain species of the fungus *Penicillium*. It is bacteriostatic for many microorganisms and useful in the treatment of infections caused by most of the gram-positive bacteria. Frequent adverse gastrointestinal effects are epigastric distress, nausea, vomiting, sore mouth or tongue, diarrhea, and colitis. It may reduce vitamin K synthesis and folate utilization, and inactivate pyridoxine. It may also cause electrolyte imbalances, especially in patients with impaired renal function. Brand names include Bicillin®, Pen-Vee® K, Tegopen®, and Wycillin®.

Pentosan. Also called "pentan"; a homopolysaccharide or a complex carbohydrate of pentose units that is widely distributed in wood, straw, gums, hulls, and corncobs. It is not digested by humans, but it is hydrolyzed by acid to yield pentoses. Examples are arabans and xylans.

Pentose. A 5-carbon monosaccharide. Pentoses important to nutrition are arabinose, xylose, ribose, 2-deoxyribose, xylulose, and ribulose. The first four are aldoses and do not occur free in nature; the last two are ketoses. Arabinose is obtained by hydrolyzing various gummy substances; xylose is derived from wood and straw; ribose and deoxyribose are constituents of *nucleic acids;* xylulose is often the sugar found in the urine in *pentosuria;* and ribulose is found only as an intermediate metabolite in the *hexose monophosphate shunt*. See also *Pentosan* and *Riboflavin*.

Pentosuria. Excretion of pentose in the urine. *Alimentary pentosuria* occurs temporarily in normal individuals after ingestion of large amounts of prunes, cherries, grapes, and other foods rich in pentoses (mainly arabinose and xylose). *Congenital pentosuria* is due to an inborn error of metabolism; the body cannot metabolize L-xylulose.

Pepsin. Digestive enzyme in gastric juice that converts protein to peptones and proteoses. It occurs in the stomach as an inactive precursor, *pepsinogen,* which is converted to the active pepsin by hydrochloric acid.

Peptamen®. Brand name of an isotonic, elemental formula containing enzymatically hydrolyzed whey protein, maltodextrin, starch, MCT oil, and sunflower oil. Also available as Peptamen VHP™ (very high protein) and Peptamen® Oral. See also Appendix 39.

Peptic ulcer disease (PUD). Ulceration on the mucous membranes of the esophagus, stomach, or duodenum caused by hyperacidity, lowered cellular resistance, insufficient mucus secretion, local trauma, and predisposing factors. The exact etiology is not known, although individuals prone to peptic ulcer include those with a family history of recurrent episodes; emotional, highly stressed persons; chronic users of alcohol and of certain drugs, such as aspirin and corticosteroids; and possibly those with susceptibility to the bacterium *Campylobacter pyloridis*. The main symptom of peptic ulcer is epigastric pain 1 to 3 hours after meals, characterized as a burning, gnawing, or sharp pain. This is relieved by eating or by using alkalis or antacids. Clinical signs include weight loss, low plasma protein, anemia, and hemorrhaging. *Nutrition therapy:* with modern drug therapy, which is now the treatment of choice, dietary restrictions are minimal. The traditional bland diet is no longer used because of the wide variations among individuals as to which foods are irritating to them. The emphasis now is on the frequency and volume of feeding per meal or snack. Except for alcohol, coffee (regular and decaffeinated), carbonated beverages, and "hot" spices such as red and black pepper, the only foods avoided are those not tolerated. Soft and soluble fibers are encouraged to prevent constipation, which is a side effect of antacid use. Cimetidine may decrease vitamin B_{12} absorption; the status of this vitamin should be monitored. Dietary changes consist largely of eliminating foods that worsen symptoms and are not tolerated. The diet and meal size are individualized. Emphasize regularity of meals.

Peptide. Compound formed when two or more amino acids are linked together by the *peptide linkage,* which is the main bond in protein structure. Peptide-based enteral solutions may be beneficial in hypermetabolic states and diseases of the gastrointestinal tract. Available data suggest that peptides have a higher biologic value than intact protein and are better absorbed than intact protein or amino acids. Other advantages attributed to peptides are maintenance of gut integrity and prevention of bacterial translocation, improved secretion of trophic gut hormones, improved visceral protein synthesis and wound healing, and improved liver function.

Peptone. A secondary protein derivative. When protein is hydrolyzed, it is changed to proteoses, then to peptones, peptides, and finally amino acids. Peptones, as distinguished from proteoses, are not precipitated by ammonium sulfate or tannic acid. *Peptonization* is the conversion of protein to peptones, using suitable proteolytic enzymes such as those found in *pancreatin* or pancreatic juice.

PER. Abbreviation for *protein efficiency ratio*.

Perative®. Brand name of a specialized tube feeding product for metabolically stressed patients with injuries such as multiple fractures, wounds, burns, and surgery. Supplemented with arginine and may be given orally or by tube. See Appendix 39.

Pericarditis. Inflammation of the pericardium; may be caused by trauma, infection, uremia, myocardial infarction, rheumatic fever, tuberculosis, or collagen disease. Symptoms are chest pains radiating to the shoulder and neck, dyspnea, dry cough, anxiety, mild fever, difficulty in breathing, and rapid pulse rate. *Nutrition therapy:* initially, parenteral fluids may be required. As the patient recovers, oral feeding is given, following the dietary modifications for *myocardial infarction*.

Periodontal disease. Any pathologic condition of the supporting tissues surrounding the teeth. The main characteristic is the inflammation of the gums due to plaque and microbial flora buildup, caused by poor dental hygiene and *pyorrhea alveolaris*. Periodontal diseases are classified as inflammatory (gingivitis and periodontitis), dystrophic (trauma and periodontosis), and other abnormalities of the areas surrounding the teeth. *Nutrition therapy:* preparation for and after periodontal surgery requires adequate nutrients to regenerate tissue and prevent infection. Ensure adequate vitamin C, vitamin A, zinc, and protein. If the procedure or the wound prevents normal dietary intake for an extended period, supplemental liquid nutritional products should be given

Peripheral parenteral nutrition (PPN). Intravenous administration of nutrients via a peripheral vein; used as a temporary route (3 to 5 days) for nutritional support of patients who are unable to consume adequate meals via enteral feeding. See *Parenteral feeding (nutrition)*.

Peristalsis. Normal wavy movement of the gastrointestinal tract characterized by alternate contraction and relaxation along the walls from the esophagus to the intestine, forcing the contents toward the anus. *Reverse peristalsis* is an abnormal backward movement of the intestine, as seen in pyloric obstruction and *diverticulitis*.

Peritoneal cavity. Region bordered by the parietal layer of the peritoneum. It contains all the abdominal organs except the kidneys.

Peritoneal dialysis (PD). Process of removing excess fluid and metabolic waste products from the body by filtering the blood artificially using a hyperosmolar solution. The peritoneal membrane acts as the semipermeable membrane for the exchange of fluids. There are three methods of PD: intermittent PD (IPD), continuous ambulatory PD (CAPD), and automated or continuous cycler PD (APD or CCPD). IPD is not commonly used because it takes about 40 hours each week. In CAPD, the exchanges are performed at least four times a day and once at night. The dialysate is available in varying volumes from 1 to 3 liters/bag to suit individual needs. In APD, a machine or cycler introduces the dialysate in and out of the peritoneal cavity at more rapid intervals than with CAPD. It is done at night which allows more freedom during the day. Compared with CAPD, there is a minimal risk of peritonitis with CCPD. The most common dialysates are 1.5%, 2.5%, or 4.25% dextrose in 1.5 to 2 liters of solution. Protein is lost with each exchange of dialysis. As much as 30 g of protein may be lost per 24-hour treatment; protein loss increases with peritonitis. An estimation of protein loss is about 0.5 g/L dialysate with a normal peritoneum and 1.0 g/liter with peritonitis. *Nutrition therapy:* the general guidelines are to restore and maintain nutritional status; replace protein, vitamin, and mineral losses; and maintain acceptable electrolyte and fluid balances. Recommended daily nutrient intakes are as follows: calories, 35 kcal/kg body weight for IPD and 25 to 35 kcal/kg for CAPD or APD, 35 to 50 kcal/kg for repletion, and 20 to 25 kcal/kg for weight reduction; protein, 1.2 to 1.3 g/kg for maintenance and 1.5 g/kg for repletion (IPD, CAPD, or APD), supplying 50% protein of high biologic value; sodium, 2 g/day for IPD and 3 to 4 g/day for CAPD or APD; potassium, 2 to 2.5 g/day for IPD and 3 to 4 g/day for CAPD or APD; phosphorus, 15 mg/g of protein for all types of PD; calcium, 1.0 to 1.5 g/day for all types of PD; zinc, 10 to 50 mg/day if deficient; fluid, output plus 3 cups for IPD and generally unrestricted for CAPD or APD; 60 to 100 mg/day of ascorbic acid; 0.4 to 1.0 mg folate; 5 to 10 mg/day of pyridoxine; and other water-soluble vitamins at the RDA level. Weight control and restriction of alcohol, simple carbohydrates, cholesterol, and saturated fat are recommended to control hyperlipidemia. For infants, children, and adolescents, general dietary guidelines based on dry weight, height, and age are as follows: total calorie requirements are 105 to 115 kcal/kg for infants; 100 kcal/kg, 1 to 3 years; 85 kcal/kg, 4 to 10 years; 60 kcal/kg, 11 to 14 years male; 49 kcal/kg, 11 to 14 years female; 42 kcal/kg, 15 to 18 years male; and 38 kcal/kg, 15 to 18 years female. To determine dietary calories, subtract kilocalories from dialysate from the daily total kilocalories. Grams of protein per kilogram of dry body weight per day: birth to 1 year, 3 to 4 g/kg; 1 to 5 years, 3 g/kg; 5 to 10 years, 2.5 g/kg, 10 to 12 years, 2 g/kg, and >12 years, 1.5 g/kg. Fluids are generally not restricted. Infants may require sodium supplementation; restrict sodium to 2 to 3 g/day (no extra salt diet) in children and adolescents. Use high-potassium foods in moderation and restrict potassium if serum level is elevated. Avoid foods high in phosphorus, except meat; limit milk and milk products to ½ to 1 cup/day. Supplement with calcium to meet RDA levels and with multivitamins containing 1 mg of folic acid. Emphasize complex carbohydrates if hypertriglyceridemia exists. Use unsaturated fats and restrict cholesterol in hypercholesterolemia.

Peritonitis. Inflammation of the peritoneum due to the introduction of bacteria or irritating substances into the abdominal cavity by a wound or perforation of an organ in the gastrointestinal or reproductive tract. It is accompanied by fever, abdominal pain, vomiting, and constipation. Basal metabolic rate is elevated by about 10% to 15%. There may be fluid and electrolyte imbalances, increased creatinine clearance, and proteinuria which continues for several days after antibiotic therapy. These losses, combined with increased protein needs due to infection or fever, put the patient at risk for protein malnutrition. Absorption of insulin may also increase in diabetic patients, necessitating dose changes and dietary adjustments. *Nutrition therapy:* initially, the patient is given intravenous feedings and nothing by mouth. When gastrointestinal tract or bowel sounds are present, resume feeding by mouth, starting with clear liquids and progressing to a nutritional formula to supplement a full liquid diet. Monitor fluid and electrolyte balances. Provide soft to regular diets as the patient's condition improves, and adjust protein and calorie levels, according to the assessed needs and the disorder that caused the peritonitis.

Permeability. Property or state of being penetrated. *Capillary permeability* is the property of the capillary walls that allows filtration and diffusion. This is important in the exchange of substances between the blood and tissue fluids. See also *Osmosis*.

Pernicious. Causing injury or death; destructive or incurable. *Pernicious anemia* has become a misnomer since the advent of liver extract and vitamin B_{12} therapy. However, the term is retained traditionally to designate a clinical entity.

Pernicious anemia. A chronic macrocytic anemia found mostly in middle-aged and elderly persons. The red blood cells cannot be supplied by the bone marrow as rapidly as needed because of vitamin B_{12} deficiency conditioned by the lack of intrinsic factor. This factor is a glycoprotein found normally in gastric juice and is essential for the absorption of vitamin B_{12}. Pernicious anemia is characterized by disturbances in the gastrointestinal, nervous, and blood-forming systems. The symptoms are achlorhydria, atrophy of the gastric mucosa, diarrhea, glossitis, paresthesia, ataxia, degeneration of the lateral and posterior tracts of the spinal cord, macrocytosis that is often hyperchromic, and hyperplastic bone marrow. Treatment consists of intramuscular or subcutaneous injection of vitamin B_{12}. This usually starts at an initial dose of 50 μg/day for 2 weeks, followed by 100 μg twice weekly for the next 2 months, then 100 μg monthly thereafter. See also *Vitamin* B_{12}.

Peroxidation. Formation of peroxides as a result of the action of oxygen on polyunsaturated fatty acids. Moderate intake of Vitamin E reduces lipid peroxidation in cells by interrupting free radical reactions that otherwise can cause membrane damage. See *Free radicals*.

Perspiration. Also called "sweat"; secretion and exudation of fluid by the sweat glands of the skin, averaging about 700 mL/day. See also *Insensible perspiration*.

Petechiae. Small pinpoint-sized, nonraised, purplish-red spots on the skin formed by a subcutaneous effusion of blood. It is seen in vitamin C deficiency.

PGA. Abbreviation for *pteroylglutamic acid*.

pH. Symbol commonly used in expressing hydrogen ion concentration; a measure of alkalinity and acidity. It is the logarithm of the reciprocal of the hydrogen ion concentration. Examples of pH values of body fluids are as follows: blood serum, 7.35 to 7.45; cerebrospinal fluid, 7.35 to 7.45; gallbladder bile, 5.4 to 6.9; pure gastric juice, 0.9; pancreatic juice, 7.5 to 8.0; saliva, 6.35 to 6.85; and urine, 4.5 to 7.5.

Phagocyte. Cell that can engulf particles or cells that are foreign or harmful to the body. Phagocytes are present in the blood, lymph, lungs, liver, and spleen.

Phagocytosis. Also called "pinocytosis." Process of ingesting a moving or foreign particle through the cell membrane. It is one of the body's defense mechanisms against bacteria and other harmful substances. In this process, a portion of the cellular membrane envelops the foreign body by forming a small pocket outside the cell and then pinches it off from the surface to create an intracellular vacuole. Phagocytosis is also an important process in the absorption of certain nutrients. See *Absorption, nutrients*.

Pharmacologic nutrients. Certain nutrients have pharmacologic effects when given in excess of amounts needed to prevent or correct nutritional deficiencies. Most of the effects are related to the immune system. Among the nutrients that have pharmacologic properties are arginine; glutamine; ribonucleic acid; vitamins A, C, and E; and the omega-3 and omega-6 fatty acids. A high dose of arginine increases the resistance to infection and improves T cell immunity and wound healing; it also inhibits the immunosuppressive effect of blood transfusion. Glutamine improves the barrier function of the gastrointestinal tract and may increase the resistance to infection by improving phagocytic function. Ribonucleic acid may also improve resistance to infection. Vitamins A, C, and E act primarily as antioxidants. Vitamins C and

E are especially important in critically ill and traumatized patients. High doses of vitamin C will prevent the development of edema related to thermal injury. The omega-3 and omega-6 fatty acids produce eicosanoids and strengthen the immune system of the body.

Phaseolin. A simple protein of the globulin type occurring in kidney beans.

Phenelzine. A monoamine oxidase inhibitor antidepressant agent. See *Monamine oxidase*.

Phenex. Brand name of a phenylalanine-free dietary product for infants and toddlers (Phenex™-1) and children and adults (Phenex™-2) with phenylketonuria or hyperphenylalaninemia.

Phenobarbital. A barbiturate used as a sedative, hypnotic, and anticonvulsant. It may cause vitamin D deficiency, especially in children, by inactivation of 25-hydroxyvitamin D. It may also decrease thiamin absorption; increase vitamin C excretion in urine; and lower serum levels of folacin, vitamin B_{12}, pyridoxine, magnesium, and calcium. It is excreted in breast milk. Brand names are Luminal® and Solfoton®.

Phenolphthalein. A stimulant laxative; a diphenylmethane derivative. Abuse of this drug can cause potassium depletion and decreased absorption of vitamin D, glucose, calcium, and other minerals. Brand names include Agoral®, Alophen®, Correctol®, Evacugen,® Ex-Lax,® and Phenolax.®

Phenylalanine (Phe). Alpha-amino-beta-phenylpropionic acid, an *essential amino acid* rarely limiting in protein foods. It is easily converted to *tyrosine,* but the reaction is not reversible. As a precursor of tyrosine, phenylalanine is also involved in reactions in which tyrosine plays a direct role, as in melanin formation. Phenylalanine participates in *transamination* and can be ketogenic as well as glycogenic. Deficiency results in weight loss, impaired nitrogen balance, aminoaciduria, decreased serum globulin, mental retardation, and anemia. See also *Phenylketonuria*.

Phenylbutazone. A pyrazole derivative that is a nonsteroidal anti-inflammatory agent used in the treatment of gout and rheumatoid arthritis. It may decrease serum folate, increase urinary protein and uric acid, and cause weight gain due to sodium and fluid retention. Adverse gastrointestinal effects include abdominal and epigastric distress, stomatitis, dry mouth or throat, anorexia, gastritis, peptic ulcer, and gastrointestinal bleeding. Available by nonproprietary name.

Phenyl-free®. Brand name of a phenylalanine-free supplement for older children with *phenylketonuria;* contains all the essential amino acids except phenylalanine, sucrose, corn syrup solids, modified tapioca, and corn oil. Its use allows a greater variety of natural foods containing phenylalanine. See Appendix 37.

Phenylketonuria (PKU). An inborn defect in phenylalanine metabolism due to a lack of the liver enzyme phenylalanine hydroxylase, which is needed in the conversion of phenylalanine to tyrosine. As a result, phenylalanine and its metabolite, phenylpyruvic acid, accumulate in the blood and tissues and are eventually excreted in the urine. Abnormal amounts of unmetabolized phenylalanine interfere with normal brain development and may cause severe mental retardation. There is also poor growth, nonspecific neurologic abnormalities, minor seizures, an eczema-like skin eruption, and reduced pigment production resulting in fair skin and blond hair. The treatment requires dietary restriction of phenylalanine. *Nutrition therapy:* the diet is highly individualized and adjusted for growth and serum phenylalanine level. Monitored blood levels of phenylalanine are used as guidelines for ongoing assessment, which allows a limited amount of phenylalanine that will maintain an acceptable serum phenylalanine level of 2 to 10 mg/dL (120 to 600 μmol/L). Women with PKU should restrict phenylalanine intake and maintain blood phenylalanine levels between 2 and 6 mg/dL (120 to 360 μmol/L) before becoming pregnant. The normal range is 0.7 to 2.0 mg/dL (40 to 120 μmol/L). See *Diet, phenylalanine-restricted*.

Phenytoin. A hydantoin-derivative anticonvulsant used in the management of seizures. A high intake of folic acid (greater than 5 mg/day) can interfere with seizure control. However, long-term drug administration can induce folate deficiency, with the development of megaloblastic anemia; vitamin D deficiency and a decrease in bone density, which may result in rickets in children and osteomalacia in adults; and vitamin K deficiency with bleeding, especially in the infants of mothers who have been taking the drug. The drug is excreted in breast milk. Brand names are Dilantin®, and Diphenylan®.

Pheochromocytoma. A rare disease caused by a vascular tumor of the adrenal medulla that causes the production of abnormal amounts of *catecholamines*. Various cardiovascular and metabolic disturbances are associated with this disorder, such as hypertension, palpitations, hyperglycemia, and heart failure. *Nutrition therapy:* modifications focus on the clinical signs and symptoms. Encourage intake of fluids without caffeine and give small, frequent feedings. If the patient has to undergo surgery, the standard surgical care procedures are observed. See also *Vanillylmandelic acid* and *Diet, VMA test*.

Phlorizin. Also spelled "phlorhizin"; a bitter-tasting glycoside from the root bark of apple, cherry, plum, and pear trees. It causes glycosuria by blocking the tubular reabsorption of glucose. It is used as a test for kidney function and to examine the formation of glucose from other ingredients of the diet.

Phosphatase. Enzyme that hydrolyzes monophosphoric esters, with the liberation of inorganic phosphate. It is found in practically all cells, body fluids, and tissues. Phosphatases constitute a large and complex group of enzymes, some of which appear to be highly specific.

Phosphate bond energy. Energy trapped in the phosphate bond of "phosphate carriers," which are of two categories: the relatively inert types, or those that release 3000 calories on hydrolysis, such as triose phosphates, and the active phosphate carriers that yield 5000 to 12,000 calories for each energy-rich bond hydrolyzed. To the latter group belongs creatine phosphate and adenosine di- and triphosphate. The energy released by the energy-rich bonds is used mainly for biologic oxidation in muscular work and maintenance of cell potential.

Phosphocreatine. Also called "creatine phosphate"; a constituent of the muscles that acts as a phosphate donor when hydrolyzed into creatine and phosphate. For its biologic role, see *Creatine*.

Phosphoinositide. Inositol-containing phosphatides. The synthesis and degradation of these compounds occur mainly in the brain. On hydrolysis, phosphoinositide yields glycerol, L-myoinositol, fatty acids, and phosphoric acid. These compounds are important in the transport processes of the cells and in hormone action.

Phospholipid. Also called "phosphatide"; substituted fat containing a phosphoric acid residue, nitrogenous compounds, and other constituents in addition to fatty acids and glycerol. Phospholipids include *lecithins, sphingomyelins, cephalins,* and *plasmalogens*. Sphingomyelins contain 4-sphingenine in place of glycerol. Phospholipids are essential components of the cell membrane. They play a role in electron transport, stimulate protein synthesis in special systems, affect cell permeability in ion transport, affect fat absorption and transport, and induce blood coagulation with the formation of thromboplastin.

Phosphoprotein. Conjugated protein compounds with a phosphorus radical other than nucleic acid (nucleoproteins) or lecithin. To this group belong the ovovitellin of egg yolk and the casein of milk.

Phosphorus. An essential mineral that comprises 22% of the total minerals in the body. Of this, 85% is in bones and teeth as insoluble calcium phosphate (apatite) crystals, and the rest occurs in soft tissues, cells, and in combination with enzymes, proteins, carbohydrates, and lipids and other compounds. Practically all biologic reactions need phosphorus to form adenosine triphosphate and in enzyme systems for energy metabolism. Phosphorus, together with calcium, gives rigidity to bones and teeth; is a constituent of a buffer system and helps to regulate the pH of blood; transports fatty acids as phospholipids; is a component of nucleic acids; and regulates osmotic pressure. Phosphorus is present in nearly all foods; hence, dietary lack is unlikely. Foods rich in protein and calcium are also good sources of phosphorus, such as milk, cheese, meats, whole grain cereals, and legumes. Phosphate is also abundant in cola beverages and other soft drinks. Phosphorus deficiency may occur with intravenous administration of glucose or total parenteral nutrition without sufficient phosphorus and with excessive use of a aluminum hydroxide antacid, which binds phosphorus and makes it unavailable for absorption. A high intake of phosphate relative to calcium, resulting in a high serum P/Ca ratio, will enhance phosphate loss. Characteristic symptoms of deficiency are similar to those of calcium deficiency and include weakness, malaise, pain, anorexia, and bone loss. Phosphorus deficiency may also be seen in small premature infants fed human milk exclusively. Such infants need more phosphorus than is contained in human milk for the rate of bone mineralization required; without additional phosphorus, hypophosphatemic rickets may develop. There is no evidence of phosphorus toxicity. See Appendix 1 for the recommended dietary allowances. See also *Hyperphosphatemia* and *Hypophosphatemia*.

Photophobia. Abnormal sensitivity to light as seen in riboflavin deficiency.

Phylloquinone. Vitamin K_1; homolog of *vitamin K* synthesized by plants. It is especially abundant in alfalfa and green leafy vegetables, and probably accounts for most of the vitamin obtained from the diet.

Phytanic acid. An oxidation product of *phytol* which is found in a large variety of foods. An enzymatic defect in the metabolism of phytanic acid results in *Refsum's disease,* a rare autosomal recessive disorder.

Phytochemicals. Plant compounds with biological activity; these include phytoestrogens, carotenoids, indoles, and flavonoids. Phytochemicals may help prevent certain cancers. The phytoestrogens bind to estrogen receptors, reduce estrogen synthesis, and lower circulating levels of free estradiol. Because many human breast cancers are estrogen-dependent, cancer cell growth could result. Indoles increase the activity of an enzyme that converts estradiol to antiestrogenic metabolites. Carotenoids also exert a protective effect against breast cancer. The flavonoids function as antioxidants, metal chelators, and inducers of enzymes that affect chemical carcinogenicity. Phytoestrogens are found primarily in soybeans, soy products, flaxseed, whole grain products, and legumes. Carotenoids are found in deeply pigmented fruits and vegetables, including carrots, sweet potatoes, tomatoes, spinach, broccoli, cantaloupe, and apricots. Indoles are found in cruciferous vegetables such as broccoli, brussels sprouts, and cabbage. Flavonoids are found in citrus fruits.

Pica. An abnormal craving for unusual articles such as hair, clay, chalk, laundry starch, and dirt. Although this is rarely seen in humans, it presents a nutritional problem when it occurs. It has been associated with iron and zinc deficiencies. The cause is not well understood. It has been observed in hysteria, mentally defective children, and some pregnant women.

PID. Abbreviation for *pelvic inflammatory disease*.

PIH. Abbreviation for *pregnancy-induced hypertension*.

Pinocytosis. 1. Process whereby a cell absorbs liquids. The membrane invaginates and then closes to form a liquid-filled vacuole. It is similar to *phagocytosis,* whereby a cell engulfs solid particles.

Pituitary gland. See *Hypophysis*.

PKU. Abbreviation for *phenylketonuria*.

Placenta. A highly vascular organ within the uterus that is the means of communication between the fetus and mother through the umbilical cord. The maternal and fetal blood supplies are separated by two thin membranes that eventually fuse. All fetal nourishment must pass through this barrier. The placenta also secretes hormones needed for normal pregnancy.

Plasma. 1. The liquid portion of the blood in which corpuscles are suspended. (When fibrinogen is separated from the plasma by blood clotting, the remaining fluid is called *serum*.) 2. The lymph without its cells. 3. The cytoplasm or protoplasm.

Plasma amino acids (PAA). The amino acid levels in the plasma which can be used as a measure of protein quality. It is expressed as follows:

$$\text{PAA} = \frac{\text{B} - \text{A}}{\text{Amino acid needs}} \times 100$$

where A and B are plasma amino acid concentrations measured immediately before and after the meal. See also *Protein quality*.

Plasma protein. Any of the proteins in the blood plasma (e.g., albumin, globulin, fibrinogen, prothrombin). Plasma proteins constitute about 7% of the blood plasma in the body. They help maintain the immune function, water balance, blood pressure, and blood viscosity. Fibrinogen and prothrombin are important in blood coagulation. Measurements of plasma or serum albumin, transferrin, thyroxine-binding prealbumin or transthyretin, and retinol-binding protein, which are constituents of the visceral protein compartment, are useful in evaluating protein status.

Plasma volume. The total volume of plasma in the body. The normal plasma volume in males is about 39 mL/kg of body weight; for females, it is about 40 mL/kg of body weight. Plasma volume is decreased in dehydration, Addison's disease, and shock. It is increased in vitamin C deficiency and in liver disorders.

Plumbism. Chronic *lead* poisoning characterized by poor appetite, vomiting, headache, pain in the joints, pallor, colic, and increased restlessness. In severe cases, lead poisoning affects the bone marrow, kidneys, and nervous system, which may result in convulsions and coma. Lead interferes with the activity of several enzymes involved in hematopoiesis; the anemia resulting from lead toxicity can be severe. Damage to the kidneys leads to increased excretion of phosphates and amino acids and interferes with the formation of the active metabolite of vitamin D. Lead and lead compounds are found in lead-based paints in many old homes and homemade toys and furniture, and in ceramic glazes, leaded gasoline, and industrial processing. The absorption of lead is 5 to 10 times higher in children than in adults.

PNI. Abbreviation for *prognostic nutritional index*.

Poliomyelitis. An infectious disease caused by one of three polioviruses. The nonparalytic form lasts for periods ranging from a few days to a week and is characterized by fever, malaise, nausea and vomiting, and back stiffness. The paralytic form has all the clinical signs and symptoms of the nonparalytic type plus paralysis. The large proximal muscles of the limbs are often affected, and in the bulbar type of poliomyelitis, the brain stem and spinal cord are affected. *Nutrition therapy:* nutrition intervention is very important due to the high fever, nausea, and vomiting. Tube feeding may be needed in some cases. Foods that tend to produce mucus, usually milk and cream, are generally not tolerated and should be avoided. The paralytic patient needs assistance in feeding. See *Dysphagia* and *Fever*.

Polycose®. Brand name of a carbohydrate supplement consisting of glucose polymers in liquid or powder form. It is tasteless, mixes readily with foods and beverages, and requires little or no amylase for digestion. See also Appendix 39.

Polydipsia. Excessive thirst due to loss of body fluids, especially from the urine, as seen in diabetes mellitus.

Polyneuritis. A condition that involves the inflammation of many nerves. It occurs in thiamin deficiency.

Polyneuropathy. Multiple noninflammatory degeneration of nerves. Nutritional polyneuropathy is caused by a lack of nutrients, particularly the B-complex vitamins. It is characterized by bilateral lesions of the legs, with cramps, ataxia, foot drop, loss of vibratory and positional sense, and paresthesia. It is seen in beriberi, chronic alcoholism, and starvation. *Nutrition therapy:* adequate diet to meet assessed needs, with supplementation of the B-complex vitamins and in particular, a daily dose of 50 to 200 mg of thiamin. Discontinue thiamin when a therapeutic response is achieved. Deficiencies in pyridoxine, niacin, pantothenic acid, biotin, and vitamin B_{12} are associated with peripheral neuropathies. Supplementation with the deficiency vitamin is necessary. Avoid alcohol in alcoholic neuropathy. See also *Beriberi* and *Alcohol dependence*.

Polypeptide. A long chain of amino acids (usually about 100 units) linked by the peptide bond. Polypeptides are smaller than proteins, but larger than proteoses.

Polyphagia. 1. Swallowing abnormally large amounts of food at a meal. 2. Excessive appetite as in diabetes mellitus.

Polysaccharide. A carbohydrate containing ten or more monosaccharide units. Those of nutritional significance are glycogen, starch, and dietary fiber. See Appendix 8.

Polyunsaturated fatty acid (PUFA). A fatty acid that has more than one unsaturated linkage in its carbon chain. See *Fatty acid*.

Polyuria. Excessive secretion and discharge of urine as seen in diabetes mellitus.

POMR. Abbreviation for *problem-oriented medical record.*

Portagen®. Brand name of a nutritionally complete product that contains 85% of the fat as *medium-chain triglycerides,* which are more readily hydrolyzed and absorbed than conventional long-chain fatty acids. It is used in infants and children with cystic fibrosis, pancreatic insufficiency, biliary atresia, and other disorders of fat absorption. See also Appendix 37.

Potassium. An essential mineral that is the chief cation in intracellular fluids. It is important in the maintenance of acid–base and water balances, osmotic equilibrium, muscle and nerve irritability, and normal blood pressure. Potassium is believed to block the ability of sodium to raise blood pressure. Potassium is widely distributed in many foods but is especially abundant in nuts, whole grains, meats, and fruits. Dietary potassium deficiency does not occur under normal circumstances. However, secondary deficiency (hypokalemia) may occur when there is excessive loss of potassium with prolonged vomiting, chronic diarrhea, laxative abuse, and use of certain diuretics; some forms of renal disease, diabetic acidosis, and other metabolic disturbances may also lead to severe potassium loss. Symptoms of deficiency are lack of appetite, nausea, muscle weakness, mental disorientation, nervous irritability, and cardiac irregularities. Elevated blood potassium (hyperkalemia) is seen when there is tissue damage, as in myocardial infarction, burns, and surgery, or inadequate excretion as in renal failure. See also *Hyperkalemia* and *Hypokalemia.*

PPN. Abbreviation for peripheral parenteral nutrition. See *Parenteral nutrition.*

Prader–Willi syndrome. A rare congenital disorder characterized by hypotonia, hyperphagia, obesity, and mental retardation. Obesity appears after infancy and becomes progressive and intractable, reaching morbid proportions in some individuals. Endocrine studies indicate that malfunction of the hypothalamus is a fundamental part of the disorder. The disorder is associated with hyposecretion of gonadotropic hormones by the pituitary gland. *Nutrition therapy:* weight control and prevention of obesity should begin during early childhood, with counseling in dietary reduction of calorie intake, alteration of eating patterns by some form of behavior modification, and increase in physical activity. Provide 10 to 11 kcal/cm height for weight maintenance, and 8 to 9 kcal/cm height for weight reduction. The use of a very-low-calorie diet may impair growth and is not recommended. More drastic measures such as jaw wiring and gastric by-pass are also not recommended, except in older adolescents with life-threatening complications of morbid obesity.

Prazosin. A derivative of quinazoline, an adrenergic blocking agent used in the treatment of hypertension. It may cause weight gain due to fluid retention; it may also cause nausea, vomiting, diarrhea, constipation, and abdominal discomfort. Brand names are Minipress® and Minizide®.

Prealbumin. A plasma albumin that moves ahead of albumin on paper electrophoresis. It is synthesized by the liver and serves as a carrier of retinol and thyroxine in the blood. Prealbumin is an early indicator of protein nutritional status; it has a short half-life (2 to 3 days). Prealbumin responds rapidly to refeeding, with a significant increase noted within 4 days. Normal plasma concentration is 15 to 30 mg/dL. Values of 10 to 15 mg/dL indicate mild depletion; 5 to

10 mg/dL, moderate depletion; and < 5 mg/dL, severe depletion. The serum level is reduced in acute catabolic states, after surgery, and with infection, stress, trauma, liver disease, and hyperthyroidism; it is increased in patients with chronic renal failure on dialysis.

Pre-Attain®. Brand name of a half-strength starter formula for tube feeding. It is lactose-free and contains sodium caseinate, maltodextrin, and corn oil. See also Appendix 39.

Prednisone. A synthetic corticosteroid used for its anti-inflammatory and immunosuppressant effects. It is excreted in breast milk. See *Corticosteroid* for its adverse effects. Brand names are Deltasone®, Meticorten®, and Orasone®.

Preeclampsia. A *toxemia* of pregnancy characterized by hypertension, albuminuria, and edema of the lower extremities. See *Pregnancy-induced hypertension.*

Preemie SMA®. Brand name of a formula for premature and low-birth-weight infants. Contains whey, casein, glucose polymers, lactose, MCT oil, and coconut and soy oils. It has a whey/casein ratio of 60:40 (compared to 18:82 in standard infant formulas) and gives a better distribution of amino acids. See also Appendix 37.

Pregestimil®. Brand name of a lactose-free formula for infants sensitive to intact protein and recovering from prolonged diarrhea. Contains casein hydrolysate, free amino acids, glucose polymers, modified tapioca starch, MCT oil, and corn oil. See also Appendix 37.

Pregnancy. Also called "gestation"; the condition of having a developing embryo or fetus in the body after the union of an ovum and a spermatozoon. In the mother's womb, the period of pregnancy is about 266 to 280 days. It is divided into three main phases: *implantation,* the first 2 weeks of gestation, during which the fertilized ovum becomes embedded in the wall of the uterus and the placenta develops; *organogenesis,* the next 6 weeks, during which the developing fetal tissue undergoes differentiation and is the most sensitive period for nutrition-induced birth defects; and *growth,* the remaining 7 months of pregnancy, characterized by rapid cell division and development. See also *Nutrition, pregnancy.*

Pregnancy-induced hypertension (PIH). High blood pressure developed by some pregnant women with serious complications such as edema, convulsions, and kidney failure. It may even be fatal to the mother and fetus. The exact etiology is not known, but nutritional intervention reduces the seriousness of this disorder. Formerly called "preeclampsia" (presence of hypertension and proteinuria) and "eclampsia" (with convulsive seizures). *Nutrition therapy:* sodium intake is mildly restricted to 2 to 3 g of sodium (87 to 130 mEq) per day. Calories and protein are controlled to provide the optimal needs for pregnancy, avoiding excesses. A strict sodium restriction does not significantly alter blood pressure and is, therefore, not necessary. Supplemental vitamins and minerals are recommended. Monitor the weight gain desirable for the gestational period.

Premature infant. A liveborn infant, regardless of birthweight, born before the 37th week of the gestational period. There are many causes of premature birth, some of which are unknown. Undoubtedly, maternal malnutrition is a significant cause. See *Nutrition, prematurity* and *Low-birth-weight infant.*

Premenstrual syndrome (PMS). A disorder seen in some women a few days just before the menstrual period; characterized by headache, bloating, depression and "moodiness," breast pain, back pain, and changes in metabolism. The appetite increases, particularly for carbohydrate foods. Researchers have studied the relationship between nutrition and PMS, but the results are not conclusive. Some of the nutrients implicated are vitamin B_6, vitamin E, and magnesium. Treatment consists of alleviating the symptoms, which disappear once the menstrual period has begun. *Nutrition therapy:* small meals with moderate salt restriction, high protein, limited sugars and concentrated sweets, and avoidance of caffeine may help alleviate some symptoms of PMS.

Preschool Nutrition Survey (PNS). A study conducted in 1966 to 1970 to determine the nutritional status of a cross-sectional sample of 3400 preschool children, ages 1 to 6 years, from all income levels in the United States.

Pressure sore. Also called decubitus ulcer or bed sore. An inflammation, sore, or ulceration around bony prominences due to local interference with circulation, usually caused by lack of oxygen. Common among patients who are malnourished and have circulatory disorders, diabetes mellitus, renal failure, spinal injury, fractures, and other conditions that keep them immobilized or bedridden. Morbid obesity is also a risk factor. A pressure sore is graded by stage of severity into: stage I for persistent erythema; stage II for skin blister, abrasion, or shallow ulcer; stage III for ulceration extending through the skin and subcutaneous fat; and stage IV for deep ulceration extending to muscle or bone. The extent and depth of pressure ulcers are related to hydration status, anemia, and undernutrition, especially hypoproteinemia. Wound drainage from the lesion contains about 400 mg of nitrogen, corresponding to almost 2.5 g of protein. Up to 30 g of protein can be lost through drainage from a single large open wound in 24 hours. Closely related to protein is the level of ascorbic acid which is necessary for the synthesis and maintenance of collagen, the chief protein component of connective tissue. *Nutrition therapy:* wound healing is an energy-intensive process. For stage I pressure sore, provide maintenance calories [basal energy expenditure (BEE)] adjusted for activity; increase BEE calories by an injury factor of 1.2 (or 30 kcal/kg body weight) for stage II sores, and by 1.5 (or 35 kcal/kg body weight) for stages III and IV. Adjust total calories to attain the weight goal, especially if the individual is underweight or there was recent significant weight loss. Estimate protein needs based on desirable body weight as follows: 1.0 to 1.1 g/kg for stage I pressure sore; 1.2 to 1.4 g/kg for stage II; 1.5 to 1.7 g/kg for stage III; and 1.8 to 2.0 g/kg for stage IV. Use the upper range for multiple sores, if there is drainage from the lesion, and if there is anemia or low serum albumin. It may be necessary to provide up to 2.5 g of protein/kg/day in a stage IV pressure sore with a large amount of drainage and in multiple stage III/IV sores with underlying low serum albumin. Supplement the diet with vitamin C (500 mg/day) and zinc (15 to 20 mg/day). Ensure adequate intakes of other nutrients necessary for wound healing, such as vitamins A and E, and iron. Correct the anemia with iron supplement. Fluid intake should be at least 35 mL/kg body weight per day.

Primidone. A structural analog of phenobarbital used in the management of seizures. Its adverse effects are the same as those of *phenobarbital*. It is also excreted in breast milk; infants should be watched for drowsiness. The brand name is Mysoline®.

ProBalance™. Brand name for a nutritionally complete liquid formula for older adults. It exceeds the recommended dietary allowances (RDA) for normal healthy adults to meet the enhanced needs of geriatric adults. Contains slightly higher levels of protein, vitamins, and minerals and has moderate fiber content to assist in bowel management. See Appendix 39.

Probenecid. A sulfonamide-derivative uricosuric agent used in the treatment of gout. It increases the urinary excretion of riboflavin, calcium, magnesium, sodium, phosphorus, and chloride. It may also cause anorexia, sore gums, anemia, nausea, vomiting, and abdominal discomfort. Probenecid increases the concentration of uric acid in the renal tubules, which may crystallize and form stones if the urine is acidic, or in diets producing an acid-ash residue. Brand names are Benemid®, ColBenemid®, and Probalan®.

Problem-oriented medical record (POMR). See *Nutritional care.*

Probucol. An antilipemic agent used as an adjunct to diet therapy to decrease elevated serum cholesterol in Type IIa hyperlipoproteinemia. The drug is structurally unrelated to other currently available antilipemic agents. Its adverse effects are generally mild to moderate in severity and of short duration. There may be diarrhea, nausea, vomiting, flatulence, and abdominal pain. The brand name is Lorelco®.

Procalamine®. Brand name of a 3% amino acid and a 3% glycerin solution with electrolytes for intravenous injection. It supplies 29 g of protein equivalent (4.6 g of nitrogen) and 130 nonprotein calories per 1000 mL. See Appendix 40.

Procarbazine. An antineoplastic hydrazine derivative; used in the treatment of advanced Hodgkin's disease and other lymphomas. Severe nausea and vomiting occur frequently following its administration, and there may be anorexia, stomatitis, dry mouth, and dysphagia. The drug is a pyridoxine antagonist and may cause depletion of the vitamin. Because it has some monoamine oxidase inhibitory activity, the drug should not be taken with alcohol and tyramine-containing foods to avoid a disulfiram-like reaction. The brand name is Matulane®.

Prochlorperazine. A derivative of phenothiazine used for the control of severe nausea and vomiting of various etiologies. It is excreted in breast milk; avoid the use of the drug, if possible, when breast-feeding. It may also cause dry mouth, constipation, and difficulty in swallowing. Brand name is Compazine®.

Product 3200AB. A special dietary product that is low in phenylalanine (11 mg/dL) and tyrosine (6 mg/dL) for use in infants with hereditary tyrosinemia II. Contains enzymatically hydrolyzed and specially treated casein to reduce phenylalanine and tyrosine, corn syrup solids, modified tapioca starch, and corn oil.

Product 3200K. A low-methionine diet powder for use in infants with homocystinuria. Made of soy protein isolate, corn syrup solids, corn oil, and coconut oil.

Product 3232A. A mono- and disaccharide-free diet powder used in infant formulas for the nutritional management of disaccharidase deficiencies (lactase, sucrase, and maltase) and for impaired glucose transport and fructose utilization. The desired carbohydrate is added to the formula base, which contains casein hydrolysate, modified tapioca starch, medium-chain triglycerides, and corn oil.

Product 80056. A protein-free diet powder for use in the dietary management of amino acid disorders, such as hyperalanemia, propionic aciduria, hyperlysinemia, malonic aciduria, isovaleric acidemia, argininosuccinic aciduria, and maple syrup urine disease. Contains corn syrup solids, modified tapioca starch, and corn oil, plus vitamins and minerals.

Professional Standards Review Organizations (PSRO). Established in 1972, the aim of this agency is to evaluate the quality of patient care, including dietary services. Nutritional screening, assessment, monitoring, and education are examples of professional standards of practice needing review to ensure that they meet professional standards.

Profiber®. Brand name of a high-fiber, isotonic, lactose-free tube feeding formula. Contains sodium caseinate, hydrolyzed cornstarch, corn oil, and 13 g of soy fiber per liter. See also Appendix 39.

Prognostic nutritional index (PNI). Predictive model relating morbid complications to four variables of nutritional status. Calculated from the following formula:

$$\text{PNI (\% risk)} = 158 - (16.6 \times \text{ALB g/dL}) - (0.78 \times \text{TSF mm}) - (0.2 \times \text{STR mg/dL}) - 5.8\ (\text{HST})$$

where ALB is serum albumin, TSF is triceps skinfold thickness, STR is serum transferrin, and HST is hypersensitivity skin test which is assigned a value of 1 for anergy and 2 for reactive. The risk of morbidity and mortality, using the PNI, is divided into three categories: low risk, PNI $< 30\%$; intermediate risk, PNI 30% to 50%; and high risk, PNI $\geq 50\%$. See also *Hospital prognostic index (HPI)*.

Project Head Start. Program started in 1965 by the Office of Economic Opportunity to help disadvantaged preschool children from poverty-stricken areas attain their potential in growth and physical and mental development before entering school. Nutrition is an important part of this project. Meals and snacks are provided at the Head Start centers. In addition, meal planning and preparation classes are held for the parents of these children.

Project LEAN. A project conducted by organizations or state and local agencies to educate the public on low-fat eating. The goal is to reduce the total fat consumption of Americans to less than 30% of calories by purchasing and preparing the right food choices. Broadcast and print media help disseminate accurate information. Also, food industries and retailers (supermarkets) promote the use of low-fat products by marketing strategies and giving away recipes. Food manufacturers and fast food and other restaurants have modified their products to include items low in fat.

Prolamine. Simple protein insoluble in water, neutral solvents, and absolute alcohol but soluble in 70% to 80% alcohol. Examples are *zein* (corn) and *gliadin* (wheat).

Proline. A heterocyclic nonessential amino acid. See *Amino acid.*

Promix RDP. Brand name of a powdered dietary supplement containing rapidly dispersing protein (whey) that is easily suspended in liquids.

ProMod®. Brand name of a powdered protein supplement made of whey; provides 75 g of protein/100 g of powder. Can be mixed with food or liquids. See Appendix 39.

Promote®. Brand name of a high-protein, lactose-free nutritional product for tube feeding or oral supplementation. See Appendix 39.

Propac™. Brand name of a powdered protein supplement for enteral feeding; made of whey protein. Provides 75 g of protein/100 g of powder. See Appendix 39.

Pro-Phree®. Brand of a protein-free dietary product for infants and toddlers on protein restriction but who need extra energy, minerals, and vitamins.

Propimex®. Brand name of a methionine- and valine-free, low-isoleucine and low-threonine dietary product for infants and toddlers (Propimex®-1) and children and adults (Propimex®-2) with propionic or methylmalonic acidemia.

Propionic acidemia. An inherited metabolic disorder due to the failure of the body to metabolize the amino acids threonine, isoleucine, methionine, and valine. It is characterized by acidosis, lethargy, mental retardation, hyperammonemia, and seizures. Acidosis occurs as a result of excess propionic acid in the blood due to defective propionyl-CoA reductase. *Nutrition therapy:* restrict isoleucine, methionine, threonine, and valine; provide greater than normal energy intake and, if necessary, give drinks sweetened with sugar or with added Polycose® or Moducal® to maintain energy intake at 150% RDA for age; use long-chain unsaturated fatty acids, large fluid intake to promote urinary excretion of abnormal metabolites, and supplemental carnitine and biotin. A dose of 10 mg/day of biotin has been suggested.

Propranolol. A beta-adrenergic blocking agent used in the management of hypertension, angina pectoris, and cardiac arrhythmia. Uptake is enhanced when the drug is given with food. It may increase blood urea nitrogen and decrease carbohydrate intolerance; adverse gastrointestinal effects include nausea, vomiting, diarrhea, epigastric distress, abdominal cramping, flatulence, and constipation. Excreted in breast milk. Brand names are Inderal® and Inderide®.

Prosobee®. Brand name of a milk-free formula made with soy protein isolate, corn syrup solids, soy oil, and coconut oil. It is used for infants with protein hypersensitivity, lactose or sucrose intolerance, and galactosemia. See also Appendix 37.

Prostaglandins. New family of hormone-like compounds that are long-chain polyunsaturated fatty acids with a 5-carbon ring. Several forms exist, varying slightly in structural formula; designated as PGD, PGE, PGF, PGG, PGH, and PGI, with numerical subscripts to indicate the number of double bonds. The polyunsaturated fatty acids serve as precursors for these compounds. Prostaglandins perform a variety of functions in the body, acting, for example, as blood pressure depressants, smooth muscle stimulants, and antagonists to several hormones.

Prosthetic group. Nonprotein molecule that confers characteristic properties on the complex protein. The binding to protein is very firm, as in the heme of hemoglobin. See also *Coenzyme* and *Cofactor*.

Protain XL™. Brand name of a high-protein tube-feeding formula with fiber (8 g/liter). Fortified with zinc; vitamins A, C, and other vitamins and minerals associated with wound healing. See Appendix 39.

Protamine. A simple protein soluble in water or in ammonium hydroxide but not coagulated by heat. This basic polypeptide contains a relatively small number of amino acids. Examples are salmine (in salmon) and sturine (in sturgeon). See also Appendix 9.

Protease inhibitor. A substance that has the ability to inhibit the proteolytic activity of certain enzymes. It is found throughout the plant kingdom, particularly among the legumes. *Trypsin inhibitor* is found in soybeans, lima beans, and mung beans. *Chymotrypsin inhibitor* is found in cereal grains and potato. Protease inhibitors are destroyed by heating.

Protein. Complex organic compound essential to all living organisms. It is a polymer of *amino acids* linked together by peptide bonds. This forms the *primary structure,* and the sequencing or order of the amino acid chain determines its function and tertiary structure. Hydrogen bonds and disulfide, ester, and salt bridges between polypeptide chains result in the *secondary structure* of proteins. The folding of the amino acid chain and its shape (globular, coiled, or spiral, etc.) is its *tertiary structure*. The grouping of units of protein molecules not joined by peptide bonds is its *quaternary structure*. An amino acid which is the backbone of proteins always contains carbon, hydrogen, oxygen, and nitrogen, and occasionally phosphorus, sulfur, zinc, copper, and iron. The nitrogen content distinguishes protein from fat and carbohydrate and is responsible for its unique physiologic functions (See under *Protein, functions*). Proteins are classified according to *composition and chemical properties* (simple, conjugated or compound, and derived), *nutritional quality* (complete, partially complete, and incomplete), *structure* (fibrous or globular), and *solubility* (in water, acid, or alcohol). Native protein in solution is colloidal because of its high molecular weight (6000 to several million). Protein is easily denatured, acts as an electrolyte, forms zwitterions, hydrates with water, and precipitates at its isoelectric point. Enzymes, which are protein in nature, are inactivated by heat. See also Appendix 9.

Protein-bound iodine (PBI). Thyroxine-binding globulin. This is the form in which the thyroid hormone, bound to a plasma protein, is transported in the blood. The normal concentration is about 5 μg/dL of plasma or serum. It increases to about 8 μg/dL in hyperthyroidism. The PBI determination provides a good index of thyroid function and is useful in measuring basal metabolism.

Protein-calorie malnutrition (PCM). See *Marasmus* and *Protein-energy malnutrition*.

Protein catabolic rate (PCR). An indication of the amount of protein that is catabolized per kilogram of body weight in 24 hours. Calculated from the amount of urea produced per minute and can be used to assess nitrogen balance. In a nutritionally stable patient, PCR is equal to the dietary protein intake (DPI). A person is in nitrogen equilibrium when DPI is equal to PCR, in positive nitrogen balance when DPI is greater than PCR, and in negative nitrogen balance when DPI is less than PCR.

Protein deficiency. Lack of protein results in stunted growth and development, poor musculature, skin lesions, thin and fragile hair, hormonal imbalances, edema due to hypoalbuminemia, and diminished immune response. In developing countries where good-quality protein is scarce and food intake is insufficient, prolonged protein deficiency accompanied by low intake of calories results in *protein-energy malnutrition*.

Protein-efficiency ratio (PER). Biologic method of evaluating *protein quality* in terms of weight gain per amount of protein consumed by a growing animal. Feeding, usually for a month, is ad libitum, with a diet that contains 10 g of protein/100 g of food. Casein is often used as a control protein. PER is calculated as follows:

$$\text{PER} = \frac{\text{Weight gain (g)}}{\text{Protein consumed (g)}}$$

Protein-energy malnutrition (PEM). Preferred name for protein-calorie malnutrition (PCM). A complex disorder caused by lack of protein and caloric intakes. Characterized by a broad array of clinical conditions resulting from varying combinations and degrees of protein and energy deficiency. The three major forms of PEM are *marasmus,* in which the deficiency is primarily of energy; *kwashiorkor,* in which the deficiency is primarily protein; and *marasmic kwashiorkor,* in which protein and energy are both deficient. Marasmus is associated with starvation or prolonged inadequate dietary intake; kwashiorkor is associated with catabolic stress and is seen in patients whose diets contain adequate carbohydrate and fat but not adequate protein; and the combined form occurs when a marasmic patient experiences a catabolic stress. This is characterized by body fat and skeletal muscle wasting; decreased serum albumin, transferrin, and total lymphocyte count; and the skin test reaction to common antigens may be depressed or absent. Risk factors for PEM in hospitalized patients include: major surgery, burns, radiation or chemotherapy, ventilator dependent, renal failure, metabolic stress, sepsis, and trauma. Prolonged PEM impairs immune function, leading to increased infections and delayed wound healing. See *Kwashiorkor* and *Marasmus*. See also *Malnutrition*.

Protein food mixtures. Products that are specially formulated to provide cheap sources of protein in areas of the world where animal protein sources are expensive or scarce. Made from local vegetables such as legumes, nuts, cereals, and leaves, and may have added skim milk powder, vitamins, and minerals.

Protein, functions. The biologic role of protein is so unique that it merits its name, which originated from a Greek word, *proteios,* meaning "to take first place." An important function of protein is tissue synthesis. It is a *structural component* of all living cells and is found in muscles, nerves, bone, teeth, skin, hair, nails, blood, and glands. Almost all body fluids contain protein, with the exception of urine, sweat, and bile. All enzymes and some hormones are composed of protein. Thyroid hormones (triiodothyronine and thyroxine) and insulin are examples. Many inborn errors of metabolism are caused by lack of specific enzymes. As a *source of energy,* protein yields 4 kcal/g; this function should be spared by fat and carbohydrate so that protein is used for building and repairing tissues, which is its primary role. Protein is a *regulator* of blood pH, osmotic pressure, and water balance. As a buffer, protein can accept or donate hydrogen ions so that blood pH is maintained within 7.35 to 7.45. Because of their larger size, protein particles cannot move from capillary beds into the tissues, which accounts for the oncotic force (osmotic pressure) exerted by protein, thereby keeping the body fluid in the bloodstream and counteracting the force of blood pressure. If a person has reduced blood albumin and globulin levels, as in protein deficiency, osmotic pressure is reduced and fluid is drawn out from the blood vessels to the extracellular or interstitial spaces, resulting in edema. Beta-lymphocytes form specific proteins called *antibodies* which aid in combating infections by producing antibody–antigen complexes. Other specific functions of protein are attributable to certain component amino acids such as *tryptophan* for vitamin synthesis, *methionine* as a lipotropic agent, and *phenylalanine* and *tyrosine* for neurotransmitter formation. Protein can form glucose, a process called "gluconeogenesis," which is an important process during starvation. This is critical for the brain, which uses glucose as its energy source. The brain uses about one-third of the total body needs for energy at rest (resting energy expenditure) in the form of glucose. On the topic of protein function, recent attention has focused on the so-called "transport proteins." These are body proteins specialized in moving molecules into and out of cells. Acting like a pump, a transport protein maintains equilibrium in the surrounding fluids by picking up substances on one side and carrying them to the other side, without leaving the membrane. See also *Acute-phase proteins*.

Protein hydrolysate. A predigested protein that contains a mixture of amino acids and peptides. It is useful for oral or parenteral feeding of patients with poor or impaired digestion, as seen in pancreatic diseases, postoperative cases, and severe burns.

Protein measurement. The quantity of protein can be estimated by a number of techniques, including colorimetry, microbiologic assay, isotope dilution, countercurrent distribution, the enzymatic method, and chromatography. For practical purposes, the *Kjeldahl method* for nitrogen determination is most commonly used, since proteins contain an average of 16% nitrogen. Thus % nitrogen × 6.25 = % protein. See also *Proximate analysis*.

Protein quality. An attribute of a protein that depends on the kinds and amounts of *amino acids* present relative to body needs. No two food proteins are alike in their efficiency for tissue synthesis. In general, plant proteins are lacking or "limiting" in the essential amino acids: lysine, methionine, threonine, and tryptophan. Animal proteins are of high quality or are said to be complete proteins. A *complete* protein contains all the essential amino acids in amounts sufficient for growth and life maintenance, e.g., casein and egg albumin. A *partially complete* protein can maintain life but cannot support growth, e.g., gliadin. An *incomplete* protein is deficient in one or more of the essential amino acids and cannot support life or growth, e.g., zein and gelatin. The last two classes of protein can be used effectively for anabolic processes by combining them with small amounts of complete proteins, by adding the limiting amino acid in synthetic form, or by mixing incomplete proteins to obtain a complete assortment of amino acids in the amounts needed for tissue synthesis. For different methods of evaluating protein quality, see *Biologic value, Chemical score, Net protein utilization,* and *Protein-efficiency ratio*.

Protein requirement. Protein needs vary among individuals according to age, body size, physiologic condition (growth, pregnancy, and lactation), state of health, type of protein in the diet, and adequacy of caloric intake. The minimum requirement for the maintenance of nitrogen balance is between 0.4 and 0.5 g/kg. See Appendix 1 for the recommended dietary allowance for protein. A practical guide for meeting daily protein needs is provided by the Food Guide Pyramid. See Appendix 4. For obese patients, protein needs are calculated on the basis of 1.0 times the adjusted body weight. (See *Weight, adjusted*.) Most hospitalized patients can be adequately maintained on protein intakes between 1.0 and 1.5 g/kg of body weight per day. Fever, sepsis, surgery, trauma, and burns increase protein catabolism and, therefore, greater amounts of protein may be needed. For estimating protein requirements in special conditions and selected disorders, see Appendix 24.

Protein reserve. The body does not store protein in the same way or to the same extent that it stores fat and carbohydrate. To a limited extent, all organs and tissues store protein, which is more appropriately termed "labile tissue protein." The latter designation signifies that the protein reserve is in a dynamic state, being constantly broken down or resynthesized, and is at equilibrium with the metabolic pool of amino acids. Muscle protein is not as readily available as is protein stored in the liver, kidney, and other organs. The determination of plasma amino acid levels or the regeneration of blood constituents and protein in the liver is indicative of the extent of protein storage or depletion. See also *Serum albumin, Total lymphocyte count,* and *Transthyretin.*

Protein-sparing action. The ability of a substance to save protein from being catabolized to supply energy. Protein can then be used for its unique function of anabolism, or tissue synthesis. Fats and carbohydrates are protein sparers.

Protein-sparing modified fast (PSMF). A *very-low-calorie diet* that provides a liberal amount of protein to preserve lean body tissue while one is fasting in order to lose weight. See *Diet, protein-sparing modified fast.*

Protein supplementary value. The ability of a protein to provide or add the "limiting" or missing amino acid of another protein so that the combination approximates a *complete protein,* e.g., the combination of rice or corn with nuts and legumes. In general, plant foods contain less of the essential amino acids than animal foods. However, plant proteins from a variety of sources, when combined in appropriate amounts, will provide a mixture of all the essential amino acids in amounts similar to those found in proteins of animal origin. See *Protein quality.*

Protein synthesis. Building of protein as opposed to catabolism. A specific protein is synthesized according to the genetic code as governed by the DNA. The DNA in the nucleus uncoils and a strand of mRNA is formed. A complement mRNA carries the genetic message from DNA to the *ribosome,* the site of protein biosynthesis. Transfer RNA (tRNA) picks up amino acids that have been activated by adenosine phosphate in the cytoplasm and brings them to a specific site on the mRNA. The amino acids are added one at a time by means of a peptide bond until the complete protein is synthesized. Protein synthesis follows the *all-or-none-law:* all amino acids (both essential and nonessential) needed according to the genetic code must be present at the same time at the site of protein synthesis. The absence of a single amino acid will prevent synthesis. See also *Nucleic acid.*

Proteinuria. Also called "albuminuria." Presence of protein (mainly albumin) in the urine as a result of faulty reabsorption from a damaged tubule or leakage of an excessive amount of protein through a damaged or inflamed glomerulus of the kidney. Proteinuria is usually a sign of renal disease or renal complication of another condition, such as hypertension, heart failure, or shock. However, it can be orthostatic proteinuria associated with the upright position or a result of heavy muscular exertion. Test results and corresponding amounts of protein present per liter of urine in 24 hours are as follows: trace (0.1 to 0.5 g); 1+ (0.5 to 1.0 g); 2+ (3.0 g); 3+ (5.0 g); and 4+ (10 g or more). Heavy proteinuria (above 4 g) are seen in nephrotic syndrome and glomerulonephritis.

Proteolysis. Enzymatic or hydrolytic conversion of protein into simpler substances such as polypeptides, proteoses, peptones, and amino acids. See Appendix 11.

Prothrombin. Protein in blood plasma needed for blood clotting. It is converted to thrombin during the clotting process. Prothrombin time is the length of time it takes for the plasma to clot. This reflects the intake of vitamin K.

Protoplasm. The organized colloidal material in the living cell that is the seat of vital functions such as digestion, metabolic activities, growth, and reproduction. The protoplasm contains water, organic compounds, and inorganic salts.

ProViMin®. Brand of a protein-base powder with added vitamins and minerals; used for infants and children with chronic diarrhea and other malabsorptive disorders requiring restricted carbohydrate and fat intake.

Provitamin. A vitamin *precursor,* e.g., carotene is the provitamin of vitamin A. A provitamin is chemically related to the preformed vitamin, but it has no vitamin activity unless it is converted to the biologically active form.

Proximate analysis. Also called the "Weende scheme"; named after the Weende Experimental Station in Germany, which established the analytic method in 1865. A method of analyzing food and biologic materials according to their molecular components, such as water, ash or minerals, protein, fat, and carbohydrate, by difference. The first four are determined as follows: the sample is dried, and the difference in weight before and after drying represents *moisture content* or water; a subsample of the dried material is extracted with ether, and the ether extract represents *crude fat;* another subsample is analyzed for nitrogen content using the *Kjeldahl method,* and the percentage of nitrogen multiplied by the factor 6.25 represents *crude protein;* the third subsample is boiled for 30 minutes in dilute sulfuric acid, filtered, and the residue boiled in sodium hydroxide. The insoluble residue is made of crude fiber and ash; when ignited, the remaining residue is *ash,* representing minerals, and the loss or difference in weight is taken as *crude fiber* or indigestible carbohydrate. The proximate analysis of food is the basis of food composition tables. It is the starting point for analyzing individual nutrients. See *Carbohydrate by difference.*

PSMF. Abbreviation for *protein-sparing modified fast,* a variety of very-low-calorie diet (VLCD).

Psoriasis. Skin disease with many varieties. It is characterized by scaly red patches on the extensor surfaces of the body, scalp, ears, genitalia, and bony prominences. Treatments include corticosteroid and methotrexate, a folic acid antagonist. *Nutrition therapy:* A taurine-restricted diet may be beneficial. See *Diet, taurine-restricted.*

P/S ratio. Ratio of polyunsaturated to saturated fatty acids. A high P/S ratio (at least 1.0) is considered beneficial. Foods in this category include the oils of corn, cottonseed, linseed, safflower, sesame, soybean, and sunflower; almonds, and walnuts. A medium P/S ratio (0.5) is provided by glandular organs, hydrogenated shortenings, solid margarines, peanut oil, peanut butter, and pecans. A low P/S ratio (0.3) is found in lard, olive and palm oils, pork fat, and veal. A very low P/S ratio (0.1) occurs in beef, butter, egg yolk, milk, and mutton.

PSRO. Abbreviation for *Professional Standards Review Organizations.*

Psychodietetics. The interrelationship between psychology and nutrition. It deals with the study of the connotative meanings of food, the attitudes and habits of people, the interaction of diet and behavior, and the emotional aspects of eating. See *Nutrition, emotional stability* and *Nutrition, psychiatry.*

Psyllium. Dried, ripe seed of *Plantago psyllium* and related species. It is a rich source of soluble fiber which is effective in regulating the blood sugar level and in lowering blood cholesterol, especially low-density-lipoprotein (LDL) cholesterol. Psyllium is also a bulk-forming laxative but it requires a large amount of fluid to be effective. Undesirable side effects of psyllium are anorexia, abdominal discomfort, flatulence, and early satiety. The absorption of riboflavin, iron, and other minerals may also be impaired. Brand names include Effersyllium®, Fiberall®, Hydrocil®, Konsyl®, and Metamucil®.

Pteroylglutamic acid (PGA). Chemical name for *folic acid,* one of the B-complex vitamins. It is a conjugated product containing paraaminobenzoic acid (PABA), pteridine, and one to seven molecules of glutamic acid. The metabolically active forms have reduced pteridine rings and several glutamic acids attached.

Public health. Science and art of preventing diseases, promoting health, and prolonging life through organized community efforts. See *Nutrition, public health.*

PUFA. Abbreviation for polyunsaturated fatty acids. See *Fatty acid.*

Pulmocare®. Brand name of a high-fat, low-carbohydrate product designed to reduce carbon dioxide production in the dietary management of respiratory insufficiency. It is also lactose-free and may be given as an oral supplement or in tube feeding. Contains sodium and calcium caseinates, sucrose, hydrolyzed cornstarch, and corn oil. See also Appendix 39.

Pulmonary embolism. A life-threatening disorder caused by a blood clot developed in another part of the body which has found its way to the lung. Main symptoms are cyanosis, shortness of breath, pallor, fainting, and sometimes edema. *Nutrition therapy:* restore fluid balance and restrict sodium to 2 g/day if edema is present. Follow general dietary principles for *chronic obstructive pulmonary disease.*

Purine. A nitrogenous ring structure widely distributed in nature, especially in *nucleic acids.* Examples are xanthine, hypoxanthine, adenine, guanine, and uric acid. Purines in the body come from the breakdown of nucleic acids (endogenous source) and from purines ingested in food (exogenous source). See *Diet, purine-restricted* and *Gout.*

Pyelonephritis. Inflammation of both the kidney and its pelvis. It is caused by bacterial invasion from the urinary tract, bloodstream, or periureteral lymphatics. *Nutrition therapy:* an *acid-ash diet* is beneficial to increase the acidity of the urine and inhibit bacterial growth. See under *Diet, ash.*

Pyrazinamide. A synthetic antituberculosis agent. It can induce vitamin B_6 deficiency. Neuritis and anemia due to pyrazinamide have been reported. This can be prevented by giving a vitamin B_6 supplement. Excreted in breast milk. Available by nonproprietary name.

Pyridoxal. The aldehyde form of vitamin B_6.

Pyridoxal phosphate (PLP). Also called "codecarboxylase"; a coenzyme that contains vitamin B_6. It is important in many reactions, including amino acid metabolism involving transamination, deamination, decarboxylation, and desulfuration; formation and metabolism of tryptophan and the conversion of tryptophan to niacin and serotonin; hemoglobin synthesis and antibody formation; carbohydrate metabolism, particularly the breakdown of glycogen to glucose; fatty acid metabolism, particularly the conversion of linoleic acid to arachidonic acid; and synthesis of cholesterol and other sterols.

Pyridoxamine. The amine form of *vitamin* B_6.

Pyridoxamine phosphate (PMP). Coenzyme of vitamin B_6 that participates only in transamination reactions.

Pyridoxic acid. Metabolite of pyridoxine that is excreted in the urine. It is used to assess vitamin B_6 nutriture.

Pyridoxine. The alcohol form of vitamin B_6. It is also the collective name given to the three chemically related forms of vitamin B_6: aldehyde (pyridoxal or PL), amine (pyridoxamine or PM), and alcohol (pyridoxine, pyridoxol, or PN). These forms are converted in the liver, erythrocytes, and other tissues to *pyridoxal phosphate* and *pyridoxamine phosphate*. See also *Vitamin B_6*.

Pyridoxol. Former name given to the alcohol form of *vitamin B_6*. The new accepted name is *pyridoxine*.

Pyrimethamine. An antimalarial drug that inhibits dehydrofolate reductase and can produce folic acid deficiency and megaloblastic anemia when given in large doses. It may also decrease serum vitamin B_{12} and may cause loss of appetite, nausea, vomiting, and stomatitis. Excreted in breast milk. Brand names are Daraprim® and Fansidar®.

Pyrimidine. A six-membered ring compound with the two nitrogen atoms separated by a carbon atom. Pyrimidine bases are found in nucleic acids. Examples are cytosine, thymine, and uracil. See also *Nucleic acid*.

Pyrithiamin. Pyridine analog of *thiamin* that is antagonistic to the vitamin.

Pyruvic acid. Also called "ketopropionic acid"; an important compound in the intermediary metabolism of carbohydrate, protein, and fat. Pyruvate is the negatively charged ion and is the key substance or starting point of many reactions, including (1) complete oxidation to carbon dioxide and water, providing energy via the Krebs cycle; (2) oxidative decarboxylation, supplying acetyl CoA, the starting point for lipid metabolism; (3) reversible reduction to lactic acid, i.e., the Pasteur effect; (4) transamination to alanine, an amino acid; and (5) glycogenesis, or the storage of glycogen in the liver and other tissues.

PZI. Abbreviation for protamine zinc insulin.

Q

QAP. Abbreviation for *Quality Assurance Program.*

q.d. Or quotid. Abbreviation for the Latin phrase *quaque die,* meaning "every day."

Q enzyme. Factor isolated from potatoes that catalyzes the formation of branching linkages of the alpha-1,6 type in starches.

q.h. Abbreviation for the Latin phrase *quaque in hore,* meaning "every hour."

q.i.d. Abbreviation for the Latin phrase *quater in die,* meaning "four times a day."

Q. Symbol for quantity or blood volume.

QQ_2. Oxygen consumption per milligram of tissue (dry weight); the amount of oxygen is expressed in microliters at standard pressure and temperature. It is a measure of the rate of respiration of different tissues.

Q substance. Term applied to a giant molecule complex of cholesterol with proteins and lipids that may be associated with the development of atherosclerosis.

Quackery. The actions, claims, or methods of a quack. A *quack* is an untrained person who practices with deception. *Food quackery,* like food faddism, leads people to believe that a particular food has miraculous properties to cure diseases. It makes use of vague, meaningless terms such as "health foods" and "cure-alls" that have scientific overtones and emotional appeal to gullible people.

Quality Improvement (QI). Formerly known as quality assurance (QA). In healthcare facilities, QI refers to any activity designed to evaluate patient care, identify and analyze any deficiency, and plan/implement corrective measures to improve patient care. The Joint Commission on Accreditation for Healthcare Organizations (JCAHO) has a QI standard that requires that "there should be evidence of a well-defined, organized program to enhance patient care and the correction of identified problems." QI is an effective system of monitoring nutrition assessment and care. The 10 steps of a QI process are: (1) develop criteria for optimum care; (2) assign responsibility; (3) delineate the scope of care; (4) identify indicators; (5) establish thresholds for evaluation; (6) collect and organize data; (7) evaluate data related to quality of care; (8) take action to improve care; (9) assess action and document improvements; and (10) communicate findings or outcomes. A more complete QI process is reflected in the term "continuous quality improvement" or CQI, sometimes called "total quality management" or TQM. This more comprehensive process indicates a continuous improvement and systems prevention. The overall objective is focused on customer satisfaction and using resources effectively. A special committee or department is organized to monitor CQI activities in a facility, with a chairperson and department heads as members of the team. However, it is emphasized that CQI is the responsibility of every employee.

Quercetin. A yellow pigment that is a flavone derivative; found in onion skin, tea, red rose, asparagus, the bark of the American oak, and lemon juice. It is used to reduce abnormal capillary fragility.

Quetelet index. The original terminology for the w/h^2 ratio, now more popularly called body mass index. Calculated as weight (kg) divided by height $(m)^2$. See *Body mass index*.

Quinic acid. An acid that is incompletely oxidized in the body; it forms hippuric acid and is excreted in the urine. Found in large amounts in plums, prunes, and cranberries.

Quinidine. An antiarrhythmic agent used to treat atrial fibrillation or flutter. Diarrhea, anorexia, bitter taste, abdominal cramps, nausea, and vomiting occur commonly with quinidine therapy. An alkaline urine delays excretion and may cause toxicity. May cause vitamin K deficiency when taken with anticoagulants. Trade names include Cardioquin®, Cin-Quin®, Duraquin®, and Quinaglute®.

Quinolinic acid. One of the intermediary products formed in the metabolism of tryptophan. The level of its excretion in the urine is a measure of the extent of tryptophan utilization in the body.

Quinone. 1. A substance obtained by the oxidation of quinic acid. 2. Any benzene derivative in which two hydrogen atoms are replaced by two oxygen atoms. Vitamin K is a quinone derivative.

R

Rachitic. Pertaining to or affected with rickets. *Rachitic rosary* is a descriptive term used for the costochondral beading (appearance of nodules or bead-like swelling) on the ribs at the junction with the cartilage often seen in children with rickets. See also *Rickets*.

Radiation. Emission of electromagnetic waves such as those of light or particulate rays, e.g., alpha, beta, and gamma rays. Radiation treatment (e.g., for patients with buccal, esophageal, and gastric cancers) produces loss of taste acuity, decreased appetite, nausea, and vomiting. See *Radiotherapy*.

Radioallergosorbent test (RAST). Diagnostic test for food allergy. It detects specific antibodies against foods by indicating the presence of a Type I allergic reaction (IgE-mediated). It is useful for patients in whom the skin test technique is not recommended. A positive test result must be confirmed by an adverse reaction to a food challenge.

Radiotherapy. The use of radiation (x-rays or gamma rays) in the treatment of neoplastic disease. Adverse side effects of radiation vary according to the area affected. Radiation to the head and neck causes sore throat, mucositis, dry mouth (xerostomia), gum destruction, and altered taste and smell. Radiation to the thorax induces esophagitis and esophageal stricture with accompanying dysphagia. Radiation to the abdomen may produce acute gastritis or enteritis with nausea, vomiting, anorexia, and diarrhea. There may be gastrointestinal damage, malabsorption of glucose, lactose, fats, and electrolytes. Radiation enteritis can develop into a chronic form, with symptoms of ulceration or obstruction intensifying the risk of malnutrition. Anorexia is common, and weight loss is a major problem. As with chemotherapy, radiation depresses immune function and nutritional status.

Raffinose. Naturally occurring trisaccharide composed of a unit each of glucose, galactose, and fructose. It is found in sugar beets, roots and underground stems, cottonseed meal, and molasses. It is only partially digestible and not well utilized by humans, but it can be hydrolyzed by enzymes of the gastrointestinal bacteria in herbivorous animals.

Ranitidine. A histamine H_2 inhibitor used for relief of symptoms associated with active peptic ulcer. Its action reduces gastric acid secretion under daytime and nocturnal basal conditions and also when stimulated by food. It has a bitter taste and may cause transient diarrhea. It may also decrease iron absorption and induce vitamin B_{12} depletion, especially if taken by vegans for an extended period of time. It also increases the risk of bleeding associated with warfarin-induced vitamin K deficiency. The brand name is Xantac®.

RAST. Abbreviation for *radioallergosorbent test*.

RBC. Abbreviation for *red blood cell*. See *Erythrocyte*.

RBP. Abbreviation for *retinol-binding protein.*

RCF®. Ross Carbohydrate Free soy formula base. Used in the dietary management of infants unable to tolerate the type or amount of carbohydrate in milk or conventional infant formulas. The type and amount of carbohydrate that can be tolerated is added to the formula before feeding. See Appendix 37.

RDA. Abbreviation for *recommended dietary allowances.* See Appendix 1.

RDI. Abbreviation for *reference daily intake.*

RDS. Abbreviation for *respiratory distress syndrome.*

RE. Abbreviation for *retinol equivalent.*

Reabilan®. Brand name of a monomeric liquid formula containing peptides, maltodextrin, tapioca starch, soy oil, and MCT oil. Also available in a higher calorie and protein concentration as Reabilan® HN. See Appendix 39.

Recombinant human erythropoietin (r-HuEPO). A product of biotechnology that functions like native erythropoietin and stimulates red blood cell production. Primarily used for the treatment of anemia associated with renal failure; also used to correct anemia induced by injury, infection or inflammation, cancer, AIDS, and rheumatoid arthritis. Administration of r-HuEPO requires an adequate supply of iron for red cell production.

Recommended dietary allowances (RDA). These are "levels of intake of essential nutrients that, on the basis of scientific knowledge, are judged by the *Food and Nutrition Board* to be adequate to meet the nutrient needs of practically all healthy persons" in the United States. Expressed as average daily intakes over time, the RDA is intended to be met by a diet consisting of a wide variety of foods from different food groups. The latest RDA, revised in 1989, is given in Appendix 1. The term "allowance" is used to avoid the implication that these are absolute standards and to emphasize that the levels of nutrient intake recommended are based on a consensus of scientific opinion that should be reevaluated periodically as new information becomes available. Dietary allowances are designed to maintain good nutrition in healthy persons and are based on the average body sizes of adult men and women at different levels of activity, pregnant and lactating women, and children and adolescent boys and girls grouped according to age. It should be noted that dietary allowances are higher than physiologic requirements in order to allow for a safety factor that considers bioavailability of the nutrients and to allow for individual variations. The amount added differs for each nutrient because of variability in the body's ability to absorb and store the nutrient, the range of observed requirements among U.S. population groups, the criteria established for assessing requirements, and the possible hazards of excessive intake of certain nutrients.

Rectal surgery. Any surgical operation done in the rectum, as in rectal cancer or hemorrhoidectomy. *Nutrition therapy:* a clear liquid diet that has no residue immediately before and after surgery. After 2 days, the postsurgical regimen is a monomeric formula or minimal-residue diet to permit wound healing and avoid infections. Depending on the type of surgery and the progress of recovery, gradual resumption of the normal diet is encouraged to provide fiber.

Red blood cell (RBC). See *Erythrocyte.*

REE. Abbreviation for *resting energy expenditure.*

Refeeding syndrome. Also called "nutritional recovery syndrome." Severe fluid and electrolyte imbalances and other complications resulting from the rapid refeeding (either orally, enterally, or parenterally) of individuals who have been malnourished or starved. The characteristic symptoms are severe hypophosphatemia, hypokalemia, hypomagnesemia, thiamin deficiency, hyperglycemia, fluid overload, cardiac dysfunction, and pulmonary and neurologic complications. Patients at great risk for refeeding syndrome include those with chronic alcohol abuse, anorexia nervosa, cachexia due to starvation or marasmus, prolonged fasting, and patients who have not been fed for 7 to 10 days. *Nutrition therapy:* slow refeeding with gradual increases in calories and protein to allow the body to adjust to changes in metabolic load. Estimate the previous intake of calories and start refeeding with this amount or allow about 20 kcal/kg of body weight. Gradually increase calories and protein every day during the first week. If there are no complications, continue with modest food increases to attain the desired weight and to replenish body protein stores. Allow 35 kcal/kg and 1.2 to 1.5 g of protein/kg of body weight.

Reference daily intake (RDI). The standard value for protein, vitamins, and minerals used in nutrition labeling. See Appendix 6.

Reference individuals. The Recommended Dietary Allowances shown in Appendix 1 are expressed in terms of designated weights and heights for "reference individuals" in different age and sex classes. The weights and heights of reference adults are actual medians for the U.S. population of the designated age, as reported by NHANES II. The median weights and heights of those under 19 years of age were based on longitudinal growth studies.

Reference values (RVs). Label reference values for use in nutrition labeling of food. Two sets of reference values have been established, namely: *Reference Daily Intakes (RDIs),* which are values based on the Recommended Dietary Allowances (RDAs) for protein, vitamins, and minerals, and *Daily Reference Values (DRVs),* which are values for eight other nutrients, including fat, cholesterol, and fiber, that are not included in the RDA but are important to health. Previously called U.S. RDA, the RDIs are the same values for vitamins and minerals established in 1973 by the Food and Drug Adminstration (FDA). See Appendix 6.

Reflux esophagitis. Esophageal inflammation and irritation that results from the backflow of the stomach contents into the esophagus. See *Esophagitis* and *Gastroesophageal reflux disease.*

Refsum's disease. Rare genetic disorder due to a defect in the enzyme system for the metabolism of *phytanic acid,* which accumulates in the cerebrospinal fluid, liver, kidney, and blood. The main clinical features are peripheral neuropathy, cerebral ataxia, cataracts, nerve deafness, and skin changes. *Nutrition therapy:* since phytanic acid comes from exogenous sources, the treatment involves restriction of foods containing phytanic acid and its precursors. See *Diet, phytanic acid-restricted.*

Regain®. Brand name of a solid nutritional supplement for patients on chronic hemodialysis with protein-energy malnutrition. It is low in sodium, potassium, phosphorus, and magnesium. Each bar provides 330 kcal and 15 g of protein.

Regional enteritis. One of the two types of inflammatory bowel disease (IBD). See *Crohn's disease.*

Rehydralyte®. Brand name of an oral electrolyte solution for rehydration of fluids and electrolytes lost in moderate to severe diarrhea. See Appendix 38.

Relapse. Return of symptoms after a disease seems to have been cured. The relapse–response–relapse method is one of the criteria for determining the nutritional essentiality of a substance. For example, deficiency signs of a specific nutrient disappear when the nutrient is administered; withdrawal of the nutrient leads to recurrence of the signs, and therapeutic supplementation again alleviates the deficiency signs. Thus the proof of essentiality is a series of alternate relapse–response–relapse reactions to the nutrient.

Renal acidosis. Reduction or lack of alkali reserve caused by the inability of the kidney to conserve base while excreting acid. See *Acidosis*.

Renal calculi. See *Urolithiasis*.

Renal diabetes. Type of diabetes characterized by the presence of sugar in the urine even with a normal or below-normal blood sugar level. The condition is not associated with a disturbance in carbohydrate metabolism. It is ascribed either to a low renal threshold for sugar or to a defective reabsorption process in the kidney tubules, allowing glucose to find its way into the urine. This condition is more appropriately called *renal glycosuria*. See also *Phlorizin*.

Renal dialysis. The removal of toxic materials from metabolism from the blood and body fluids by mechanical means. Peritoneal dialysis uses the peritoneal membrane, while hemodialysis uses an artificial kidney and an extracorporeal dialysis method. Self-dialysis methods used at home are *continuous ambulatory peritoneal dialysis (CAPD)* and *continuous cycler peritoneal dialysis (CCPD)*. The comparative nutritional effects of hemodialysis (HD) and peritoneal dialysis (PD) are as follows: HD has no protein loss into the dialysate, while in CAPD, about 9 g of protein/day is lost; however, there is less loss of amino acids in PD (2 to 4 g/day) compared to 5 to 8 g/day in HD. With both methods, water-soluble vitamins are lost. For further details on dietary management, see *Hemodialysis* and *Peritoneal dialysis*.

Renal erythropoietic factor (REF). An enzyme released by the kidney that acts on the plasma protein globulin to split off erythropoietin, which in turn acts on the bone marrow to stimulate red blood cell production.

Renal failure. The inability of the kidneys to carry out their many functions, such as excretion of more than 200 waste products, excess fluid and drugs or poisons, regulation of blood pressure, maintenance of acid–base balance and bone health, and production of the hormone erythropoietin that stimulates red blood cell production. Renal impairment results in edema, uremia, anemia, hypertension, bone disease, and metabolic acidosis. The types of renal failure are acute, chronic, and end-stage.

Acute renal failure (ARF). Characterized by sudden onset of symptoms and rapid deterioration of renal function. It may be due to shock or sudden loss of blood supply to the kidneys, trauma, bacterial infection, effects of toxic drugs, or may be part of multiorgan system failure. Symptoms include uremia, metabolic acidosis, and fluid and electrolyte imbalance. It is often associated with oliguria or anuria, but can occur with normal urine flow. The condition is life-threatening and needs immediate aggressive therapy. The kidneys usually recover with nutrition therapy and treatment by peritoneal or hemodialysis. *Nutrition therapy:* in acute renal failure without dialysis, allow 0.5 to 0.6 g of protein/kg actual body weight, but not less than 40 g per day. Increase protein up to 0.8 to 1.0 g/kg/day as the glomerular filtration rate returns to normal. Critically ill, hypercatabolic patients may require 1.0 to 1.3 g/kg/day of protein. If the patient is on dialysis, provide 1.0 to 1.5 g of protein/kg body weight. Insulin may be required to control blood glucose. Provide 30 to 35 kcal/kg body weight for weight maintenance, and 40 to 50 kcal/kg for repletion or if the patient is hypercatabolic to prevent use of protein for energy. Supplement with nonprotein calories from fats and simple carbohydrates. In the anuric–oliguric phase, allow 500 to 1000 mg (20 to 40 mEq) sodium, 1200 to 2000 mg (30 to 50 mEq) potassium, and fluids equal to output plus 500 mL. Increase fluid in fever and other fluid losses. In the diuretic phase, sodium is restricted to 1500 to 2000 mg (60 to 90 mEq), and potassium is about 2500 mg (65 mEq) per day if hyperkalemia exists. Adjust phosphorus intake according to laboratory values. Give parenteral nutrition to supplement inadequate oral intake or if the patient is unable to eat or tolerate oral feedings. Continuous arteriovenous hemofiltration is being increasingly used in critical care units to facilitate nutritional support of patients with acute renal failure.

Chronic renal failure (CRF). Results from the progressive deterioration of kidneys over a period of months or years, with permanent impairment of renal functions. Causative factors include glomerulonephritis, nephrosclerosis, pyelonephritis, obstructive renal diseases, diabetes mellitus, systemic lupus erythematosus, and drug abuse. CRF progresses until treatment with dialysis or transplantation is required. Dietary restriction is generally not recommended if the glomerular filtration rate (GFR) is greater than 70 mL/minute. However, patients showing progressive decline in GFR will benefit from early reductions in protein and phosphorus intake to slow down the rate of renal deterioration. It may be prudent to keep protein intake at the RDA level of 0.8 g/kg body weight and limit foods high in phosphorus. Further restriction is necessary when the glomerular filtration rate or creatinine clearance approaches 30 mL/minute. *Nutrition therapy:* for CRF without dialysis, energy intake is individualized per kilogram of body weight per day at 35 kcal/kg for weight maintenance, 45 kcal/kg if the patient is underweight or catabolic, and 20 to 30 kcal/kg if the patient is obese. For obese patients who are greater than 125% of desirable body weight, calculate energy needs using an adjusted body weight. Protein is from 0.6 to 0.8 g/kg actual body weight (corrected for edema) plus the amount lost in the urine if proteinuria is present. Fat contributes about 30% of total calories, with a polyunsaturated-to-saturated fatty acid ratio of 1:1. The remainder of calories are supplied by carbohydrates. Sodium is restricted to 1.5 to 2.0 g (65 to 90 mEq) per day if edema is present. Potassium is not restricted but if serum level is high, limit to 2700 mg/day (70 mEq). Balance fluid intake with urine output in patients with edema or congestive heart failure. Recommended amounts for minerals and vitamins are: 1200 to 1600 mg/day elemental calcium (including supplements); 8 to 12 mg/kg/day phosphorus; 100 mg/day elemental iron if needed with erythropoietin therapy; 20 mg of niacin equivalents, 2 mg thiamin, 2 mg riboflavin, 10 mg of pantothenic acid, 5 mg of pyridoxine, 200 μg of biotin, 0.8 to 1 mg folic acid, 3 μg vitamin B_{12}, and 60 mg vitamin C. Supplement with vitamin K (10 mg every other day) if the patient is on an antibiotic and not eating. Vitamin A should be avoided; there is evidence of increased storage in uremia. As renal failure progresses, supplementation with vitamin D and the use of phosphate binders, in addition to dietary phosphorus restriction, may be necessary. Dietary recommendations for infants, children, and adolescents are the following: calories per kilogram of body weight for infants, 105 to 115 kcal; 1 to 3 years, 100 kcal; 4 to 10 years, 85 kcal; 11 to 14 years, 60 kcal for boys and 48 kcal for girls; 15 to 18 years, 42 kcal for boys and 38 kcal for girls. Grams of protein per kilogram of actual body weight per day: birth to 1 year, 2 to 3 g/kg; 1 to 2 years, 2 g/kg; 2 to 6 years, 1.2 g/kg; 7 years to adolescence, 1 g/kg. Fluid intake is not restricted unless severe edema is present; increased fluid may be necessary for children with a renal concentrating defect. Restrict sodium to 2 to 4 mEq/kg/day in infants and 2 to 3 g/day in children and adolescents, if needed to control hypertension and edema. Avoid foods high in phosphorus and potassium; if needed, restrict potassium to 2 to 4 mEq/kg. Calcium, iron, and vitamins are at the RDA levels.

End-stage renal failure (ESRF). More commonly called end-stage renal disease (ESRD). The last phase of renal failure when the glomerular filtration rate decreases to 10 mL/minute and the kidneys lose 90% or more of their functional ability. Treatment modalities include conservative management, kidney transplantation, hemodialysis, and peritoneal dialysis. *Nutrition therapy:* without dialysis treatment, dietary protein is restricted to 0.4 to 0.5 g or lower per kilogram of body weight, or an alternative diet that provides about 0.25 g of essential amino acids (EEAs) or a mixture of essential and nonessential amino acids (NEAAs), which allows a slightly higher protein intake of up to 0.6 g/kg/day. Use protein of high biologic value and provide sufficient calories to spare protein. Amino acid solutions may be beneficial when dialysis is delayed; however, benefits obtained from long-term use have not been clearly established. Without dialysis treatment or kidney transplantation, severe protein restriction for an extended period will eventually produce negative nitrogen balance and malnutrition. See *Kidney transplantation, Hemodialysis,* and *Peritoneal dialysis.*

Renal hypertension. Elevation in blood pressure as a result of reduction of blood flow to the kidney, as in ischemia. The ischemic kidney liberates into the blood an enzyme, *renin,* that splits angiotensin I from angiotensinogen, an alpha globulin formed in the liver. An enzyme present in the plasma acts on angiotensin I to form angiotensin II, which is a powerful pressor agent. However, the kidney tissues also contain a dipeptidase enzyme, angiotensinase, that destroys angiotensin II.

Renal insufficiency. Partial or mild kidney failure characterized by less than normal urine excretion. Hypoalbuminemia, albuminuria, and a decrease of glomerular filtration rate (GFR) are laboratory findings indicating early renal damage. Renal insufficiency is associated with Type I diabetes mellitus. If uncontrolled, it leads to diabetic nephropathy. *Nutrition therapy:* a low-protein, low-phosphorus diet decreases albuminuria and the progression of renal insufficiency. Allow 0.8 g of protein per kilogram of body weight, 1 g phosphorus, and 1 g calcium per day. Calculate energy needs to maintain weight, providing 12% to 20% of kilocalories from protein, 50% to 60% from carbohydrates, and less than 30% from fat. For persons with diabetes, control blood glucose (70 to 140 mg/dL) with proper diet and adequate medication. Monitor hypertension; sodium restriction may be beneficial.

Renal osteodystrophy. Also called renal bone disease. A generic term for bone lesions associated with renal failure. It develops in the early stages of renal insufficiency. There are four types of lesions:

osteitis fibrosa, osteomalacia, adynamic bone disease, and mixed lesions. The latter is the most common renal osteodystrophy, occurring in about 45% to 85% of patients undergoing dialysis. Main characteristics are high levels of PTH (parathyroid hormone), impaired bone formation, demineralization, and vitamin D deficiency. *Nutrition therapy:* treatment includes control of calcium and phosphorus, use of vitamin D, and avoidance of aluminum. Calcium intake is kept high and phosphorus intake is kept low. Because most of the high-calcium foods are also high in phosphorus, calcium is increased by giving calcium supplements, in addition to the 300 to 500 mg of calcium provided in the diet. Phosphate intake is restricted to 1200 mg or less. In the case of a high phosphorus blood value, restrict phosphorus further to 800 to 900 mg/day and/or use phosphate binders. Calcium-containing binders are preferred to aluminum-containing ones. The latter tend to cause aluminum overload with adverse side effects. Vitamin D (10,000 to 30,000 IU/day) is given only when the hypocalcemia of renal failure is severe or causing osteomalacia. See *Diet, phosphorus-restricted.*

Renal solute load (RSL). The amount of urea, sodium, potassium, and chloride in the urine. These solutes require water for urinary excretion. In pediatric formula feeding, the renal solute load is related to the fluid requirement and the protein and mineral content of the formula. About 1 mL of free water is required to excrete 1 mOsm of RSL. If the renal solute load is too high and fluids are restricted, hypertonic dehydration will occur. The kidneys of a normal adult have the ability to concentrate urine to 1200 to 1400 mOsm/liter. In kidney disease or in immature infants, the kidneys need a higher fluid requirement for an equivalent RSL. A person with hypercatabolism has an increased RSL. Each gram of dietary protein contributes about 4 mOsm of renal solute, and each milliequivalent (mEq) of Na, K, and Cl contributes 1 mOsm. To calculate the RSL, the following equation is used for children and anabolic patients:

$$\text{RSL} = (\text{g of protein} \times 4) + [\text{Na(mEq)} + \text{K (mEq)} + \text{Cl (mEq)}]$$

To calculate the RSL for adults, multiply grams of protein by 5.7 in the above equation.

Renal threshold. Concentration of a substance in plasma above which the substance appears in the urine. Various substances in plasma, such as glucose, do not appear in the urine until their plasma concentrations rise to certain values. Such substances are referred to as *threshold substances*. The renal threshold of glucose in the adult varies between 140 and 170 mg/dL.

Renal transplantation. See *Kidney transplantation*.

RenAmin®. Brand name of a 6.5% amino acid solution for total parenteral nutrition; has no electrolytes and contains 60% essential and 40% nonessential amino acids. Designed for use in renal disease and hepatic encephalopathy. See Appendix 40.

Re/Neph™. Brand name for a shelf-stable formula for potassium, phosphorus, and fluid-restricted diets. It is also low in sodium, calcium, and saturated fats. Also cholesterol-free and lactose-free. See Appendix 39.

Renin. Proteolytic enzyme formed in the kidney and released into the blood. It liberates angiotensin I from its inactive precursor, angiotensionogen. See *Renal hypertension*.

Rennin. Enzyme present in the gastric juice of infants and young animals that is primarily responsible for the coagulation of milk (casein). It is capable of clotting about 10 million times its weight of milk at 37°C in 10 minutes. The process of clotting involves a change in the casein molecules to *paracasein,* which then forms calcium paracaseinate, or the milk clot.

Replete®. Brand name for a high-protein, lactose-free formula for patients with pressure ulcer, burns, multiple trauma, and other healing needs. Also available with added fiber (14 g/1000 mL). See Appendix 39.

RES. Abbreviation for *reticuloendothelial system.*

Reserpine. A rauwolfia alkaloid used in the treatment of hypertension. It may cause sodium and fluid retention with edema and weight gain, especially if the drug is not taken with a diuretic. It may also cause dry mouth, increased hunger, increased gastric secretion, and intestinal cramping. Brand names are Releserp®, Sandril®, Serpate®, Serpalan®, and Serpasil®.

Residue. The remainder; the portion remaining after a part has been removed. In nutrition, it refers to the amount of bulk remaining in the intestinal tract following digestion. It is composed of undigested and unabsorbed food, as well as metabolic and bacterial products. Sloughed gastrointestinal cells and intestinal bacteria make up a large part of the residue. See also *Diet, residue-modified.*

Resol®. Brand name of an oral electrolyte solution for infants. See Appendix 38.

Resource®. Brand name of a liquid lactose-free, low-residue formula for oral supplementation or tube feeding. Contains sodium and calcium caseinates, soy protein isolates, maltodextrin, sucrose, and corn oil. Also available in a 1.5 kcal/mL concentration (Resource Plus®). See Appendix 39.

Respalor®. Brand name of a specialized formula designed for respiratory patients, for oral or tube feeding. It is lactose-free and provides 1.5 kcal/mL. See Appendix 39.

Respiratory carriers. Group of electron carriers, including coenzymes I and II (nicotinamide adenine dinucleotide and nicotinamide adenine dinucleotide phosphate), the flavoproteins (flavin mononucleotide and flavin adenine dinucleotide), coenzyme Q, and the cytochrome systems (cytochrome a_1, a_3, b, and c). They convey electrons from the dehydrogenated substrates to oxygen, harnessing free energy in the course of the reaction for the synthesis of adenosine triphosphate (ATP), a form of energy utilized in the endergonic processes of a living cell.

Respiratory distress syndrome (RDS). A secondary lung condition resulting from pulmonary edema, sepsis, shock and other trauma, or critical illnesses. Specifically among infants, RDS refers to an acute lung disorder caused by a deficiency of pulmonary surfactant resulting in inelastic lungs, distended alveoli, and severe hypoxemia, hence its other name, "live membrane disease." RDS occurs in about 1% of all live births, 10% of all preterm babies, 60% of low-birth-weight infants weighing below 1000 g, and five times more often in infants of diabetic mothers. *Nutrition therapy:* see *Nutrition, prematurity.*

Respiratory enzymes. Enzymes found in the *mitochondria* that catalyze a series of reactions involved in the cellular oxidation of substrates, resulting in their complete oxidation to carbon dioxide and water. Electrons removed from the substrates are passed on to a highly ordered array of *electron* or *respiratory carriers* and thence to oxygen, forming water. The principal types of respiratory enzymes are the *oxidases* and *dehydrogenases*.

Respiratory failure. Inability of the pulmonary system to maintain an adequate exchange of oxygen and carbon dioxide in the lungs. Oxygenation failure is characterized by hyperventilation and occurs in diseases involving the alveoli. It is a common feature of critical illness, acute pulmonary infection, and severe chronic obstructive pulmonary disease. Ventilatory failure is due to increased arterial tension of carbon dioxide. Mechanical ventilation may be required. A healthy nutritional status is linked to strong respiratory muscles and endurance. With respiratory failure, extra work for breathing is increased by 30%. Malnutrition may lead to respiratory failure and respiratory failure may be aggravated by malnutrition.

Acute respiratory failure. Sudden absence of respiration with failure of gas exchanges in the lungs. May be caused by drugs, muscular dystrophy, gross obesity, asthma, pneumonia, pulmonary embolism, Guillain-Barré syndrome, intracranial disorder, and trauma. The patient is often confused and may be unresponsive. *Nutrition therapy:* adequate supply of nutrients, but with decreased carbon dioxide production. Provide a high-calorie diet (35 to 40 kcal/kg/day for anabolism), with equal amounts of carbohydrate and fat. Tube feeding of low osmolality is initially given. Special care should be taken to avoid aspiration. Continuous feedings and elevation of the head of bed are recommended. When enteral feeding is not possible, parenteral nutritional support is required, especially when the patient is malnourished. Introduce oral feeding gradually in frequent small feedings. If pulmonary edema occurs, restrict sodium to 2 g/day. Mineral and vitamin supplements may be required.

Chronic respiratory failure. Or chronic respiratory insufficiency, which includes pulmonary diseases such as chronic obstructive pulmonary disease (COPD) and emphysema, or is secondary to congestive heart failure. Chronic respiratory failure may lead to cor pulmonale. Several factors may result in decreased nutrient intake in chronic respiratory insufficiency, such as: intestinal distress due to COPD or effects of drugs; chronic sputum production, which may alter desire for and taste of food; a full stomach makes breathing difficult after a large meal; and shortness of breath may decrease food intake. *Nutrition therapy:* afebrile, mechanically ventilated patients need additional calories of 30% to 35% above basal energy expenditure (BEE) and protein intake of 1.2 g/kg of body weight. In malnourished patients, energy intake should be 1.4 to 1.6 × BEE and protein is 1.5 g/kg/day. Distribute calories from carbohydrate, protein, and fat to minimize carbon dioxide production, especially during weaning from mechanical ventilation. Fat may be as high as 40% of total calories and a ratio of 150:1 nonprotein calories per gram of nitrogen is recommended. Give small meals of nutrient-dense foods, but avoid overfeeding. Use of commercial supplements and modular products may be helpful to patients who are unable to eat adequately.

Respiratory quotient (RQ). Ratio of the volume of carbon dioxide eliminated to the volume of oxygen used. One gram molecule of carbon dioxide has the same volume as one gram molecule of oxygen. The complete oxidation of carbohydrate gives a respiratory quotient of 1. Oxidation of fat and oxidation of protein give approximate respiratory quotients of 0.7 and 0.8, respectively. Because the three foodstuffs are metabolized simultaneously, the respiratory quotient is always a result of the three. It is about 0.85 on an ordinary mixed diet. See also *Metabolic cart*.

Resting energy expenditure (REE). The amount of energy expended by a person in 24 hours when at rest and 3 to 4 hours after a meal. The REE includes some energy expended for digestion and absorption, which is about 10% higher than the basal energy expenditure (BEE). This difference is not considered significant in clinical practice. The two terms, REE and BEE, are often used interchangeably. REE correlates directly with measures of lean body mass and is the largest component of total energy expenditure in normal activities.

Resting metabolic rate (RMR). The resting energy expenditure expressed as kcal/kg body weight/hour. The RMR includes some energy expended for digestion and absorption. Unlike the *basal metabolic rate (BMR)* which is taken at complete rest and after at least a 12-hour fasting, the RMR is measured in a nonfasting state but not less than 2 hours after a light meal and after resting in the supine position for more than 30 minutes. Because the RMR includes some energy expended for digestion and absorption, it is about 10% higher than the BMR.

Reticuloendothelial system (RES). Group of cells (except for leukocytes) with phagocytic properties. They are distributed throughout the body, particularly in the spleen, bone marrow, liver, and lymph nodes. These cells have the property of engulfing and digesting foreign particles or cells harmful to the body. The reticuloendothelial system also removes red blood cells on their destruction. Protein and iron are recovered for formation of new erythrocytes, and the heme portion is converted to *bile pigments*. See also *Hemopoiesis*.

Retinaldehyde. Name now used for vitamin A aldehyde, retinal, or retinene.

Retinoic acid. Name for vitamin A_1 acid, formed by the oxidation of vitamin A aldehyde. It has no activity in the visual process or the reproductive system and cannot be stored in the body.

Retinoids. A group of compounds consisting of four isoprenoid units and containing five conjugated double bonds. It is the collective term for the various forms of vitamin A activity (*retinol, retinaldehyde,* and *retinoic acid*) and a large number of synthetic analogs, with or without vitamin A activity. Retinoids vary qualitatively as well as quantitatively in vitamin A activity.

Retinol. Vitamin A_1 alcohol. Its corresponding aldehyde and acid are called "retinaldehyde" and "retinoic acid," respectively. Retinol and retinaldehyde can be reversibly oxidized and reduced, but retinoic acid cannot be converted back to the other two. Retinol circulates in the blood as a complex with retinol-binding protein (RBP) and transthyretin (TTR). The Recommended Dietary Allowances (RDAs) for vitamin A are given in retinol equivalents (REs), which include the preformed vitamin A and its precursor, beta-carotene, and other provitamin A carotenoids. See Appendix 1.

Retinol-binding protein (RBP). A plasma protein that binds and transports vitamin A, in the form of *trans*-retinol, from the liver to extrahepatic tissues. The binding of vitamin A to the protein serves to solubilize vitamin A and protect it against oxidation. RBP has a half-life of only 12 hours. The normal range in plasma is 30 to 70 mg/liter. It is decreased in vitamin A deficiency, liver disease, trauma, acute catabolic states, hyperthyroidism, and after surgery; it is increased in renal failure due to impaired glomerular filtration and decreased renal metabolism.

Retinol equivalent (RE). A measure of the total vitamin A activity. It takes into account the amount of absorption of the carotenes, as well as the degree of conversion to vitamin A. One RE equals 1 μg of retinol, 6 μg of beta-carotene, and 12 μg of other provitamin A carotinoids. The RDA standards in retinol equivalent for vitamin A include the preformed vitamin A and its precursors, beta-carotene and other provitamin carotenoids. See *Vitamin A* and also Appendix 1.

Retinopathy. Noninflammatory disease of the retina; characterized by retinal detachment, a waxy discharge, and a tiny amount of hemorrhaging. The nonproliferative type of retinopathy is more common and is confined to the retina. The proliferative type, which may occur in about 10% of diabetics, is characterized by the retinal capillaries extending into the vitreous, whereby the person's sight is impaired, eventually leading to blindness.

Retrolental fibroplasia (RLF). Also called "retinopathy of prematurity." A form of vitamin E deficiency to which premature or low-birth-weight infants are prone, with an incidence of about 10% in these infants. The occurrence is reduced with daily doses of vitamin E. The anemia seen among premature babies due to lack of body stores can be treated with vitamin E, iron, and folic acid.

Rheumatic fever. Acute or chronic inflammatory process that comes as a sequel to hemolytic streptococcal infection, usually after 3 to 4 weeks. It occurs most frequently in children and tends to recur. The inflammatory process initially affects connective tissues, but may spread to many organs, and when pronounced, conditions such as myocarditis and arthritis occur. *Acute rheumatic fever* is characterized by a sudden onset, with high fever and swelling and pain in the joints. Recurring episodes of rheumatic fever lead to damage of the heart muscle and heart valves, a disorder called "rheumatic heart disease". *Nutrition therapy:* follow guidelines for *fevers*. Calorie and protein intakes are individualized for the febrile condition and to recover from weight loss. If edema is present and/or steroids are used, restrict sodium intake to 2 g/day. Supplement with vitamins and minerals, especially vitamins A and C, potassium, magnesium, and zinc.

Rheumatoid arthritis. Also called "arthritis deformans" and "atrophic arthritis"; a painful, chronic autoimmune disorder common among women over the age of 30 and characterized by swelling, stiffness, and eventual deformity of the joints. Manifestations vary in severity and can begin at any age. Some patients have fever and other systemic manifestations, leading to anorexia and weight loss. Patients with rheumatoid arthritis are generally underweight. Small joints of the hands are frequently affected and this hinders independence with eating; involvement of the temporomandibular joint limits opening and closing of the mouth; there may be dysphagia and decreased salivary secretion; necessary medications may cause gastritis and mucosal damage, resulting in malabsorption of nutrients; and urinary excretion of vitamin C is increased. *Nutrition therapy:* the main goal is to prevent malnutrition and encourage the person to eat adequately. Provide adaptive feeding devices to assist with self-feeding. For calories, give basal energy expenditure (BEE) × 1.3 to 1.4 for patients with limited activity receiving physical therapy, and BEE × 1.5 to 1.6 for those receiving intensive daily physical therapy. Protein is at the RDA level; increase protein to 1.5 g/kg body weight if the person is malnourished and during the acute inflammatory phase. Restrict sodium to 2 to 3 g/day if there is fluid retention because of immobility or secondary to medication, especially steroid therapy. Ensure adequate intake of vitamins and minerals, particularly vitamin C, pyridoxine, vitamin D, potassium, calcium, selenium, and zinc. The hypochromic anemia frequently associated with arthritis is not related to dietary iron deficiency and does not respond to iron therapy; iron supplementation is not indicated and may exacerbate the arthritis. If joint pain is a manifestation of food allergy, eliminate the offending food from the diet. About 10% to 15% of patients respond to a milk-free diet. An increase in the consumption of seafoods and fish oils rich in omega-3 fatty acids may help diminish joint inflammation and morning stiffness. Monitor the side effects of drugs and avoid gastrointestinal side effects by giving food or milk with drugs.

Rheumatoid cachexia. Extreme weight loss and severe muscle wasting associated with rheumatoid arthritis. The changes are identical to those seen in many chronic illnesses. The resting energy expenditure (REE) is higher than normal by 12% and increased cytokine production is associated with loss of body-cell mass.

Rhodopsin. Formerly called "visual purple"; the pigment in the rods of the retina that contains vitamin A. On exposure to light, it is bleached through a series of products, forming at the end opsin and another pigment known as "retinaldehyde (visual yellow)." As a result of these changes, images are transmitted to the brain through the optic nerve. Vitamin A is required for the regeneration of rhodopsin.

Riboflavin. Also called "vitamin B_2"; one of the water-soluble B-complex vitamins. It was isolated from milk (lactoflavin), eggs (ovoflavin), and liver (hepatoflavin) and identified with Warburg's yellow enzymes. Riboflavin is associated with the health of the skin and eyes. As an essential component of two flavoprotein coenzymes, flavin mononucleotide (FMN) and flavin adenine dinucleotide (FAD), it participates in biologic oxidation as a hydrogen acceptor in aerobic dehydrogenases. Riboflavin is also involved in the activation of vitamin B_6 and the conversion of folic acid to its coenzymes. Deficiency in the vitamin results in *ariboflavinosis,* which is characterized by burning and itching of the eyes, photophobia, corneal vascularization, generalized seborrheic dermatitis, cheilosis, angular stomatitis, and glossitis with a characteristic magenta color of the tongue. It is rare to find riboflavin deficiency as an isolated and pure deficiency state. Secondary riboflavin deficiency usually results from malabsorption; diseases such as acute and chronic infection, burns, dialysis, and malignancy, which increase riboflavin requirement and cause tissue depletion; and use of drugs such as phenothiazine, tricyclic antidepressants, barbiturates, and some antibiotics, which increase urinary excretion of the vitamin. Because riboflavin is essential to the functioning of vitamins B_6 and niacin, some symptoms attributed to riboflavin deficiency are due to the failure of systems requiring these vitamins. Manifestations of riboflavin deficiency in animals vary with the species. Riboflavin is widely distributed in foods, but in small amounts. The richest food sources are milk, eggs, meats, poultry, and fish; green vegetables such as broccoli, spinach and turnip greens; and enriched grains and cereals are other good sources. It occurs in bound form in plant and animal tissues and is not made available unless binding is liberated by cooking. The vitamin is stable in heat and only slightly soluble in water but is readily destroyed in the presence of light and alkali. It is absorbed in the proximal small intestine and transported either attached to protein or linked with a phosphate molecule. There is little storage in the body, and the amount stored depends on the saturation of protein. No cases of toxicity from ingestion of riboflavin have been reported; toxicity may occur if it is given in massive doses by injection. See Appendix 1 for the recommended dietary allowances.

Ribonucleic acid (RNA). Also called "ribose nucleic acid"; one of the two main types of *nucleic acid,* with ribose as the pentose constituent. It is present in the cell cytoplasm and nucleolus and plays an important role in protein synthesis. Three distinct types of RNA are involved in *protein synthesis,* with varying size, shape, origin, and function.

Messenger RNA (mRNA). Also called "template RNA" or "informational RNA"; conceived to be a complementary copy of DNA, thus containing the genetic "information" for protein synthesis. mRNA functions at the site of protein synthesis and carries the genetic message from DNA to the *ribosome,* the site of protein synthesis.

Ribosomal RNA (rRNA). Ribonucleic acid in the ribosome, which is believed to direct the arrangement of the amino acids of proteins into their proper sequence within the polypeptide chain.

Transfer RNA (tRNA). Also called "soluble RNA" or "acceptor RNA." It occurs free in the cytoplasmic fluid and transfers the activated amino acid to a specific site on the ribosomal RNA template, resulting in an alignment of amino acids in a particular sequence to form the primary structure of a protein.

Ribose. Pentose sugar of significant physiologic importance. It is a constituent of ribonucleic acid, of the coenzymes nicotinamide adenine dinucleotide (NAD) and nicotinamide adenine dinucleotide phosphate (NADP), and of adenosine triphosphate (ATP). Any glycoside containing ribose as the sugar component is called a *riboside.*

Ribosome. Site of protein synthesis within the cells; contains 80% to 90% of the ribonucleic acid within the cell. As seen by electron microscopy, it appears as a delicate network of membranous tubules attached to numerous dense, spherical granules with diameters of 100 to 150 Å.

Ricelyte®. Brand name for an oral electrolyte solution made with rice syrup solids; may be used for maintenance of water and electrolytes lost during diarrhea and vomiting. See Appendix 38.

Ricinoleic acid. A monohydroxy monounsaturated fatty acid containing 18 carbon atoms; found in castor oil.

Rickets. A nutritional deficiency disease occurring in infancy and early childhood due to a lack of vitamin D or a disturbance in calcium–phosphorus metabolism. It is more likely to develop in dark, overcrowded sections of large cities where the ultraviolet rays of sunshine, especially in the winter months, cannot penetrate the fog, smoke, and soot. Rickets is essentially a disease of defective bone formation. The characteristic symptoms are delayed closure of the fontanelles; poor muscle tone, resulting in a "pot-belly" appearance of the abdomen; soft, fragile bones, leading to bowing of the legs; costochondral beading at the junction of the rib joints, forming the "rachitic rosary"; and projection of the sternum, giving the appearance of a "pigeon breast." Conditioned rickets may occur as a secondary result of other diseases, as in fat malabsorption of celiac disease, in renal failure, and in certain genetic disorders such as familial hypophosphatemia or vitamin D-refractory rickets. *Nutrition therapy:* in vitamin D-deficiency rickets, treatment is with 4000 IU (100 μg) of vitamin D daily for 4 to 8 weeks, which is about 10 times the RDA. After the initial period of high vitamin D therapy, ensure a daily intake of at least 400 IU (10 μg) vitamin D and adequate intakes of calcium and phosphorus at the RDA level to prevent recurrence. Human milk is deficient in vitamin D and contains only 22 IU (0.55 μg) per liter. Breast-fed infants, therefore, require an additional source of the vitamin, such as casual exposure to sunlight. Fortified cow's milk and most infant formulas contain 400 IU (10 μg) per quart. Several types of vitamin D-resistant rickets require large and variable amounts of vitamin D and high intakes of calcium and phosphate to achieve a satisfactory response.

Rifampin. An antibacterial used in the treatment of tuberculosis. The most frequent adverse effects include anorexia, altered taste sensation, abdominal cramps, epigastric distress, and diarrhea. Rifampin is distributed into milk and interferes with tests for serum folate and vitamin B_{12}. Its absorption is reduced by food; give 1 hour before or 2 hours after eating food to ensure maximum absorption. Trade names are Rifadin®, Rifamate®, and Rimactane®.

RLF. Abbreviation for *retrolental fibroplasia*.

RMR. Abbreviation for *resting metabolic rate*.

RNA. Abbreviation for *ribonucleic acid*. Abbreviations for the three types of RNA are mRNA, rRNA, and tRNA for messenger RNA, ribosomal RNA, and transfer RNA, respectively.

Ross SLD®. Brand name of a powdered mixture of eggwhite solids, sucrose, and hydrolyzed cornstarch; readily soluble in water and suitable as a protein source for patients on a clear liquid diet. See also Appendix 39.

Roughage. Now called "insoluble dietary fiber." Indigestible carbohydrate material in plants; it passes through the intestines unchanged but absorbs and holds water, thus acting as a laxative. Usually composed for the most part of cellulose, an indigestible polysaccharide. See also *Dietary fibers* under *Fiber*.

R-protein. Protein produced by the salivary glands found to enhance the absorption of vitamin B_{12}.

RQ. Abbreviation for *respiratory quotient*.

RV. Abbreviation for *reference value*.

S

S. Chemical symbol for sulfur.

Saccharin (Saccharine). Artificial sweetener about 400 times sweeter than cane sugar or sucrose. It has no caloric value, but some find a bitter aftertaste; possibly a carcinogenic substance, but the risk from ingestion of moderate levels is considered extremely small. Sold under the brand names Sweet & Low® and Sugar Twin®.

Saliva. Secretion of the *salivary glands* in the mouth. It serves to moisten and hold particles of food together, thus aiding chewing and swallowing. It contains an enzyme, *salivary amylase,* that helps to digest starch. The salivary secretion also has some bacteriostatic properties.

Salivary amylase. Formerly called "ptyalin." It is the principal enzyme of human saliva, acting on starches, with oligosaccharides and maltose as the end products. It requires chloride ion and an optimum pH of 6.6 to 6.8 for its action. Digestion in the mouth is not appreciable, as food stays in contact with the enzyme for only a short time. Salivary amylase is easily inactivated at a pH of 4.0 or less, and its action ceases when food enters the stomach.

Salt substitutes. These are mineral bases consisting of potassium chloride, calcium chloride, and ammonium chloride. Potassium-based salt substitute is a good source of potassium for patients on diuretics requiring potassium supplementation but it could be harmful to patients with renal insufficiency. "Low-sodium" salt substitutes contain half as much sodium as regular table salt; they can be used in mild sodium-restricted diets. See Appendix 36.

Saturation. 1. Point beyond which a solution can no longer dissolve a given substance. 2. Property of having all the chemical affinities satisfied. For example, in *saturated fatty acids,* all the carbons in the chain have all the hydrogen linked to carbon.

Saturation test. See *Load test.*

SBS. Abbreviation for *short bowel syndrome.*

SCFA. Abbreviation for short-chain fatty acid. See under *Fatty acid.*

Schilling test. Procedure used in the differential diagnosis of macrocytic anemias to determine an individual's ability to absorb vitamin B_{12}. After an oral dose of radioactive vitamin B_{12} is given, the urinary excretion of the vitamin is low in patients with *pernicious anemia.* When the same test is repeated with intrinsic factor also given orally, the urinary excretion becomes almost normal in these patients because the vitamin is absorbed; it remains low in vitamin B_{12} deficiency due to other causes, such as malabsorption syndrome and intestinal resection.

School Breakfast Program. Established in 1966 as part of the *Child Nutrition Act.* It provides nutritious low-cost breakfasts to children in participating schools. All public schools and voluntary private schools may participate, but the qualifying family income must be between 130% and 185% of poverty level.

Administered by the U.S. Department of Agriculture.

School Lunch Program. More appropriately called the "National School Lunch Program." The National School Lunch Act became a law in 1946 (and is the largest of the Child Nutrition Programs). See *National School Lunch Program.*

Scleroprotein. Group of simple proteins that are insoluble in water and neutral solvents and resistant to digestive enzymes. They include the *collagen* of skin, tendons, and bones and the elastic proteins known as *elastin* and *keratin.* Scleroproteins have protective and supportive functions in bones, cartilage, ligaments, tendons, and other tough parts of the animal body. See Appendix 9.

Sclerosis. Hardening of a part of the body due to the growth of tough, fibrous tissues. The term is more commonly used to describe a disorder of the nervous system characterized by hardening of the tissues as a result of hyperplasia.

Sclerotherapy. Treatment procedure using a sclerosing solution for cauterizing ulcers, arresting hemorrhage, and treating hemangiomas. It is the primary treatment of choice for esophageal variceal bleeding. Performed by introducing the sclerosing chemical, such as sodium tetradecyl sulfate, with the use of an endoscope. *Nutrition therapy:* start with clear liquids at room temperature. Progress to full liquids, then a soft diet as tolerated. As the patient improves, give a regular diet or a modified diet appropriate for the underlying medical condition.

SCP. Abbreviation for *single-cell protein.*

Scurvy. Deficiency disease caused by a lack of vitamin C. It tends to affect the very young, the elderly, and the chronically ill. In *infantile scurvy,* the onset is usually in the second half of the first year. There is pain, tenderness, and swelling of the thighs and legs; the infant assumes the "pithed frog" position, with the legs flexed at the knees. Enlargement of the costochondral junction produces the scorbutic rosary, which has a sharper feel than that due to rickets (rachitic rosary). If the teeth have erupted, the gums may be swollen, spongy, and prone to bleeding. *Adult scurvy* may occur after several months on a diet devoid of vitamin C. Early symptoms are weakness, easy fatigue, and listlessness, followed by shortness of breath and aching in bones, joints, and muscles, especially at night. Perifollicular hemorrhages are common in the thorax, forearms, thighs, legs, and abdomen. In advanced scurvy the slightest injury produces excessive bleeding, and large hemorrhages may be seen beneath the skin; the gums are friable and bleed easily. *Nutrition therapy:* give 200 to 300 mg/day vitamin C for 1 week (in synthetic form or from foods), followed by 100 mg daily for several weeks to replenish tissue stores. Good food sources of vitamin C are green and red peppers, broccoli, tomatoes, potatoes, strawberries, kiwi, oranges, and other citrus fruits.

SDA. Abbreviation for *specific dynamic action.*

Se. Chemical symbol for selenium.

Seborrheic dermatitis. Skin lesion characterized by greasy scaling, especially on the nasolabial folds and around the eyes and ears. Hard sebaceous plugs also form over the bridge of the nose. The condition may be a result of various factors, such as oily skin, hormonal imbalance, emotional disturbance, and nutritional deficiency (as in riboflavin and pyridoxine deficiencies).

Secretin. Hormone secreted by the intestinal mucosa when gastric acid or chyme reaches the intestine. It is carried by the bloodstream to the pancreas and stimulates it to secrete the pancreatic juice.

Selenium (Se). Trace mineral essential to humans, animals, and plants. It is an active component of glutathione peroxidase, an enzyme that catalyzes the breakdown of hydroperoxides. In this manner, selenium protects membrane lipids from oxidant damage. A functional interrelationship exists between selenium and vitamin E; a deficiency in one can be partially corrected by supplementation with either nutrient. Selenium also protects against the toxicity of mercury, cadmium, and silver, and there is some evidence suggesting that selenium may reduce the incidence of cancer. It also functions in thyroid hormone metabolism. Deficiency in humans has been described in severely malnourished children and in patients maintained on prolonged total parenteral nutrition unsupplemented with selenium. Characteristic signs are a low blood selenium level, abnormal nail beds, growth retardation, and muscle cramps; heart enlargement and varying degrees of heart insufficiency are seen in severe deficiency. Seafoods, liver, meats, and whole grains are good food sources; fruits and vegetables generally contain little selenium. The average daily intake is about 50 to 200 μg/day, and absorption is about 80% or more. Selenium is transported to various tissues bound to very-low-density and low-density lipoproteins and is taken up by the red blood cells, liver, heart, spleen, nails, and tooth enamel. Selenium is toxic beginning at levels 20 to 30 times the requirement. Signs of intoxication include loss of hair and nails, dental caries, dermatitis, peripheral neuropathy, irritability, and fatigue. Chronic selenium toxicity occurs in many seleniferous areas in the Americas, South Africa, Australia, New Zealand, and China. Intoxication from dietary selenium supplements has also been reported. The indiscriminate use of selenium self-supplementation should be avoided. See Appendix 1 for the recommended daily allowances. See also *Keshan disease*.

Self-feeding problems. Obstacles encountered by patients who are either blind or have problems in chewing, swallowing, and coordination, as well as physical handicaps with feeding. *Nutrition therapy:* independence in self-feeding, where appropriate, should be encouraged. Provide adaptive feeding equipment such as special plates and utensils, special straws, plate guards, and tray rails. Alter feeding position if necessary. If there is a problem with chewing and swallowing, provide pureed, ground, or chopped foods. Soft foods such as cottage cheese, eggs, milk, and mashed fruits and vegetables are recommended. See also *Dysphagia* and *Diet, supraglottic*.

Senna. An anthraquinone glycoside stimulant laxative obtained from the dried leaves of *Cassia angustifolia*. Abuse of this drug can cause potassium depletion and malabsorption with weight loss. Brand names are Gentlax®, Senokot®, and Senolax®.

Sepsis. Infection that may be the result of fungal or bacterial agents. When it has spread from one part of the body to other areas via the circulatory system, it is called *septicemia* or, more colloquially, "blood poisoning." The presence of pathogenic bacteria and their toxins in the blood is accompanied by chills, excessive perspiration, intermittent fever, and weakness. In severe form, the body is in a hypercatabolic state which needs immediate nutrition intervention. Sepsis stimulates a hypermetabolic state that causes rapid mobilization of lean tissues from the somatic muscle compartment. *Nutrition therapy:* because of the high fever, the resting energy expenditure is increased as much as 50% to 75% which requires an increase in the patient's caloric intake to about 40 to 45 kcal/kg/day. Increased protein losses due to hypercatabolism should be replaced by providing protein of high biologic value and a daily protein intake of 1.7 to 2.0 g/kg/day. Branched-chain amino acids, arginine, and taurine may be beneficial in supplementing enteral feedings. Increased fluids (up to 3 to 4 liters/day) may be required to achieve optimal hydration. Supplemental vitamins and minerals, especially thiamin, riboflavin, folic acid, niacin, potassium, magnesium, and zinc are recommended. Supplemental iron is not indicated; the anemia seen in infection is not caused by deficient intake of iron but by increased hepatic uptake.

Septic encephalopathy. A clouding of consciousness common in sepsis due to hepatic insufficiency or failure. *Nutrition therapy:* same as in *hepatic encephalopathy*.

Serine (Ser). Alpha-amino-beta-hydroxypropionic acid, a nonessential amino acid first obtained from the silk protein sericin. It is converted to glycine and one carbon (1-C) fragment, with *tetrahydrofolic acid* as the acceptor. The 1-C fragment becomes the source of methyl groups needed in the biosynthesis of many compounds.

Serotonin. 5-Hydroxytryptamine (5-HT), a *tryptophan* derivative. It is found in the serum and in a number of tissues, including the gastrointestinal tract, blood platelets, brain, and nerve tissues. Serotonin is a powerful vasoconstrictor and plays a role in brain and nerve function, gastric secretion, and intestinal peristalsis. It is also believed to influence the regulation of food intake by the brain. Metabolism of serotonin yields 5-hydroxyindoleacetic acid (5-HIAA). See also *Diet, serotonin test*.

Serum. Clear liquid left after protein has clotted; refers to both blood and milk serum. The serum from milk is called *whey;* it contains lactose, proteins, water-soluble vitamins, and minerals. Blood serum is plasma *without* its fibrinogen. See also *Plasma*.

Serum albumin. See *Albumin*.

Serum creatinine. See *Creatinine*.

Serum enzymes. Enzymes present in the blood plasma resulting from the breakdown of body cells. Examples are serum glutamic pyruvic transaminase (SGPT), serum glutamic oxalacetic transaminase (SGOT), and lactic dehydrogenase (LDH). The levels of subgroups of these enzymes provide diagnostic and clinical information on specific tissue damage.

SGA. Abbreviation for *subjective global assessment*.

SGOT. Abbreviation for serum glutamic oxalacetic transaminase; also called "aspartate aminotransferase (AST)." This *serum enzyme* is widely distributed in all body tissues (except bone), but its levels are highest in the heart muscle, skeletal muscle, liver, kidney, and brain. The normal level ranges from 5 to 40 units. Activity is high in myocardial infarction, infectious hepatitis, extrahepatic biliary obstruction, and liver damage from toxic agents. It is moderately increased in rheumatic fever and in disorders involving necrosis of the heart, liver, or muscle.

SGPT. Abbreviation for serum glutamic pyruvic transaminase; also called "alanine aminotransferase (ALT). A *serum enzyme* found in higher concentration than SGOT in the liver. It is also found in heart muscle, but in lower concentration than SGOT. This concentration difference between the two enzymes provides a more accurate diagnosis of myocardial infarction and liver disease.

Short-bowel syndrome (SBS). Set of symptoms resulting from massive resection of the small intestine which is life-threatening, especially if over 50% of the organ is removed. The patient undergoes three post surgical phases: (1) severe diarrhea, fluid losses, and electrolyte imbalance; (2) two months or more of anorexia, mild diarrhea, steatorrhea, and weight loss; and (3) anemia, osteomalacia, gallstone formation, gastric hyperacidity, and nutritional deficiencies. *Nutrition therapy:* during the first phase, intravenous and total parenteral nutrition feedings are needed. As the patient improves, slowly advance to a transitional diet consisting of clear liquids, elemental formulas, and oral electrolyte solutions, followed by gradual introduction of lactose-free polymeric formulas and other more complex foods. Taper TPN as oral intake increases and intestinal adaptation occurs. The transition from parenteral to complete oral feeding may take several weeks. Provide 1.5 to 2.0 g of protein/kg/day and 40 to 45 kcal/kg/day or a daily calorie intake of basal energy expenditure × 1.5; restrict fat to 30% of total calories and give medium-chain triglyceride oil if there is steatorrhea; restrict lactose if there is intolerance; restrict foods high in oxalate if urinary oxalate levels are high or when renal stones occur; avoid concentrated sweets and alcohol; and ensure adequte fluid intake (35 mL/kg body weight). The recommended amounts for supplemental vitamins and minerals are: water-soluble form of vitamin A (25,000 to 50,000 IU), vitamin D (50 to 100 μg of 25-hydroxycholecalciferol or 50,000 IU cholecalciferol), and vitamin K (4 to 12 mg of menadione); folic acid (5 mg daily for 1 month; 1 mg daily maintenance), vitamin B_{12} (initial 1000 μg loading dose; 100 to 300 μg/month maintenance), calcium carbonate (500 mg t.i.d.), magnesium gluconate (1 to 4 g), ferrous sulfate (325 mg), copper sulfate (7.5 mg), and zinc sulfate (220 mg). Slowly progress from a low residue to a regular diet by gradually adding fiber as tolerated. There is current evidence that fiber may enhance gut function by promoting mucosal growth. Prevent and correct fluid and electrolyte imbalances.

Short-chain fatty acid (SCFA). See under *Fatty acid (FA).*

Si. Chemical symbol for silicon.

Sialic acid. Acetyl derivative of an amino sugar acid; present in saliva, glycoproteins, lipids, and polysaccharides.

Siderophilin. Also called "transferrin," an iron–carbonate–protein complex circulating in the blood plasma.

Siderosis. The presence of excess iron in the body. This may be the result of hemolysis, excess intake of iron, multiple blood transfusions, or failure to regulate iron utilization.

Silicon (Si). An essential trace element in chickens and rats. Silicon deficiency results in growth retardation and incomplete development of the skeleton. It is believed to function in the metabolism of connective tissue, formation of collagen, calcification of bones, and maintenance of elastic tissue integrity. There is some evidence indicating that silicon may also be essential in humans. Of interest is its role in the development of atherosclerotic vascular disease. There is an inverse relationship between the silicon content of the arterial wall and the degree of atherosclerosis present, and some studies have reported blood lipid-lowering effects of silicates in the drinking water. Silicon may also be involved in osteoarthritis, hypertension, and the aging process. The minimum requirement for silicon has not been ascertained in animals, and nothing is known about the possible human requirement. Silicon is supplied by many foods, especially unrefined grains, cereal products, and root vegetables; foods of animal origin are low in silicon. When taken orally, silicon is essentially nontoxic. Magnesium silicate, an over-the-counter antacid, and other silicates in food additives used as anticaking or antifoaming agents have been used by humans without obvious harmful effects.

Similac®. Brand name of a line of milk formulas for infant feeding; contains skim milk with added lactose, soy oil, coconut oil, and vitamins, with or without added iron. Similac® 20 is designed for normal, full-term infants; Similac® 13 for infants who have not been fed for several days or weeks and for whom a dilute formula is desired; Similac® 24 for premature infants with a limited intake or for infants with increased energy needs; and Similac 27 for premature infants with increased growth needs; Similac® PM 60/40 for infants who have impaired renal or cardiovascular function; and Similac® Special Care for low-birth-weight and premature infant's before switching to a standard formula. See also Appendix 37.

Simplesse. A fat substitute made from egg white and milk protein with only 1.5 kcal/g. It is usually used in dressings, butter, ice cream, and other foods that do not require cooking. Simplesse produces allergic reactions in individuals allergic to eggs or milk.

Single-cell protein (SCP). Protein produced by the growth of single-cell organisms such as algae, bacteria, yeasts, and fungi.

Sitosterol. Plant sterol similar to cholesterol but having an extra methyl group. Large doses can lower blood cholesterol and beta-lipoprotein levels. High intakes, however, produce toxic effects such as anorexia, diarrhea, and cramps.

SI unit. Abbreviation for *le Systéme International d'Unités*. It is an international system of "base units" and "derived units" for weights and measures to allow interchange of scientific information among different nations. Examples of SI "base units" are meter, kilogram, second, Celsius, and mole. Examples of SI "derived units" are square meter (m^2), kilogram per cubic meter (kg/m^3), moles per liter (mol/L), and joule. The SI units for laboratory values are given in molar concentrations, with the liter as the reference volume. See Appendix 30. In this dictionary, both conventional and SI units (enclosed in parentheses) are given when laboratory values are referenced in the definition.

Skatole. The substance that gives the characteristic foul odor to feces. It is a product of tryptophan deamination in the intestines.

Skin. Also called "integument"; the outermost covering of the body, consisting of a double-layered, tough, resilient epithelium averaging 1.7 m^2 of surface area. The skin protects the underlying tissues from mechanical injury, helps regulate body temperature, synthesizes vitamin D, and is sensitive to sensations of pain, touch, and temperature. The general condition of the skin is taken as an index of health and the state of nutrition (e.g., pallor of anemia, petechiae of vitamin C deficiency, follicular keratosis of vitamin A deficiency, and pellagrous dermatitis of niacin deficiency). See also *Nutrition, dermatology*.

Skinfold measurement. Also called the "pinch test"; measurement of the thickness of a fold of skin at selected body sites where adipose tissue is normally deposited, as in the biceps, triceps, subscapular, suprailiac, thigh, and calf muscles. The skinfold is measured with a *caliper* and gives an estimate of the degree of fatness of an individual. The most commonly measured skinfold is the triceps. See *Triceps skinfold* and Appendix 23.

SMA®. Brand name of an infant formula that is low in sodium, potassium and renal solute load; used for infants with impaired renal or cardiovascular function. Contains demineralized whey, nonfat cow's milk, lactose, coconut oil, and soy oil. The protein and mineral contents are comparable to those of breast milk. See also Appendix 37.

Sn. Chemical symbol for tin.

SOAP charting. The acronym for a style of charting in the medical record, as follows: *Subjective* (includes information provided by the patient, caretaker, or family; nutritional/dietary history; significant socioeconomic, cultural, and emotional factors; physical activity and other factors that affect the lifestyle); *Objective* (includes age, height, weight, and other anthropometric measurements; laboratory or clinical data; current diet order and medications); *Assessment* (includes evaluation of the condition based on the subjective and objective data; estimation of nutritional needs; projection of weight loss or gain; assessment of diet order); *Plan* (includes goals of nutrition therapy; recommendations for care and follow-up; referrals for consultation to other healthcare professionals; plans for counseling and nutrient supplementation).

Society for Nutrition Education (SNE). Professional organization formed to promote good nutrition among the public by making sound nutrition education more available and effective; provides technical assistance and consultation for interested groups; develops extensive bibliographies with evaluated annotations of content; and maintains the National Nutrition Education Clearing House. It publishes the *Journal of Nutrition Education*.

Sodium (Na). A major mineral essential to life. About 50% of the body's sodium is in extracellular fluids, 40% in bones, and the rest inside the cells. Sodium is the chief cation in the extracellular fluids. It regulates water and acid–base balance, osmotic pressure, contraction of muscles, and conduction of nerve impulses. Almost all foods contain sodium, either naturally or as an ingredient added during processing or cooking. The main sources are sodium chloride and sodium bicarbonate, and the average daily intake is about 4 g of sodium (10 g of salt). The American Heart Association recommends an intake not exceeding 3 g/day. Excessive sodium may increase the risk of heart disease, stroke, and kidney damage when the blood pressure is high. The association between sodium and hypertension has been firmly established; blood pressure goes down in sodium-sensitive individuals when they reduce dietary sodium intake. Normally, the quantity of sodium ingested daily equals the amount excreted, so that a state of sodium balance is maintained. Aldosterone, a hormone secreted by the adrenal cortex, controls the regulation of sodium balance. When sodium intake is high, the aldosterone level decreases and the urinary sodium level increases. When the sodium level is low, the aldosterone level increases and urinary sodium excretion decreases. Dietary sodium deficiency does not normally occur, even among those on very-low-sodium diets. Sodium depletion is usually the result of excessive loss from diuretic therapy, persistent diarrhea or vomiting, profuse sweating, and other disorders marked by loss of body fluids or inability to retain sodium. Symptoms of deficiency include weakness, muscle cramps, fatigue, and dizziness. In severe cases, there may be a drop in blood pressure, leading to confusion, fainting, and palpitations. See *Hypernatremia* and *Hyponatremia*. See also *Diet, sodium-restricted*.

Sodium pump. Mechanism that mediates the active transport of sodium across cell membranes whereby sodium is pumped out in exchange for potassium. The operation of this pump requires cellular adenosine triphosphate and adenosine triphosphatase.

Somatic. Pertaining to the body framework as distinguished from the viscera.

Somatomedin C. Also called insulin-like growth factor I (IGF-I). One of a family of insulin-like peptides that have anabolic actions on fat, muscle, cartilage, and cultured cells. It has a very short half-life (3 to 7 hours) and is a sensitive indicator of recent protein and calorie intake and deprivation. Plasma levels (normal range, 0.55 to 1.4 IU/mL) fall rapidly with fasting and quickly recover during refeeding; low values are also seen in hypothyroidism and with estrogen administration.

Somatostatin. A hormone secreted by the delta cells of the pancreas that inhibits the release of both *insulin* and *glucagon*. It is also secreted by the hypothalamus to inhibit the release of growth hormone and thyroid stimulating hormone.

Somatotropin (somatotropic hormone). Growth hormone secreted by the anterior lobe of the *pituitary gland*. Its main action is to stimulate the growth of the epiphyseal cartilages of long bones. It also increases nitrogen retention, facilitates the transfer of amino acids from extracellular to intracellular compartments of the body, influences carbohydrate metabolism by its insulin-like effect, and causes lowering of the fat content in the body. Commercial bovine somatotropin is used in dairy farms to increase the efficiency of milk production.

Sorbitol. A 6-carbon sugar alcohol formed by the reduction of glucose or fructose. It is a nutritive sweetener having the same caloric value as glucose (4 kcal/g). However, sorbitol is slowly absorbed and delays the onset of hunger. It is only 50% as sweet as sucrose and is used in dietetic candies, gums, and ice creams. It can be converted to utilizable carbohydrate in the form of glucose. Excessive use may cause gastrointestinal distress and diarrhea.

Special Milk Program. Administered by the U.S. Department of Agriculture, this program provides milk to children of school age in child care centers and in schools and other approved institutions where lunch is not served and the institutions do not participate in a meal service program.

Specific dynamic action (SDA). See *Thermal effect of food*.

Specific fuel factor (value). Coefficient of digestibility and particular caloric contribution of foods coming from similar sources. For example, the coefficient of digestibility of proteins in milk, eggs, and meat is 97%, whereas that of cornmeal protein is only 60%. Thus the caloric value per gram of protein would be much less for corn protein than for proteins in milk, egg, and meat. Specific fuel factors (values) for estimating calories from various foods have been established.

Specific gravity. Weight of a substance compared with that of an equal volume of another substance taken as a standard (water for liquids and solids, hydrogen for gases). The specific gravity of water is 1 (i.e., 1 g/mL); it is less than 1 for fats.

Specificity. The ability of an enzyme to catalyze a single reaction or a limited range of reactions. Specificity is the main distinction between enzymes and inorganic catalysts such as minerals, which are nonspecific.

Sphingomyelin. Complex *phospholipid* composed of 4-sphingenine (sphingosine), fatty acids, phosphoric acid, and choline. It is a part of cell structures and is found primarily in brain and nervous tissue as a constituent of the myelin sheath.

Spinal cord injury. Traumatic disruption of the spinal cord, often associated with musculoskeletal involvement. Muscle function is lost below the level of injury. Immobilization brings about negative nitrogen balance, urinary loss of calcium and nitrogen, and low serum albumin. Energy needs are lower because of the decreased metabolic activity of denervated muscle. The higher the injury to the spinal cord, the lower the energy expenditure. Up to 65% reduction in energy expenditure is seen in chronic paraplegia and quadriplegia. Excess weight gain is common. *Nutrition therapy:* the aims are to prevent protein depletion without overfeeding and to maintain body weight slightly below the standard. In acute SCI and immediately after the injury, provide 1.5 to 1.7 g of protein/kg body weight and 80% of basal energy expenditure via nasogastric tube feeding, if tolerated, or parenteral nutrition if there is vomiting. Gradually introduce oral feeding and monitor weight. The goal is 10 to 15 lb below standard body weight for paraplegics, and 15 to 20 lb below standard for quadriplegics. In the rehabilitation stage after the injury, provide 28 kcal/kg body weight for paraplegics and 23 kcal/kg body weight for quadriplegics and adequate protein (1.1 to 1.2 g/kg body weight) to maintain muscle mass and prevent negative nitrogen balance and pressure sore formation. Supplementation with vitamins and minerals, especially calcium, is recommended. Encourage plenty of fluids (2 to 3 liters/day) to dilute the urine and prevent urinary tract infections and formation of renal stones.

Spirometer. Also called "respirometer"; an apparatus that measures air taken into and from the lungs. It is used in indirect calorimetry. See *Respirometer* and *Calorimetry*.

Spironolactone. A synthetic steroid aldosterone antagonist; a potassium-sparing diuretic used for edema and hypertension. It may cause hyperkalemia, especially in patients with renal insufficiency. Dehydration and hyponatremia manifested by low serum concentration, dry mouth, thirst, and mental confusion may occur. Brand names are Aldactone® and Spiractone®.

Spleen. Ductless organ situated in the upper part of the abdomen just below the diaphragm and to the left of and behind the stomach. It is composed of a mass of sinuses with various openings. The spleen is an organ of the *reticuloendothelial system*. It also serves as a reservoir for the storage of blood and is capable of increasing and decreasing its volume to maintain normal blood cell levels in active circulation.

Sports anemia. Also called "runners' anemia." Observed in some athletes who have decreased ability to carry oxygen (in the red blood cells) caused by excessive iron loss through perspiration or increased red blood cell destruction from intense physical activity.

Sprue. Disorder characterized by malabsorption of nutrients, diarrhea, weight loss, and abnormalities in the mucous membrane lining the digestive tract. It occurs in both nontropical and tropical forms. Nontropical sprue is due to an intolerance to gluten. (See *Gluten-sensitive enteropathy.*) Tropical sprue is of unknown cause. It is endemic in the tropics and subtropics and is characterized by protein malnutrition, multiple nutrient deficiencies, and a megaloblastic anemia due to lack of folic acid and vitamin B_{12}. *Nutrition therapy:* a high-protein diet with supplemental vitamins A, D, K; the B-complex; folic acid; intramuscular vitamin B_{12}; calcium; and iron.

Squalene. Unsaturated hydrocarbon formed by four molecules of acetic acid. It is an intermediate step in the synthesis of *cholesterol*.

Stachyose. Tetrasaccharide containing glucose, fructose, and two molecules of galactose. It is found in tubers, peas, lima beans, and beets.

Stapling, gastric. See under *Gastric bypass (stapling)*.

Starch. Storage form of carbohydrates in plants. It is a polysaccharide composed of many glucose units linked in a straight line (amylose) or with branches (amylopectin). Starch is the principal source of energy and the basic staple of the daily diet. Chief food sources are cereals and cereals products, root crops such as potatoes and yams, tapioca, legumes, and starchy vegetables. See Appendix 8.

Starvation. A condition resulting from prolonged deprivation of food, hence the lack of calories and nutrients. The starved body suffers from physiologic malfunctions. Energy expenditure continues during starvation; thus the body mobilizes nutrients from body fuel reserves. Glycogen stores are depleted fairly early in starvation, after which the body adjust by hydrolyzing skeletal muscle protein and using the amino acids as sources of glucose. In the early days of starvation, the urinary excretion of nitrogen is about 12 g/day (equal to 75 g of protein or 360 g of lean tissue). As much as 500 g of protein, or 5% of total body intracellular protein, may be lost in 7 days of starvation. During prolonged starvation, the adipose tissue becomes the principal source of energy. Production of ketone bodies from fatty acids is accelerated, and the brain and other tissues begin to use ketones as an energy source. Muscle protein continues to be catabolized but at a reduced rate, with protein loss decreasing to 20 g/day. At the same time, the body reduces its energy needs by slowing the metabolic rate. *Nutrition therapy:* rehabilitation of the starved, hypometabolic patient should be gradual. The aim is to avoid *refeeding syndrome*. Start with low fluid volume and salt load. Suggested daily caloric progressions are as follows: basal energy expenditure (BEE) × 0.8 on days 1 and 2; BEE × 1.0 on days 3 and 4; BEE × 1.2 on days 5 and 6; and BEE × 1.5 on day 7 and thereafter. Increase the calories to BEE × 2 if weight gain is desired. Protein intake should be 1.2 to 1.5 g/kg of body weight, depending on the degree of repletion required. Monitor weight, albumin and other indicators of protein status, and electrolytes. The resynthesis of lean tissues requires large amounts of potassium, magnesium, phosphorus, and other nutrients. Supplemental vitamins and minerals are recommended.

Sta-Slim. Brand name of a fat substitute made from modified potato starch and that has the texture and mouth feel of fats and oils. Used partially to replace fats and oils in baked products, cheesecake, dips, frosting, frozen dessert, and salad dressing. Provides 4 kcal/g.

Stature. The natural height in an upright position. When direct measurement of height cannot be taken, stature can be estimated using other means. See *Armspan measurement, Knee height,* and *Total arm length*.

Stearic acid. Long-chain saturated *fatty acid* with 18 carbon atoms. It is present in most animal and vegetable fats as the triglyceride *stearin*.

Steatorrhea. Presence of fat in stool. It may be caused by defective fat absorption, lack of bile, or lack of lipase. Fatty stools are seen in celiac disease and other malabsorption syndromes. As much as 60 g of fat may be lost in the stool, in contrast to the 2 to 5 g normally excreted each day. The underlying cause of the fat malabsorption must be determined and treated. *Nutrition therapy:* reduce dietary fat intake to 40 to 60 g/day. Give as much fat as can be tolerated without an increase in steatorrhea. If weight gain is desired, use medium-chain triglycerides (MCTs) for additional calories. MCT oil can be substituted for fat in cooking and some enteral products contain MCT oil. Alcohol should be avoided. Supplemental minerals, especially calcium, magnesium, and zinc, and the water-soluble forms of vitamins A, D, E, and K may be necessary.

Steroids. Large group of cyclic lipid compounds. Included in this group are the *sterols, sex hormones, adrenocortical hormones, bile acids, vitamin D, saponins,* and *sterol glycosides*.

Sterols. Class of *steroids* that are complex monohydroxy alcohols universally found in both plants and animals. Mycosterols are found in yeasts and fungi; the most important one is *ergosterol,* a precursor of vitamin D. Phytosterols, or plant sterols, include *sitosterol,* which is found in oils of higher plants, especially wheat germ oil. *Cholesterol* is the most familiar sterol; it is present only in animal sources.

STH. Abbreviation for *somatotropic hormone*. See *Somatotropin*.

Stomach. Also called the "gastric gland"; the most dilated part of the alimentary canal, situated below the diaphragm. It is composed of three parts: the *cardia,* or the upper part; the *fundus,* which secretes digestive juices and stores food temporarily; and the *antrum,* which provides powerful mixing movements. Three types of cells in the stomach secrete the gastric juice: *mucous cells,* which secrete mucin; *parietal cells,* which secrete hydrochloric acid; and *zymogenic* or *chief cells,* which secrete the enzymes pepsin, the intrinsic factor, rennin, and lipase. See also *Gastrectomy, Gastric bypass, Gastric emptying, Gastric surgery, Gastritis,* and *Gastroparesis*.

Stomatitis. Inflammation of the oral mucosa or soft tissues of the mouth. In *angular stomatitis,* the inflammation occurs at the angles of the mouth. Stomatitis may result from an infection, side effects of drugs, or vitamin deficiency (riboflavin, niacin, and pyridoxine). See also *Nutrition, tongue and mouth conditions*.

Stress. 1. Time of extreme pressure or a trying period. 2. Any stimulus that disrupts the homeostasis of the organism. Stress factors that alter nutrient needs are *physiologic stresses,* as in growth, pregnancy, and lactation; *pathologic stresses* such as fever, infection, and disease; *physical stresses* such as heavy labor, strenuous exercise, and severe environmental conditions; and *psychological stresses* such as anorexia nervosa and psychic overeating. The body responds to stress in a variety of ways, and these are often accompanied by hypermetabolism and hypercatabolism. *Nutrition therapy:* see Appendix 24 for estimating energy and protein requirements in various stress conditions. The dietary goal is to provide sufficient energy, protein, fluid, and electrolytes to maintain the metabolic demands and immune responses. Reversal of catabolism may not be possible in severe stress but it can be curtailed. See *Nutrition, stress*.

Stroke. 1. A sudden and severe attack of a disease. 2. Common term for *apoplexy,* a symptom complex caused by hemorrhage of the brain or thrombosis of the cerebral vessels. See *Cerebrovascular accident.*

Strontium. A metallic element that is present in various compounds, seawater, marine plants, and food. It is also found in the body, although it is not established as essential to humans. It is metabolized in a manner similar to that of calcium and has the ability to replace calcium in bone formation. There are several radioisotopes of the element, of which strontium-90 is of public health importance. It is produced in nuclear fission reactions and is present in the fallout from nuclear bomb explosions. Of major concern is the possibility of ingesting radioactive strontium-90 by drinking milk from cows fed grass and hay that have absorbed the element from the soil or the atmosphere as radioactive fallout. Strontium-90 emits radiation for a long time (half-life of 28 years) and accumulates in the bone, where the radiation may cause leukemia and/or bone tumors.

Struvite. Compound of magnesium ammonium phosphate, a common constituent of urinary tract stones. Usually associated with chronic urinary tract infections, neurogenic bladder, and paraplegia.

Subjective global assessment (SGA). An alternative method for assessing nutritional status based on clinical parameters taken from a routine history and physical examination. The routine history determines weight loss over both the long term (past 6 months) and the short term (past 2 weeks); decreased or unusual food intake; presence of gastrointestinal symptoms such as anorexia, diarrhea, nausea, and vomiting; and functional capacity. The physical examination is noted as normal, mild, moderate, or severe based on a subjective impression on loss of subcutaneous fat (triceps and chest), muscle wasting (quadriceps and deltoid), presence of edema (ankle or sacral), and mucosal lesions including glossitis and skin rashes. Without reference to any other values and without a specific weighting scale, the examiner derives a "global" assessment of nutritional status and assigns the following rating: A for well nourished, B for moderately malnourished, and C for severely malnourished.

Substance abuse. Chronic use of alcohol and drugs resulting in physiologic, social, emotional, and occupational problems. Recent studies show that persons with substance abuse have abnormal metabolism of dopamine, serotonin, and norepinephrine. See *Alcoholism* and *Nutrition, chemical dependence.*

Sucaryl®. Brand name for *cyclamate,* an artificial sweetening agent for diet soft drinks. Its use was banned in the United States under the Delaney Amendment. See also *Alternative sweeteners.*

Succus entericus. The intestinal juice. It is slightly alkaline and contains mucin, amylase, lipase, peptidases, disaccharidases, and other enzymes. Its composition and volume throughout the intestines vary.

Sucralose. Also called "chlorosucrose." A noncaloric sweetener under study; has high-intensity sweetening power (600 times that of sucrose). Extremely heat stable. Its use is approved in Canada but pending approval in the United States.

Sucrose. Table sugar; made from cane or beet sugar. Also found in molasses, maple syrup, honey, as well as fruits and vegetables. It is a *disaccharide* consisting of glucose and fructose. Sucrose is easily hydrolyzed by acid or the enzyme invertase to form *invert sugar.* Intestinal *sucrase* readily splits sucrose into glucose and fructose.

Sucrose intolerance. Condition characterized by watery diarrhea associated with the malabsorption of sucrose. It is due to a genetic defect in which the enzyme sucrase-isomaltase, required in the hydrolysis of sucrose, is absent or deficient. See *Diet, sucrose-restricted*.

Sucrose polyester (SPE). A fat substitute under study, made by replacing the glycerol of the fat with sucrose. It is totally nonabsorbable and noncaloric.

Sugar. 1. Any sweet, soluble, crystalline organic compound belonging to the carbohydrates. 2. Specifically refers to sucrose extract from sugar cane and sugar beet. See *Sweetener*.

Sulfasalazine. A sulfonamide derivative used in the treatment of ulcerative colitis and Crohn's disease. It inhibits intestinal transport of folate and may induce folate deficiency. It may also decrease the intestinal synthesis of vitamin K and increase the urinary excretion of protein and ascorbic acid. Adverse gastrointestinal effects include anorexia, nausea, vomiting, gastric distress, and diarrhea. It is excreted in breast milk and known to have adverse effects on the infant. Brand names are Azaline® and Azulfidine®.

Sulfinpyrazone. A uricosuric agent used in the treatment of gouty arthritis and tophaceous gout. It reduces the amount of uric acid in the blood by increasing the amount excreted in the urine. Maintenance of a large volume of alkaline urine reduces the risk of stone formation in the kidneys. The brand name is Anturane®.

Sulfites. Oxides of sulfur widely used in the food industry and in restaurants; as sanitizing agents for food containers and equipment; in wine and beer to stop fermentation; on seafoods, vegetables, and fruits (fresh or dried) to prevent discoloration or spoilage; and in salad bars to keep vegetables and fruits looking fresh. Some asthmatics experience acute reactions to foods treated with sulfites; these include wheezing, flushing, weakness, and tightness in the chest. Severe reactions, which have resulted in death, prompted the Food and Drug Administration to issue a labeling regulation requiring manufacturers to declare sulfite on the label of any food containing sulfite at a level of 10 ppm or more. Sulfite used as a preservative must be declared on the label regardless of the amount in the finished product. There is another regulation that bans the use of sulfites on fruits and vegetables intended to be served or sold raw to consumers. See also *Diet, sulfite-restricted*.

Sulfonamides. Also called "sulfa drugs"; a group of antibacterial drugs used mainly to treat urinary tract infections. Sulfonamides interfere with the utilization of para-aminobenzoic acid and thus inhibit the biosynthesis of folic acid by colonic bacteria. Possible adverse effects include diarrhea, anorexia, stomatitis, and gastritis. Brand names include Bactrim®, Gantanol®, Gantrisin®, Renoquid®, and Thiosulfil®.

Sulfonylurea. Class of chemical compounds that includes the oral hypoglycemic agents *acetohexamide, chlorpropamide, glipizide, glyburide, tolazamide,* and *tolbutamide*. These drugs stimulate the synthesis and release of insulin from the beta cells of the pancreas and are used to treat patients with non-insulin-dependent diabetes mellitus. They have no hypoglycemic effects on patients with insulin-dependent diabetes mellitus who have nonfunctional beta cells and cannot produce insulin.

Sulfur (S). Also spelled "sulphur." Mineral that is present in all cells, especially in cartilage and keratin of skin and hair. It occurs principally as a constituent of the amino acids cystine, cysteine, and methionine; it is also a constituent of insulin, thiamin, biotin, heparin, glutathione, coenzyme A, and other coenzymes. All protein foods provide sulfur, and the need for this mineral is met when the protein supply is adequate.

Sumacal®. Brand name of a powdered carbohydrate (maltodextrin) supplement that is protein-free and low in electrolytes; provides 3.8 kcal/g. See Appendix 39.

Summer Food Service for Children. This program provides one nutritious meal to school children as a substitute for the National School Lunch and School Breakfast programs. Public and nonprofit residential institutions of local, municipal, and county governments may participate. Administered by the U.S. Department of Agriculture (USDA).

Suplena®. Brand name of a specialized liquid nutrition for nondialyzed renal patients requiring protein, electrolytes, and fluid restrictions. Can be taken orally or by tube feeding. See Appendix 39.

Supplementary feeding. The giving of food in addition to the regular meals to increase or supplement nutrient intake.

Suprarenal glands. The *adrenal glands*. The term "suprarenal" means lying above the kidneys.

Surgery. Treatment of disease by manual or instrumental operations. The patient's nutritional status is affected directly or indirectly, depending on the part of the body that undergoes surgery. For example, gastric resection results in decreased gastric acid production and the dumping syndrome; intestinal resection causes general malabsorption of all nutrients; and pancreatic resection leads to diabetes mellitus and malabsorption; other surgical procedures place the patient under physiologic and psychological stresses. The patient's weight, blood glucose, albumin, blood count, and electrolytes should be assessed and monitored regularly. *Nutrition therapy:* if surgery is elective and the patient is overweight or obese, a weight reduction regimen is recommended. The undernourished patient should be rehabilitated with a diet high in protein, carbohydrate, and calories to build up tissue and glycogen reserves. Immediately before and after surgery, nothing is given by mouth to avoid vomiting and aspiration during surgery and postoperatively while recovering from anesthesia. Initially, intravenous glucose is the main source of energy. Depending on the patient's recovery and type of surgery, progressive oral feeding of clear to full liquids, followed by soft to regular foods, is often tolerated within a few hours or the next few days. Provide 35 kcal/kg body weight; increase to 40 kcal/kg if the person is undernourished. Protein allowances per kilogram of body weight per day are as follows: 1.2 to 1.3 g in minor surgery, 1.4 to 1.5 g in major surgery, and 1.7 to 1.9 g if the patient is nutritionally depleted. When enteral feeding is not possible, parenteral nutrition may be given for brief periods by peripheral vein or for longer periods by central vein or total parenteral nutrition (TPN). The postoperative nutritional approach depends on whether the patient is hypometabolic or hypermetabolic. Because of the risk of hypophosphatemia and heart failure on refeeding, nutritional support for the nonstressed, hypometabolic, starved patient should be increased cautiously and gradually, taking up to a week to reach the final caloric goal. This allows the patient to adapt to the high caloric and glucose loads. If the intravenous route is required, one-third or more of the calories should be provided as lipid to reduce the glucose load. With either type of feeding, blood phosphorus levels should be assessed before repletion and monitored daily during the initial period of refeeding. In the hypermetabolic, stressed patient, nutritional support should be more aggressive. Provide 40 to 45 kcal and 1.8 to 2.0 g of protein per kilogram body weight. It is possible to reach the goal for calorie and protein intake within 2 to 3 days of initiating enteral or parenteral feeding. See *Refeeding syndrome*.

Sustacal®. Brand name of a high-protein, lactose-free product for tube feeding or oral supplementation. Also available as Sustacal® Plus (high-calorie) and with fiber (5.6 g of dietary fiber/liter). Sustacal® Pudding and Sustacal® Nutritional Powder are brand names for products containing milk and lactose, available in different flavors, for use as high-protein oral supplements. See also Appendix 39.

Sustagen®. Brand name of a high-protein, high-calorie, low-fat oral supplement; contains whole and nonfat milk, calcium caseinate, corn syrup solids, and dextrose. See Appendix 39.

Sweetener. A substance that gives a sweet taste; may be *nutritive* (supplies calories) or *nonnutritive* (supplies no calories). Sucrose is the most common sweetener and is used as the standard (100%) for comparing the sweetness of other agents. The relative sweetnesses of natural agents are as follows: fructose, 173%; glucose, 74%; maltose, 33%; lactose, 16%; glycerol, 60%; sorbitol, 60%; and glycine, 70%. See also *Artificial sweeteners*.

Sympathetic nervous system. One of two parts of the autonomic nervous system. Its actions are opposite to those of the other part, the parasympathetic nervous system. For example, heart action is accelerated by the sympathetic system but decelerated by the parasympathetic system; intestinal peristalsis is decreased by the sympathetic system and increased by the parasympathetic system.

Symptom. The manifestation or expression of a disease as the patient experiences it, in contrast to the *sign,* or the manifestation of a disease as the examiner perceives it. Headache is a symptom; rapid pulse is a sign.

Syndrome. Set of symptoms and signs that occur together and characterize a particular disease or condition.

Synergism. Joint action of agents so that their combined effect is greater than the sum of their individual effects. Malnutrition lowers resistance to infection, and infectious diseases tend to magnify an existing malnutrition. The simultaneous presence of malnutrition and infection results in an interaction with an enlarged effect that is more serious than would be expected if malnutrition or infection acted separately.

T

T. Abbreviation for temperature.

T_3. Abbreviation for *triiodothyronine,* a thyroid hormone. See *Thyroid gland.*

T_4. Abbreviation for *tetraiodothyronine* or *thyroxine*. See *Thyroid gland* and *Thyroxine*.

Tachycardia. Rapid heartbeat. The term is usually applied to a pulse rate above 100/minute; the rapid stimulus of heart action is associated with several causes, varieties, and sites in the heart.

Tachysterol. An isomer of ergosterol produced by irradiation, as is calciferol. It has no antirachitic activity unless reduced to dihydrotachysterol.

TAG. Abbreviation for *total available glucose*.

Tangier disease. A rare inherited disorder of lipid metabolism characterized by enlarged, orange-colored tonsils and storage of large amounts of cholesterol esters in foam cells; the mucosa of the colon and rectum also frequently has an orange color. The most striking feature is the almost complete absence of high-density lipoprotein. Plasma triglycerides may be elevated by the presence of chylomicrons and raised levels of low-density-lipoprotein. The condition is probably due to the absence of an apoprotein involved in the metabolism of lipoproteins. There is no specific treatment. *Nutrition therapy:* partial alleviation of symptoms occurs with a low-cholesterol diet, thereby reducing the storage of dietary cholesterol and cholesteryl esters. A low-fat diet is unnecessary, as fat transport appears to be normal. Because diabetes mellitus may develop in old age, weight control from an early stage is advisable.

Tartrazine. An alternate name for the yellow color FD & C Yellow No. 5, which is approved for use in foods, drugs, and cosmetics. Tartrazine may cause allergic reactions, including bronchial asthma, in susceptible individuals.

Taurine. Or beta-aminoethanesulfonic acid. It is unique among other natural amino acids in that it is not incorporated into proteins or dissociated to give its side groups and sulfur; hence, it is referred to as a free amino acid. It is found in most tissues and fluids, especially the central nervous system, bile acids, muscles, blood platelets, and lymphocytes. As glutaurine, it acts as a neurotransmitter. It is also involved in many biological processes, such as cell differentiation, sperm motility, retinal photoreceptor activity, detoxification, and as an oxidant for free radicals that cause cellular damage. Although taurine is synthesized from methionine and cysteine, humans have poor synthetic ability. Dietary sources are meats, seafoods, and milk. Human milk, particularly colostrum, is very high in taurine (contains 40 mg/L or 300 μmol/L), compared to cow's milk with only 4 mg/L or 30 μmol/L. Under normal physiologic conditions, taurine is not considered a dietary essential, but is classified with the conditionally essential nutrients. There is concern that infants, particularly premature babies, and persons receiving long-term parenteral nutrition have low plasma taurine. Reduced body pools of taurine are associated with retinal degeneration. Thus, most infant formulas are supplemented with taurine to provide concentrations similar to that in human milk. Two parenteral solutions containing taurine are *TrophAmine®* and pediatric *Aminosyn.®* Taurine supplementation is recommended in preterm as well as small-for-gestational age neonates who are not on human milk.

TBA. Abbreviation for *thyroxine-binding albumin.*

TBI. Abbreviation for *traumatic brain injury.*

TBPA. Abbreviation for *thyroxine-binding prealbumin.*

α-TE. Abbreviation for *α-tocopherol equivalent.*

TEE. Abbreviation for *total energy expenditure.*

TEF. Abbreviation for *thermal effect of food.*

Temporary Emergency Food Assistance Program (TEFAP). A U.S. Department of Agriculture (USDA) program that provides commodity foods to low-income households through public or private nonprofit agencies who are approved emergency providers.

Temporomandibular joint dysfunction (TMJD). A broad array of disorders of the masticatory structures in the temporomandibular region. The most common complaints are pain or muscle spasm in the temple and cheek, limited jaw motion, and clicking or popping of the joint. *Nutrition therapy:* eat soft foods that are easy to chew and avoid gum chewing and chewy foods such as caramel and gummy rolls. Cut food into bite-size pieces to avoid opening the mouth widely. Ensure an adequate intake of foods and fluids. Nutrient supplementation is not necessary unless there is a diagnosed nutrient deficiency. There is no scientific evidence to support the effectiveness of nutrients claimed to improve TMJ function; these include vitamins A, B-complex, C, and E; and the minerals calcium, iron, magnesium, and zinc.

Ten-State Nutrition Survey. Originally called the National Nutrition Survey; conducted between 1968 and 1970, covering about 40,000 persons in California, Kentucky, Louisiana, Massachusetts, Michigan, New York, South Carolina, Texas, Washington, and West Virginia. Nutritional status was evaluated using physical and dental examinations, anthropometric measurements, biochemical analysis for hematocrit and hemoglobin, and dietary data. Details of the findings are available from the U.S. Department of Health and Human Services or have been published in books and journals. See also *National Health and Nutrition Examination Survey (NHANES).*

Tetracycline. A group of antibiotics used to treat specific conditions including acne, bronchitis, syphilis, gonorrhea, and certain types of pneumonia. Milk and dairy products and foods high in calcium, iron, magnesium, and zinc inhibit drug absorption and should not be taken 2 hours before or after oral drug administration. Tetracycline decreases the synthesis of vitamin K by intestinal bacteria and increases the urinary excretion of vitamin C, riboflavin, folic acid, and niacin. Frequent gastrointestinal effects include nausea, vomiting, diarrhea, bulky loose stools, anorexia, stomatitis, glossitis, dysphagia, and epigastric burning. Tetracycline is excreted in breast milk and can have adverse effects; it may also discolor developing teeth. Brand names are Achromycin®, Declomycin®, Minocin®, Panmycin®, Sumycin®, Terramycin®, and Vibramycin®.

Tetrahydrofolic acid (THFA). Designated name for the compound tetrahydropteroylglutamic acid or tetrahydrofolacin; the most active form of *folic acid*. It acts as a carrier of 1-carbon fragments that are important for the synthesis of *purines* and *pyrimidines* and for methylation reactions.

Tetraiodothyronine (T_4). See *Thyroxine*.

Textured vegetable protein (TVP). Fabricated food product made from vegetable protein sources such as peanuts, sesame seeds, soybeans, and wheat and suitably flavored, colored, and textured to simulate commonly used foods such as bacon, beef, and chicken. Useful for select vegetarians and for food allergies.

Theine. Alkaloidal stimulant in tea chemically identical to caffeine.

Theobromine. Alkaloidal stimulant in cocoa beans; also occurs in tea leaves and cola nuts. It is closely related to caffeine and used as a diuretic, an arterial dilator, and a myocardial stimulant.

Theophylline. A bronchodilator used to treat bronchial asthma, bronchitis, and emphysema. Anorexia, epigastric pain, and abdominal cramps are common side effects. High fat intake promotes rapid absorption and increased risk of side effects. Char-broiled meats increase rate of drug metabolism, thus reducing duration of effect. High-carbohydrate, low-protein diets reduce rate of drug metabolism and can increase risk of adverse effects (dizziness, flushing, headache) and toxicity. Brand names include Bronkodyl®, Slo-bid®, Somophyllin®, and Theo-dur®.

Thermal effect of food (TEF). Also known as "thermogenic" or "thermic effect of food." Formerly called "specific dynamic action (SDA) of food." Energy expended when food is digested, absorbed, and metabolized. It refers to an increase in metabolism after eating, which varies with the quantity and type of food consumed. The thermic effect of protein is about 30%; of fat, 13%; and of carbohydrate, 5%. For a mixed diet, the thermic effect is about 5% to 10% of total calories needed for basal metabolism and physical activity. See *Energy requirement*.

Thermogenesis. The reaction of the body to produce extra heat by shivering or by increasing the metabolic rate.

THFA. Abbreviation for *tetrahydrofolic acid*.

Thiamin (thiamine). Vitamin B_1, one of the water-soluble B-complex vitamins. It was formerly called "aneurine," "antiberiberi vitamin," "antineuritic factor," "vitamin F," "oryzamine," and "morale vitamin." As a component of the coenzyme *thiamin pyrophosphate,* it functions in oxidative decarboxylation of keto acids and in the transfer of glycolaldehyde in the pentose phosphate pathway. Thiamin also helps maintain normal nervous system activity and regulates the muscle tone of the gastrointestinal tract. Early signs of thiamin deficiency include loss of appetite, irritability, depression, gastrointestinal disturbances, and easy fatigability. A severe deficiency is clinically recognized as *beriberi,* the primary symptoms of which involve the nervous and cardiovascular systems. The causes of the deficiency are several and include inadequate intake due to diets dependent on milled and unenriched grains such as rice and wheat; chronic ingestion of raw fish containing microbial *thiaminase,* which destroys the vitamin; chronic alcoholism, in which there is not only a low intake of thiamin (and other B vitamins) but also impaired absorption and increased requirement; and thiamin-responsive inborn errors of metabolism. Other persons at risk are patients undergoing long-term renal dialysis or intravenous feeding and those with chronic febrile infections. Good food sources are brewer's yeast, unrefined or enriched cereal grains, organ meats, lean pork, legumes, and nuts. The vitamin is very soluble in water. Cooking losses are high if the cooking liquid is discarded and if high temperature and prolonged heating are employed. It is absorbed primarily from the duodenum. The body is incapable of storing the vitamin; excess quantities are excreted in the urine. There is no evidence of thiamin toxicity by oral administration, although there is some toxicity from large doses given parenterally. See Appendix 1 for the recommended dietary allowances.

Thiaminase. An enzyme that splits the thiamin molecule, thereby inactivating it and causing a thiamin deficiency. It is present in bracken fern, raw fish, and a variety of fruits and vegetables. Cooking inactivates thiaminase.

Thiamin pyrophosphate (TPP). Also called "cocarboxylase"; the thiamin-containing coenzyme that participates in the oxidative decarboxylation of alpha keto acids and in the formation of alpha ketols.

Thioridazine. A phenothiazine antipsychotic agent. It may induce riboflavin depletion, alter glucose tolerance, and increase serum cholesterol. It may also cause weight gain due to fluid retention. Liquid preparations are incompatible with enteral formulas and can block nasogastric tubes if given during formula administration. Adverse gastrointestinal effects include anorexia, constipation, dry mouth, and stomatitis. The brand name is Mellaril®.

Thiouracil. Antithyroid pyrimidine derivative that interferes with the formation of thyroxine; used in the treatment of thyrotoxicosis.

Thiourea. Also called "thiocarbamide"; an antithyroid substance used in the treatment of thyrotoxicosis. It inhibits the production of thyroxine by interfering with the incorporation of inorganic iodine into the organic form.

Thirst. A sensation or desire for water or other fluids. Thirst is usually an adequate guide for water intake, except in infants, the sick, and the elderly in whom thirst sensation is diminished, and in extreme heat or excessive sweating when thirst cannot keep pace with the actual requirement for water.

Threonine (Thr). Alpha-amino-beta-hydroxybutyric acid. An *essential amino acid* that participates in many of the reactions involving glycine and is important in purine synthesis and methylation reactions. Its metabolism is similar to that of serine, and both act as phosphate carriers in phosphoproteins. Deficiency results in arrested weight gain, glossitis, reddening of the buccal mucosa, and decreased serum globulin.

Thyrocalcitonin. Also called "calcitonin." Thyroid hormone having a significant effect on the calcium content of blood and bone. It is secreted in response to an elevated plasma calcium level and acts principally on bone, causing inhibition of bone resorption. Thyrocalcitonin is a polypeptide composed of 32 amino acids.

Thyroglobulin. Gelatinous iodine-containing protein synthesized by the *thyroid gland*. Hydrolysis of thyroglobulin yields *thyroxine* and other iodinated amino acids.

Thyroid gland. Butterfly-shaped endocrine gland consisting of two major lobes connected by a central isthmus. It is located in the neck just below the larynx. The thyroid gland has a unique ability to remove and concentrate blood iodide. This activity is influenced largely by the *thyrotropic hormone* and other chemical substances such as thiouracil, thiourea, thiocyanates, sulfonamides, and goitrogens. The chief function of the thyroid gland is to elaborate the thyroid hormones thyroxine and mono-, di-, and triiodothyronine. Thyroxine and triiodothyronine are the most active biologically. The thyroid hormones regulate metabolism by stimulating oxygen consumption. For the effects of hypo- and hyperfunction of the thyroid gland, see *Cretinism, Goiter, Hyperthyroidism, Hypothyroidism,* and *Myxedema.*

Thyronine. An amino acid that occurs in proteins only in the form of iodinated derivatives (iodothyronines), such as thyroxine. See *Thyroid gland.*

Thyrotropic hormone. Thyroid-stimulating hormone (TSH) or thyrotropin; a hormone secreted by the anterior lobe of the pituitary gland. It stimulates the thyroid gland to oxidize iodide to iodine and to release the thyroid hormones into the circulation.

Thyroxine. Also called "tetraiodothyronine"; the principal hormone of the thyroid gland. It is secreted into the blood bound to plasma protein (PBI) for transport to the tissues. Thyroxine regulates the rate of oxygen consumption in cells. It is also involved in growth and differentiation of the tissues.

Thyroxine-binding albumin (TBA). A thyroid hormone (triiodothyronine and thyroxine), a carrier protein of blood plasma.

Thyroxine-binding prealbumin (TBPA). A protein that binds the retinol-binding protein complex in the plasma and serves to carry vitamin A to the eye. It also binds the thyroid hormones thyroxine and triiodothyronine. See *Transthyretin.*

Tin (Sn). An ultra trace element; possibly essential based on its growth-enhancing effect when added to the purified diet of rats. The human body contains about 0.2 ppm of tin, or about 12 mg/60 kg of body weight, with the highest concentrations in liver and spleen. Naturally occurring tin deficiency is unknown either in animals or in humans. The average intake is about 3 to 4 mg/day, mostly in inorganic form. Ingested tin is poorly absorbed and is excreted mainly in the feces. There is no evidence of human toxicity from inorganic tin in foods; however, the widespread use of unlacquered tin and tin foil in cans and in packaged foods presents a potential hazard. Very high intakes of tin in experimental animals have produced changes in zinc and iron metabolism, with decreased hematocrit, hemoglobin, and serum iron. In humans, inhalation from the industrial environment may cause a mild lung condition called "pneumoconiosis."

TLC. Abbreviation for *total lymphocyte count.*

T lymphocytes. Type of white blood cells, or lymphocytes, traced from thymus-derived cells, which in turn come from the bone marrow. They are important cells for the immune system of the body.

TMJD. Abbreviation for *temporomandibular joint dysfunction.*

TNF. Abbreviation for *tumor necrosis factor.*

TOBEC. See *Total body electrical conductivity.*

Tocopherol. A generic name for all mono-, di-, and trimethyl tocols, which are complex alcohols of the chromanol type. Several tocopherols have been isolated, but only four forms have vitamin E activity (alpha-, beta-, gamma-, and delta-tocopherol). Alpha-tocopherol is the most potent biologically, and delta-tocopherol is the most active antioxidant.

Tocopherols occur naturally in certain plant oils, particularly in wheat germ; they can also be produced synthetically. See *Vitamin E.*

α-Tocopherol equivalent (α-TE). Since vitamin E occurs in foods in several forms and with varying biologic activity, the vitamin E activity for dietary purposes is expressed as RRR-α-tocopherol equivalent (α-TE). One α-TE is the activity of 1 mg of RRR-α-tocopherol or natural α-tocopherol (formerly termed *d*-α-tocopherol). To estimate the total α-TEs of mixed diets containing only natural forms of vitamin E, multiply the number of milligrams of β-tocopherol, γ-tocopherol, and alpha-tocotrienol respectively, by 0.5, 0.1, and 0.3. If all-*rac*-α-tocopherol is present, the number of milligrams is multiplied by 0.74.

Tolazamide. A *sulfonylurea* oral hypoglycemic agent used in the management of non-insulin-dependent diabetes mellitus. It is five times as potent as *tolbutamide*. Concomitant use with alcohol may cause disulfiram-like reactions. The drug is bound to plasma proteins. In protein malnutrition with hypoalbuminemia, more drug is available for a hypoglycemic effect. Hypoglycemia may also occur with inadequate food intake, with prolonged exercise without caloric supplementation, or with alcohol consumption. Excreted in breast milk. The brand names are Tolamide® and Tolinase®.

Tolbutamide. A *sulfonylurea* oral hypoglycemic agent used in the treatment of non-insulin-dependent diabetes mellitus. Concomitant use with alcohol may cause disulfiram-like reactions. There is a risk of hypoglycemia, although less than with *chlorpropamide* and *tolazamide*. Excreted in breast milk. The brand names are Oramide®, and Orinase®.

Tolbutamide test. Blood sugar determination before and 20 minutes after the intravenous administration of a solution containing 1 g of tolbutamide. A fall in blood sugar level by more than 89% in 20 minutes is diagnostic of diabetes mellitus.

Tolerance. 1. Limit to which substances can be ingested, absorbed, and metabolized with no deleterious physiologic effect. 2. Maximum limit established by the Food and Drug Administration to which additives may be incorporated in food.

Tolerex®. Brand name of an elemental formula consisting of free amino acids, glucose oligosaccharides, and safflower oil. See Appendix 39.

Tonsillectomy and adenoidectomy (T&A). Surgery performed to remove diseased tonsils and/or to correct adenoidal impairment. *Nutrition therapy:* postoperatively during the first 24 hours, give ample cold liquids, including gelatin, ice cream, and sherbet. Avoid milk and milk beverages if not tolerated. Gradually progress to soft foods on the second day and introduce warm foods and regular texture as healing occurs and as tolerated. The usual diet can be given within the next few days. See also *Cold semi-liquid diet* under *Diet, liquid.*

Tooth. One of the calcified structures supported by the gums of both jaws. Several nutrients are essential for proper tooth formation and calcification. Protein influences matrix formation in the enamel and dentin of the developing tooth. The presence of vitamin A affects the formation of the enamel matrix and the maintenance of the epithelium of the periodontal tissue. Vitamin C influences the formation of the collagen matrix in dentin, cementum, and the periodontal membrane. Vitamin D, calcium, and other minerals are needed for the calcification of enamel, dentin, and cementum. Deficiencies and excesses of several minerals (fluoride, calcium, phosphorus, etc.) affect the composition of the calcified tissues. See *Nutrition, dental health.*

TOPS. Abbreviation for *Take Off Pounds Sensibly*. A noncommercial self-help group concerned with the management and problems of obesity. See *Nutrition, weight control*.

Total arm length (TAL). Useful measurement to estimate the height of bedridden patients or persons with kyphosis or bone deformities (bowlegged, curved spine, bent knee gait, etc.). Using a nonstretchable tape, measure from the tip of the acromial process of the scapula at the shoulder to the end of the arm at the styloid process of the ulna at the wrist.

Total available glucose (TAG). A precise method of predicting the amount of glucose from foods. TAG values are based on the assumption that the metabolism of 100% carbohydrate, 58% protein, and 10% fat in foods is converted to blood glucose. TAG is useful in determining the amount of exogenous insulin needed to metabolize food, which is expressed as glucose to insulin ratio (G/I ratio). Generally, 10 TAGs require one unit of regular insulin.

Total body electrical conductivity (TOBEC). A method of measuring body composition based on the principle of differences in electrical conductivity of fat, muscle, and bone. Used to monitor weight loss, fitness of athletes, and to analyze fat content of tissues and food. See also *Bioelectrical impedance analysis (BIA)*.

Total energy expenditure (TEE). The sum total of the basal or resting energy expenditure (BEE or REE), the energy expended for physical activity, the thermal effect of food (TEF), and the energy associated with stress or pathologic conditions, minus the energy savings in sleep. Because TEF and energy savings during sleep are about the same, they are generally not included in the calculation. To estimate TEE, see Appendix 24.

Total iron binding capacity (TIBC). A measure of the concentration of serum iron that is equal to the sum of the iron bound to *transferrin* and the unsaturated binding capacity. TIBC is often increased in iron deficiency, in the third trimester of pregnancy, and in hypoxia. It is decreased during infections and in iron overload, cancer, protein-calorie malnutrition, chronic diseases, and conditions associated with loss of protein. The normal range is 250 to 400 μg/dL (45 to 72 μmol/liter). See also *Iron-binding capacity*.

Total lymphocyte count (TLC). An indicator of immune function and visceral protein stores. Determined from the formula:

$$\text{TLC} = \text{Total white blood cell count} \times \%\ \text{lymphocytes}.$$

A TLC of 1200 to 2000/mm^3 indicates depressed immune competence associated with mild protein depletion; 800 to 1200/mm^3, moderate depletion; and less than 800/mm^3, severe protein depletion. Values less than 2500/mm^3 may be abnormal in infants less than 3 months of age. TLC is not always useful in detecting malnutrition. It shows marked fluctuations daily and is affected by other factors such as blood dyscrasias, infection, chemotherapy, and immunosuppressive therapy.

Total parenteral nutrition (TPN). A system of feeding that bypasses the gastrointestinal tract. A nutritionally adequate solution is administered through the veins. In TPN, calories are supplied by carbohydrate in the form of dextrose, fat as a lipid emulsion, and protein as a crystalline amino acid solution. Vitamins, minerals, and trace elements are added together with certain medications as needed. There are two routes: through the subclavian or internal jugular vein that leads to the large central vein (called "central parenteral nutrition or CPN") and through the peripheral vein (called "peripheral parenteral nutrition or PPN"). Generally, PPN is indicated when short-term nutritional support is necessary (<7 days). It is a useful supplement in patients who are unable to meet nutritional requirements completely by the oral or enteral feeding. When a standard solution for PPN administration is not adequate to meet the increased demands for more nutrients and calories, hypertonic solutions are given by CPN, which is sometimes called "parenteral hyperalimentation." Typically, TPN is infused continuously over the entire day. It is usually started at a slow rate of 25 to 50 mL/hour. The rate is increased at 12- to 24-hour intervals until the appropriate calorie needs and fluid volume are met. Cyclic TPN refers to the intermittent infusion of intravenous solution over a specified period of time, usually at night. It is useful for patients who need home TPN so that they can resume normal activities during periods when the TPN is not administered. Cyclic TPN is especially useful for cases of fatty liver. The decision to begin TPN is based on the presence or absence of malnutrition and on the attending medical problems. Indications when TPN should be used include: patients whose gastrointestinal tract are totally unavailable, those undergoing high-dose chemotherapy, radiation, or bone-marrow transplantation; persons who are unable to absorb nutrients via the gastrointestinal tract (e.g., severe diarrhea, massive small bowel resection, severe enteritis, acute pancreatitis); severely malnourished patients in the presence of a nonfunctional gastrointestinal tract; patients who cannot receive adequate nutrition by enteral feedings within 7 days (e.g., severe mental state or comatose state); and those who require increased nutrients due to severe catabolism and whose gastrointestinal tract has been nonfunctional for 5 to 7 days (e.g., severe trauma, major surgery, extensive body surface burns). The benefits of TPN in cancer patients remain unclear. Clinical settings where TPN should not be used include: patients who have functional gastrointestinal tract capable of absorbing adequate nutrients; patients whose prognosis does not need aggressive nutritional support; and cases when the risks of TPN outweigh potential benefits, and the patient or legal guardian is against aggressive nutrition support.

The main problems attending TPN are sepsis or infection and metabolic complications. The latter group includes hyper- or hypoglycemia, electrolyte imbalance, excess or lack of certain trace elements or vitamins, anemia, amino acid imbalance, essential fatty acid deficiencies, and many other complications that need nutrition intervention. Blood values that need monitoring are prothrombin time, glucose, BUN/creatinine, electrolytes, liver enzymes, CBC with differential count, serum triglycerides if lipid emulsions are used, bilirubin, and visceral proteins. Weight gain should not exceed ½ lb a day. See also *Parenteral nutrition*.

TPN. Abbreviation for (1) *total parenteral nutrition*. (2) Archaic term for *triphosphopyridine nucleotide* or coenzyme II, currently called "nicotinamide adenine dinucleotide phosphate."

TPP. Abbreviation for *thiamin pyrophosphate*.

Trace mineral. An element present in the body at less than 0.01% of total body weight or one part in 10,000; required by humans in minute amounts measured in mg or mcg/day. (see Appendix 2.) Trace minerals are essential to metabolic processes as components of enzyme systems (i.e., metalloenzymes) or as cofactors to activate enzymes.

Tracer technique. Research method that uses a radioactive element to follow the fate of a substance or its reactions. Compounds containing tracer elements are said to be "tagged" or "labeled."

Transamination. Transfer of an amino group from one compound to another, with *pyridoxal phosphate* acting as the intermediate amino carrier. The reaction is catalyzed by the enzyme *transaminase*. By this process the body is able to use ammonia and synthesize the nonessential amino acids.

Transcobalamin. Vitamin B_{12} bound to a protein; the transport form of the vitamin. There are three different cobalamin-binding proteins: transcobalamin I, II, and III.

Trans-fatty acid. See under *Fatty acid*.

Transferrin. A beta-globulin that transports iron in plasma. It is synthesized by the liver and has a half-life of 8 to 10 days. Transferrin transports iron from the intestine into the bloodstream and distributes iron where it is needed throughout the body. The percentage of transferrin saturation indicates the amount of iron in blood available for use by the bone marrow for erythropoiesis. The normal range for serum transferrin is 250 to 300 mg/dL (2.5 to 3.0 g/liter); values of 150 to 200 mg/dL (1.5 to 2.0 g/liter) indicate mild depletion; 100 to 150 mg/dL (1.0 to 1.5 g/liter), moderate depletion; and less than 100 mg/dL (1.0 g/liter), severe depletion. Transferrin is decreased in acute catabolic states, protein-losing enteropathy and nephropathy, chronic infections, and uremia. It is increased during pregnancy, estrogen therapy, iron deficiency, and acute hepatitis. Serum transferrin is measured by radial immunodiffusion, but a close estimate can be obtained from the more widely available measurement of total iron-binding capacity (TIBC), using the following formula:

$$\text{Transferrin} = (0.68 \times \text{TIBC}) + 21$$

Transitional feeding. Nutritional support during the time when the feeding method is being changed to another method. The transition can be from tube feeding to oral feeding or from parenteral nutrition to oral or tube feeding. The transition should be done gradually. For example, when changing from tube to oral feeding, start by stopping the tube feeding for 1 hour before and after each meal. Stop the tube feeding during the day when oral intake is consistently about 500 to 750 kcal/day (or 30% of total kcal/day) to give the patient a chance to regain appetite and increase oral intake. Supplemental tube feeding can be given overnight. The feeding tube can be removed if the patient has consistently consumed two-thirds of the nutrient needs orally for 3 to 5 days.

Transketolase. Enzyme found in blood cells, liver, and other tissues; necessary for the synthesis of the 5-carbon sugars found in DNA and RNA. It requires thiamin pyrophosphate as a coenzyme.

Transmanganin. Protein carrier that transports manganese in the blood. See *Manganese*.

Transmethylation. Transfer of a methyl radical ($-CH_3$ group) from one compound to another. This reaction is important in intermediary metabolism, particularly in fat, sulfur, and creatine metabolism. Vitamin B_{12} and folic acid are involved in the synthesis of methyl groups. Methionine is considered the primary methyl donor in transmethylation reactions. Choline and betaine are also methyl donors.

Transplantation. The transfer of a graft (organ or tissue) from one body area or person to another site or person to replace a diseased organ or tissue. Skin, whole organs, and bone marrow may be transplanted. Solid organ transplantation specifically refers to the heart, kidney, liver, lung, or pancreas, when the organ has reached the end-stage and is refractory to other medical treatment. Rejection and infection are two major complications. Immunosuppressive drugs are given in large doses as a prophylactic measure. These drugs have many side effects that adversely affect nutrient intake and contribute to severe depletion of protein and other nutrients. Organ transplant recipients undergo three phases of care: pretransplantation phase, immediate post-transplantation phase, and chronic post-transplantation phase. *Nutrition therapy:* the main goal is to promote anabolism and prevent infections. Precautions for a low-bacterial (microbial) diet are followed in order to minimize infections from exposure to foodborne pathogens. Prior to transplantation, nutriture should be improved in order to decrease surgical risk and postsurgical complications. Give nutritional supplements as needed. In the second phase or about 8 weeks after transplantation, the goal is to support the hypercatabolic demands of postsurgery and immunosuppressive therapy. Enteral or parenteral feedings may be necessary if oral intake is inadequate to meet the needs for energy and protein. Nutritional support may be provided by a combination of oral, enteral, and parenteral feedings. Most patients are usually able to resume oral intake within a few days after surgery. Energy needs are based on dry weight. On the average, heart, liver, kidney or lung transplants require 35 kcal/kg/day; pancreatic transplant needs 30 to 35 kcal/kg. For most post-transplants, an average range for protein is 1.3 to 1.5 g/kg dry weight per day. Increased protein intake is recommended due to the effects of surgical stress, muscle catabolism, and glucocorticoids. Protein restriction is not needed when a patient undergoes dialysis. Monitor serum proteins, which tend to be decreased, until organ function is normalized. Albumin is commonly depressed in all organ transplants, but in renal failure, prealbumin and retinol-binding proteins are usually increased. Total lymphocyte count is reduced, especially due to the effects of immunosuppressive medications. Carbohydrates should provide 50% to 60% of nonprotein energy needs; limit to 50% only when there is glucose intolerance or when hyperglycemia persists. Fat provides 30% to 50% of nonprotein calories, but restrict if hyperlipidemia occurs. Depending on renal function, post-transplant patients experience fluid overload. Nutrition support is concentrated to a low volume, but in amounts to prevent dehydration. The requirements for vitamins and minerals vary according to serum levels and kind of organ transplantation. Nutritional rehabilitation is continued to the third phase until the patient is back to normal. For more specific nutrition therapy, see *Bone marrow transplantation, Heart transplantation, Kidney transplantation, Liver transplantation,* and *Pancreas transplantation.*

Transport protein. See *Protein, functions.*

Transthyretin. Also called "thyroxine-binding prealbumin (TBPA)." A plasma protein that serves as a secondary carrier of thyroxine and exists as a complex with *retinol-binding protein*. The very short half-life of this protein (2 days) makes it extremely sensitive to a decreased protein and energy intake of a few days duration. It declines in 3 days in response to a lowered energy intake, even when protein intake is adequate. Normal ranges are between 16 and 36 mg/dL. Values of 5 to 10 mg/dL are indicative of moderate depletion; less than 5 mg/dL, severe protein depletion.

Trauma. Severe injury caused by an accident or injury, violent disruption, or ingestion of a toxic substance. Trauma is also classified according to etiology as thermal (burns), neurologic (central nervous system or brain), severe emotional shock, chemical (toxic agent), and physical (as in multiple fractures and major surgery). Internal risk factors include the nutritional status of the person. The immediate reaction to trauma is a high plasma level of catecholamines, glucocorticoids, and glucagon. Blood glucose is elevated due to impaired utilization or insulin resistance by muscles; protein and fat catabolism is increased; gut absorption is impaired; and the electrolyte/water balance is affected. *Nutrition therapy:* specialized nutrition support is needed, usually by *tube feeding,* if the gastrointestinal tract is functional. In severe cases, as in serious head injuries, intravenous feeding is a more effective method. Return to oral feedings only when there is no vomiting and the patient can gradually resume normal digestion and absorption. The goal is to maintain body cell mass, meet metabolic needs, and support the immune system while avoiding the complications of overfeeding. Energy and protein requirements vary throughout the various stages of trauma. Depending on the extent of injury, the protein requirement may range from 1.8 to 2.4 g/kg/day and may be higher (2.5 to 3.0 g/kg/day) in septic and severely traumatized patients. Maintain a calorie-to-nitrogen ratio of 100:1. Provide adequate hydration and monitor serum levels of protein, glucose, and electrolytes. The body becomes anabolic between the fifth and tenth days post injury, and albumin usually starts to increase in the second week. Repletion of body fat follows restoration of lean body mass, and normal body weight is regained about 50 days after the initial trauma, providing there are no complications. Supplemental vitamins and minerals are recommended, especially vitamin C, zinc, and vitamin A to promote healing. See also *Traumatic brain injury*.

Traumacal®. Brand name of a high-nitrogen, restricted-carbohydrate formula for stressed patients with hyperglycemia. Contains branched-chain amino acids (23%) plus essential and nonessential amino acids, corn syrup, sucrose, MCT oil, and soy oil. See also Appendix 39.

Traum-Aid® HBC. Brand name of a chemically defined formula high in branched-chain amino acids (50%) plus essential amino acids. For oral or tube feeding of hypercatabolic patients and in hepatic insufficiency. See also Appendix 39.

Traumatic brain injury (TBI). Or head trauma; a sudden blow or impact to the head causing severe injury to the brain. This activates the systemic metabolic responses, resulting in hypermetabolism with increased energy expenditure, increased protein catabolism and urinary nitrogen excretion, and depletion of lean body mass. *Nutrition therapy:* immediate nutritional support by either the enteral or parenteral route is critical to meet the hypermetabolic demand for increased energy and protein. Provide 45 to 50 kcal/kg body weight (1.8 to 2.0 × basal energy expenditure) and 2.0 to 2.5 g protein/kg of body weight. Start tube feeding at half strength or at a slower rate if the solution is hypertonic. (Start TPN if enteral feeding is not well tolerated after 3 days.) Most patients are able to tolerate the full volume required by the second week. Fluid restriction may be necessary to minimize brain edema. Monitor weight, serum albumin, glucose, and electrolytes. If the patient is conscious and able to swallow, gradually introduce oral feeding and watch for any problem with aspiration. Modify food consistency as tolerated. See *Dysphagia*.

Travasol®. Brand name of an amino acid solution for total parenteral nutrition. Contains both essential and nonessential amino acids in 3.5%, 5.5%, 8.5%, and 10% concentrations, with and without electrolytes. See also Appendix 40.

Travasorb®. Brand name of a line of products for specialized nutritional support by tube feeding or as an oral supplement. Travasorb® Hepatic contains amino acids as a protein source, high in branched-chain amino acids (50%) and low in aromatic amino acids (2%) for oral supplementation in liver failure; Travasorb® MCT contains medium-chain triglycerides, with a high ratio of medium- to long-chain triglycerides (80:20 w/w); Travasorb® Renal is a low-protein (from amino acids), high-calorie, electrolyte-free powder for use in renal failure; Travasorb® STD and Travasorb® HN contain peptides from enzymatically hydrolyzed lactalbumin in soluble powder form. See also Appendix 39.

Triamterene. A potassium-sparing diuretic that is structurally related to folic acid. Used in the treatment of edema associated with congestive heart failure, liver cirrhosis, or nephrotic syndrome. It is a folic acid antagonist and can cause folate depletion; it may cause hyperkalemia, decrease serum vitamin B_{12}, and increase urinary excretion of calcium, sodium, and chloride. Possible gastrointestinal effects are dry mouth, nausea, vomiting, diarrhea, and gastric distress. Excreted in breast milk. The brand name is Dyrenium®.

Tricarboxylic acid (TCA) cycle. Krebs cycle or citric acid cycle; the final common pathway of energy metabolism for carbohydrate, protein, and fat. See *Krebs cycle*.

Triceps skinfold (TSF). Anthropometric measurement to estimate subcutaneous fat indirectly. Together with midarm muscle circumference, it also gives an estimate of skeletal muscle mass. Measurement of a skinfold is taken at the midpoint between the acromion of the scapula (bony protrusion on the posterior of the upper shoulder) and the olecranon of the ulna (bony part of the elbow). A decrease in triceps skinfold thickness reflects a chronically inadequate nutritional intake. Measurements greater than the 50% percentile are acceptable; those in the 40th to 50th percentile indicate mild depletion; in 25th to 39th percentile, moderate depletion; and below the 25th percentile, severe fat depletion. See Appendix 22.

Triglyceride. Also called "triacylglycerol." *Fat* in which the glycerol molecule has three fatty acids attached to it. Chemically, triglycerides comprise about 95% of dietary fats. Triglycerides are transported on very-low-density lipoproteins (VLDL) and chylomicrons to tissues for use as fuel or to adipose tissue for storage. Circulating triglyceride levels are influenced by genetics, dietary factors (calories, fat, simple sugars, alcohol intake), and disease (diabetes mellitus, obesity, pancreatitis, fatty liver). Elevated serum triglyceride levels are generally associated with low levels of high-density lipoprotein (HDL)-cholesterol which is an independent risk factor for coronary heart disease (CHD). See *Hypertriglyceridemia*. See also *Hyperlipoproteinemia* and *Diet, fat-modified*.

Trigonelline. An inactive form of *niacin* that is found in seeds and nuts. Roasting coffee beans activates this substance.

Triidothyronine (T_3). A thyroid hormone that exerts the same effects as thyroxine, but is present in much smaller amounts. See *Thyroid gland.*

Trimethoprim. A synthetic folate antagonist antibacterial used in the treatment of urinary tract infections caused by susceptible organisms. High doses for prolonged periods (more than 6 months) can cause folate deficiency, especially in geriatric, malnourished, alcoholic, pregnant, or debilitated patients; in patients receiving other folate antimetabolites, such as the phenytoin anticonvulsants; and when the dietary folate intake is low. Supplementation with folic acid may be necessary. Brand names are Bactrim®, Proloprim®, Septra®, and Sulfatrim®.

Triose. A sugar that contains 3 carbon atoms. It is an intermediate product of metabolism and does not occur naturally. *Glyceraldehyde* is a triose.

Trisaccharide. An oligosaccharide containing three monosaccharide units. Examples are *raffinose* (has fructose, glucose, and galactose) and *melezitose* (has two molecules of glucose and one molecule of fructose). See Appendix 8.

TrophAmine®. Brand name of a 6% amino acid solution especially formulated for pediatric total parenteral nutrition. Contains taurine; is high in branched-chain amino acids and low in phenylalanine, methionine, and glycine. See also Appendix 40.

Trypsin. Proteolytic enzyme of the pancreas secreted as the inactive precursor *trypsinogen.* It is an endopeptidase and catalyzes the hydrolysis of peptide linkages containing the carboxyl group of lysine and arginine, yielding polypeptides with C-terminal lysine and arginine groups.

Trypsin inhibitor. A substance capable of reducing the activity of the proteolytic enzymes in the digestive juices; it can also slow down the absorption of some amino acids, either by reducing the utilization of nitrogenous material in food or by increasing the needs of the organism for certain amino acids. It is found in raw egg white, soybeans, peanuts, peas, beans, and lentils.

Tryptophan (Try). Alpha-amino-beta-indolylpropionic acid. An *essential amino acid* for humans and animals. It is frequently a limiting amino acid for tissue synthesis. Tryptophan is the only amino acid with an indole nucleus; it can be converted to nicotinic acid, serotonin, and melatonin. Deficiency results in weight loss, impaired nitrogen retention, and decreased plasma cholesterol.

TSF. Abbreviation for *triceps skinfold.*

TSH. Abbreviation for *thyroid-stimulating hormone.* See *Thyrotropic hormone.*

Tube feeding (TF). Delivery of enteral nutrition via a feeding tube into the gastrointestinal tract as a nutritional supplement when oral intake is inadequate, as transitional feeding when parenteral support is being discontinued, or as the only source of nourishment for patients who are unable or unwilling to eat. The feeding tube can be inserted, by a surgical or nonsurgical procedure, at various sites along the gastrointestinal tract for a nasogastric, nasoduodenal, nasojejunal, esophagostomy, gastrostomy, and jejunostomy placement. The nasoenteric route is usually indicated for short-term feeding, whereas the feeding enterostomies are for long-term use. Risk for aspiration, gastric emptying, reflux, obstruction, and other problems also determine tube placement. Tube feeding formulas vary from blenderized mixture of foods to a wide selection of liquid nutritional products that are monomeric, polymeric, fiber-containing, lactose-free, nutrient-dense, modular, or specialized for specific therapeutic needs. (See Appendix 39.) The formula may be administered by bolus feeding, by gravity via continuous drip or intermittent feeding, or by cyclic continuous infusion. In general, continuous feeding is used for head-injured or acutely ill patients. For patients who have not eaten for several days, the initial feeding rate is usually started at 50 mL/hour, with increases of 25 mL/hour every 8 to 12 hours until the volume goal is reached. A higher starting rate of up to 150 mL/hour can be given to those who were eating previously. Gastric residual is checked every 4 hours; if it exceeds 150 cc, the feeding is interrupted and given later. Other solutions for some of the common tube feeding problems are as follows: flush the tube with 30 to 50 cc of water before and after each feeding or every 4 hours during continuous feeding (tube obstruction or clogging); elevate the head of the bed 30 to 45 degrees during and for 2 hours after tube feeding (aspiration pneumonia, gastric retention); use soft, small-bore feeding tubes and keep the mouth and lips moist (tube discomfort); lower the infusion rate, dilute the formula or use isotonic strength if gastric residuals are high, increase the rate and/or concentration over several days (nausea, vomiting, cramping, and distention); use a formula with fiber and add free water if intake is not greater than output by 500 to 1000 mL/day (constipation); dilute the formula or reduce rate of administration if formula is hypertonic, or consider using an isotonic formula (hyperosmolar dehydration); restrict fluids or use a concentrated formula (fluid overload, overhydration); use an isotonic formula, slow down the rate or decrease bolus volume per feeding, check infection control for tube feeding administration (diarrhea). Low serum albumin and certain medications, especially antibiotics, are the usual cause of diarrhea in tube-fed patients. Drug interaction with the formula may degrade or inactivate nutrients or alter the absorption and availability of the drug. Certain drugs such as ibuprofen, psyllium, theophylline, and hydrochlorothiazide are incompatible with enteral formulas and cause clumping of the formula and clogging of the tube. The following are general guidelines for drug administration via the feeding tube: do not crush enteric coated tablets and sustained-release capsules; reconstitute capsules with 10 to 15 mL of water; avoid mixing drugs with enteral formulas in the feeding bag; flush the feeding tube with 15 to 30 mL of water before and after drug administration; administer each drug separately and flush the feeding tube with at least 15 mL of water between each drug; contents of hard gelatin capsules may be poured and mixed in water; compressed tablets may be crushed and thoroughly mixed in 15 to 30 mL of water; dilute with 15 to 30 mL of water prior to administration of viscous liquid drugs, hypertonic or irritating drugs, and drugs that are usually administered with meals to avoid gastrointestinal irritation.

Tuberculosis. An infectious disease caused by the tubercle bacillus, *Mycobacterium tuberculosis,* which invades the lungs. It is usually transmitted by airborne droplets produced by a person with untreated tuberculosis of the lung or larynx. Other organs may become infected by other modes of transmission. Active tuberculosis is characterized by weight loss and tissue wasting, anorexia, fever, night sweats, coughing and expectoration, and hemoptysis. Tuberculosis is a public health problem in the United States. It is common among persons with HIV infection. The incidence has increased with the development of multiple drug resistant strains of tuberculosis (MDRTB). Thus the infection remains active for a longer period of time. *Nutrition therapy:* provide a liberal intake of protein (1.2 to 1.5 g/kg to restore plasma protein and promote wound healing) and a high caloric level (1.5 times more than the individual's needs) due to fever. Supply adequate minerals, especially calcium to help in the calcification of the lesions, and iron, plus other hematopoietic nutrients in case of hemorrhaging. If the drug used is isoniazid, increase vitamin B_6 (isoniazid is an antagonist for pyridoxine) and folic acid.

Tumor necrosis factor (TNF). A hormone-like protein produced in the body and released during illness and in response to toxic substances; it has anticancer effects. TNF kills tumor cells, induces fever, influences inflammatory responses, accelerates lipolysis, alters glucose metabolism in muscle, induces anorexia, and contributes to fat depot depletion and cachexia in chronic illness.

TwoCal® HN. Brand name of a high-calorie, high-protein supplement for severely hypermetabolic and stressed patients on fluid restriction. Provides 2 kcal/mL and contains 84 g of protein per liter. See also Appendix 39.

Type A Lunch. The Food and Nutrition Services under the U.S. Department of Agriculture conduct this program using federal funds and government surplus commodities. Type A lunch must provide one-third of a child's recommended dietary allowances (RDA). See also *National School Lunch Program*.

Tyramine. A pressor amine similar in action to epinephrine; found mainly in foods and beverages that have undergone bacterial decomposition during the process of fermentation, aging, pickling, or spoiling. It is normally degraded in the body by the enzyme monoamine oxidase (MAO). Inhibitors of this enzyme interfere with this process, causing the release of norepinephrine, a powerful vasoconstrictor. The concomitant ingestion of tyramine-rich foods and MAO inhibitors can trigger a hypertensive crisis. As little as 6 mg of tyramine may cause increased blood pressure, nausea, and vomiting; intakes of 10 to 25 mg may induce severe headaches with possible intracranial hemorrhage; and intakes exceeding 25 mg may result in severe hypertensive crisis. See *Monamine oxidase* and *Diet, tyramine-restricted*.

Tyrex®. An amino acid-modified dietary product without phenylalanine and tyrosine. Used in feeding children and adults with Type II tyrosinemia.

Tyromex™. An amino acid-modified dietary product free of phenylalanine, tyrosine, and methionine. Used in feeding infants and toddlers with Type I tyrosinemia.

Tyrosinase. Copper-containing enzyme that oxidizes tyrosine and other phenolic compounds, forming brown to black pigments. Lack of this enzyme results in *albinism*.

Tyrosine (Tyr). Alpha-amino-beta-hydroxyphenylpropionic acid. A nonessential amino acid that has some sparing action on the essential amino acid phenylalanine. It participates in transamination reactions and is the starting material for the synthesis of melanin, thyroxine, and epinephrine. Deficiency results in impaired nitrogen retention and low levels of catecholamine and thyroxine. Tyrosine is an essential amino acid for children with phenylketonuria (PKU).

Tyrosinemia. An inherited disorder characterized by an elevated blood tyrosine level due to an enzyme deficiency in tyrosine metabolism. Type I tyrosinemia, or *tyrosinosis*, affects predominantly the liver and kidneys; it is due to a deficiency in the enzyme fumarylacetoacetase. Type II tyrosinemia affects the eye and skin,

U

U. Abbreviation for Unit, as in IU (International Unit.)

u. Symbol for micro-; often written as μ.

UAC. Abbreviation for *upper arm circumference*.

UAMA. Abbreviation for *upper arm muscle area*.

Ubiquinone. General term for a group of related quinones with variable numbers of isoprene residues. Ubiquinone is found in the mitochondria and serves as an electron transport agent. See *Coenzyme Q*.

UFA. Abbreviation for unsaturated fatty acid.

UKM. Abbreviation for *urea kinetic modeling*.

Ulcer. An eroded lesion or an excavated sore. See *Peptic ulcer disease, Pressure sore,* and *Ulcerative colitis*.

Ulcerative colitis. A chronic inflammatory disease of the large intestine, or colon and rectum, characterized by diarrhea, fever, abdominal cramps, anemia, and weight loss. The mucosal and submucosal tissue layers of the large intestine are usually inflamed for brief periods. The diarrhea may contain varying amounts of mucus, pus, and blood. Anemia may be present as a result of blood loss. Surgery of the affected area may be required if treatment with corticosteroids and anti-inflammatory drugs is not effective. *Nutrition therapy:* in the acute stage, use a minimal-residue diet to avoid undue irritation to the colon. In intractable cases, total parenteral feeding is needed. As the patient improves, oral feeding is resumed, using a fiber-restricted diet high in protein (1.5 g/kg) and calories (40 to 50 kcal/kg). Use preparations high in medium-chain triglycerides for better fat digestion and utilization. Restrict lactose or avoid during flare-ups if there is intolerance; reintroduce during periods of remission. Vitamin and mineral supplements, particularly folate (if on sulfasalazine), calcium, chromium, copper, selenium, and zinc, are recommended. See also *Crohn's disease*.

Ultra-. A combining form meaning "beyond a certain limit or more than the normal range."

Ultracal®. Brand name of a fiber-containing formula for tube feeding; has a fiber content of 14.4 g/liter. It is lactose-free and isotonic, and contains MCT oil, and a blend of oat and soy fibers. See Appendix 39.

Ultralan™. Brand name of a lactose-free, nutrient-dense formula for patients requiring a hypercaloric regimen with fluid restriction. See Appendix 39.

Ultratrace minerals. Elements with estimated dietary requirements of less than 1 mg per day. These include chromium, fluoride, iodine, manganese, molybdenum, and selenium. Ultratrace minerals known to be essential to animals and probably also essential for humans are arsenic, boron, cadmium, nickel, silicon, tin, and vanadium.

Undernutrition. A form of *malnutrition*. A deficiency state due to lack of calories and/or one or more of the essential nutrients. For examples of undernutrition, see *Anorexia nervosa, Kwashiorkor, Marasmus,* and *Nutrition, starvation.*

Underweight. Term applied to individuals whose body mass index (BMI) is below 20 kg/m^2 or body weight is more than 10% below the established standard for individuals of the same age, sex, and height. Underweight may be caused by inadequate dietary intake, excessive activity, poor absorption and utilization of food, a wasting disease such as cancer or hyperthyroidism, and psychologic or emotional stress. Malnutrition may accompany the underweight. A weight that is 30% below the standard (BMI of 15 kg/m^2) is life threatening starvation. *Nutrition therapy:* theoretically, an excess of 500 kcal/day results in a weekly gain of 1 lb. A gradual gain of 1 to 2 lb/week is a desirable rate. A high-calorie, high-protein diet is best given in six to eight feedings per day. Adjust protein level according to the degree of protein depletion. Supplemental vitamins and minerals may be necessary.

UNICEF. Abbreviation for United Nations International Children's Emergency Fund. An organization that aims to help children all over the world by eradication of disease, improvement of health, and provision of emergency relief rations by milk distribution and school feeding programs and through the establishment of maternal and child health care centers.

United States Department of Agriculture (USDA). A department of the federal government whose goals are to improve farm income and develop markets for agricultural products; to reduce and cure poverty, hunger, and malnutrition; to ensure standards of food quality through inspection and grading services; and to conduct research and nutrition programs. Examples of the last are the Food Stamp Program, the National School Lunch Program, the Child Care Food Program, the WIC Program, the Special Milk Program for Children, the School Breakfast Program, and the Commodity Supplemental Food Program. The divisions directly concerned with nutrition include the Extension, Human Nutrition Center, Food and Nutrition Service (FNS), Food Safety and Quality Service (FSQS), Agriculture Research, Cooperative Research, and Technical Information Service.

United States Department of Health, Education and Welfare (USDHEW). Created on April 11, 1953, it was renamed the Department of Health and Human Services (DHHS) on October 17, 1979.

United States Department of Health and Human Services (USDHHS). The goals of this department are to serve people ranging from newborn infants to the oldest citizens; to advise the president on health, welfare, and social security plans; and to form policies related to human development, general health, and welfare. Some of the offices *directly* concerned with nutrition are the Administration on Aging (AOA); Administration for Children, Youth, and Families (ACYF); Administration for Native Americans (ANA); Public Health Service (PHS); President's Council on Physical Fitness and Sports; National Center for Health Statistics; Centers for Disease Control (CDC); Food and Drug Administration (FDA); National Institutes of Health (NIH); National Cancer Institute (NCI); National Heart, Lung, and Blood Institute (NHLBI); and the Alcohol, Drug Abuse, and Mental Health Administration.

United States Pharmacopeia (USP). The standard weight reference for nutrients in the United States. The USP standards for ascorbic acid, calcium pantothenate, choline, chloride, nicotinamide, nicotinic acid, pyridoxine hydrochloride, riboflavin, thiamin hydrochloride, essential amino acids, and vitamins A and D are available to the public. When an international standard exists, the USP standard is compared and brought as closely as possible into agreement. A USP unit is therefore equal to an international unit.

Unsaturated fatty acid (UFA). See under *Fatty acid.*

Upper arm circumference (UAC). Also called "midarm circumference (MAC)." Combination of muscle mass, bone size, and subcutaneous fat deposits. Measurement is taken midway between the tip of the acromial process of the scapula and the olecranon process of the ulna. UAC measures skeletal mass and fat stores but is not a sensitive method by itself; it is used in formulas for other anthropometric measurements.

Upper arm muscle area (UAMA). A reliable indicator of lean body mass and skeletal protein reserves. Useful in evaluating protein-energy malnutrition. It is determined from the following formula:

$$\text{UAMA (mm}^2) = \frac{(\text{MAC, mm} - \text{TSF mm} \times 0.314)^2}{4 \times 0.314}$$

where MAC is *midarm circumference* and TSF is *triceps skinfold* measurement. To measure the bone-free arm muscle area, subtract 10 (for males) and 6.5 (for females) from the results of this formula.

Urate. A salt of uric acid. An increased amount of urates in the urine is called "uraturia." Urates may be deposited as crystals in body joints or as calcareous deposits in tissues. See also *Gout.*

Urea. The diamide of carbonic acid. It is the major end product of human nitrogen (protein) metabolism and the chief nitrogenous constituent of *urine*. Urea formation occurs chiefly in the liver. See *Blood urea nitrogen.*

Urea clearance test. Test that measures the quantity of blood "cleared" of urea per minute to determine renal function.

Urea cycle. The overall reactions of the urea cycle proceed as follows: the combination of CO_2, NH_3, and adenosine triphosphate forms carbamyl phosphate, which combines with ornithine to form citrulline. The latter combines with aspartic acid to form argininosuccinic acid, which is broken down to arginine and fumaric acid. The last reaction of the urea cycle is the hydrolytic cleavage of arginine to yield urea and ornithine.

Urea kinetic modeling (UKM). A computerized tool for the clinical assessment of protein metabolism and nitrogen balance; an important aid in monitoring the adequacy of dialysis and protein intake. UKM is used in dialysis units to select treatment parameters and optimize the diet. It is based on the relationship between the production of urea (a major metabolite of protein degradation) and removal of urea (by residual kidney function and dialyzer).

Urease (urase). Specific enzyme that decomposes urea, forming ammonia and carbon dioxide.

Uremia. Excessive retention of urinary constituents in the blood. It is a toxic condition and the terminal manifestation of *renal failure* or end-stage renal disease. The clinical features include nausea and vomiting, anorexia, dizziness, anemia, convulsions, and coma. *Nutrition therapy:* protein is restricted to 15 to 20 g/day, supplied mainly by one egg and 4 to 6 oz milk. Provide sufficient calories to prevent breakdown of body tissues. Nonprotein calories should come from fats, sugars, fruits and vegetables, and special low-protein, high-calorie commercial products. See *Diet, minimal-protein,* under *Diet, protein-modified*.

Ureterolithiasis. Formation of a stone or calculus in the ureter, which is a long, narrow tube that conveys the urine from the pelvis of the kidneys to the bladder. The surgical removal of a stone in the ureter is called *ureterolithotomy*. See also *Urolithiasis*.

Uric acid. The end product of purine metabolism in humans and of protein metabolism in birds and some reptiles. In humans, uric acid is excreted in the urine in the free state and as the urates of sodium, potassium, and ammonium. It is formed in part from purines taken in food (exogenous uric acid) and in part from body purines as a result of the breakdown of nucleic acids (endogenous uric acid). Abnormal metabolism of uric acid is characteristic of *gout*.

Urico-. A combining form meaning "pertaining to uric acid."

Uridine. Nucleoside that consists of uracil and ribose. This is obtained by the removal of phosphate from uridylic acid.

Uridine diphosphate glucose (UDPG). The glucose ester of uridine diphosphate formed from glucose-1-phosphate in the presence of a pyrophosphorylase. It is the prosthetic group of the enzyme responsible for the conversion of galactose to glucose.

Uridylic acid. Nucleotide containing ribose, phosphoric acid, and the pyrimidine uracil. The nucleotides of uridine monophosphate, uridine diphosphate, and uridine triphosphate function as coenzymes in a wide variety of reactions.

Urinalysis. Physical, chemical, and microscopic analyses of the urine. These include a description of color and clarity or turbidity; the determination of pH and specific gravity; and the observation of the presence or absence of abnormal constituents such as proteins (albumin), sugar, ketone bodies, casts, bacterial cells, pus, and blood cells. See Appendix 31.

Urinary calcium. Normally <300 mg (7.6 mmol) per 24 hours on a usual diet; elevated in hypercalciuria and may be as high as 800 to 900 mg (20.2 to 22.3 mmol) per 24 hours in hypercalcemia associated with metastatic tumor.

Urinary calculi (urolithiasis). See *Urolithiasis*.

Urinary chloride. Helps in differentiating between types of metabolic alkalosis. Values vary with intake. It is usually less than 10 mEq/liter (10 mmol/liter) when alkalosis is due to vomiting, gastric suction, or diuretic use; usually greater than 20 mEq/liter (20 mmol/liter) when metabolic alkalosis is due to profound potassium depletion.

Urinary osmolality. Elevated in fluid volume deficit (the kidneys conserve needed fluid, causing urine to be more concentrated). Decreased in fluid volume excess (kidneys excrete unneeded fluid causing urine to be dilute). After a fast of 12 hours, the urine osmolality should be at least three times the serum osmolality.

Urinary pH. Usually increased (more alkaline) in alkalosis; however, urine may be acidic when alkalosis is accompanied by severe hypokalemia. Decreased (more acidic) with use of acidifying agents such as ascorbic acid. In kidney stones, acidification of the urine is recommended for calcium phosphate and struvite stones, and alkalinization is recommended for uric acid and cystine stones. See *Diet, ash*.

Urinary potassium. Values vary with intake. Increased in hyperaldosteronism (Na/K ratio of 1:2), and decreased in adrenal insufficiency (Na/K ratio may be 10:1). The normal Na/K ratio is 2:1.

Urinary sodium. Less than 10 mEq/L (10 mmol/L) in hypovolemia and hyponatremia. Greater than 20 mEq/L (20 mmoL/L) in hypovolemia associated with adrenal insufficiency, osmotic diuresis, and salt-wasting renal disease.

Urinary specific gravity. Elevated in fluid volume deficit; decreased in fluid volume excess. Specific gravity persistently below 1.015 is a sign of significant renal disease.

Urinary tract infection (UTI). Infection usually caused by gram-negative bacteria in one or more of the organs in the urinary tract (kidneys, ureter, bladder, urethra). Characterized by a burning sensation during urination, pain, frequent urination, and sometimes visible blood and pus in the urine. Treatment is by drug therapy, liberal fluid intake, and control of urinary pH. See also *Diet, ash*.

Urinary urea nitrogen (UUN). About 80% to 90% of urinary nitrogen is comprised of urea. A 24-hour urine collection is analyzed to obtain the UUN. The value is variable, depending on the amount of protein in the diet. Thus, UUN is useful in assessing the adequacy of protein intake. See *Nitrogen balance*.

Urine. Fluid excreted by the kidneys. The quantity excreted in 24 hours varies with the amount of fluid consumed but averages between 1000 and 1500 mL. It is slightly acidic in reaction (pH 4.6 to 7) and has a specific gravity of 1.005 to 1.030. The amount of solids varies with the diet and with renal function, although urine collection normally contains 40 to 75 g of solids/24 hours. Urine formation is the result of three processes that occur in the nephron: filtration through the glomerular capillaries; reabsorption of fluid and solutes in the proximal tubule, the loop of Henle, and the distal tubule; and secretion into the lumen of the distal tubule. Urinalysis is an important biochemical method of *nutritional assessment* and aids in diagnostic testing. See also Appendix 31.

Urobilin. Also called "stercobilin"; a brownish pigment derived from the oxidation of urobilinogen, a derivative of the bile pigment bilirubin. This is found in the feces and sometimes in the urine after exposure to air. It is primarily responsible for the brown color of the feces.

Urochrome. The chief yellow pigment of the urine. Other pigments are uroerythrin and uroporphyrin.

Urogastrone. A substance found in the urine similar to enterogastrone. It also inhibits gastric secretion and motility.

Urokinase. An enzyme produced in the kidney and found in the urine. As a drug preparation, it is used to treat pulmonary embolism.

Urolithiasis. Formation of urinary calculi or insoluble constituents in the urine that precipitate as stones in the urinary passages. Variable in composition, these may contain urates, cystine and calcium oxalates, phosphates, carbonates, and *struvite*. Urinary calculi formation is the result of a number of factors, including hyperfunction of the parathyroid glands; vitamin A deficiency; systemic infections; inadequate fluid intake; metabolic disturbances; prolonged bed rest; and obstruction in the renal flow, producing stasis of the urine. Excessive intakes of sodium, calcium, oxalate, and protein tend to increase the risk of future stone formation. About 80% of calculi seen are calcium-containing stones. Excess vitamin C intake can contribute to formation of oxalate stones. *Nutrition therapy:* a liberal fluid intake to dilute the urine and prevent concentration of "stone-forming" substances is the main guideline. Control of urinary pH and foods to avoid will depend on the composition of the stone. Ensure a minimum urine output (about 2 liters/day for women and 2.5 liters/day for men) by having the person consume approximately 3 liters of fluids daily. Increase fluids in warm climates and in case of heavy physical activity. The patient should drink fluids throughout the day with half of the total volume coming from water. Moderate sodium restriction (2 to 3 g/day) might be helpful for patients with hypercalciuria to decrease urinary calcium excretion. Calcium restriction is not indicated in idiopathic hypercalciuria with normal intestinal absorption of calcium, although excessive intake should be avoided (≥1000 mg/day). In Type II diet-dependent absorptive hypercalciuria, a mildly restricted calcium intake is recommended (600 to 800 mg/day for adult men; up to 1000 mg/day for premenopausal women and 1200 mg/day for postmenopausal and pregnant and lactating women). Restriction of calcium to less than these levels yields no additional benefit and is not recommended. Calcium is not restricted in Type I diet-independent absorptive hypercalciuria because urinary calcium is elevated irrespective of dietary calcium intake. Avoid foods high in oxalate to prevent precipitation of calcium oxalate stones. A high protein intake will increase the urinary excretion of calcium, oxalate, and uric acid. A high insoluble fiber diet has been shown to increase fecal calcium excretion by binding calcium and preventing its absorption. Bran fiber provides phytic acid, which combines with dietary calcium in the intestines to form calcium phytate. Encourage vegetable protein sources and decrease animal protein, especially muscular protein sources rich in purines, in uric acid stones. See *Diet, calcium-modified* and *Diet, oxalate-restricted* (for calcium oxalate stones and hypercalciuria); *Diet, methionine-restricted* (for cystinuria); and *Diet, purine-restricted* (for urinary uric acid lithiasis). See also *Acid-ash* and *Alkaline-ash diets* under *Diet, ash*.

Uroporphyrin. A porphyrin found in small amounts in the urine. It is excreted in abnormally large amounts in lead poisoning, congenital porphyria, and porphyrinuria.

USDA. Abbreviation for *United States Department of Agriculture*.

USP. Abbreviation for *United States Pharmacopeia*.

UTI. Abbreviation for *urinary tract infection*.

V. Chemical symbol for vanadium.

Vagotomy. Cutting of certain branches of the vagus nerve, often accompanied by gastrectomy. Usually performed to reduce the recurrence of a gastric ulcer. The vagus nerve is the longest cranial nerve originating from the brain and carrying impulses to the head, neck, chest, and abdomen. *Nutrition therapy:* see *Gastric surgery*.

Valine (Val). Alpha-aminoisovaleric acid; an *essential* amino acid necessary for growth and maintenance of tissues. Deficiency causes poor appetite, weight loss, decreased plasma albumin, hyperesthesia, and muscular incoordination.

Valproic acid. A derivative of carboxylic acid that is used alone or with other anticonvulsants in the management of seizures. It may cause hyperglycinemia and carnitine deficiency with hyperammonemia; it may also cause nausea, vomiting, diarrhea, and stomach cramps. It is excreted in breast milk; infants must be observed for drowsiness. Brand names are Depakene® and Depakote®.

Vanadium (V). Trace element found in the human body; the concentration in human blood is about 100 μg/mL. It is probably essential and may have a role in the regulation of the sodium pump and the metabolism of bones, glucose, and lipids. Vanadium at pharmacologic levels can inhibit cholesterol synthesis in the liver and lower plasma cholesterol and triglycerides. It can also influence the shape of erythrocytes, and stimulate glucose oxidation and glycogen synthesis in the liver. There has been no report of vanadium deficiency. It is found in small amounts in most foods. The typical American diet probably provides 25 μg daily. Vanadium can also be absorbed through inhalation. Potential toxicity from exposure in industry has been a concern because vanadium is rapidly absorbed through the lungs and skin. The use of vanadium in various industrial processes and its release into the environment have resulted in many cases of vanadium toxicity in humans. Signs of intoxication include sore eyes, diarrhea, dermatitis, depressed food intake, elevated tissue vanadium level, and death. An interesting aspect of vanadium toxicity is its depletion of ascorbic acid and the counteraction of vanadium toxicity by this vitamin. Toxicity from the diet is rare, although vanadium is toxic in relatively low doses. Dietary substances that were found to reduce vanadium toxicity are aluminum hydroxide, chloride, chromium, ferrous iron, protein, and vitamin C. Vanadium toxicity can be completely prevented by administration of ethylenediamine tetraacetic acid, which interferes with vanadium absorption.

Vanillylmandelic acid (VMA). Also spelled "vanylmandelic acid." An end product in the metabolism of epinephrine and norepinephrine. It is excreted in urine in the free form and is frequently increased in *pheochromocytoma*. The reference range for VMA is 1 to 8 mg/24 hours (5 to 40 μmol/24 hours).

Vascular. Relating to blood or lymphatic vessels and ducts. Most often refers to blood vessels. The vascular bed is the total blood supply (i.e., arteries, capillaries, and veins) of an organ or region.

Vasoconstriction. Constriction of the blood vessels, especially the arterioles, leading to a decrease in the caliber of the vessels and a reduced blood supply. This is brought about by a *vasoconstrictor agent*.

Vasodilation. Dilation or increase in the caliber of a blood vessel, especially the arterioles, leading to an increased blood supply. This effect is brought about by a *vasodilator agent*.

Vasomotor. Regulating the movements of the walls of the blood vessels, i.e., their contraction (vasoconstriction) and expansion (vasodilation).

Vasopressin. A posterior pituitary hormone that exerts both pressor and antidiuretic actions. Its *pressor* action is a result of peripheral vasoconstriction in the systemic arterioles and capillaries. There is also constriction of the coronary and pulmonary vessels but dilation of the cerebral and renal vessels. The *antidiuretic effect* is exerted by increasing the rate of reabsorption of water in the renal tubules, resulting in relatively concentrated urine. A lack of this hormone results in *diabetes insipidus*. This is characterized by excessive renal loss of water and excessive thirst. Release of vasopressin is stimulated by a variety of neurogenic stimuli such as pain, trauma, and emotional stress.

Vegan. A *vegetarian* who excludes from the diet all foods of animal origin.

Vegetable protein mixtures. Blend of processed vegetable protein foods with or without skim milk powder and with or without added vitamins and minerals. They can be cheap sources of protein-rich food in countries where animal protein foods are expensive or unavailable. See also *Textured vegetable protein*.

Vegetarian. A person subsisting entirely or largely on foods of plant origin; intake of food of animal origin is restricted or not allowed. A vegetarian follows one of several diet patterns. A *pure vegetarian* (or vegan) is one who eats only foods of plant origin, without specific restrictions as to kind. A *fruitarian* is one who restricts the variety of plant foods eaten to fresh and dried fruits, nuts, honey, and sometimes olive oil. A *lactovegetarian* eats plant foods plus milk and other dairy products. An *ovovegetarian* includes eggs as the only source of animal protein. A *lactoovovegetarian* consumes milk, dairy products, and eggs, in addition to plant foods. A *pescovegetarian* includes fish, and a *pollovegetarian* includes poultry in the diet. Other people who also consider themselves vegetarians are the *red meat abstainers,* who eat any animal product except red meat, and the *semi-vegetarians,* who allow some animal foods but meat is usually excluded. See also *Nutrition, vegetarianism* and *Diet, Zen macrobiotic*.

Verdohemoglobin. Also called "choleglobin." Intermediate compound formed in the breakdown of hemoglobin. It is a biliverdin–iron–protein complex that has a green color.

Very-low-calorie diet (VLCD). See *Diet, calorie-modified*.

Very-low-density lipoprotein (VLDL). A plasma protein that has a density of 0.95 to 1.006 g/mL; classified as the pre-beta fraction on the basis of electrophoresis and contains about 10% protein, 18% phospholipid, 22% cholesterol, and 50% triglycerides. VLDL is involved in the transport of triglycerides from the intestinal tract and liver to adipose tissues and muscles. An increase in the concentration of VLDL is believed to be linked to an increase in the incidence of atherosclerosis.

Villi. Small, finger-like projections on the surface of a mucous membrane, as in the walls of the small intestine, where absorption takes place.

Vinblastine. Antineoplastic drug used for treating breast and testicular cancer, lymphomas, and Kaposi's sarcoma. It may cause nausea, vomiting, anorexia, diarrhea, stomatitis, anemia, and constipation. Brand name is Velban®.

Vincristine. An antineoplastic drug that interferes with amino acid metabolism and inhibits nucleic acid and protein synthesis. It is used alone or in combination with other chemotherapeutic agents in the treatment of Hodgkin's disease and other lymphomas. Its adverse effects are dose related and may include nausea, vomiting, diarrhea, stomatitis, oral ulcerations, hyponatremia, anorexia and weight loss. The brand name is Oncovin®.

Viral hepatitis. Inflammatory liver disease caused by a hepatitis virus. Five human hepatitis viruses have been identified: hepatitis A virus (HAV), hepatitis B (HBV), hepatitis C (HCV), hepatitis D (HDV), and hepatitis E (HEV). The two most common are HAV, which is spread by direct contact or fecal-contaminated food or water, and HBV, which is transmitted in contaminated serum in blood transfusion or by the use of contaminated needles and instruments. Clinical signs include anorexia, fever, nausea and vomiting, jaundice, malaise, diarrhea with clay-colored stools, and pain over the liver area. Serum bilirubin and SGOT levels are elevated. The severity of symptoms and the course of the illness depend on the kind and strain of the virus. Hepatitis A (also called "infective hepatitis") is characterized by slow onset of symptoms. Hepatitis B (also called "serum hepatitis") has a rapid onset; the infection may be severe and result in prolonged illness, destruction of liver cells, cirrhosis, and even death. *Nutrition therapy:* a diet that is high in protein (1.5 to 2.0 g/kg body weight), high calories (basal energy expenditure × 1.5), liberal carbohydrates (55% to 60% of calories), and moderate fat (25% to 30% of calories). For further details, see *Hepatitis*.

Virus. Disease-producing agent smaller than the ordinary germ; consists of a nucleic acid, either RNA or DNA, enclosed in a protein layer. It is a living pathogen that can multiply only in the presence of living, healthy host cells. Some viruses are visible under the ordinary microscope; others, the ultraviruses, are visible only under the ultramicroscope. Some can pass through porcelain filters; the nonfilterable viruses, however, cannot.

Viscera. Organs enclosed within the four great cavities: the cranium, thorax, abdomen, and pelvis. Pertains most commonly to the digestive organs within the abdominal cavity.

Visceral protein. Although most of the body's protein is found in the skeletal muscle, a smaller but significant amount is found in the visceral pool. Visceral proteins are found in the circulating serum proteins, liver, kidneys, pancreas, and heart. The liver is the main site of serum protein synthesis and is the first organ to be affected by a limited supply of protein. The serum protein level is affected by deficiencies in protein and energy and is decreased in liver and kidney diseases, infections, stress, burns, wounds, and protein-losing enteropathies. Assessment of the visceral protein status includes measurement of serum *albumin, prealbumin, retinol-binding protein, thyroxine-binding prealbumin, total iron binding capacity,* and *transferrin.* Also useful in assessing short-term effects on nutritional status are other body proteins such as *somatomedin-C* and *fibronectin*. See also Appendix 24.

Viscosity. Resistance of a liquid to flow. The viscosity of a liquid is measured by an instrument called a *viscosimeter*.

Visual process. The eyes contain two kinds of light receptors (rods and cones) located in the retina. The rods are involved in vision in dim light, and the cones function in vision in bright light. The light sensitivity of these two receptors comes from two photosensitive pigments, *rhodopsin* (in the rods) and *iodopsin* (in the cones), which are protein complexes of vitamin A. The prosthetic group in both of these pigments is vitamin A in the form of retinaldehyde, although the proteins to which the aldehyde is attached are different. On exposure to light, rhodopsin is bleached through a series of intermediate compounds, giving as end products *opsin* (a protein) and retinaldehyde (vitamin A aldehyde). These changes initiate a nerve impulse that is transmitted to the brain by way of the optic nerve. Regeneration of rhodopsin occurs in the dark, but some retinaldehyde is lost in each cycle, so that a constant supply must be present in the blood to recombine with opsin to regenerate rhodopsin.

Visual purple. Also called "rhodopsin"; a conjugated protein containing vitamin A. It is a photosensitive pigment in the retina of the eye that is bleached to visual yellow by light.

Visual threshold. Minimal light intensity required to evoke a visual sensation.

Visual violet. Also called "iodopsin"; a photosensitive pigment of the cones in the retina of the eye that is important for vision. It contains vitamin A.

Visual yellow. Colorless substance formed in the retina when visual purple (rhodopsin) is exposed to light. It is a mixture of retinaldehyde and a protein, opsin.

Vital® High Nitrogen. Brand name of a low-residue product for patients with limited digestion and/or absorption. It is lactose-free, low in electrolytes, and contains partially hydrolyzed whey. For tube feeding or oral feeding. See Appendix 39.

Vital statistics. Figures on births, deaths, longevity, disease rates, and other data that indicate the state of health of a population.

Vitamer. 1. Substance structurally related to a vitamin and capable of producing the same biologic activity. 2. One of the early names given to vitamins.

Vitamin. General term given to a group of organic substances that are present in food in minute quantities but are distinct from carbohydrates, lipids, and proteins; essential for normal health and growth; cause a specific deficiency disease when not adequately supplied by the diet or improperly absorbed from the food. See *Vitamin nomenclature*.

Vitamin A. Formerly called "axerophthol" and "antixerophthalmic vitamin"; a group of fat-soluble compounds that occur in several isomeric forms and occur preformed only in foods of animal origin. Vitamin A exists in three forms: alcohol (retinol), aldehyde (retinaldehyde), and acid (retinoic acid). As a group, the various forms of vitamin A are called retinoids. The vitamin is present in yellow and green leafy plants as provitamin A, of which there are several forms; the most important ones in human nutrition are the *carotenoids* alpha- and beta-carotene and cryptoxanthin. These are converted to the active vitamin in the intestinal wall and liver. The richest sources of preformed retinol are fish liver oils, egg yolk, and fortified milk. Biologically active carotenoids are found in dark green leafy vegetables and yellow fruits and vegetables such as squash and carrots. The color intensity of a fruit or vegetable is not a reliable indication of its content of provitamin A, because many other yellow and orange carotenoids in plants are not readily converted to the vitamin. Vitamin A is necessary for normal growth and development, maintenance of normal epithelial tissue structure, integrity of the immune system, and other physiologic functions, including vision and reproduction. Vitamin A, in the form of retinoic acid, acts like a hormone in regulating the expression of a number of genes. Vitamin A deficiency is usually due to inadequate dietary intake; deficiency also occurs as a result of chronic fat malabsorption. Symptoms of deficiency vary with the animal species. In humans the most common signs are poor growth, lowered resistance to infection, night blindness, and rough, scaly skin. Severe deficiency leads to *keratomalacia* and *xerophthalmia*. Vitamin A is not normally excreted and can accumulate in the body. Excessive intake (hypervitaminosis A) causes headache, dry skin, loss of hair, softening of bones, and liver damage. A high incidence of spontaneous abortions and birth defects has been reported in women ingesting large doses of *isotretinoin* (13-*cis*-retinoic acid) during the first trimester of pregnancy. Carotenoids are not toxic but can color adipose tissue stores and make the skin look yellow when taken in large doses over several weeks; the yellow coloration disappears gradually when the high intake is discontinued. See Appendix 1 for the recommended dietary allowances and *Retinol equivalent (RE)*.

Vitamin B complex. Group of water-soluble vitamins generally found together in nature and somewhat related in function, although unrelated chemically. These include vitamin B_1 (thiamin), vitamin B_2 (riboflavin), the vitamin B_6 group (pyridoxine, pyridoxal, and pyridoxamine), the vitamin B_{12} group (the cobalamins), nicotinic acid (niacin), folic acid (pteroylglutamic acid or PGA), pantothenic acid, and biotin.

Vitamin B_1. Also called *thiamin;* originally designated "water-soluble vitamin B" or the "antiberiberi vitamin" or "antineuritic factor" found in rice polishings. It was given the subscript number 1 when the B vitamin was discovered to be not a single factor but a complex composed of several factors. The vitamin was identified and synthesized by R. R. Williams and co-workers, who coined the word "thiamine" to indicate its structure, containing both sulfur (thio group) and an amino group. See *Thiamin*.

Vitamin B_2. Also called *riboflavin;* name given to the heat-stable fraction of vitamin B to differentiate it from the heat-labile fraction, designated "vitamin B_1." It was recognized to be a yellowish-green fluorescent pigment belonging to a group of compounds known as "flavin" and given the names Warburg's "yellow enzyme," "vitamin G," and "lactoflavin," "ovoflavin," "hepatoflavin," or "verdoflavin," since it was isolated from milk, egg, liver, and grass. The compound was later called "riboflavine" because it contains a ribose conjugated to a protein plus a pigment, flavin. The final "e" was dropped from the spelling because the vitamin is not really an amine. See *Riboflavin*.

Vitamin B_6. A group of chemically related compounds that are metabolically interchangeable: pyridoxine (alcohol form), pyridoxal (aldehyde form), and pyridoxamine (amine form). The three forms are all widely distributed in low concentrations in all animal and plant tissues. Good food sources are muscle meats, liver, pork, egg, whole grain cereals, and soybeans. The vitamin is readily and completely absorbed from the gut, but the level in the blood is extremely low and storage is very limited. The biologic activity of the three forms is about equal; all can be converted in the liver, erythrocytes, and other tissues to *pyridoxal phosphate* and *pyridoxamine phosphate,* which function as coenzymes in transamination reactions. Pyridoxal phosphate is also involved in the degradation, decarboxylation, and racemization of amino acids; in the conversion of tryptophan to nicotinic acid; in urea production; and in the metabolism of essential fatty acids. It is used therapeutically to control nausea and vomiting of pregnancy and to alleviate the peripheral neuritis associated with isonicotinyl hydrazide medication. Vitamin B_6 deficiency in adults rarely occurs alone, but it is seen in persons with multiple vitamin deficiencies and during long-term use of certain drugs, such as oral contraceptives, isoniazid, levodopa, and hydralazine. Signs of deficiency include seborrheic dermatitis, nausea and vomiting, stomatitis, and depression. In infants, vitamin B_6 deficiency has been induced with milk formulas deficient in the vitamin, causing hyperirritability, poor growth, and convulsions. The vitamin is relatively safe in oral doses of up to 50 times the recommended dietary allowance. A transient dependency state, consisting of ill-defined symptoms including nervousness and tremulousness, is induced when pharmacologic doses are given for several weeks and then suddenly withdrawn. Toxicity in humans has been seen with 1000 to 3000 times the RDA (2 to 6 g/day) used for treatment of carpal tunnel syndrome, which resulted in a peripheral neuropathy. See Appendix 1 for the recommended dietary allowances.

Vitamin B_{12}. Also called "cobalamin"; the antipernicious anemia factor found to be identical to the extrinsic factor of Castle, erythrocyte maturation factor, animal protein factor, and zoopherin. It has a characteristic red color and contains cobalt as an essential mineral constituent. As a constituent of two coenzymes, methylcobalamin and 5-deoxyadenosylcobalamin, vitamin B_{12} functions in the stimulation of red blood cell formation; synthesis of nucleic acids and nucleoproteins; and metabolism of nervous tissue, folate, sulfur-containing amino acids, carbohydrate, fat, and protein. It is present only in animal foods; plant foods are practically devoid of the vitamin. Good food sources are liver, eggs, milk, meat, and fish. The average intake is about 0.6 μg/day. Cyanocobalamin is the commercially available form of vitamin B_{12} used in pharmaceuticals and vitamin pills. The primary absorption of the vitamin involves cleavage from dietary protein by gastric acid and binding to the intrinsic factor of Castle secreted by the gastric mucosa. A small amount of the vitamin (1% to 3%) may be absorbed by simple diffusion. The vitamin is present in blood bound to three different proteins as transcobalamins and is transported to the liver and other tissues. There is unusual storage in the liver; the amount stored may last for 3 to 4 years in the absence of any additional supply. Dietary deficiency is rare but may be seen in strict vegetarians. However, symptoms of deficiency have been observed in some breast-fed infants of women who are strict vegetarians. Vitamin B_{12} deficiency is nearly always the result of inadequate absorption due to absence of intrinsic factor, which may be an inherited condition or a result of surgical resection of the stomach or the absorbing surfaces of the ileum. Small bowel diverticula, intestinal infestations, sprue, and other malabsorption syndromes may also induce a deficiency state. Manifestations of deficiency include sore tongue, general weakness, macrocytic anemia, and neurologic symptoms due to demyelination of the spinal cord, brain, and optic and peripheral nerves. Correction of a deficiency resulting from inadequate absorption requires injection of 100 μg/month of vitamin B_{12}. There has been no report of toxicity with large oral intakes. Adverse effects from vitamin B_{12} injection include hypersensitivity reaction, vascular thrombosis, and local pain from the injection site. See Appendix 1 for the recommended dietary allowances.

Vitamin B_{13}. A compound from distillers' dried solubles that was provisionally called "vitamin B_{13}"; later identified as orotic acid, an intermediate in pyrimidine metabolism. Orotic acid is not recognized as a vitamin, as all amino acids are capable of contributing to the orotic acid pool.

Vitamin B_{14}. Crystalline compound isolated from wine and originally thought to be a metabolite of xanthopterin; claimed to check the growth of cancer cells.

Vitamin B_{15}. Pangamate (pangamic acid), a preparation marketed as a vitamin but not recognized as such by United States drug authorities. See *Pangamic acid*.

Vitamin B_{17}. A term used to describe laetrile and/or amygdalin. Not recognized as a vitamin by United States drug authorities. See also *Laetrile*.

Vitamin C. Formerly called "antiscorbutic vitamin" and "cevitamic acid"; a water-soluble vitamin that exists in several forms, of which the two most active are L-*ascorbic acid* and L-*dehydroascorbic acid*. It is synthesized from glucose or galactose by most animals except the human, monkey, guinea pig, Indian fruit-eating bat, and red-vented bulbul bird. Vitamin C functions in a wide variety of roles, such as formation and maintenance of the intercellular cementing substance; metabolism of phenylalanine, tyrosine, folic acid, and histamine; conversion of ferric to ferrous iron to facilitate absorption; and wound healing and immune response. It also acts as a good antioxidant and may prevent the formation of carcinogenic nitrosamines by reducing nitrites. The ingestion of fruits and vegetables rich in vitamin C has been associated with a reduced incidence of some cancers. Deficiency of the vitamin results in *scurvy*, anemia, delayed or incomplete wound healing, and reduced resistance to infection. The richest food sources are acerola and camu-camu; other good sources are green and red peppers, collard greens, broccoli, spinach, tomatoes, potatoes, strawberries, kiwi, oranges, and other citrus fruits. The dietary vitamin C is generally much lower than the calculated amount in foods because of its destruction during processing and loss in cooking water. It is the most unstable vitamin, and is easily oxidized on exposure to air and light and destroyed by high temperature, alkali, and copper. Absorption in the intestines is almost complete, although tissue storage is quite limited and deficiency can result readily when intake is inadequate. Deficiency may develop rapidly in 24 to 48 hours following a major burn. Much of the body's store is concentrated in cells and is found in greatest concentration in the adrenals, eye lens, and liver. The vitamin is used up more rapidly in persons who smoke tobacco or are under stress. Excess intake is readily eliminated in the urine after tissue saturation. Pharmacologic doses of ascorbic acid have been reported to reduce the frequency and severity of symptoms of the common cold and other respiratory illnesses and to lower serum cholesterol in some hypercholesterolemic individuals. However, there is no general agreement on the benefits obtained from large doses of the vitamin for these conditions. Large doses of vitamin C may cause diarrhea and renal stones (cystine, oxalate, or urate) and may interfere with the anticoagulant effects of heparin. Sudden cessation of massive intakes could produce a "rebound scurvy." See Appendix 1 for the recommended dietary allowances.

Vitamin C_2. See *Bioflavonoids*.

Vitamin D. Also called "calciferol"; a fat-soluble vitamin necessary for the formation of the skeleton and for mineral homeostasis. It was previously called the "sunshine vitamin," "antirachitic factor," and "rachitamin." Vitamin D is a group of several related sterols, but the two most important are vitamin D_2 (ergocalciferol or ercalciol), which is obtained from irradiation of the provitamin ergosterol found in plants, and vitamin D_3 (cholecalciferol or calciol), which is produced from the provitamin 7-dehydrocholesterol found underneath the skin on exposure to ultraviolet light from the sun. The vitamin is metabolized in the liver into 25-hydroxyvitamin D (25-hydroxycholecalciferol, 25-(OH)D, 25-HCC, or calcidiol) and then further hydroxylated in the kidney to the active metabolite 1,25-dihydroxyvitamin D (1,25-dihydroxycholecalciferol, 1,25-$(OH)_2$D, 1,25-DHCC, or calcitriol). The active metabolite then returns to the intestinal mucosal cells, where it initiates the production of a calcium-binding protein that assists in the absorption of calcium from the intestines. It also helps to maintain plasma calcium regulation by increasing bone resorption synergistically with parathormone and by stimulating reabsorption of calcium by the kidney. Because of this action, vitamin D acts more like a hormone than a cofactor for an enzyme. Vitamin D is thus considered both a vitamin and a prohormone. The requirement for vitamin D can be met under normal conditions of exposure to sunlight. When sun exposure is restricted or limited, dietary intake may be important, especially in the elderly, whose capacity to synthesize the vitamin is approximately half that of younger people. Fish liver oil and foods fortified with vitamin D are the major dietary sources; smaller amounts are found in liver, egg yolk, sardines, and salmon. Vitamin D deficiency is characterized by inadequate mineralization of the bone. Severe deficiency in children results in *rickets;* a deficiency in adults leads to *osteomalacia*. Primary nutritional deficiency of vitamin D is seen in certain population groups in the United States. It is relatively common in institutionalized elderly who lack casual exposure to sunlight, especially in Eastern states; it also occurs in black children on prolonged breast-feeding without vitamin D supplementation. However, vitamin D deficiency may occur secondary to malabsorption syndromes, prolonged use of anticonvulsants or corticosteroids, hypoparathyroidism, cholestatic disease, chronic renal disease, and certain metabolic disorders. The vitamin is potentially toxic; excessive intake (hypervitaminosis D) causes anorexia, nausea, calcification of soft tissues, and renal damage. Although the toxic level has not been established for all ages, consumption of as little as 45 μg (1800 IU) of cholecalciferol per day has been associated with signs of hypervitaminosis D in young children. Toxicity should be monitored when large doses of vitamin D (25 μg or more) are given for an extended period. See Appendix 1 for the recommended dietary allowances.

Vitamin E. Originally known as the "antisterility vitamin." It is a fat-soluble vitamin required by humans and several species of animals. A derivative of chromanol, vitamin E includes eight naturally occurring forms of the tocopherol and tocotrienol compounds found in plants. The most active form, and the one most widely distributed in nature, is alpha-tocopherol. Vitamin E is a biologic antioxidant capable of protecting cellular membranes from oxidative damage by preventing lipid oxidation, especially the peroxidation of polyunsaturated fatty acids (PUFAs) and cholesterol; it can also inhibit accumulation of ceroid pigment granules. Together with selenium, carotenoid, and ascorbic acid, vitamin E is the primary defense against potentially harmful oxidation. Vitamin E appears to be important for the development and maintenance of nerve and muscle function. Major food sources are the vegetable and seed oils, especially those with polyunsaturated fatty acids, such as soybean, corn, cottonseed, and safflower oils; however, various amounts of the vitamin in these oils are lost during production, storage, and refining processes. Meats, fish, animal fats, and most fruits and vegetables have little vitamin E. Absorption of the vitamin varies, depending on total lipid absorption and the presence of bile salts and pancreatic enzymes. The efficiency of absorption decreases as large amounts of tocopherol are consumed. Once absorbed, the vitamin is transported in plasma with the lipoproteins and taken up by most tissues, including the liver, lung, heart, skeletal muscle, and adipose tissue. But unlike other tissues, fat tissue can sequester and accumulate the vitamin. The results of vitamin E deficiency in animals are varied and include reproductive failure, muscular dystrophy, liver necrosis, and neurologic abnormalities. In humans the occurrence of vitamin E deficiency is rare, except in susceptible individuals such as premature and very-low-birth-weight infants and those with steatorrhea and malabsorption problems associated with a variety of conditions such as biliary atresia, cystic fibrosis, chronic pancreatitis, and abetalipoproteinemia. Newborn infants delivered prematurely develop vitamin E deficiency because of limited storage at birth, intestinal malabsorption, and rapid growth rates that increase nutrient requirements in general. Some infant formulas also have marginal or low tocopherol contents relative to the amount of PUFAs and iron present. The requirement for vitamin E depends on the amount of PUFAs consumed. A subclinical deficiency of vitamin E in humans is considered when the plasma or serum tocopherol level is below 0.5 mg/dL, accompanied by a low ratio of serum tocopherol to lipid and/or hemolysis of erythrocytes incubated in 2% hydrogen peroxide. Pharmacologic doses of vitamin E (200 to 800 mg/day) can significantly increase peripheral blood circulation, may be beneficial in the recovery from ischemic conditions such as after coronary bypass or brain surgery, and may improve mobility and diminish pain associated with osteoarthritis. Limited studies have not shown consistent evidence of toxicity. Adverse effects include fatigue, weakness, blurred vision, headache, and delayed clotting time in patients taking warfarin. See Appendix 1 for the recommended dietary allowances.

Vitamin K. Formerly called "antihemorrhagic factor" and "Koagulation vitamin"; a group of fat-soluble 2-methyl-1,4-naphthoquinone derivatives necessary for the prevention of hemorrhagic conditions. The naturally occurring vitamins K are vitamin K_1 (phylloquinone), found in plants, and vitamin K2 (menaquinone), formed by bacterial synthesis in the intestine. Vitamins K_3 to K_7 are synthetic preparations, of which the most active is vitamin K_3 (menadione). Compounds with vitamin K activity are essential for the formation of prothrombin and other coagulation factors involved in the regulation of blood clotting. Vitamin K is also required for the biosynthesis of certain proteins, found in plasma, bone, and kidney, that bind calcium ions and probably function in bone crystal formation, photosynthetic phosphorylation, and the respiratory enzyme system. Deficiency of the vitamin results in a delayed blood clotting time and a hemorrhagic tendency. Because of its wide distribution in plants and intestinal synthesis by bacteria, deficiency resulting from dietary lack in human adults is rare. It may, however, occur in the presence of conditioning factors such as fat malabsorption, biliary obstruction, liver dysfunction, and prolonged treatment with broad-spectrum antibiotics and anticoagulants. Infants are susceptible to vitamin K deficiency because of small prenatal vitamin K storage and inadequate intestinal flora for its synthesis. Although vitamin K is fat-soluble, there are no known cases of toxicity in humans, except for the synthetic form, menadione, which can produce hemolytic anemia, hyperbilirubinemia, and kernicterus in the newborn with immature liver function. See Appendix 1 for the recommended dietary allowances.

Vitamin P. Name originally given to a group of factors that decrease capillary fragility. They were later identified as flavonoids but are now more commonly called *bioflavonoids* to indicate that they have biologic activity. They are no longer considered vitamins.

Vitamine. Original spelling of *vitamin,* as proposed by Funk in 1911 to indicate that the *accessory food factor* necessary for life is a vital amine. The final letter, "e," was dropped when analysis of the chemical nature of several of these factors showed that not all of them are amines.

Vitamin-like substances. Substances that, on the basis of current information, fail to meet all the criteria necessary to be classified as vitamins but still have some properties of vitamins. In some cases, they are present in larger amounts than vitamins; in others, the body can synthesize sufficient amounts to meet its needs if precursors are present. Examples of these substances are *carnitine, inositol,* and *bioflavonoids*.

Vitamin nomenclature. As suggested by Osborne and Mendel and McCollum and Davis, vitamins were originally classified, according to their solubility, as fat-soluble A and water-soluble B. Successive letters were assigned to new vitamins as they were characterized and isolated. Later it became evident that vitamin B was not a single vitamin but a group of vitamins, and subscripts were added for identification. The nomenclature became confusing when new factors thought to be new vitamins were so named, only to be found to be duplicates of other vitamins already named. The present trend is to call the vitamins by their chemical names. To date we recognize 13 vitamins essential to human nutrition: four fat-soluble vitamins (A, D, E, and K) and nine water-soluble vitamins (ascorbic acid, thiamin, riboflavin, nicotinic acid, pyridoxine, cobalamin, pantothenic acid, folic acid, and biotin). The *fat-soluble vitamins* have the following general properties: soluble in fat and fat solvents; not absolutely necessary in the diet every day; have precursors or provitamins; intake in excess of daily need is not excreted but stored in the body; and deficiencies are slow to develop. The *water-soluble vitamins* have the following general properties: soluble in water; must be supplied every day in the diet; generally do not have precursors; intake in excess of daily need is excreted in the urine, with minimal storage in the body; and deficiency symptoms often develop rapidly.

Vitaneed®. Brand name of a lactose-free blenderized formula containing fiber (8 g/liter) for tube feeding and not intended for oral use. Made with beef puree, sodium and calcium caseinates, maltodextrins, pureed fruits and vegetables, and soy oil. See also Appendix 39.

Vitellin. Phosphoprotein found in egg yolk.

Vitreous humor. Transparent gel-like fluid that fills the posterior chamber of the eye; consists of hyaluronic acid within a protein framework called "vitrein."

Vivonex®. Brand name of monomeric (elemental) formulas containing crystalline amino acids and glutamine; lactose-free and low in fat. Vivonex® T.E.N. has 10 g of free glutamine per liter. Vivonex® Plus contains 2% of total calories from arginine. For oral or tube feeding of stressed patients with gastrointestinal impairment. See Appendix 39.

VLCD. Abbreviation for very-low-calorie diet. See *Diet, calorie-modified*.

VLDL. Abbreviation for *very-low-density lipoprotein*. See also *Lipoprotein*.

Vomiting. Process of "throwing up" or expelling materials from the stomach through the esophagus and out of the mouth. Prolonged or persistent vomiting leads to nutritional deficiencies and is one factor considered in the initial screening and nutritional assessment of patients. If food cannot be retained and liquid intake is not adequate, dehydration and undernutrition can be avoided by using methods other than oral feeding. See *Parenteral feeding*. See also *Hyperemesis gravidarum*.

von Gierke's disease. Type I glycogen storage disease; an inborn error of metabolism due to a deficiency in the enzyme glucose-6-phosphatase, which is required for the conversion of glycogen to glucose. This results in the accumulation of glycogen in the liver, which becomes enlarged, and in hypoglycemia, which may be frequent, severe, and even fatal. Excessive glycogen accumulation, accompanied by acute and chronic hypoglycemia, causes liver damage and growth retardation. *Nutrition therapy:* small, frequent feedings of glucose as the carbohydrate source, including interruption of sleep for feeding, to prevent hypoglycemia. Nasogastric or gastrostomy feeding or even intravenous administration of glucose may be necessary at times. Both sucrose and lactose should be avoided because their end products, fructose and galactose, are readily converted to glycogen in the liver. Thus, milk and sugar-containing foods are to be avoided. Starches are not restricted unless they contain forbidden sugars. Acceptable infant formulas and oral supplements include Isomil®, Nutramigen®, and Pregestimil®.

Vulnerability. Susceptibility to injury or contagion. In nutrition, the phrase *vulnerable group* refers to infants, children, pregnant or lactating women, and elderly people, groups particularly prone to develop nutritional disorders.

W

Waist/hip ratio (WHR). An anthropometric measurement to determine body fat distribution and abdominal obesity. The ratio is calculated by dividing the waist circumference, where it is smallest, by the hip circumference, where it is largest:

$$\text{WHR} = \frac{\text{smallest waist measurement (cm or inch)}}{\text{largest hip measurement (cm or inch)}}$$

Ratios above 0.8 for women and 0.95 for men are linked to greater risk for several diseases, such as ischemic heart disease, hypertension, stroke, and Type II diabetes mellitus. The WHR is also called abdominal/gluteal ratio, which differentiates between android and gynoid obesity. See also *Obesity, Weight,* and Appendix 20 A.

Warfarin. An anticoagulant used in the treatment of thrombosis and embolism. Nutritional supplements and foods high in vitamin K may reduce its anticoagulant effect by increasing the synthesis of clotting factors. Vitamin E and herbal teas containing coumarin may increase drug effects and promote bleeding. Brand names are Coumadin®, Panwarfin®, and Sofarin®.

Water (H_2O). One of the major nutrients needed by the body. It comprises about 50% to 80% of total body weight, depending on body fat content. It performs varied functions and is second only to oxygen in maintaining life. The body can live for several days, even months, without food but dies within 5 to 10 days without water. Loss of 20% of body water results in death. All chemical reactions in the body take place in the presence of water. It acts as a solvent for products of digestion, as a lubricant of moving parts, and as a regulator of body temperature. Blood is 90% water and urine is 97% water. Water is also important for the proper elimination of waste products. See also *Water, sources*.

Bound w. Portion of water in food and body tissues that is attached to the colloids and is therefore more difficult to release than *free water*.

Endogenous w. Also called "metabolic water"; water derived from the metabolism of food in the body.

Exogenous w. Water in the body coming from dietary sources either as liquid or as a food component.

Free w. The portion of the water in the body or food that is not closely bound by attachment to the colloids.

Metabolic w. Also called "water of combustion"; water in the body that is provided by the combustion of foodstuffs (i.e., carbohydrate, protein, and fat). Oxidation of 100 g of carbohydrate, protein, and fat yields approximately 55, 41, and 107 g of water, respectively. On the average, water produced by the body's metabolic activity is about 200 to 300 mL/day.

Water balance. Balance between water intake and output. Water intake comes from fluids and beverages (free water), as a component of food, or as a product of the oxidation of foods in the body (metabolic water). The channels of water output are through the kidney (urine), the skin (sweat and insensible perspiration), the lungs (expired air), and the gastrointestinal tract (saliva and feces). Insensible water loss from lungs and skin accounts for 50% of the turnover of water, even without visible perspiration. Water intake must equal output, the difference resulting in edema or dehydration, depending on whether intake is greater or less than output. Water intake is controlled by the thirst center in the hypothalamus. Water output is controlled by the hormone *vasopressin* (also called "antidiuretic hormone or ADH"), which is secreted by the pituitary gland. Release of this hormone decreases water excretion by the kidney by increasing the rate of water reabsorption from the tubules. The urine is an important medium for the elimination of excess water. Abnormal losses of water, however, may occur in diarrhea, excessive vomiting, and severe burns. The following formula can be used to calculate water status:

$$\text{Water deficit} = (\%\ \text{TBW} \times \text{BW}) \times \left(1 - \frac{\text{Na predicted}}{\text{Na measured}}\right)$$

where, % TBW is percent total body water which estimated for adults to be 60% for normal male and female, 70% for lean male, 50% for obese male, and 42% for lean female; BW is actual body weight in kilogram; Na predicted is the constant average serum sodium of 140 mEq/L; and Na measured is the actual measurement of serum sodium. See also *Balance study* and *Dehydration*.

Water compartment, body. Water inside the body exists in two main compartments: within the cells (intracellular water) or outside the cells (extracellular water). Water outside the cells is found within the blood vessels (intravascular water) or between the vascular spaces and the cells (interstitial water). Smaller amounts are found in cerebrospinal fluid, synovial fluid, aqueous and vitreous humors, and lymph.

Water determination, body. The two general methods of measuring body water are direct and indirect. The direct method is obviously feasible only in human autopsy studies. The indirect methods of estimating the volume of each of the various water compartments are essentially the same. A material, previously found to be distributed almost exclusively within the compartment to be measured, is given intravenously in known amounts. After a sufficient time for mixing, a plasma sample is obtained and the concentration of the administered material is measured. *Total body water* can be measured by the use of a substance such as heavy water or antipyrine, which passes freely through capillary walls, cell membranes, and the blood–brain barrier. Determination of this material in any available fluid (e.g., plasma or urine) indicates the amount of total body water. Determination of total *extracellular fluid* requires a substance such as inulin, thiocyanate, or thiosulfate that can cross capillary walls and distribute itself uniformly in the plasma and interstitial fluid without entering the cells. Estimation of plasma volume requires the administration of dyes or iodine-131 attached to albumin which will be retained only within the vascular spaces. The volume of the *interstitial water* is the difference between the volume of total extracellular fluid and plasma volume.

Water intoxication (water toxicity). Condition that results from excessive intake of fluids without an equivalent amount of salt, as in intravenous glucose administration to persons with inadequate renal function. The kidney cannot excrete the extra load, and the accumulated water enters all the fluid compartments, including the cells and tissues, which become waterlogged. Serious symptoms develop, including confusion, convulsions, coma, and even death.

Water (fluid) requirement. Two vital needs of the body demand a continual expenditure of water: removal of body heat by vaporization of water through the skin and lungs, and excretion of urea and other products of metabolism in the urine. There is an obligatory daily loss of approximately 1500 ml water. Of this amount, about 600 mL is lost through the skin as insensible perspiration, 400 mL in expired air, and 500 mL in urine. Any excess in water intake over this obligatory water loss appears as an increase in urine volume, and any deficit under the obligatory water loss is taken at the expense of total body water, resulting in dehydration. The need for water is increased in hot climates; with excessive exercise because of the loss of water through sweat; and in burns, fever, infection, diarrhea, vomiting, draining fistula or open wound, and other pathologic conditions that increase the need for water above the normal requirement. Water needs increase 100 mL to 150 mL/day for each degree rise in body temperature over 37°C (98.6°F). A concentrated or high-protein diet also requires a higher fluid intake. An estimate of the daily water (fluid) requirement can be determined from the total calorie intake/day or body weight. An allowance of 1 mL/kcal for adults and 1.5 mL/kcal for infants and children are generally considered adequate under normal conditions. Based on body weight, the estimated daily fluid requirement for infants, children, and adults are the following:

Premature, <1000 g	200 mL/kg
Premature, 1000 to 1500 g	175 to 200 mL/kg
Premature, 1500 to 2500 g	150 to 180 mL/kg
Newborn	140 to 150 mL/kg
1 year (9 kg average weight)	120 to 135 mL/kg
2 years (12 kg average weight)	115 to 125 mL/kg
3 years (16 kg average weight)	100 to 110 mL/kg
6 years (20 kg average weight)	90 to 100 mL/kg
16 to 25 years	40 mL/kg
25 to 55 years	35 mL/kg
56 to 65 years	30 mL/kg
>65 years	25 mL/kg

In pregnancy, the water requirement is increased compared to that of the nonpregnant state in order to supply the needs of the growing fetus, the increased extracellular fluid space, and the amniotic fluid. The extra water needed for lactation is for milk secretion.

Water sources. The main form of water ingested to meet bodily requirements is water as such. The second category is in the form of beverages such as fruit juices, milk, coffee, tea, soft drinks, soups, and alcoholic beverages. For Americans, most of the water sources come from beverages and fewer from water as such. The rest of their water needs are supplied by the so-called solid foods, which vary in their water content from 95% in succulent fruits and vegetables to 65% in meats and fish. Butter and nuts contain 16% and 5% water, respectively. Water from exogenous sources, referred to as "preformed water," averages about 2000 mL/day. Water produced internally by oxidation or from metabolic processes amounts to about 300 mL/day.

WBC. Abbreviation for *white blood cell*. See *Leukocyte*.

Wean. Gradual or abrupt stoppage of breast feeding when the infant is capable of taking substantial nourishment from sources other than breast milk. Early weaning may be indicated when there is insufficient breast milk or when the mother is in poor health or when she becomes pregnant again.

Weanling diarrhea. Diarrhea associated with weaning of malnourished infants. It is commonly seen in developing countries where the incidence of food-borne infections is high. Malnutrition lowers body resistance to disease, thus making these infants more susceptible to such infections.

Weight. Force with which a body is attracted to the earth. In reference to body weight, the word "weight," if used without qualification, means the *actual* body weight as measured on a weighing scale. *Standard weight* is the average weight for each sex for different statures at various ages. *Ideal weight* (taken as the standard weight at age 25) is the weight associated with the most favorable in terms of lower mortality. Standard weight tables imply that average weights are normal (see Appendix 19). Currently, the concept and usage of ideal body weight have been replaced by that of *healthy* body weight. A healthy weight depends on how much of the body weight is fat, where fat is located in the body, and whether there is a weight-related medical problem. See Appendix 20-A. The National Institutes of Health (NIH) have defined *desirable* weight as the midpoint of the recommended weight range at a specified height for persons of medium build, according to the 1983 Metropolitan Life Insurance table. See Appendix 20-B. This criterion for desirable weight corresponds to a body mass index (BMI) of 22.0 for men and 21.5 for women. The BMI corrects body weight for height and is useful for assessing obesity and health risk. The *Hamwi formula* gives a quick estimate of ideal body weight for men and women taller than 5 feet. See also Appendix 25 for interpretations of body weight status.

Weight, adjusted. Because body fat is less metabolically active than lean body mass, an adjusted weight is used to determine energy and protein requirements for individuals whose body weight is greater than 125% of the desirable body weight (DBW). Adjustment in body weight is calculated as follows:

$$\text{Adjusted weight (kg)} = (\text{ABW} - \text{IBW}) \times 0.25 + \text{IBW}$$

where ABW is actual body weight, IBW is ideal body weight, and 0.25 represents 25% of body fat tissue metabolically active.

Weight, computed. Useful equations to estimate body weight for persons who have bone fractures that need traction or casting; also for persons whose weight cannot be measured directly due to some specific illnesses. The formulas are given below:

Body weight for men (in kg)

$$= (0.98 \times \text{calf C}) + (1.16 \times \text{knee H}) + (1.73 \times \text{MAC}) + (0.37 \times \text{subsc SF}) - 81.69$$

Body weight for women (in kg)

$$= (1.27 \times \text{calf C}) + (0.87 \times \text{knee H}) + (0.98 \times \text{MAC}) + (0.4 \times \text{subsc SF}) - 62.35$$

The recumbent measurements are in centimeters for the calf circumference (calf C), knee height (knee H), midarm circumference (MAC), and subscapular skinfold thickness (subsc. SF), which is measured just posterior to the left scapula or shoulder blade.

Weight control. Process of adjusting variable factors that affect body weight, aiming at a desirable level considered optimum. As a result of the prevalence of obesity, weight control programs concentrate on reducing body weight using a multifaceted approach. See *Nutrition, weight control, Obesity, Overweight,* and *Underweight.*

Weight for stature index. Used to identify infants and children with acute protein-energy malnutrition (PEM) and obesity. It is calculated by dividing the actual weight by the 50% weight for length.

An index below 0.9 (90% of standard) requires further assessment for acute PEM, and index above 1.1 (110% of standard) requires evaluation for overweight or obesity.

Weight loss. Weight loss is usually due to increased energy expenditure, decreased energy intake, or both. With uncomplicated mild trauma, weight loss among adults may be 250 g or less per day. In severe trauma, the loss may be up to 500 g per day. Weight loss is more rapid and prolonged following serious infection or injury. Losses greater than 500 g per day are associated with loss of lean body mass, in addition to water loss. The rate and percentage of weight change are important factors in assessing nutritional status. See Appendix 25 for interpretation of weight change.

Wernicke-Korsakoff syndrome. Disorder combining *Wernicke's encephalopathy* and *Korsakoff's* neurologic signs and symptoms. In severe alcoholism, both syndromes occur. Lack of vitamins, especially thiamin, is the main cause, in addition to the direct toxic effects of alcohol on the brain and peripheral nerves.

Wernicke's encephalopathy. Disease caused by an acute biochemical lesion in the brain due to *thiamin deficiency*. The syndrome was originally described by Wernicke among alcoholics and may also occur in hyperemesis gravidarum. It can be considered as the human counterpart of the encephalopathy produced in animals by acute deprivation of thiamin. It is characterized by paralysis or weakness of eye movements, ataxia, and mental disturbances. *Nutrition therapy:* Up to 1 g of thiamin administered intravenously for acute control, followed by 25 to 100 mg every 12 hours orally. See *Thiamin*.

Wetzel grid. Technique for measuring and evaluating a child's growth. It consists of nine physique channels that range from the very fat to the very thin. The height–weight data of a child are plotted throughout the growth period. The rate of growth is thought to be a more accurate measure of nutritional status than one weight measurement. The grid can be used to detect children who are failing to develop normally.

White blood cell (WBC). See *Leukocyte*.

WHO. Abbreviation for *World Health Organization*.

WHR. Abbreviation for *Waist/hip ratio*.

WIC. Abbreviation for *Women, infants,* and *children*.

Wilms' tumor. A highly malignant embryonoma of the kidney occurring mostly in young children. Symptoms include weight loss, anorexia, abdominal pain, anemia, fever, and hypertension. *Nutrition therapy:* aimed at correcting these symptoms. A high-calorie, high-protein diet with mineral supplements is indicated. Sodium intake is controlled if hypertension exists.

Wilson's disease. Also called "hepatolenticular degeneration." Rare inherited disorder of copper metabolism characterized by a decrease in plasma *ceruloplasmin* concentration and excessive accumulation of copper in the liver, brain, cornea, and kidney. If untreated, it causes liver damage and progressive neurologic changes. Treatment of Wilson's disease is aimed at reducing copper absorption and mobilizing copper in the liver and other organs. *Nutrition therapy:* A diet low in copper (1 to 2 mg/day) may be of value, although the mainstay of treatment is D-penicillamine or other drugs that inhibit copper absorption or increase urinary copper excretion. Adequacy of vitamin B_6 intake should be monitored with D-penicillamine administration, as it is an antagonist of pyridoxine. Oral vitamin B_6

and zinc supplementation of about 25 mg of zinc/day may be necessary. Zinc promotes copper binding to intestinal cells and subsequent excretion in the stools. See Diet, copper-restricted.

Women, Infants, and Children (WIC) Program. An assistance program for women, infants, and children providing nutrition education and supplemental foods. Qualifications of participants are children under 5 years of age of mothers who meet certain income guidelines; women who are pregnant or have recently delivered a baby; and mothers, infants, and children who are identified nutritional risks by a health professional. They must live in a local WIC agency's service area to receive vouchers for purchasing specific nutritious foods.

World Health Organization (WHO). International organization that aims to eliminate all kinds of diseases. In the field of nutrition, WHO has been involved in developing and testing new protein-rich foods; combating protein-calorie malnutrition, nutritional anemias, vitamin A deficiency, endemic goiter, and rickets; assessing nutritional requirements; and developing coordinated applied nutrition programs and training personnel for them. WHO was created in 1948 with about 100 member countries and has its headquarters in Geneva, Switzerland.

Wound. Physical injury to the body tissues disrupting the normal continuity of structures. *Wound healing* involves tissue synthesis and occurs in two phases. Initial wound healing occurs readily during a period of negative energy balance; subsequent healing occurs between the fifth and fifteenth day after surgery or trauma. Sufficient calories, protein, zinc, essential fatty acids, vitamin A, and vitamin C are necessary for continued wound healing. If exogenous nutrients are not provided at this stage, wound healing will cease. Other nutrients required in wound healing include arginine, histidine, magnesium, and selenium.

Wrist circumference. A simple anthropometric measurement to estimate frame size. Determined by measuring the smallest part of the wrist distal to the styloid process of the ulna and radius. Frame size is based on a specific "R" value for a small, medium, or large body frame for men and women. See Appendix 21.

WSB. Wheat-soy blend; a mixture of 74% wheat flour and 24% soy flour with a protein concentrate, vitamins, and minerals. Distributed by United States government programs, it is used in developing countries as a dietary supplement.

Xanthine. An intermediate product in the catabolism of adenine and guanine. It is formed by the oxidation of hypoxanthine or by the hydrolytic deamination of guanine. Oxidation of xanthine by *xanthine oxidase* yields uric acid, the major end product of purine metabolism in humans and higher apes.

Xanthine oxidase. Flavoprotein enzyme found in liver and milk. It contains iron, molybdenum, and flavin mono- or dinucleotide as a prosthetic group. It catalyzes the oxidation of hypoxanthine to xanthine and of the latter to uric acid. It also catalyzes the oxidation of various aldehydes.

Xanthinuria. Hereditary disorder in humans resulting from a lack of xanthine oxidase and characterized by the excretion of xanthine in the urine. *Nutrition therapy:* purine restriction and plenty of fluids. See *Diet, purine-restricted*.

Xanthoma. A deposit of lipid in the skin, often around the eyes, elbows, and knees; usually yellow-orange in color due to the presence of lipid-soluble pigments such as carotene. Related to hyperlipemia.

Xanthomatosis. Condition characterized by yellow lipoid deposits in the skin, tendon sheaths, and internal organs. The condition is associated with lipemia and hypercholesterolemia.

Xanthophyll. Yellow plant pigment occurring with carotene in green leaves and other vegetables. It is one of the *carotenoids* but has no vitamin A activity.

Xanthoproteic test. Test for the presence of tyrosine and tryptophan. A yellowish derivative formed with concentrated nitric acid turns orange on the addition of ammonia.

Xanthopterin. Yellow pigment present in the liver and urine that has folic acid activity in several animal species. It has antianemic activity in large doses.

Xanthosis. Yellowish discoloration of the skin due to the deposition of carotenoid pigment; often results from excessive intake of yellow fruits and vegetables such as carrots and squash. The skin discoloration is reversible.

Xanthurenic acid. Compound formed in the metabolism of tryptophan to niacin. It is excreted in large amounts in the urine in pyridoxine deficiency. Thus the *xanthurenic acid index* in the urine after administration of a standard dose of tryptophan is used to determine the degree of pyridoxine deficiency. A high index in pregnancy can be reduced by pyridoxine administration.

Xeroderma. Sometimes called "fish skin" or "alligator skin"; the skin becomes dry and rough and comes off in fine or branny scales. Sometimes a layer of "lacquer" appears on the surface that, on drying, breaks up into individual "islands" or patches of varying sizes. There is often desquamation from the borders of each island, while the intervening gaps become fissured. Xeroderma is often associated with vitamin A deficiency, although exposure to dirt and alternative heat and moisture often contribute to its causation.

Xerophthalmia. Dry, lusterless condition of the eyeball characterized by atrophy of the paraocular glands, hyperkeratosis of the conjunctiva, and finally, involvement of the cornea, which becomes dry and opaque. This is followed by cloudiness and infection, leading to ulceration, softening, and blindness. Xerophthalmia is associated with severe vitamin A deficiency; it may also follow chronic conjunctivitis and other diseases of the tear-producing gland.

Xerosis. Abnormal dryness of the skin, mucous membranes, or conjunctiva. In *conjunctival xerosis,* the bulbar conjunctiva is dry, thickened, wrinkled, and pigmented, due to a failure to shed the epithelial cells and consequent keratinization. The pigmentation gives the conjunctiva a peculiar "smoky" appearance. In *corneal xerosis,* the dryness spreads to the cornea, which takes on a dull, hazy, and lusterless appearance; ulceration and infection may occur, leading to softening of the cornea (keratomalacia) and blindness. The condition is associated with a deficiency in vitamin A but may also be due to other causes, such as diuretic use, long periods of exposure to glare and dust, and infections.

Xerostomia. Abnormal dryness of the mouth and lack of saliva due to a salivary gland dysfunction, affecting food intake. Associated with certain drugs, radiation therapy in the head and neck, paralysis of facial nerves, diabetes, acute infections, and Sjögren's syndrome. *Nutrition therapy:* encourage the intake of liquids at mealtime and between meals. Use foods that are moist or served with sauces or gravy. Use of sugar-free gums and artificial saliva may be beneficial.

Xylan. A hemicellulose of the pentosan type occurring in woody tissues, corncobs, peanut shells, and straw. Yields xylose on hydrolysis.

Xylitol. A sugar alcohol derived from *xylose* and commercially produced from birchwood chips, berries, leaves, and mushrooms. It tastes almost as sweet as sucrose (table sugar) and can be used as a sugar substitute. Xylitol is less cariogenic and less insulin dependent than sucrose.

Xylose. Wood sugar; a pentose obtained from the hydrolysis of *xylan*. It is used as a diagnostic aid for the detection of malabsorption. Urinary excretion of less than 4.5 g in 5 hours following ingestion of a 25-g load suggests decreased absorptive capacity.

Y

Yeast. Microscopic, unicellular, fungal plant extensively used by humans for fermentation processes and for bread making. In nutrition, yeast extract is a rich source of the B vitamins; it also contains protein and considerable ergosterol.

Yo-yo dieting. Layman's description of a dietary regimen resulting in fluctuations of body weight, usually due to calorie restriction periodically between periods of overeating or binging. Many individuals reduce and gain weight several times in their lifetime (i.e., yo-yo effect). The resting metabolic rate (RMR) decreases with weight reduction. Thus, it will take longer to lose the same amount of fat and less time to regain weight. Usually, more weight is regained than was lost and the regained weight is higher in fat in proportion to lean body mass (LBM).

Z

Zalcitabine. Antiviral drug for concurrent use with zidovudine in patients with advanced HIV infection. Adverse side effects include mouth and esophageal ulceration, stomatitis, anemia, pancreatitis, and gastrointestinal disturbances. Brand name is Hivid®.

Zeaxanthin. One of the carotenoid pigments in corn and egg yolk; used as a coloring agent and exhibits no vitamin A activity.

Zein. Protein from corn; an incomplete protein deficient in lysine and low in tryptophan. As the sole source of protein, zein can neither support growth nor maintain life.

Zen macrobiotic. See *Diet, Zen macrobiotic*.

Zidovudine. Antiretroviral drug used in the treatment of HIV infections and AIDS. Adverse side effects include anorexia, anemia, mouth ulcers, edema of tongue, bleeding gums, dysphagia, nausea, vomiting, abdominal pain, dyspepsia, diarrhea, and fatigue. Breast feeding is not recommended because of unknown risk. Brand name is Retrovir®.

Zinc (Zn). An essential trace mineral for humans, animals, and plants. As a cofactor in more than 100 different enzymes, zinc is involved in carbohydrate and energy metabolism, protein synthesis and degradation, nucleic acid synthesis, acid–base balance, carbon dioxide transport, and many other reactions; it is also important in wound healing, metabolism of vitamin A and collagen, cellular immunity, maintenance of taste acuity, and development of the reproductive organs. The normal zinc intake is about 10 to 15 mg/day. Good food sources are meat, liver, eggs, and seafoods; cereals and whole grains provide moderate amounts. Almost 30% is absorbed from the gastrointestinal tract; absorption is influenced by the level of zinc in the diet and by the presence of interfering substances like phytate, calcium, fiber, and chelating agents. Zinc also competes with copper and ferrous iron for absorption. In the body, zinc is concentrated in the liver, kidney, bone, retina, prostate, and muscle; it is mainly bound to albumin in the plasma. Zinc deficiency is usually the result of poor intake associated with conditions that decrease absorption or increase loss, such as malabsorption syndrome, inflammatory bowel disease, liver cirrhosis, thermal burn and loss from open wounds, fistulas, and exudates, and use of antimetabolites and antianabolic drugs. During stress, as much as 8000 μg of zinc may be lost in the urine per day because of catabolism of skeletal muscle which contains high concentrations of zinc. Primary zinc deficiency is also seen when inadequate zinc is supplied in solutions for total parenteral nutrition. Manifestations of deficiency are diverse, including hair loss, dermatitis and skin changes, growth retardation, impaired taste acuity, delayed wound healing, decreased dark adaptation, impaired immune function, and delayed sexual maturation. Zinc toxicity has occurred as a consequence of severe pollution, inhalation of zinc oxide fumes, and prolonged storage of food and drink in galvanized containers that allow zinc to leach out. Contamination of dialysis water stored in a galvanized tank has been reported to increase plasma zinc concentrations and to produce nausea, vomiting, fever, and anemia. Prolonged intake of dietary zinc supplements also has adverse effects, such as induction of copper deficiency anemia, depressed levels of white blood cells, increased low-density lipoprotein and decreased high-density lipoprotein cholesterol, and decreased serum ferritin and hematocrit levels. Prolonged ingestion of zinc supplements exceeding 15 mg/day is therefore not recommended. See Appendix 1 for the recommended dietary allowances.

Zn. Chemical symbol for zinc.

Zollinger–Ellison syndrome. Tumor of the delta cells of the *islets of Langerhans*. Characterized by hypersecretion of the gastric acid and ulceration of the esophagus, stomach, and upper small intestine. Malabsorption and diarrhea are the main problems. *Nutrition therapy:* correct diarrhea and electrolyte imbalance; restrict fat and use MCT oil; modify protein and fiber intakes according to the patient's needs.

Zwitterion. A dipolar ion containing both a positive and a negative charge and hence electrically neutral. Amino acids may form zwitterions in solution by migration of a hydrogen ion from the carboxyl group to the basic nitrogen atom of the amino group.

Appendix 1

Food and Nutrition Board, National Academy of Sciences—National Research Council Recommended Dietary Allowances,[a] Revised 1989

Designed for the maintenance of good nutrition of practically all healthy people in the United States

		Weight[b]		Height[b]			Fat-Soluble Vitamins				Water-Soluble Vitamins							Minerals						
Category	Age (years) or Condition	(kg)	(lb)	(cm)	(in)	Protein (g)	Vitamin A (μ RE)[c]	Vitamin D (μg)[d]	Vitamin E (mg α-TE)[e]	Vitamin K (μg)	Vitamin C (mg)	Thiamin (mg)	Riboflavin (mg)	Niacin (mg NE)[f]	Vitamin B_6 (mg)	Folate (μg)	Vitamin B_{12} (μg)	Calcium (mg)	Phosphorus (mg)	Magnesium (mg)	Iron (mg)	Zinc (mg)	Iodine (mg)	Selenium (μg)
Infants	0.0–0.5	6	13	60	24	13	375	7.5	3	5	30	0.3	0.4	5	0.3	25	0.3	400	300	40	6	5	40	10
	0.5–1.0	9	20	71	28	14	375	10	4	10	35	0.4	0.5	6	0.6	35	0.5	600	500	60	10	5	50	15
Children	1–3	13	29	90	35	16	400	10	6	15	40	0.7	0.8	9	1.0	50	0.7	800	800	80	10	10	70	20
	4–6	20	44	112	44	24	500	10	7	20	45	0.9	1.1	12	1.1	75	1.0	800	800	120	10	10	90	20
	7–10	28	62	132	52	28	700	10	7	30	45	1.0	1.2	13	1.4	100	1.4	800	800	170	10	10	120	30
Males	11–14	45	99	157	62	45	1000	10	10	45	50	1.3	1.5	17	1.7	150	2.0	1200	1200	270	12	15	150	40
	15–18	66	145	176	69	59	1000	10	10	65	60	1.5	1.8	20	2.0	200	2.0	1200	1200	400	12	15	150	50
	19–24	72	160	177	70	58	1000	10	10	70	60	1.5	1.7	19	2.0	200	2.0	1200	1200	350	10	15	150	70
	25–50	79	174	176	70	63	1000	5	10	80	60	1.5	1.7	19	2.0	200	2.0	800	800	350	10	15	150	70
	51+	77	170	173	68	63	1000	5	10	80	60	1.2	1.4	15	2.0	200	2.0	800	800	350	10	15	150	70
Females	11–14	46	101	157	62	46	800	10	8	45	50	1.1	1.3	15	1.4	150	2.0	1200	1200	280	15	12	150	45
	15–18	55	120	163	64	44	800	10	8	55	60	1.1	1.3	15	1.5	180	2.0	1200	1200	300	15	12	150	50
	19–24	58	128	164	65	46	800	10	8	60	60	1.1	1.3	15	1.6	180	2.0	1200	1200	280	15	12	150	55
	25–50	63	138	163	64	50	800	5	8	65	60	1.1	1.3	15	1.6	180	2.0	800	800	280	15	12	150	55
	51+	65	143	160	63	50	800	5	8	65	60	1.0	1.2	13	1.6	180	2.0	800	800	280	10	12	150	55
Pregnant						60	800	10	10	65	70	1.5	1.6	17	2.2	400	2.2	1200	1200	320	30	15	175	65
Lactating	1st 6 months					65	1300	10	12	65	95	1.6	1.8	20	2.1	280	2.6	1200	1200	355	15	19	200	75
	2nd 6 months					62	1200	10	11	65	90	1.6	1.7	20	2.1	260	2.6	1200	1200	340	15	16	200	75

[a]The allowances, expressed as average daily intakes over time, are intended to provide for individual variations among most normal persons as they live in the United States under usual environmental stresses. Diets should be based on a variety of common foods in order to provide other nutrients for which human requirements have been less well defined. See text for detailed discussion of allowances and of nutrients not tabulated.

[b]Weights and heights of Reference Adults are actual medians for the U.S. population of the designated age, as reported by NHANES II. The median weights and heights of those under 19 years of age were taken from Hamill et al. (1979) (see pages 16–17). The use of these figures does not imply that the height-to-weight ratios are ideal.

[c]Retinol equivalents. 1 retinol equivalent = 1 μg retinol or 6 μg of β-carotene. See text for caculation of vitamin A activity of diets as retinol equivalents.

[d]As cholecalciferol. 10 μg cholecalciferol = 400 IU of vitamin D.

[e]α-Tocopherol equivalents. 1 mg D-α tocopherol = 1 α-TE. See text for variation in allowances and calculation of vitamin E activity of the diet as α-tocopherol equivalents.

[f]1 NE (niacin equivalent) is equal to 1 mg of niacin or 60 mg of dietary tryptophan.

Appendix 2
Estimated Safe and Adequate Daily Dietary Intakes of Selected Vitamins and Minerals[a]

Category	Age (years)	Vitamins: Biotin (μg)	Vitamins: Pantothenic Acid (mg)
Infants	0–0.5	10	2
	0.5–1	15	3
Children and adolescents	1–3	20	3
	4–6	25	3–4
	7–10	30	4–5
	11+	30–100	4–7
Adults		30–100	4–7

Category	Age (years)	Trace Elements[b]: Cu (mg)	Mn (mg)	F (mg)	Cr (μg)	Mo (μg)
Infants	0–0.5	0.4–0.6	0.3–0.6	0.1–0.5	10–40	15–30
	0.5–1	0.6–0.7	0.6–1.0	0.2–1.0	20–60	20–40
Children and adolescents	1–3	0.7–1.0	1.0–1.5	0.5–1.5	20–80	25–50
	4–6	1.0–1.5	1.5–2.0	1.0–2.5	30–120	30–75
	7–10	1.0–2.0	2.0–3.0	1.5–2.5	50–200	50–150
	11+	1.5–2.5	2.0–5.0	1.5–2.5	50–200	75–250
Adults		1.5–3.0	2.5–5.0	1.5–4.0	50–200	75–250

[a]Because there is less information on which to base allowances, these figures are not given in the main table of RDA and are provided here in the form of ranges of recommended intakes.

[b]Because the toxic levels for many trace elements may be only several times usual intakes, the upper levels for the trace elements given in this table should not be habitually exceeded.

Source: National Research Council. *Recommended Dietary Allowances*, 10th ed. Washington, D.C., National Academy Press, 1989, p. 284. Reprinted with permission.

Appendix 3
Recommended Dietary Standards for Adults in Selected Countries and FAO/WHO

Country	Sex	Age (yr)	Wt (kg)	Activity	Calories	Protein (g)	Calcium (mg)	Iron (mg)	Vitamin A (μg RE)	Thiamin (mg)	Riboflavin (mg)	Niacin (mg NE)	Ascorbic Acid (mg)
FAO/WHO[a]	M	18–30	65	MA	3000	37[b]	450[c]	7[d]	750	1.2	1.8	20[e]	30
	F	18–30	55	MA	2200	29[b]	450[c]	21[d]	750	0.9	1.3	14[e]	30
Australia[a]	M	18–35	70	FN[b]	2800	70	800	7	750	1.1	1.7	19[d]	30
	F	18–35	58	FN[b]	2000	58	800	14[c]	750	0.8	1.2	13[d]	30
Canada[a]	M	19–24	71	MA	3000	61	800	9	1000	1.2	1.5	22	40[b]
	F	19–24	58	MA	2100	50	700	13	800	0.8	1.1	15	30[b]
Caribbean[a]	M	20–39	65	MA	3000	53[b]	500	6	750	1.2	1.7	20	30
	F	20–39	55	MA	2200	41[b]	500	19	750	0.9	1.2	15	30
China[a]	M	18–44	63	MA	3000	90	800	12	800	1.5	1.5	15	60
	F	18–44	53	MA	2700	80	800	18	800	1.4	1.4	14	60
Czechoslovakia[a]	M	19–34	NS	MA	3000	105[b]	800	12	1000	1.2	1.8	20	60[b]
	F	19–34	NS	LA	2300	90[b]	800	14	900	1.1	1.6	17	50[b]
France[a]	M	Adult	NS	FN[b]	2700	81	800	10	1000	1.5	1.8	18	80
	F	Adult	NS	FN[b]	2000	60	800	18	800	1.3	1.5	15	80
Germany[a] (former Dem Rep)	M	18–35	NS	MA	3000	85	600	10	800	1.5	1.8	20	45
	F	18–35	NS	MA	2400	75	600	15	750	1.2	1.4	16	45
Germany[a] (former Fed Rep)	M	Adult	NS	SA	2600	0.9 g/kg	800	12	900	1.6	2.0	12[b]	75
	F	Adult	NS	SA	2200	0.9 g/kg	700	18	900	1.4	1.8	12[b]	75
Hungary[a]	M	19–30	70	MA	3200	94	800	12	1000	1.4	1.8	18	60
	F	19–30	60	MA	2400	70	800	18	800	1.3	1.5	15	60
INCAP[a]	M	19–40	63	MA	2900	60	450	9	750	1.2	1.6	19	30
	F	19–40	52	MA	2050	45	450	28	750	0.8	1.1	14	30

(continued)

Appendix 3 *(continued)*

Country	Sex	Age (yr)	Wt (kg)	Activity	Calories	Protein (g)	Calcium (mg)	Iron (mg)	Vitamin A (μg RE)	Thiamin (mg)	Riboflavin (mg)	Niacin (mg NE)	Ascorbic Acid (mg)
India[a]	M	Adult	NS	MA	2800	55	450[b]	24	750	1.4	1.7	19	40
	F	Adult	NS	MA	2200	45	450[b]	32	750	1.4	1.3	15	40
Indonesia[a]	M	20–39	55	FN[b]	2530	51	500	9	4000 IU	1.0	1.4	17[c]	30
	F	20–39	47	FN[b]	1880	40	500	28	3500 IU	0.8	1.3	12[c]	30
Israel[a]	M	18–25	NS	NS	33 kcal/kg	0.8 g/kg	1200	10	1000	1.5	1.7	19	60
	F	18–25	NS	NS	31 kcal/kg	0.8 g/kg	1200	15	800	1.1	1.3	15	60
Italy[a]	M	20–39	65	MA	3000	64	600	10	750	1.2	1.6	20	45
	F	20–39	54	MA	2160	53	600	18	750	0.9	1.2	14	45
Japan[a]	M	20–29	64	MA	2550	70	600	10	2000 IU	1.0	1.4	17	50
	F	20–29	52	MA	2000	60	600	12	1800 IU	0.8	1.1	14	50
Korea[a]	M	20–29	64	MA	2500	70	600	10	700	1.2[b]	1.5	16[b]	55
	F	20–49	52[c]	MA	2000	60	600	18	700	1.0	1.2	13	55
Mexico[a]	M	18–34	65	NS	2750	83	500	10	1000	1.4	1.7	25	50
	F	18–34	55	NS	2000	71	500	18	1000	1.0	1.2	18	50
Netherlands[a]	M	19–22	NS	MA	2900	79	800[b]	10	1000	1.2	1.6	NS	70
	F	19–22	NS	MA	2250	60	800[b]	16	800	1.0	1.3	NS	70
Philippines[a]	M	20–39	56	MA	2570	60	500	12	525	1.3	1.3	25	75
	F	20–39	49	MA	1900	52	500	26[b]	450	1.0	1.0	18	70
Russia[a]	M	18–60	NS	MA	3050[b]	96[c]	800	18[d]	1500	1.8[e]	2.3[f]	19[g]	72[h]
	F	18–60	NS	MA	2500[b]	80[c]	800	18[d]	1500	1.4[e]	2.0[f]	16[g]	62[h]
Spain[a]	M	20–40	NS	FN[b]	3000	54	600	10	750	1.2	1.8	20	45
	F	20–40	NS	FN[b]	2300	41	600	18	750	0.9	1.4	15	45
Sweden[a]	M	19–30	70	MA	3150	75	600	10	1000	1.4	1.7	18	60
	F	19–30	60	MA	2500	62	800	18	800	1.1	1.3	14	60
Turkey[a]	M	19–39	65	NS	2700	65	500	7	750	1.1	1.6	15	50
	F	19–39	55	NS	2200	55	500	23	750	0.9	1.3	12	50

United Kingdom[a]	M	18–34	NS	MA	2900	72	500	10	750	1.2	1.6	18	30
	F	18–34	NS	FN[b]	2150	54	500	12	750	0.9	1.3	15	20
United States[a]	M	19–24	72	FN[b]	2900	58	1200	10	1000	1.5	1.7	19	60
	F	19–24	58	FN[b]	2200	46	1200	15	800	1.1	1.3	15	60
Venezuela[a]	M	20–39	65	NS	3000	63	450	9	750	1.2	1.6	20	30
	F	20–39	55	NS	2200	48	450	28	750	0.9	1.2	14	30

FN = footnote; NS = not specified; LA = light activity; MA = medium or moderate activity; SA = sedentary activity

Explanations:

The purpose for establishing a national dietary standard is not the same in all countries. Some variation in nutrient allowances from country to country should therefore be expected. The "reference" individual varies in different countries, and even in instances where presumed similar objectives exist among countries as to the purpose and usefulness of proposed dietary standards, the table shows that there is no uniform agreement as to the nutrient allowance that may be considered desirable.

FAO/WHO (1985)

[a]Handbook on Human Nutritional Requirements. FAO Nutritional Studies No. 28, WHO Monograph Series No. 61, Rome, 1974; Energy and Protein Requirements, Report of a Joint FAO/WHO Expert Group, Technical Report Series No. 724, 1985. [b]Protein quality varies from country to country. The FAO/WHO standard is stated in terms of high-quality protein of milk or egg (protein score of 100). Protein recommendations are 46 g for males and 36 g for females for a protein score of 80; and 62 g for males and 47 g for females for a protein score of 60. [c]Average of 0.4–0.5 g of calcium recommended for both males and females. [d]Average of 5–9 mg of iron for males, and 14–28 mg iron for females. The recommendations were based on the assumption that the upper limit of iron absorption is 10% if less than 10% of kilocalories come from foods of animal origin. [e]Rounded to nearest whole number.

Australia (1987)

[a]Recommended Dietary Intakes for Use in Australia. National Health and Medical Research Council, Australian Government Publishing Service, Canberra, 1987. [b]Basic activity level. [c]Average of 12–16 mg range. [d]Average of 18–20 mg range for males, and 12–14 mg range for women.

Canada (1990)

[a]Nutrition Recommendations. From Health and Welfare Canada: The Report of the Scientific Review Committee. [b]Smokers should increase vitamin C by 50%.

Caribbean (1979)

[a]Recommended Dietary Allowances for Use in the Caribbean. Caribbean Food and Nutrition Institute, Kingston, Jamaica, 1979. [b]Adjusted to NPU = 70 for average Caribbean diet.

China (1988)

[a]Chinese Nutrition Society. Recommended Dietary Allowances, Revised October, 1988. *Acta Nutrimenta Sinica*, 12:3, March 1990.

(continued)

Appendix 3 *(continued)*

Czechoslovakia (1981)

[a]Institute of Hygiene and Epidemiology. Recommended Dietary Allowances, 1981. [b]Protein and Vitamin recommendations increase with calories for increasing degrees of activity.

France (1981)

[a]Apports Nutritionnels Conseilles du Centre National de Coordination des Etudes et Recherches sur l'Alimentation et la Nutrition (CNERNA), 1981. "Apports Nutritionnels Conseilles pour la population francaise." Technique et Documentation, Paris.

Germany, former Democratic Republic (1980)

[a]Central Institute of Nutrition, Science Academy of the GDR and Nutrition Society of the GDR. Ketz, H.A. and Moehr, M.: Durchschnittswerte des physiologischen Energie—und Nährstoffbedarfs für die Bevölkerung der Deutschen Demokratischen Republik, Zentralinstitut für Ernährung der Akademie der Wissenschaften der DDR und der Gesellschaft für Ernährung in der DDR, Auflage, 1980.

Germany, former Federal Republic (1975)

[a]German Nutrition Society (Deutsche Gesellschaft für Ernährung). *Empfehlungen für die Nährstoffzufur—Emfehlungen der Deutschen Gesellschaft für Ernährung e. V.*, 4th ed. 1979, Umschau-Verlag, Frankfurt/Main.

Hungary (1988)

[a]Biro, G. and Karoly, L.: Tapanyagtablazat, Tapanyagszukseglet es Tapanyag-Osszetetel (*Food Composition Tables. Requirements and Composition of Nutrients*), 11th revised, enlarged edition. Medicina, Konyvkiado, Budapest, 1988.

INCAP (1973)

[a]Institute of Nutrition for Central America and Panama (Guatemala, Honduras, Nicaragua, El Salvador, Costa Rica, and Panama, Instituto de Nutricion). "Recommendaciones Dieteticas Diarias para Centro America y Panama," Publicacion INCAP E-709, 1973.

India (1981)

[a]Indian Council of Medical Research, National Institute of Nutrition. Recommended Dietary Intakes of Nutrients, 1981. *Nutrition News*, 2:3, Hyberabad. [b]Average of 400–500 mg range for both males and females.

Indonesia (1980)

[a]Recommended Dietary Allowance (RDA), Indonesia. *Nutr Abs Rev, Clinical Nutrition Series A*, 53:972, 1983. [b]RDA's "for good health in Indonesia." Activity level not specified. [c]Figures were rounded: from 16.7 mg of niacin for males, and 12.4 mg for females.

Israel (1990)

[a]Recommended Dietary Allowances. State of Israel Ministry of Health, Department of Nutrition, 1990. (Based on National Academy of Sciences USA 1989 and World Health Organization.)

Italy (1978)
[a]Commissione "Ad Hoc" Della Societa Italiana di Nutrizione Umana, Istituto Nazionale Della Nutrizione, Ministero Dell 'Agricoltura e Delle Foreste (1978): "Livelli di Assunzione Raccomandati di Nutrienti per gli Italiani," Roma.

Japan (1991)
[a]Recommended Dietary Allowances, revised 1991. From the Health Promotion and Nutrition Division, Health Policy Bureau, Ministry of Health and Welfare, Tokyo, Japan.

Korea (1989)
[a]Recommended Daily Dietary Allowances. Ministry of Health and Social Affairs, Kyonggi, Korea. [b]Figures were rounded from 1.25 mg for thiamin and 16.5 mg for niacin. [c]Figure was rounded from 52.5 g of protein.

Mexico (1970)
[a]Bourges, H., Chavez, A., and Arroyo, P.: Recommendaciones de Nutrimentos para la Poblacion Mexicana. Publ. L-17, Instituto Nacional de Nutricion, 1970.

Netherlands
[a]Adviescollege van de Minister van Welzijn, Volksgezondheid en Cultuur en de Minister van Landbouw en Visserij inzake voeding en voedselvoorziening, Voedingsraad.

Philippines (1989)
[a]Food and Nutrition Research Institute, Department of Science and Technology. Recommended Dietary Allowances for Filipinos, 1989 ed. [b]Cannot be met by the usual diet; thus supplementation is recommended.

Russia (1980)
[a]Institute of Nutrition, U.S.S.R. Academy of Medical Sciences, Moscow. Voroncova, I.M. and Mazurina, A.V.: "Spravochnik po Detskoy Dietetike." Medicina, Leningrad. ("Guide in Children's Dietetics"); Vanhanen, V.D. et al.: Higiene Pitanija" (1980), Zdorovja, Kiev. ("Nutrition Hygiene"). [b–h]Dietary allowances shown are for people living "in cities with developed communal services." Figures shown are rounded averages. [b]Average for calories (2800–3300 for males, 2400–2600 for females); [c]Average for protein (92–99 g for males, 77–84 gm for females); [d]Average for iron (15–20 mg for both males and females); [e]Average for thiamin (1.7–1.8 mg for males, 1.4–1.5 mg for females); [f]Average for riboflavin (2.2–2.4 mg for males, 1.9–2.0 mg for females); [g]Average for niacin (10–18 mg for males, 15–17 mg for females); [h]Average for vitamin C (70–75 mg for males, 60–65 mg for females).

Spain (1980)
[a]Institute of Nutrition, Madrid and Spanish Nutrition Society. Instituto de Nutricion (CSIC) Facultad de Farmaciaciudad Universitaria, Madrid (1980). "Ingestas Recomendadas de Energia y Nutrientes para la Poblacion Espanola." [b]Energy for active work; subtract 10% for light work; add 20% for very active work. [c]Protein NPU = 70.

(continued)

Appendix 3 *(continued)*

Sweden (1989)

[a]Swedish Nutrition Recommendations (Svenska naringsrekommendationer, SNR), 2nd ed. Statens livsmedelsverk, National Food Administration, Uppsala, Sweden, 1989.

Turkey (1972)

[a]Institute of Nutrition and Food Sciences, Hacettepe Universitesi, Ankara. Recommended Dietary Allowances, 1972.

United Kingdom (1979)

[a]Recommended Daily Amounts of Food Energy and Nutrients for Groups of People in the United Kingdom. Report by the Committee on Medical Aspects of Food Policy, Department of Health and Social Security, London, 1979 (as per third impression, 1985).

United States (1989)

[a]Recommended Dietary Allowances, 10th ed. National Research Council, National Academy Press, Washington, DC, 1989.

Venezuela (1976)

[a]National Institute of Nutrition, National Council for Scientific Investigations and Technology. Instituto Nacional de Nutricion, Consejo Nacional de Investigaciones Cientificas y Technologicas (1976). "Requerimientos de Energi y de Nutrientes de al Poblacion Venezolana." Serie de Cuadernos Azules, publication no. 38. Republica de Venezuela.

Appendix 4
The Food Guide Pyramid (USA)[a]

The Pyramid is a general guide of what to eat daily for a healthful diet. The small tip of the Pyramid represents fats, oils, and sweets. This group provides calories and little else nutritionally. The base of the pyramid are grains, cereals, breads, rice, and pasta. On top of the base are fruit and vegetable groups. Below the tip are milk and meat groups. Details for recommended servings for each group in the Pyramid are given below:

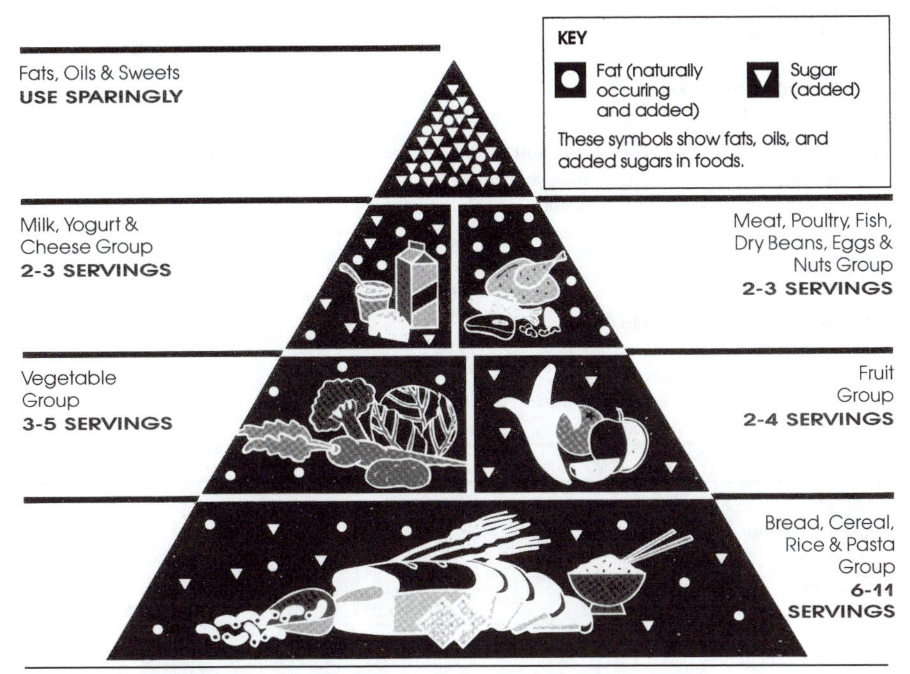

[a]Source: USDA Home and Garden Bulletin No. 249, 1992.

What Counts as a Serving?

Breads, cereals, rice and pasta (Contribute complex carbohydrates, riboflavin, thiamin, niacin, iron, protein, magnesium, and fiber.) The amount you eat may be more than 1 serving.
A dinner portion of pasta would count as 2 or 3 servings:
- ½ cup of cooked pasta or rice
- ½ cup of cooked cereal
- 1 oz. of ready-to-eat cereal
- 1 slice of bread
- ½ bagel or hamburger bun

Milk, yogurt and cheese (Contribute calcium, riboflavin, protein, vitamin B_{12}, and, when fortified, vitamins A and D.)
- 1 cup milk
- 1 cup yogurt
- 1 ½ to 2 oz. of cheese

(continued)

Appendix 4 *(continued)*

Vegetables (Contribute vitamins A and C, folic acid, potassium, magnesium, and fiber. Traces or very low in fat and cholesterol.)
- ¾ cup of vegetable juice
- ½ cup of chopped raw or cooked vegetables
- 1 cup of raw, leafy vegetables

Fruits (Contribute vitamins A and C, potassium, and fiber. Traces or very low in sodium, fat, and cholesterol.)
- 1 medium apple, banana, orange
- ¾ cup of juice
- ½ cup of canned or cooked fruit
- ¼ cup of dried fruit

Meat, poultry, fish, dry beans, eggs and nuts (Contribute protein, phosphorus, vitamins B_6 and B_{12}, thiamin, niacin, iron, magnesium, and zinc.)
- 2 or 3 oz. of cooked lean meat, poultry or fish

Count as 1 oz. of lean meat:
- ½ cup of cooked dry beans
- 1 egg
- 2 tablespoons of peanut butter

Recommended Servings Each Day

	Women and some older adults	Children, teenage girls, active women, most men	Teenage boys and active men
Calorie level	About 1500	About 2200	About 2800
Bread group	6	9	11
Vegetable group	3	4	5
Fruit group	2	3	4
Milk group	2–3	2–3	2–3
Meat group	2 (total 5 oz)	2 (total 6 oz)	2 (total 7 oz)

Appendix 5
Nutrition Labeling (USA)

As of mid-1994, food labels with nutrition information are required on almost all packaged foods. The five main features of the nutrition panel are:

1. **Nutrition Facts.** A label bearing a panel titled "Nutrition Facts" meets the January 1993 government nutrition labeling requirements.
2. **Serving Size.** Similar foods now have standard serving sizes. This makes it easier to compare foods. Always compare the label serving size with the amount you actually eat.
3. **% Daily Value.** % Daily Value shows how a food fits into an overall daily diet. The % Daily Values for total fat, saturated fat, total carbohydrate, and dietary fiber are based on a 2000-calorie diet.
4. **Vitamins and Minerals.** Only vitamins A and C, calcium, and iron have to be listed on the food label. Food companies have the option to list other vitamins and minerals that are in the food.
5. **Daily Values.** Daily Values are label reference values set by the government. Some values are recommended maximums for the day (such as 65 g total fat or less if you eat 2000 calories per day), whereas others are minimums (such as 300 g total carbohydrates or more if you eat 2000 calories per day). Many labels list Daily Values for 2000 calories per day.

Mandatory Information on Food Labels:

Every food label must prominently display and express in ordinary words the following:

1. The common name of the product
2. The name and address of the manufacturer, packer, or distributor
3. The net contents in terms of weight, measure, or count
4. The ingredient list: The FDA requires all manufactures to list the ingredients of all foods in descending order of predominance by weight and to state on the label that they have done so. Manufacturers may list the percentages of each ingredient as well. They must list all the additives used and the specific fat and oils used.

Definition of Terms on Food Labels

Calorie Terms

Calorie-free: fewer than 5 calories per servings
Light: one-third fewer calories or 50% less fat per serving
Low-calorie: less than 40 calories per serving
Reduced calorie: at least 25% calories per serving when compared with a similar food

Sugar Terms

Sugar-free: less than 0.5 g of sugars per serving
Low-sugar: may not be used as a claim
Reduced sugar: at least 25% less sugar per serving when compared with a similar food
No added sugars, Without added sugars, No sugar added: (1) no amount of sugars or any other ingredient that contains sugars that functionally substitute for added sugars is added during processing or packaging; (2) the product contains no ingredients that contain added sugars such as jam, jelly, or concentrated fruit juice; (3) the product it resembles and substitutes for normally contains added sugars; and (4) the label declares that the food is not "low calorie" or "calorie reduced" as appropriate.

(continued)

Appendix 5 *(continued)*

Fat Terms

Fat-free: less than 0.5 g of fat per serving

100% fat-free: meets requirements for fat-free (less than 0.5 g of fat per serving)

Low-fat: 3 g or less fat per serving

____**% fat free:** meets requirement for low-fat; percentage is based on amount of fat (by weight) in 100 g of food

Reduced fat: at least 25% less fat when compared with a similar food

Saturated fat free: less than 0.5 g saturated fat and less than 0.5 g *trans*-fatty acids per serving

Low saturated fat: 1 g or less saturated fat per serving and no more than 15% of kcalories from saturated fat

Reduced saturated fat: at least 25% less saturated fat per serving when compared with a similar food

Cholesterol Terms

Cholesterol-free: less than 2 mg of cholesterol per serving and 2 g or less of saturated fat per serving

Low-cholesterol: 20 mg or less of cholesterol per serving and 2 g or less of saturated fat per serving

Sodium Terms

Sodium-free: less than 5 mg of sodium per serving

Salt-free: meets requirement for sodium-free

Very-low-sodium: 35 mg or less of sodium per serving

Low-sodium: 140 mg or less of sodium per serving

Reduced sodium: at least 25% less sodium when compared with a similar food

Light in sodium: 50% less sodium per serving; restricted to foods with more than 40 kcalories per serving or more than 3 g of fat per serving

Unsalted, Without added salt, No salt added: (1) no salt is added during processing; (2) the product it resembles and substitutes for is normally processed with salt; and (3) the label bears the statement "not a sodium-free food" or "not for control of sodium in the diet" if the food is not sodium-free.

General Terms

Good source, Contains, Provides: contains 10% to 19% of the Daily Value per serving

High, Rich in, Excellent source of: contains 20% or more of the Daily Value per serving

More, Fortified, Enriched, Added: contains at least 10% more of the Daily Value for protein, vitamins, minerals, dietary fiber, or potassium per serving. May not be used as a claim on meat or poultry products.

Fiber: any food making a fiber claim must meet the requirements for a good source or high claim; must declare the level of total fat per serving if food is not low fat.

Lean: packaged seafood, game meat, cooked meat, or cooked poultry with less than 10 g total fat, 4.5 g or less of saturated fat, and less than 95 mg of cholesterol per serving (and 100 g of the food)

Extra lean: packaged seafood, game meat, cooked meat, or cooked poultry with less than 5 g total fat, less than 2 g saturated fat, and less than 95 mg of cholesterol per serving (and 100 g of the food)

Fresh: raw food that has not been frozen, heat processed, or similarly perserved

Fresh frozen, Frozen fresh: food quickly frozen while very fresh

Appendix 5 *(continued)*

Health Claims

Health messages that have scientific validity between diet and health are permitted if the relationship is clearly established. The following health claims are authorized by the Food and Drug Administration:

1. Calcium and osteoporosis—must contain 20% (200 mg) or more of the Reference Daily Intake (RDI) for calcium per serving.
2. Dietary saturated fat and cholesterol and risk of coronary heart disease—must meet descriptor requirements for "low saturated fat", "low cholesterol", and "low fat."
3. Dietary fat and cancer—must meet the requirements for a "low fat" food; fish and game meats may bear this health claim if they met requirements for "extra lean."
4. Sodium and hypertension—must meet "low sodium" descriptor requirements and must not exceed disqualifying levels for fat, saturated fat, and cholesterol.
5. Fiber-containing grain products, fruits, and vegetables and cancer—must qualify as a low-fat food and a "good source" of dietary fiber without fortification.
6. Fruits, vegetables, and grain products and risk of coronary heart disease—must be low in saturated fat and cholesterol and contain not less than 0.6 g of soluble fiber without fortification per serving.

Appendix 6
U.S. Daily Values for Nutrition Labeling

U.S. Daily Values for nutrition labeling refer to two sets standards as follows:
Reference Daily Intakes (RDI): values for protein, vitamins, and minerals based on the RDA.
Daily Reference Values (DRV): values for nutrients and food components (such as fat and fiber) that do not have RDA values, but do have important relationships with health (see text for further details)

	Reference Daily Intakes (RDI)[a]		
Nutrient	Reference Daily Intakes (RDI)	Daily Reference Values (DRV)	
Protein (g)	50	Total fat	less than 65 g
Thiamin (mg)	1.2	Saturated fat	less than 20 g
Riboflavin (mg)	1.4	Cholesterol	less than 300 mg
Niacin (mgNE)	16	Total CHO	300 g
Biotin (μg)	60	Fiber	25 g
Pantothenic acid (mg)	5.5	Sodium	less than 2400 mg
Vitamin B_6 (mg)	1.5		
Folate (μg)	180		
Vitamin B_{12} (μg)	2		
Vitamin C (mg)	60		
Vitamin A (μgRE)	875		
Vitamin D (μg)	6.5		
Vitamin E (αTE)	9		
Calcium (mg)	900		
Phosphorus (mg)	900		
Magnesium	300		
Iron (mg)	12		
Zinc (mg)	15		
Iodine (μg)	150		
Copper (mg)	2		
Manganese (mg)	3.5		
Fluoride (mg)	2.5		
Chromium (μg)	120		
Selenium (μg)	55		
Molybdenum (μg)	150		

[a]The FDA established RDI values for infants, children under 4 years, pregnant women, and lactating women.

[b]The FDA established only one set of DRV for adults and children age 4 and older. Calculations for DRV for fat, saturated fat, carbohydrate, and fiber are based on an energy intake of 2000 kcalories.

Source: Food and Drug Administration. Food labeling: Reference daily intakes and daily reference values, Federal Register 55: 29476-29486, 1991.

Appendix 7A
Dietary Guidelines for All Healthy Americans Over 2 Years Old[a]

Eat a variety of foods: Choose foods each day from the five major food groups.

Maintain healthy weight: Whether your weight is "healthy" depends on weight for height, where in your body the fat is located (potbelly puts you at added risk), and whether you have a weight-related medical problem (such as high blood pressure).

Choose a diet low in fat, saturated fat, and cholesterol: A diet that provides 30% or fewer calories from fat is suggested for individuals over age 2. These goals apply to the diet over several days, not to a single meal or food.

Choose a diet with plenty of vegetables, fruits, and grain products.

Use sugars only in moderation.

Use salt and sodium only in moderation.

If you drink alcoholic beverages, do so in moderation: Moderate drinking is described as: no more than one drink a day for women and no more than two drinks a day for men. Some people who should not drink alcoholic beverages: children; women who are pregnant or who are trying to conceive; individuals who plan to drive or engage in activities that require attention or skill.

[a]Source: *Dietary Guidelines for Americans*, 3rd ed. USDA and USDHHS, 1990.

Appendix 7B
National Nutrition Objectives for the Year 2000[a]

1. Reduce deaths from coronary heart disease to no more than 100 per 100,000 persons (age-adjusted baseline: 135 per 100,000 in 1987).
2. Reverse the increase in deaths from cancer to achieve a rate of no more than 130 per 100,000 persons (age-adjusted baseline: 133 per 100,000 in 1987).
3. Reduce the overweight population to no more than 20% among adults aged 20 years and older and no more than 15% among adolescents aged 12 through 19 years (baseline: 26% for adults aged 20 through 74 years in 1976 to 1980. 24% for men and 27% for women; 15% for adolescents aged 12 through 19 years in 1976 to 1980).
4. Reduce growth retardation among low-income children aged 5 years and younger to less than 10% (baseline: up to 16% among low-income children in 1988, depending on age and race/ethnicity).
5. Reduce dietary fat intake to an average of 30% of calories or less and reduce average saturated fat intake to less than 10% of calories among persons aged 2 years and older (baseline: 36% of calories from total fat and 13% from saturated fat for persons aged 20 through 74 years in 1976 to 1980; 36% and 13% for women aged 19 through 50 years in 1985).
6. Increase complex carbohydrates and fiber-containing foods in the diets of adults to 5 or more daily servings for vegetables (including legumes) and fruits, and to 6 or more daily servings for grain products (baseline: 2.5 servings of vegetables and fruits and 3 servings of grain products for women aged 19 through 50 years in 1985).
7. Increase to at least 50% the proportion of overweight persons aged 12 years and older who have adopted sound dietary practices combined with regular physical activity to attain an appropriate body weight (baseline: 30% of overweight women and 25% of overweight men for people aged 18 years and older in 1985).
8. Increase calcium intake so that at least 50% of youth aged 12 through 24 years and at least 50% of pregnant and lactating women are consuming 3 or more servings daily of foods rich in calcium, and at least 50% of adults aged 25 years and older are consuming 2 or more servings daily (baseline: 7% of women and 14% of men aged 19 through 24 years and 24% of pregnant and lactating women consumed 3 or more servings daily, and 15% of women and 23% of men aged 25 through 50 years consumed 2 or more servings daily in 1985 to 1986).
9. Decrease salt and sodium intake so that at least 65% of those who prepare home-cooked meals do so without adding salt, at least 80% of persons avoid using salt at the table, and at least 40% of adults regularly purchase foods modified or lower in sodium (baseline: 54% of women aged 19 through 50 years who prepared most of the meals did not use salt in food preparation, and 68% of women aged 19 through 50 years did not use salt at the table in 1985; 20% of all persons aged 18 years and older regularly purchased foods with reduced salt and sodium content in 1988).
10. Reduce iron deficiency to less than 3% among children aged 1 through 4 years and among women of childbearing age (baseline: 9% for children aged 1 through 2 years, 4% for children aged 3 through 4 years, and 5% for women aged 20 through 44 years in 1976 to 1980).
11. Increase to at least 75% the proportion of mothers who breast-feed their babies in the early postpartum period and to at least 50% the proportion who continue to breast-feed until their babies are 5 to 6 months old (baseline: 54% at discharge from birth site and 21% at 5 to 6 months in 1988).
12. Increase to at least 75% the proportion of parents and caregivers who use feeding practices that prevent baby-bottle tooth decay.
13. Increase to at least 85% the proportion of persons aged 18 years and older who use food labels to make nutritious food selections (baseline: 74% used labels to make food selections in 1988).

Appendix 7B *(continued)*

14. Achieve useful and informative nutrition labeling for virtually all processed foods and for at least 40% of fresh meats, poultry, fish, fruits, vegetables, baked foods, and ready-to-eat carry-out foods (baseline: 60% of processed foods regulated by the Food and Drug Administration had nutrition labeling in 1988; baseline data on fresh and carry-out foods are unavailable).
15. Increase the available processed food products that are reduced in fat and saturated fat to at least 5000 brand items (baseline: 2500 brand items reduced in fat in 1986).
16. Increase to at least 90% the proportion of restaurants and institutional service operations than offer identifiable low-fat, low-calorie food choices, consistent with the nutrition principles in the Dietary Guidelines for Americans.
17. Increase to at least 90% the proportion of school lunch and breakfast services and child-care food services that offer menus consistent with the nutrition principles in the Dietary Guidelines for Americans.
18. Increase to at least 80% the receipt of home food services by people aged 65 years and older who cannot prepare their own meals or are otherwise in need of home-delivered meals.
19. Increase to at least 75% the proportion of schools in the United States that provide nutrition education from preschool through 12th grade, preferably as part of quality school health education.
20. Increase to at least 50% the proportion of worksites with 50 or more employees that offer nutrition education and/or weight management programs for employees (baseline: 17% offered nutrition education activities and 15% offered weight-control activities in 1985).
21. Increase to at least 75% the proportion of primary care providers who provide nutrition assessment and counseling and/or referral to qualified nutritionists or dietitians (baseline: physicians provided diet counseling for an estimated 40% to 50% of patients in 1988).

[a]Reference: *Nutrition in Healthy People 2000*. In: National Promotion and Disease Prevention Objectives. Washington, D.C., U.S. Government Printing Office, 1991.

Appendix 8
Classification of Carbohydrates

Carbohydrate	Occurrence	Characteristics
Monosaccharides (simple sugars)		
Hexoses		
Glucose	Honey, fruits, corn syrup, sweet grapes, and sweet corn; hydrolysis of starch and of cane sugar	Physiologically the most important sugar; the "sugar" carried by the blood and the principal one used by tissues
Fructose	Honey, ripe fruits, and some vegetables; hydrolysis of sucrose inulin	Can be changed to glucose in the liver and intestine; an intermediate metabolite in glycogen breakdown
Galactose	Not found free in nature; digestive end product of lactose hydrolysis	Can be changed to glucose in the liver; synthesized in the body to make lactose; constituent of glycolipids
Mannose	Legumes; hydrolysis of plant mannosans and gums	Constituent of polysaccharide of albumins, globulins, and mucoids
Pentoses		
Arabinose	Derived from gum arabic and plum and cherry gums; not found free in nature	Has no known physiologic function in man; used in metabolism studies of bacteria
Ribose	Derived from nucleic acid of meats and seafoods	Structural element of nucleic acids, ATP, and coenzymes (NAD and FAD)
Ribulose	Formed in metabolic processes	Intermediate in direct oxidative pathway of glucose breakdown
Xylose	Wood gums, corncobs, and peanut shells; not found free in nature	Very poorly digested and has no known physiologic function; used medicinally as a diabetic food
Oligosaccharides (2–10 sugar units)		
Disaccharides		
Sucrose	Cane and beet sugar, maple syrup, molasses, and sorghum	Hydrolyzed to glucose and fructose; a nonreducing sugar
Maltose	Malted products and germinating cereals; an intermediate product of starch digestion	Hydrolyzed to two molecules of glucose; a reducing sugar; does not occur free in tissue
Lactose	Milk and milk products; formed in body from glucose nature	Hydrolyzed to glucose and galactose; may occur in urine during pregnancy; a reducing sugar
Trisaccharides		
Raffinose	Cottonseed meal, molasses, and sugar beets and stems	Only partially digestible but can be hydrolyzed by enzymes of intestinal bacteria to glucose, fructose, and galactose

Appendix 8 *(continued)*

Carbohydrate	Occurrence	Characteristics
Melizitose	Honey, poplars, and conifers	Composed of one fructose unit and two glucose units
Polysaccharides (more than 10 sugar units)		
Digestible		
Glycogen	Meat products and seafoods	Polysaccharide of the animal body, often called animal starch; storage form of carbohydrate in body, mainly in liver and muscles
Starch	Cereal grains, unripe fruits, vegetables, legumes, and tubers	Most important food source of carbohydrate; storage form of carbohydrate in plants; composed chiefly of amylose and amylopectin; hydrolyzable to glucose
Dextrin	Toasted bread; intermediate product of starch digestion	Formed in course of hydrolytic breakdown of starch
Partially digestible		
Inulin	Tubers and roots of dahlias, artichokes, dandelions, onions, and garlic	Hydrolyzable to fructose; used in physiologic investigation for determination of glomerular filtration rate
Mannosan	Legumes and plant gums	Hydrolyzable to mannose but digestion incomplete; further splitting by bacteria may occur in large bowel
Indigestible[a]		
Cellulose	Skins of fruits, outer coverings of seeds, and stalks and leaves of vegetables	Not subject to attack of digestive enzyme in man, thus an important source of "bulk" in diet; may be partially split to glucose by bacterial action in large bowel
Hemicellulose and pectin	Woody fibers and leaves	Less polymerized than cellulose; may be digested to some extent by microbial enzymes, yielding xylose

[a]Another name for indigestible polysaccharides is *dietary fiber*. There are two groups:

I. Insoluble dietary fibers (cellulose, lignin, and cutin) which are the most abundant organic compounds in the world. They help prevent constipation, colon cancer, and diverticulosis, but not hypercholesterolemia.

II. Soluble dietary fibers (hemicellulose, pectins, gums, and algal polysaccharides) which are useful in decreasing serum cholesterol and in regulating blood glucose levels.

Appendix 9
Classification of Proteins

Protein	Occurrence	Characteristics
Simple proteins		
Albumins	Blood (serum albumin); milk (lactalbumin); egg white (ovalbumin); lentils (legumelin); kidney beans (phaseolin); wheat (leucosin)	Globular protein; soluble in water and dilute salt solutions; precipitated by saturation with ammonium sulfate solution; coagulated by heat; found in plant and animal tissues
Globulins	Blood (serum globulin); muscle (myosin); potato (tuberlin); Brazil nuts (excelsin); hemp (edestin); lentils (legumin)	Globular protein; sparingly soluble in water; soluble in dilute neutral solutions; precipitated by dilute ammonium sulfate and coagulated by heat; distributed in both plant and animal tissues
Glutelins	Wheat (glutenin); rice (oryzenin)	Insoluble in water and dilute salt solutions; soluble in dilute acids; found in grains and cereals
Prolamines	Wheat and rye (gliadin); corn (zein); rye (secaline); barley (hordein)	Insoluble in water and absolute alcohol; soluble in 70% alcohol; high in amide nitrogen and proline; occur in grain seeds
Protamines	Sturgeon (sturine); mackerel (scombrine); salmon (salmine); herring (clupeine)	Soluble in water; not coagulated by heat; strongly basic; high in arginine; associated with DNA and occur in sperm cells
Histones	Thymus gland, pancreas; mucleoproteins (nucleohistone)	Soluble in water, salt solutions, and dilute acids; insoluble in ammonium hydroxide; yields large amounts of lysine and arginine; combined with nucleic acids within cells
Scleroproteins	Connective tissues and hard tissues	Fibrous protein; insoluble in all solvents and resistant to digestion
Collagen	Connective tissues, bones, cartilage, and gelatin	Resistant to digestive enzymes but altered to digestible gelatin by boiling water, acid, or alkali; high in hydroxyproline
Elastin	Ligaments, tendons, and arteries	Similar to collagen but cannot be converted to gelatin
Keratin	Hair, nails, hooves, horn, and feathers	Partially resistant to digestive enzymes; contains large amounts of sulfur, as cystine
Conjugated proteins		
Nucleoproteins	Cytoplasm of cells (ribonucleoprotein); nucleus of chromosomes (deoxyribonucleoprotein); viruses and bacteriophages	Contains nucleic acids, nitrogen, and phosphorus; present in chromosomes and in all living forms as a combination of protein with either RNA or DNA

Appendix 9 *(continued)*

Protein	Occurrence	Characteristics
Mucoprotein or Glycoprotein	Saliva (mucin); egg white (ovomucoid) Bone (osseomucoid); tendons (tendomucoid); cartilage (chondromucoid)	Proteins combined with amino sugars, sugar acids, and sulfates Containing more than 4% hexosamine, mucoproteins; if less than 4%, glycoproteins
Phosphoproteins	Milk (casein); egg yolk (ovovitellin)	Phosphoric acid joined in ester linkage to protein
Chromoproteins	Hemoglobin; myoglobin, flavoproteins; respiratory pigments; cytochromes	Protein compounds with nonprotein pigments such as heme; colored proteins
Lipoproteins	Serum lipoprotein; brain, nerve tissues, milk, and eggs	Water-soluble proteins conjugated with lipids; found dispersed widely in all cells and all living forms
Metalloproteins	Ferritin; carbonic anhydrase; ceruloplasmin	Proteins combined with metallic atoms that are not parts of a nonprotein prosthetic group
Derived proteins		
Proteans	Edestan (from elastin) and myosan (from myosin)	Results from short action of acids or enzymes; insoluble in water
Proteoses	Intermediate products of protein digestion	Soluble in water; uncoagulated by heat, and precipitated by saturated ammonium sulfate; result from partial digestion of protein by pepsin or trypsin
Peptones	Intermediate products of protein digestion	Same properties as proteoses except that they cannot be salted out; of smaller molecular weight than proteoses
Peptides	Intermediate products of protein digestion	Two or more amino acids joined by a peptide linkage; hydrolyzed to individual amino acids[a]

[a]Classification of amino acids according to essentiality:

Nonessential Amino Acids	Essential Amino Acids	Conditionally Essential Amino Acids
Alanine	Histidine	Arginine
Asparagine	Isoleucine	Cysteine
Aspartic acid	Leucine	Glutamine
Cystine	Lysine	Taurine
Glutamic acid	Methionine	Tyrosine
Glycine	Phenylalanine	
Hydroxylysine	Threonine	
Hydroxyproline	Tryptophan	
Proline	Valine	
Serine		

Appendix 10
Classification of Lipids

Lipid	Occurrence	Characteristics
Simple Lipids		
Triglycerides, neutral fats	Adipose tissue, butterfat, lard, suet, fish oils, olive oil, corn oil, etc.	Esters of three molecules of fatty acids and one molecule of glycerol; the fatty acids may all be different
Waxes	Beeswax, head oil of sperm whale, cerumen, carnauba oil, and lanolin	Composed of esters of fatty acids with alcohol other than glycerol; of industrial and medicinal importance
Compound lipids		
Phospholipids (phosphatides)	Chiefly in animal tissues	Substituted fats consisting of phosphatidic acid; composed of glycerol, fatty acid, and phosphoric acid bound in ester linkage to a nitrogenous base
Lecithin	Brain, egg yolk, and organ meats	Phosphatidyl choline or serine; phosphatide linked to choline; a lipotropic agent; important in fat metabolism and transport; used as emulsifying agent in the food industry
Cephalin	Occurs predominantly in nervous tissue	Phosphatidyl ethanolamine; phosphatide linked to serine or ethanolamine; plays a role in blood clotting
Plasmalogen	Brain, heart, and muscle	Phosphatidal ethanolamine or choline; phosphatide containing an aliphatic aldehyde
Lipositol	Brain, heart, kidneys, and plant tissues together with phytic acid	Phosphatidyl inositol; phosphatide linked to inositol; rapid synthesis and degradation in brain; evidence for role in cell transport processes
Sphingomyelin	Nervous tissue, brain, and red blood cells	Sphingosine-containing phosphatide; yields fatty acid, choline, sphingosine, phosphoric acid, and no glycerol; source of phosphoric acid in body tissue
Glycolipids		
Cerebroside	Myelin sheaths of nerves, brain, and other tissues	Yields on hydrolysis fatty acids, sphingosine, galactose (or glucose), but not fatty acid; includes kerasin and phrenosin

Appendix 10 *(continued)*

Lipid	Occurrence	Characteristics
Ganglioside	Brain, nerve tissue, and other selected tissues, notably spleen	Contains a ceramide linked to hexose (glucose or galactose), neuraminic acid, sphingosine, and fatty acids
Sulfolipid	White matter of brain, liver, and testicle; also plant chloroplast	Sulfur-containing glycolipid; sulfate present in ester linkage to galactose
Proteolipids	Brain and nerve tissues	Complexes of protein and lipids having solubility properties of lipids
Terpenoids and steroids		
Terpenes	Essential oils, resin acids, rubber, plant pigments such as carotenes and lycopenes, vitamin A, and camphor	Large group of compounds made up of repeating isoprene units; vitamin A of nutritional interest; fat-soluble vitamins E and K also related chemically to terpenes
Sterols		
Cholesterol, ergosterol, and 7-dehydrocholesterol	Cholesterol found in egg yolk, dairy products, and animal tissues; ergosterol found in plant tissues, yeast, and fungi; 7-dehydrocholesterol found in animal tissues and underneath skin	Cholesterol, a constituent of bile acids and precursor of vitamin D; ergosterol and 7-dehydrocholesterol, converted to vitamin D_2 and D_3, respectively, on irradiation
Sex hormones		
Androgens, estrogens	Ovaries and testes	
Adrenal cortical steroids	Adrenal cortex, blood	
Derived lipids		
Fatty acids[a]	Occur in plant and animal foods; also exist in complex forms with other substances	Obtained from hydrolysis of fats; usually contain an even number of carbon atoms and are straight chain derivatives

[a]Classification of fatty acids is based on the length of the carbon chain (short, medium, or long); number of double bonds (unsaturated, mono- or polyunsaturated); or essentiality in the diet (essential or nonessential). A current designation is based on the position of the endmost double bond counting from the methyl (CH_3) carbon, called the omega end. The most important omega fatty acids are:

Omega-6 fatty acids	linoleic and arachidonic acids
Omega-3 fatty acids	linolenic, eicosapentaenoic, and docosahexaenoic acids

Sample nomenclature for fatty acids according to its chemical characteristics are:

Butyric acid	4:0 (carbon length:no. of double bond)
Palmitic acid	16:0
Oleic acid	18:1 (9) indicating position of double bond
Linoleic acid	18:2 (9,12)
Linolenic acid	18:3 (9,12,15)
Arachidonic acid	20:4 (5,8,11,14)
Eicosapentaenoic acid	20:5 (5,8,11,14,17)
Docosahexaenoic acid	22:6 (4,7,10,13,16,19)

Appendix 11A
Summary of Digestive Enzymes

Source and Enzyme	Substrate	Products
Mouth		
Salivary α-amylase	Cooked starch	Dextrins, maltose, and maltriose
Stomach		
Gastric lipase	Emulsified fat	Fatty acids and glycerol
Pepsin	Proteins and polypeptides	Polypeptides and amino acids
Rennin	Casein of milk	Calcium caseinate
Pancreas		
Carboxypeptidase A	Proteins and polypeptides	Aromatic or branch chain amino acids
Carboxypeptidase B	Proteins and polypeptides	Basic side chain amino acids
Chymotrypsin	Proteins and polypeptides	Polypeptides, proteoses, and peptones
Deoxyribonuclease	DNA	Mononucleotides
Elastase	Elastin and other protein	Neutral aliphatic amino acids
Cholesterol esterase	Cholesteryl esters	Cholesterol; fatty acids
Pancreatic α-amylase	Cooked starch	Same as salivary α-amylase
Pancreatic lipase	Fat and triglycerides	Glycerides, fatty acids, and glycerol
Phospholipase A	Lecithin	Lysolecithin
Ribonuclease	RNA	Mononucleotides
Trypsin	Proteins and polypeptides	Polypeptides, peptones, proteoses
Small intestines		
Aminopeptidase	Polypeptides	N-terminal amino acids
Dipeptidase	Dipeptides	Two amino acids
Enterokinase	Trypsinogen	Trypsin
Intestinal lipase	Monoglycerides	Glycerol and fatty acids
Isomaltase	Dextrins	Glucose
Lactase	Lactose	Glucose, galactose
Maltase	Maltose	Glucose
Nucleosidase	Nucleosides	Purines, pyrimidines, and pentose
Sucrase	Sucrose	Glucose, fructose

Appendix 11B
Summary of Selected Hormones[a]

Hormone (Organ of Origin)	Main Physiologic Functions
Adrenaline	See epinephrine
Adrenocorticotropin or ACTH (anterior pituitary gland)	Stimulates the secretion of corticosteroids and other hormones. Acts on the adrenal cortex to release its hormones.
Aldosterone (adrenal cortex)	Regulates water and electrolyte balance.
Antidiuretic hormone or ADH	See vasopressin.
Calcitonin (thyroid gland)	Controls the calcium level requirement.
Cholecystokinin (duodenum and jejunum)	Stimulates the gallbladder and pancreas to release their contents into the small intestines.
Corticotropin-releasing hormones or CRH (hypothalamus)	Stimulates the release of ACTH triggered by stress; but turned off by ACTH when enough has been released.
Cortisone	See glucocorticoids.
Deoxycorticosterone (Adrenal cortex)	Maintains water and electrolyte balance.
Enterogastrone (stomach)	Inhibits gastric juice secretion.
Enterocrinin (intestinal mucosa)	Stimulates intestinal juice secretion.
Epinephrine (adrenal medulla)	Accelerates cardiac action; constricts certain blood vessels; relaxes intestines and dilates the bronchi in the trachea.
Erythropoietin (kidney)	A glycoprotein hormone; stimulates the production of red blood cells.
Estrogens (ovaries)	Responsible for female characteristics and menstrual cycle. Promotes the maturation of female ovaries and sperm formation of the testes.
Gastrin (stomach and duodenum)	Promotes the manufacture and release of acid and digestive juices in the stomach and duodenum.
GH-inhibiting hormone or somatostatin (hypothalamus)	Inhibits the release of TSH and GH; also inhibits ACTH, glucagon, and insulin.
GH-releasing hormone or GRH (hypothalamus)	Promotes the release of growth hormone; stimulated by insulin.
Growth hormone or GH (anterior pituitary gland)	Promotes linear growth of long bones. Aids in nitrogen, calcium, phosphorus, and carbohydrate metabolism.
Glucagon (alpha cells of pancreas)	Increases blood glucose levels by stimulating the breakdown of glycogen into glucose; activates gluconeogenesis.
Glucocorticoids: corticosterone, cortisone and hydrocortisone (adrenal cortex)	Maintains circulatory and vascular homeostasis. Aids in carbohydrate, protein, and lipid metabolism.
Insulin (beta cells of pancreas)	Regulates glucose metabolism by lowering blood glucose levels; promotes transport and entry of glucose into the muscle cells and other tissues; promotes glycogen synthesis, lipogenesis, and gluconeogenesis.
Motilin (small intestine)	Regulates motility of the intestines.

(continued)

Appendix 11B *(continued)*

Hormone (Organ of Origin)	Main Physiologic Functions
Oxytocin or pitocin (anterior pituitary gland)	Stimulates uterine contraction labor and bleeding after delivery. Aids in the production of milk.
Pancreozymin (pancreas)	Stimulates the secretion of pancreatic juice.
Parathyroid or PTH (parathyroid gland)	Controls the calcium content in the blood.
Progesterone (uterus and mammary glands)	Stimulates the growth of the uterus and mammary glands.
Prolactin (anterior pituitary gland)	Stimulates the growth of the mammary glands and the production of milk.
Secretin (duodenum)	Stimulates the pancreas to release acid-neutralizing bicarbonates into the small intestines.
Somatotropic hormone or STH (anterior pituitary gland)	Promotes protein synthesis in all cells; increases fat mobilization; aids in CHO metabolism.
Somatostatin or GIH (hypothalamus)	Checks and inhibits the release of growth hormones; also called growth hormone release inhibiting hormone.
Thyroid-stimulating hormone or TSH (anterior pituitary gland)	Stimulates the release of TSH; excited by low body temperature and large meals. Stimulates the thyroid gland to promote and release thyroid hormones.
Thyroxine, di- and triiodothyroxine (thyroid gland)	Stimulates oxygen consumption. Regulates carbohydrate, lipid, and protein metabolism; excessive secretion leads to glucose intolerance and increased gluconeogenesis.
Vasopressin or pitressin (posterior pituitary gland)	Also known as antidiuretic hormone; prevents water loss in urine by the contraction of the kidneys. Turned on whenever the blood volume is depleted (either the blood pressure is too low or salt concentration is too high).

[a]This is a partial list; includes hormones with direct nutritional functions.

Appendix 12
Utilization of Carbohydrates

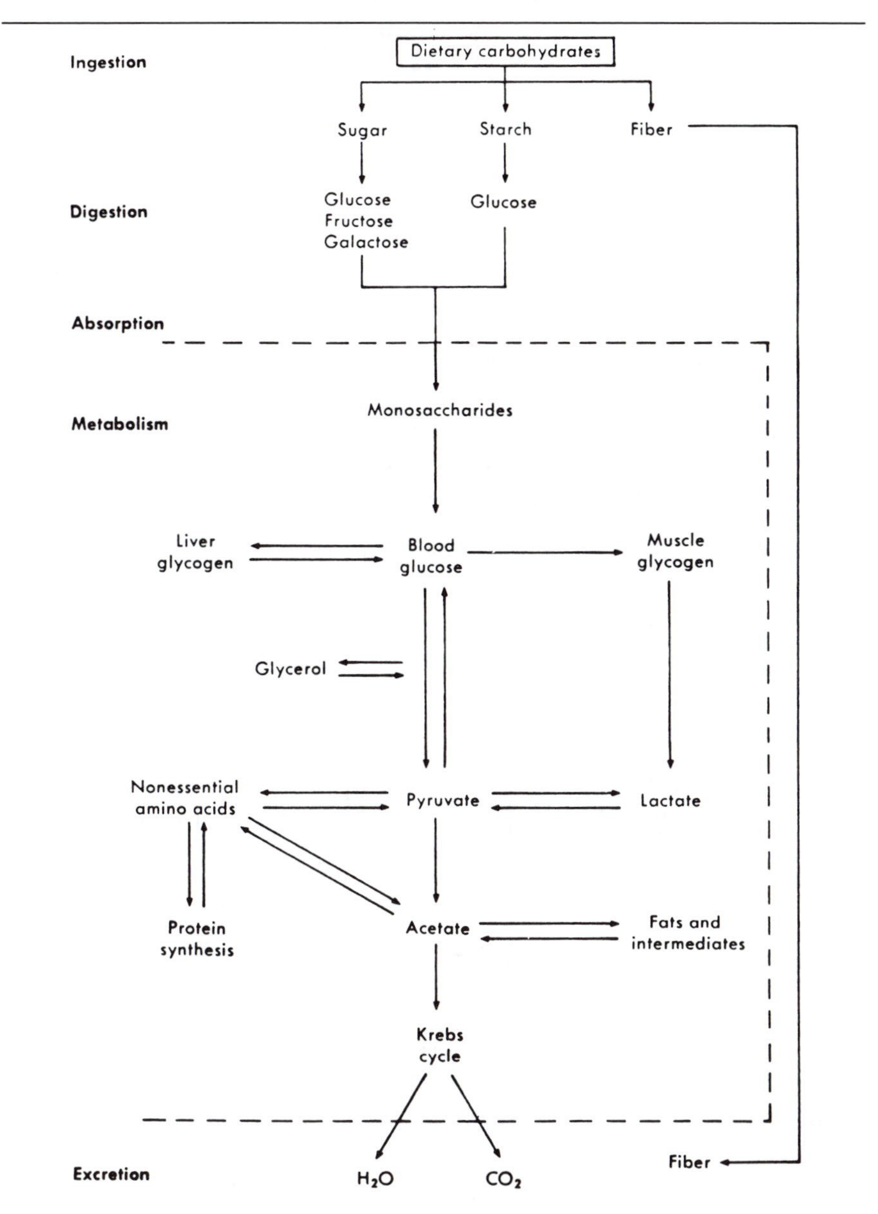

Appendix 13
Utilization of Proteins

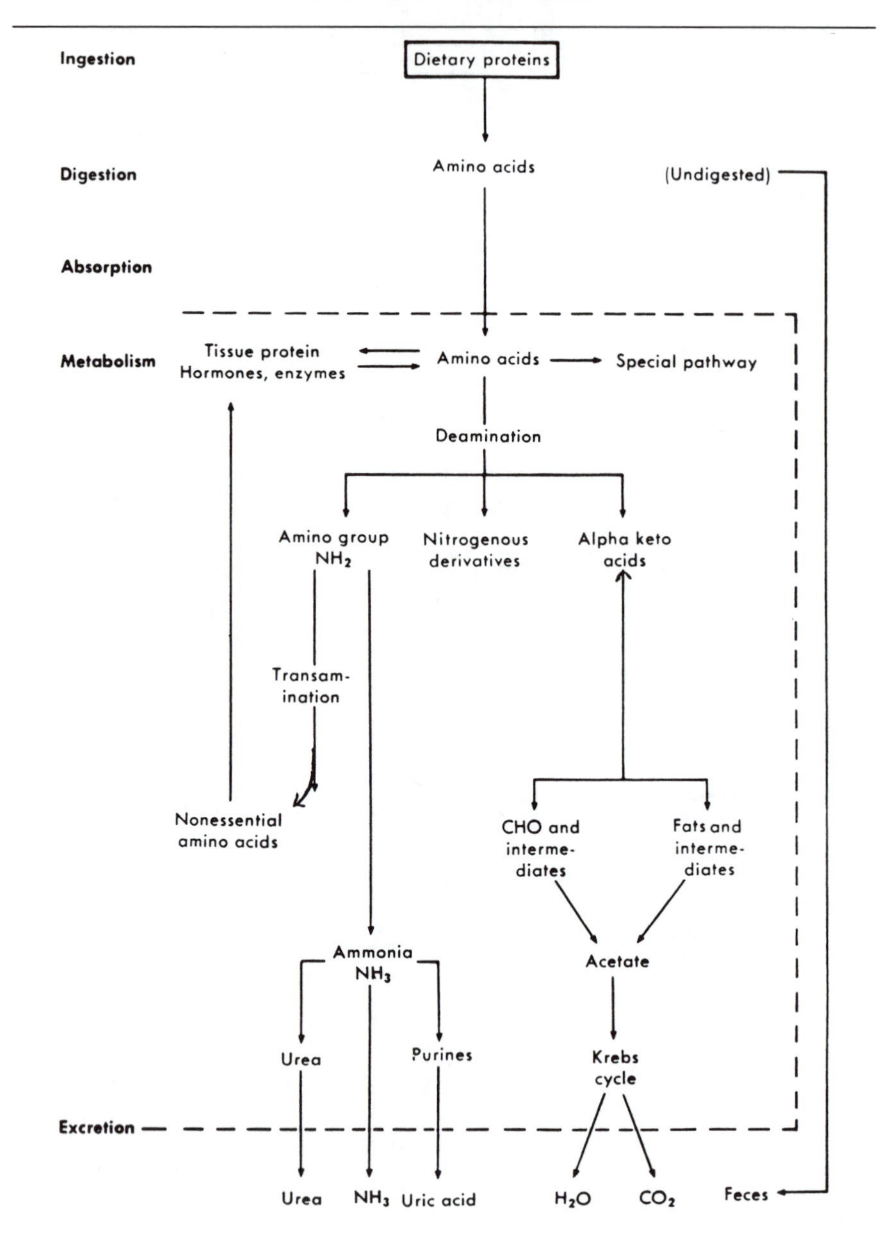

Appendix 14
Utilization of Fats

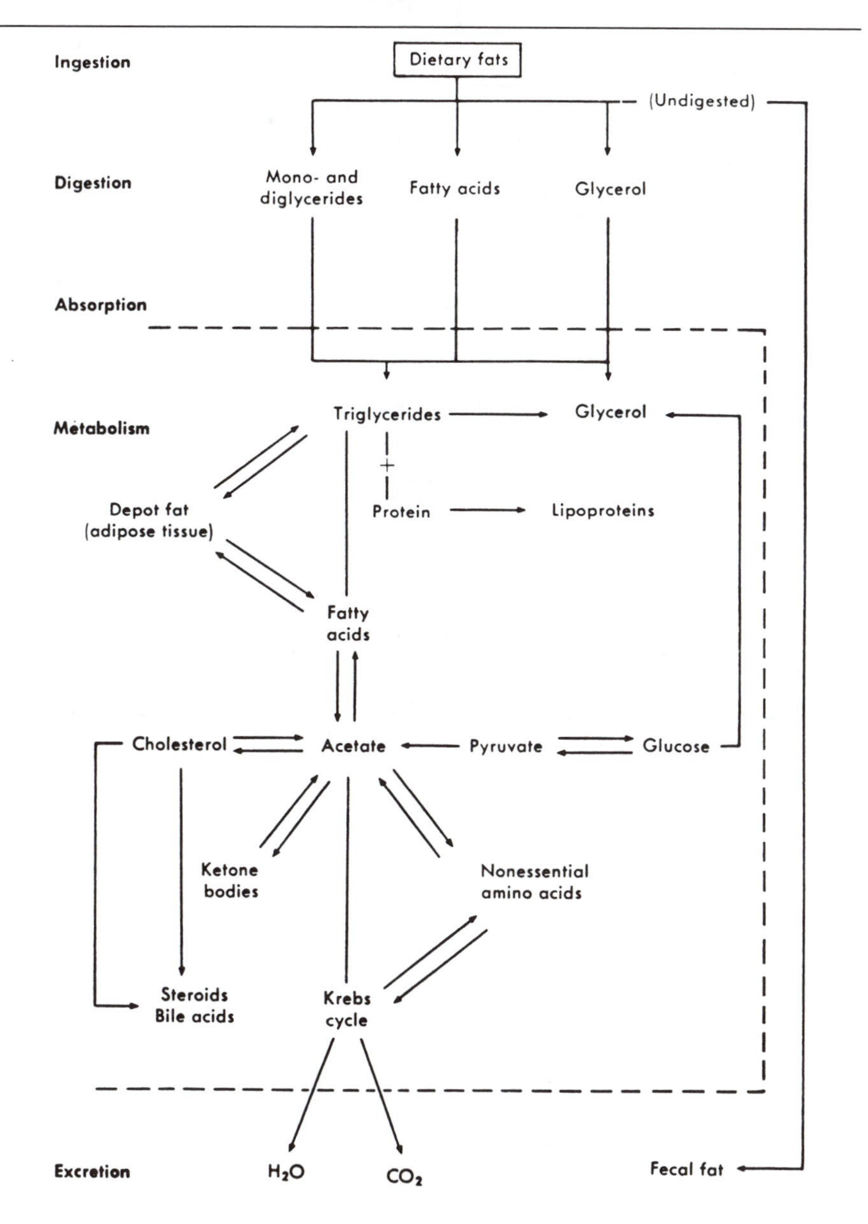

Appendix 15
Interrelationship of Carbohydrate, Protein, and Fat

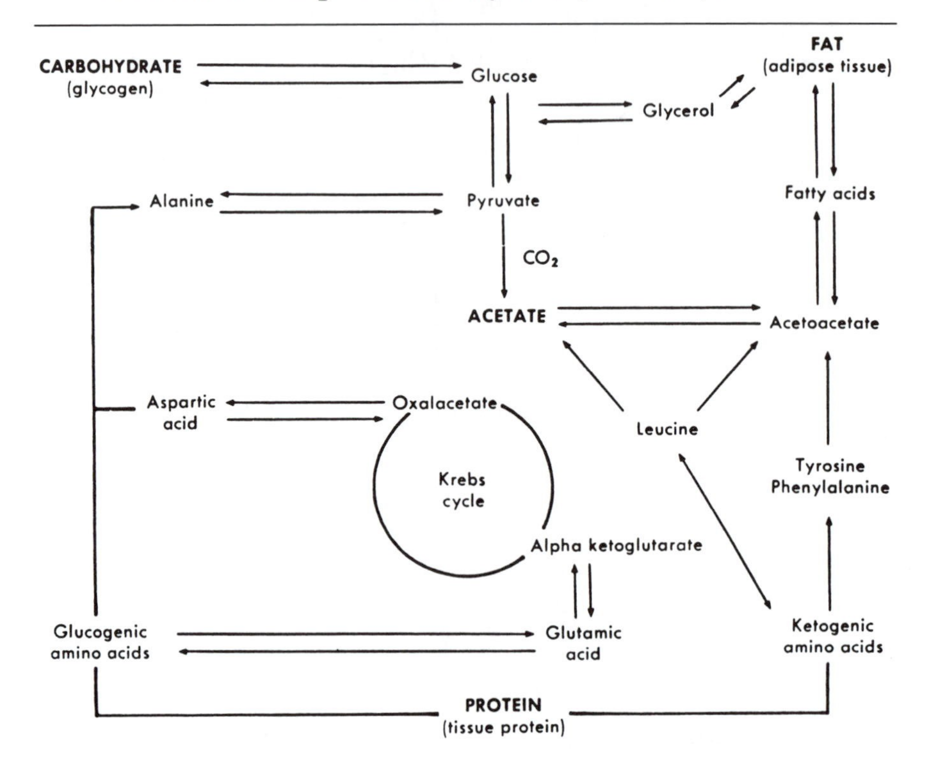

Appendix 16
Summary of Vitamins

Vitamin	Major Functions	Signs of Deficiency	Food Sources
Fat-soluble vitamins			
Vitamin A	Maintenance of skin and mucous membranes; component in visual process; immune stimulation	Poor growth; night-blindness, xerophthalmia, keratomalacia; Bitot's spots; follicular hyperkeratosis; reduced resistance to infection	Liver; dark green and deep yellow fruits and vegetables
Vitamin D	Mineralization of bones and teeth; intestinal absorption and regulation of calcium and phosphate	Rickets; costochondral beading; bowed legs; epiphyseal enlargement; muscle weakness; osteomalacia	Sunlight exposure; fortified milk and milk products; cod liver oil
Vitamin E	Antioxidant; prevents peroxidation of polyunsaturated lipids; free radical scavenger	Hemolytic anemia of newborn; increased fragility of red blood cells; ceroid pigment deposits	Oils high in polyunsaturated fatty acids; wheat germ; seeds
Vitamin K	Synthesis of prothrombin and clotting factors II, VII, IX, and X	Bleeding tendency (especially in newborns); ecchymosis; epistaxis	Green leafy vegetables; liver; vegetable oils
Water-soluble vitamins			
Vitamin C	Reducing agent; collagen synthesis; helps in wound healing and resistance to infection; iron absorption	Scurvy; easy bruising; joint tenderness; muscle ache; slow wound healing; petechiae; bleeding gums	Citrus fruits; tomatoes; papaya; raw cabbage; green pepper
Vitamin B_1 (thiamin)	Coenzyme in carbohydrate fat and protein metabolism; promotes normal functioning of the nervous system	Beriberi; peripheral edema; polyneuritis; high-output cardiac failure; anorexia; paresthesias; tender muscles	Lean pork; wheat germ; whole and enriched cereals; legumes
Vitamin B_2 (riboflavin)	Part of flavin coenzymes required in cellular oxidation; essential for growth	Cheilosis; glossitis; photophobia; angular stomatitis; seborrhea; magenta tongue; corneal vascularization	Milk and dairy products; liver; dark green vegetables; organ meats
Vitamin B_6 (pyridoxine)	Cofactor for many enzymes in metabolism of protein and amino acids; functions in hemoglobin synthesis	Anemia; irritability; convulsions; depression; skin lesions; seborrheic dermatitis; glossitis	Liver; pork; poultry; whole grain; fortified cereals; bananas; legumes

(continued)

Appendix 16 *(continued)*

Vitamin	Major Functions	Signs of Deficiency	Food Sources
Vitamin B_{12} (cobalamin)	Maintenance of nerve tissue and normal blood formation; nucleic acid synthesis; recycling of tetrahydrofolate	Megaloblastic anemia; glossitis; spinal cord degeneration; peripheral neuropathy	Liver; meat; fish; poultry; eggs; (not present in plant foods)
Niacin	Part of coenzymes for oxidation/reduction reactions, release of energy and biosynthesis of fatty acids	Pellagra; pigmented dermatitis; inflammation of mucous membranes; diarrhea; weakness; depression	Lean meats; fish; poultry; whole grains; peanuts; organ meats
Folate	Cofactor for synthesis of purine and pyrimidine; transfer of single carbon; red blood cell maturation	Megaloblastic anemia; gastrointestinal disturbances; glossitis; stomatitis; diarrhea	Dark green leafy vegetables; whole grains; legumes; nuts; organ meats
Pantothenic acid	Component of coenzyme A; functions in release of energy from carbohydrate, protein, and fat	Fatigue; malaise; insomnia; abdominal cramps; burning paresthesias; impaired coordination; depression	Liver; egg; meat; fish; poultry; whole grains; fresh vegetables
Biotin	Coenzyme for carboxylation reactions; plays a role in carbohydrate and fat metabolism	Dermatitis; neuritis; anorexia; nausea; vomiting; glossitis; insomnia; thin hair; depression	Liver; milk; egg; kidney; mushrooms; bananas; strawberries

Usual Daily Dose Ranges for Oral Vitamin Administration:

	Prophylactic	Low Dose	High Dose
Biotin, mg	—	10	—
Folic acid, mg	2.5–10	5–10	10–20
Niacin, mg	10	10–100	100–500
Pantothenate, mg	100	100–200	200–600
Vitamin A, IU	5000	5000–30,000	100,000–200,000
Vitamin D, IU	500–1000	5000	10,000–15,000
Vitamin E, mg	3–15	10–100	400–600
Vitamin K_1, mg	5–10	5–10	10–20
Vitamin B_1, mg	3–10	10–100	100–600
Vitamin B_2, mg	1–5	5–10	10–100
Vitamin B_6, mg	20	20–50	50–300
Vitamin B_{12}, μg[a]	5–50 monthly	100–250 weekly	Up to 500 weekly
Vitamin C, mg	100	100–200	Up to 3000

[a]Intramuscular

Data from Marks, J.: *The Vitamins. Their Role in Medical Practice*. Boston: MTP Press Limited, 1985.

Appendix 17
Summary of Minerals

Mineral	Major Functions	Signs of Deficiency	Food Sources
Calcium	Structure of bones and teeth; nerve transmission; blood clotting; muscle contraction	Stunted growth; rickets; osteomalacia; osteoporosis; muscle cramping; tetany; possibly hypertension	Milk; cheese; sardines with bones; mustard greens; kale
Chloride	Constituent of gastric juice; major anion of extracellular fluid; enzyme activator; acid–base balance	Mental apathy; muscle cramps; usually accompanied with sodium depletion	Table salt; seafoods; meats
Chromium	Insulin cofactor; glucose and energy metabolism; stimulates fat and cholesterol synthesis	Insulin resistance; glucose intolerance; impaired growth; elevated serum lipids	Liver; cheese; meat; whole-grain cereals
Cobalt	Constituent of vitamin B_{12}	Only as vitamin B_{12} deficiency; pernicious anemia	Organ and muscle meats; milk
Copper	Absorption and use of iron; enzyme cofactor; electron transport; myelin sheath of nerves; may be part of RNA	Anemia; disturbance of bone formation; impairment of cardiovascular system; neutropenia; kinky hair	Liver; shellfish; nuts; whole grain cereals; legumes; organ meats
Fluoride	Structure of teeth enamel; reduces dental caries	Dental caries; possibly growth depression;	Drinking water; tea; seafood
Iodine	Constituent of thyroid hormones	Goiter; depressed thyroid function; cretinism	Iodized salt; seafoods; marine fish
Iron	Constituent of hemoglobin and the cytochrome enzymes involved in oxygen and electron transport	Microcytic and hypochromic anemia; growth retardation; decreased serum iron; easy fatigability	Liver; lean meats; legumes; egg yolk; fortified cereals and breads
Magnesium	Activates enzymes; nerve impulse transmission and muscle contraction; constituent of bones and teeth	Neuromuscular irritability; weakness; spasms; apathy; growth failure; behavioral disturbances	Whole grains; nuts, legumes; green leafy vegetables; seafoods
Manganese	Constituent of enzymes in mucopolysaccharide metabolism and fat synthesis	Scaly dermatitis; retarded hair/nail growth; weight loss	Nuts; legumes; unrefined cereals dried fruits
Molybdenum	Enzyme cofactor in sulfur and purine metabolism; oxidation/reduction process	Dietary deficiency not observed in humans	Milk, organ meats; legumes; whole-grain cereals

(continued)

Appendix 17 *(continued)*

Mineral	Major Functions	Signs of Deficiency	Food Sources
Phosphorus	Structure of bones and teeth; component of phospholipids, and nucleic acids; acid–base balance; energy metabolism	Demineralization of bone; weakness; poor growth; paresthesia of hands and feet; seizures	Milk; cheese; egg; meats; whole-grain cereals; legumes
Potassium	Major cation of intracellular fluid; regulates acid–base and water balance, osmotic pressure, nerve transmission	Muscle weakness; nausea; cardiac arrhythmias; heart failure; glycogen depletion; respiratory failure	Many fruits; nuts; meats; milk; vegetables; potatoes; cereals
Selenium	Antioxidant; constituent of glutathione oxidase; associated with vitamin E and fat	Cardiac myopathy; increased fragility of red blood cells; muscle tenderness	Meat; seafoods; grains; milk; vegetables
Sodium	Major cation of extracellular fluid; regulates body fluid volume, pH, and osmolarity; influences nerve irritability and muscle contraction	Abdominal cramps; nausea; vomiting; apathy; muscle contraction; palpitation; confusion	Table salt (NaCl); processed foods; abundant in most foods except for fruits
Sulfur	Constituent of coenzyme A, certain amino acids, hair, cartilage, thiamin, biotin	No dietary deficiency if protein intake is adequate	All foods rich in protein—meat, milk, legumes
Zinc	Constituent of many enzymes; cell replication; connective tissue synthesis; immune system function	Dermatitis; impaired wound healing; taste change; depressed immunocompetence; sexual immaturity	Seafood; eggyolk; liver; meat; oysters; cereal germ

Daily oral therapeutic dosages for mineral deficiency:

Calcium	1–1.5 g b.i.d.
Chromium	200 μg (chromium chloride)
Copper	2–3 mg (copper sulfate)
Iodine	50 μg (potassium iodide)
Fluoride	1–2 mg (sodium fluoride)
Iron	325 mg t.i.d. (ferrous sulfate)
Magnesium	250–500 mg b.i.d. to q.i.d. (magnesium oxide)
Molybdenum	300 μg
Phosphorus	1–1.5 mmol/kg
Selenium	100–200 μg
Zinc	60 mg elemental zinc (zinc sulfate 220 mg)

Data from Weinsier, R. et al.: *Handbook of Clinical Nutrition*, 2nd ed., Mosby-Year Book, Inc., St. Louis, MO, 1989.

Appendix 18
Median Weights and Heights for Children from Birth to 18 Years

	Males				Females			
	Weight		Height		Weight		Height	
Age	kg	lb	cm	in	kg	lb	cm	in
Months								
1	4.29	9.4	54.6	21.5	3.98	8.8	53.5	21.1
3	5.98	13.2	61.1	24.1	5.40	11.9	59.5	23.4
6	7.85	17.3	67.8	26.7	7.21	15.9	65.9	25.9
9	9.18	20.2	72.3	28.5	8.56	18.8	70.4	27.7
12	10.15	22.3	76.1	30.0	9.53	21.0	74.3	29.3
18	11.47	25.2	82.4	32.4	10.82	23.8	80.9	31.9
Years								
2	12.34	27.1	86.8	34.2	11.80	26.0	86.8	34.2
3	14.62	32.2	94.9	37.4	14.10	31.0	94.1	37.0
4	16.69	36.7	102.9	40.5	15.96	35.1	101.6	40.0
5	18.67	41.1	109.9	43.3	17.66	38.9	108.4	42.7
6	20.69	45.5	116.1	45.7	19.52	42.9	114.6	45.1
7	22.85	50.3	121.7	47.9	21.84	48.0	120.6	47.5
8	25.30	55.7	127.0	50.0	24.84	54.6	126.4	49.8
9	28.13	61.9	132.2	52.0	28.46	62.6	132.2	52.0
10	31.44	69.2	137.5	54.1	32.55	71.6	138.3	54.4
11	35.30	77.7	143.3	56.4	36.95	81.3	144.8	57.0
12	39.78	87.5	149.7	58.9	41.53	91.4	151.5	59.6
13	44.95	98.9	156.5	61.6	46.10	101.4	157.1	61.9
14	50.77	111.7	163.1	64.2	50.28	110.6	160.4	63.1
15	56.71	124.8	169.0	66.5	53.68	118.1	161.8	63.7
16	62.10	136.6	173.5	68.3	55.89	123.0	162.4	63.9
17	66.31	145.9	176.2	69.4	56.69	124.7	163.1	64.2
18	68.88	151.5	176.8	69.6	56.62	124.6	163.7	64.4

Data from the National Health Survey: NCHS Growth Curves for Children. Birth–18 Years, United States. In: *Vital and Health Statistics*, Series 11, No. 124, DHHS Publication No. (PHS) 78-1650.

Appendix 19
Average Weights for Men and Women Aged 18–74 Years

Height	Weight in Pounds by Age Group in Years					
ft in	18–24	25–34	35–44	45–54	55–64	65–74
Men						
5 2	130	141	143	147	143	143
5 3	135	145	148	152	147	147
5 4	140	150	153	156	153	151
5 5	145	156	158	160	158	156
5 6	150	160	163	164	163	160
5 7	154	165	169	169	168	164
5 8	159	170	174	173	173	169
5 9	164	174	179	177	178	173
5 10	168	179	184	182	183	177
5 11	173	184	190	187	189	182
6 0	178	189	194	191	193	186
6 1	183	194	200	196	197	190
6 2	188	199	205	200	203	194
Women						
4 9	114	118	125	129	132	130
4 10	117	121	129	133	136	134
4 11	120	125	133	136	140	137
5 0	123	128	137	140	143	140
5 1	126	132	141	143	147	144
5 2	129	136	144	147	150	147
5 3	132	139	148	150	153	151
5 4	135	142	152	154	157	154
5 5	138	146	156	158	160	158
5 6	141	150	159	161	164	161
5 7	144	153	163	165	167	165
5 8	147	157	167	168	171	169

Data from the National Center for Health Statistics: Weight by Height and Age for Adults 18–74 Years, United States, 1971–1974. In: *Vital and Health Statistics*, Series II, no. 208, DHHS Publication No. (PHS) 79-1656.

Appendix 20A
Acceptable Weights for Men and Women

Height[a]	Weight (lb)[b]	
ft in	19–34 yrs	35 yrs and over
5 0	97–128	108–138
5 1	101–132	111–143
5 2	104–137	115–148
5 3	107–141	119–152
5 4	111–146	122–157
5 5	114–150	126–162
5 6	118–155	130–167
5 7	121–160	134–172
5 8	125–164	138–178
5 9	129–169	142–183
5 10	132–174	146–188
5 11	136–179	151–194
6 0	140–184	155–190
6 1	144–189	159–205
6 2	148–195	164–210
6 3	152–200	168–216
6 4	156–205	173–222
6 5	160–211	177–228
6 6	164–216	182–234

[a]Height without shoes

[b]Weight without clothes

Source: USDA. *Nutrition and Your Health: Dietary Guidelines for Americans*. HG-232, 1990.

Criteria for a healthy weight:

1. Weight is within the acceptable range for the height and age group.
2. Waist-to-hip ratio (WHR) is less than 0.8 for women and less than 0.95 for men, where

$$\text{WHR} = \frac{\text{Waist measurement}}{\text{Hip measurement}}$$

3. No weight-related medical problems or family history of such problems.

Appendix 20B
1983 Metropolitan Height–Weight Tables

Height ft in	Small Frame	Medium Frame	Large Frame
Men			
5 2	128–134	131–141	138–150
5 3	130–136	133–143	140–153
5 4	132–138	135–145	141–156
5 5	134–140	137–148	144–160
5 6	136–142	139–151	146–164
5 7	138–145	142–154	149–168
5 8	140–148	145–157	152–172
5 9	142–151	148–160	155–176
5 10	144–154	151–163	158–180
5 11	146–157	154–166	161–184
6 0	149–160	157–170	164–188
6 1	152–164	160–174	168–192
6 2	155–168	164–178	172–197
6 3	158–172	167–182	176–202
6 4	162–176	171–187	181–207
Women			
4 10	102–111	109–121	118–131
4 11	103–113	111–123	120–134
5 0	104–115	113–126	122–137
5 1	106–118	115–129	125–140
5 2	108–121	118–132	128–143
5 3	111–124	121–135	131–147
5 4	114–127	124–138	134–151
5 5	117–130	127–141	137–155
5 6	120–133	130–144	140–159
5 7	123–136	133–147	143–163
5 8	126–139	136–150	146–167
5 9	129–142	139–153	149–170
5 10	132–145	142–156	152–173
5 11	135–148	145–159	155–176
6 0	138–151	148–162	158–179

Weights at ages 25–59 based on lowest mortality. Weight in pounds in indoor clothing (allow 5 lb for men and 3 lb for women); shoes with 1-inch heels.

Data from 1979 *Build Study,* Society of Actuaries and Association of Life Insurance Medical Directors of America. Courtesy of the Metropolitan Life Insurance Company, 1983.

Appendix 21
Estimation of Frame Size and Stature

A. Body Frame According to Wrist Size

Wrap the fingers of one hand around the opposite wrists. If the thumb and middle finger:

Overlap by 1 cm = small frame
Touch = medium frame
Cannot touch by 1 cm = large frame

B. Body Frame According to Height/Wrist Circumference (*r*)

	Large	Medium	Small
Males, *r* values	<9.6	9.6–10.4	>10.4
Females, *r* values	<10.1	10.1–11.0	>11.0

Where r = Height (cm) divided by wrist circumference (cm)

C. Body Frame According to Elbow Breadth

	Males			Females		
Age	Large	Medium	Small	Large	Medium	Small
18–24	>7.7	6.6–7.7	<6.6	>6.5	5.6–6.5	<5.6
25–34	>7.9	6.7–7.9	<6.7	>6.8	5.7–6.8	<5.7
35–44	>8.0	6.7–8.0	<6.7	>7.1	5.7–7.1	<5.7
45–54	>8.1	6.7–8.1	<6.7	>7.2	5.7–7.2	<5.7
54–55	>8.1	6.7–8.1	<6.7	>7.2	5.8–7.2	<5.8
65–74	>8.1	6.7–8.1	<6.7	>7.2	5.8–7.2	<5.8

To measure elbow breadth, extend arm forward and bend forearm upward at 90 degrees with fingers pointing up and inside of wrist toward the body. Measure the breadth with a sliding caliper (in cm) across the elbow joint on the two prominent bones on either side.

D. Stature from Knee Height Measurement

Men Height (cm) = (knee height, cm × 2.02) − (age, yr × 0.04) + 64.19
Women Height (cm) = (knee height, cm × 1.83) − (age, yr × 0.24) + 84.88

Use a broad-blade caliper to get knee height measurement. The subject lies on the back with the knee bent to a 90 degree angle. Press the sliding blade of the caliper against the thigh about 2 inches behind the kneecap and hold the caliper shaft in line with the shaft of the tibia. Two readings should agree within ±0.5 cm.

Sources: Grant, J. P.: *Handbook of Total Parenteral Nutrition*. W.B. Saunders, 1980, p. 15; Frisancho, A. R.: New standards of weight and body composition by frame and height for assessment of nutritional status of adults and the elderly, *Am J Clin Nutr,* 40:808, 1984; Chumlea, W. C. et al.: Estimating stature from knee height for persons 60 to 90 years of age. *J Am Geriatr Soc,* 33:116, 1985.

Appendix 22
Reference Values for Triceps Skinfold Thickness

	Males (mm Percentile)					Females (mm Percentile)				
Age	10th	25th	50th	75th	90th	10th	25th	50th	75th	90th
18–74	6.0	8.0	11.0	15.0	20.0	13.0	17.0	22.0	28.0	34.0
18–24	5.0	7.5	9.5	14.0	20.0	11.0	14.0	18.0	24.0	30.0
25–34	5.5	8.0	12.0	16.0	21.5	12.0	16.0	21.0	26.5	33.5
35–44	6.0	8.5	12.0	15.5	20.0	14.0	18.0	23.0	29.5	35.5
45–54	6.0	8.0	11.0	15.0	20.0	15.0	20.0	25.0	30.0	36.0
55–64	6.0	8.0	11.0	14.0	18.0	14.0	19.0	25.0	30.5	35.0
65–74	5.5	8.0	11.0	15.0	19.0	14.0	18.0	23.0	28.0	33.0

Developed from data collected during the NHANES I, 1971–1974. From Bishop, C. W., Bowen, P. E., and Ritchey, S. J. Norms for nutritional assessment of American adults by upper arm anthropometry. *Am J Clin Nutr*, 34:2530, 1981. Adapted with permission.

Triceps skinfold: With the arm relaxed and the elbow flexed at a 90 degree angle, grasp the skin and subcutaneous tissue at the midpoint of the upper arm between the acromion and the olecranon processes of the scapula and the ulna. Measure with a caliper while the fold is still held with the hand to release skin tension. Take three measurements to the nearest 0.5 mm and average the results.

Interpretation:		
	>50th percentile:	Acceptable
	40–50th percentile:	Mild fat depletion
	25–39th percentile:	Moderate fat depletion
	<25th percentile:	Severe fat depletion

Appendix 23
Reference Values for Midarm Muscle Circumference

	Males (cm Percentile)					Females (cm Percentile)				
Age	10th	25th	50th	75th	90th	10th	25th	50th	75th	90th
18–74	24.8	26.3	27.9	29.5	31.4	19.0	20.2	21.8	23.6	25.8
18–24	24.4	25.8	27.2	28.9	30.8	18.5	19.4	20.6	22.1	23.6
25–34	25.3	26.5	28.0	30.0	31.7	18.9	20.0	21.4	22.9	24.9
35–44	25.6	27.1	28.7	30.3	32.1	19.2	20.6	22.0	24.0	26.1
45–54	24.9	26.5	28.1	29.8	31.5	19.5	20.7	22.2	24.3	26.6
55–64	24.4	26.2	27.9	29.6	31.0	19.5	20.8	22.6	24.4	26.3
65–74	23.7	25.3	26.9	28.5	29.9	19.5	20.8	22.5	24.4	26.5

Developed from data collected during the NHANES I, 1971–1974. From Bishop, C. W., Bowen, P. E., and Ritchey, S. J. Norms for nutritional assessment of American adults by upper arm anthropometry. *Am J Clin Nutr,* 34:2530, 1981. Adapted with permission.

Mid-upper arm: Allow the arm to hang relaxed at the side. Using a tape measure, get the arm circumference (in cm) at the midpoint of the upper arm. The tape should be maintained in a horizontal position touching the skin and following the contours of the limb, but not compressing underlying tissue. Calculate the midarm muscle circumference (MAMC) with the formula: MAMC (cm) = Midarm circumference, cm − [triceps skinfold (TSF) mm × 0.314]

Appendix 24
Estimation of Energy and Protein Requirements

Adult Energy Requirement:

Method I.

1. Determine basal energy expenditure (BEE) from the Harris–Benedict equation.

   Male: $BEE = 66.5 + (13.8 \times W) + (5.0 \times H) - (6.8 \times Y)$

   Female: $BEE = 655.1 + (9.6 \times W) + (1.8 \times H) - (4.7 \times Y)$

   Where W is actual weight in kilograms; Y is age in years; H is height in centimeters (height in inches × 2.54)

   Weight adjustment for obesity (>25% of IBW):

   $$(ABW - IBW) \times 0.25 + IBW = W \text{ in kilograms for BEE}$$

   Where ABW is actual body weight; IBW is ideal body weight.

2. Adjustment for activity: BEE × activity factor below

Bed rest	1.0–1.1	Moderate	1.6–1.7
Very light	1.2–1.3	Heavy	1.9–2.1
Light	1.4–1.5	Strenuous or exceptional	2.2–2.4

   (Use lower factor for females; higher factor for males)

3. Provision for illness: BEE × stress or injury factor below

No illness/nonstress	1.0
Convalescence, mild malnutrition, postoperative (no complication)	1.1
Mild illness, noncatabolic	
Confined to bed	1.2
Ambulatory/out of bed	1.3
Infections and stress, catabolic	
Mild	1.1–1.2
Moderate	1.3–1.4
Severe, hypercatabolic	1.5–1.7
Sepsis	1.8–2.0
Burns, <20% body surface	1.2–1.4
20%–40% body surface	1.5–1.7
>40% body surface	1.8–2.0
Burns, severe	2.1–2.3
Fracture, long bone	1.2–1.3
Respiratory/renal failure	1.4–1.5
COPD	1.4–1.6
Cancer with chemotherapy or radiation, cardiac cachexia	1.5–1.6
Surgery, minor/elective	1.1–1.2
Surgery, major	1.2–1.3
Trauma, skeletal/blunt	1.3–1.4
Trauma, multiple/head injury	1.5–1.6

4. Kilocalories for total energy expenditure (TEE)

   Weight maintenance TEE = BEE × activity factor × stress/injury factor

   Weight gain: Add 500 kcal/day to gain at the rate of 1 lb per week

   Weight loss: Subtract 500 kcal/day to lose at the rate of 1 lb per week

Appendix 24 *(continued)*

Method II. Quick Method. Calorie allowances per kilogram (or pound) body weight for different activity levels and stress conditions:

Activity level:	kcal/kg BW	kcal/lb BW
Bed patient	25	11
Very light activity	30	14
Light activity	35	16
Moderate activity	40	18
Active/heavy work	45	20
Strenuous/exceptional	50	23

[a]Add or subtract 5 kcal/kg if underweight or overweight, respectively.

Stress condition:		
Overweight/weight reduction	20	9
Nonstress, bed rest	25	11
Mild stress, bed rest	30	14
ambulatory	35	16
Moderate stress, bed rest	35	16
ambulatory	40	18
Severe stress, polytrauma, hypermetabolic, sepsis	45	20
Surgery, elective/minor	32	14.5
Surgery, major, bed rest	35	16
ambulatory	38	17.5
Burn, major, bed rest	45–50	20–23
ambulatory	55–60	25–27
Cancer	35–45	16–20
Predialysis	40–50	18–23
Hemodialysis	35	16
Peritoneal dialysis	30	14

Adult Protein requirement:

1. Normal requirement: 0.8–1.0 g/kg body weight
2. Requirement during illness or stress (g/kg body weight)

Mild stress	1.1–1.2	Renal failure, acute	0.3–0.5
Moderate stress	1.3–1.4	Renal failure, chronic	0.55–0.6
Severe stress	1.5–1.7	Hemodialysis	1.0–1.2
Hypermetabolic stress	1.7–2.5	Peritoneal dialysis	1.2–1.5
Polytrauma, infection	1.8–2.4	Pressure sore, stage 1	1.0–1.1
Severe sepsis, major burn	2.5–3.0	stage 2	1.2–1.4
head injury		stage 3	1.5–1.7
Surgery, minor/elective	1.2–1.3	stage 4	1.8–2.0
Surgery, major	1.4–1.5	Depleted protein stores	1.5–2.0
Cancer, malabsorption syndromes, tuberculosis	1.2–1.5	Long-bone fractures, draining wounds	1.6–1.7
Respiratory failure, acute	1.3–1.4	Hepatitis, cirrhosis	1.5–2.0

(continued)

Appendix 24 *(continued)*

Energy and protein requirements for infants and children

Age (year)	Calories (kcal/kg)	Protein (g/kg body weight)	
		Normal	Stressed
0.0–0.5	108	2.0 g	2.2–3.0 g
0.5–1.0	98	1.6 g	1.6–3.0 g
1–3	103	1.2 g	1.2–3.0 g
4–6	90	1.1 g	1.1–3.0 g
7–10	70	1.0 g	1.0–2.5 g
11–14, boys	55	1.0 g	1.0–2.5 g
girls	47	1.0 g	1.0–2.5 g
15–14, boys	45	0.9 g	1.0–2.5 g
girls	40	0.8 g	1.0–2.5 g

Appendix 25
Interpretations and Equations for Assessing Nutritional Status

Albumin

Interpretation:

>3.5 g/dL, acceptable
2.8–3.4 g/dL, mild depletion
2.1–2.7 g/dL, moderate depletion
<2.1 g/dL, severe depletion

Prealbumin

Interpretation:

>15 mg/dL, acceptable
10–15 mg/dL, mild depletion
5–10 mg/dL, moderate depletion
<5 mg/dL, severe depletion

Serum transferrin

Interpretation:

>200 mg/dL, acceptable
150–200 mg/dL, mild depletion
100–149 mg/dL, moderate depletion
<100 mg/dL, severe depletion

Transferrin, calculated from total iron-binding capacity (TIBC)

Transferrin, mg/dL = $(0.68 \times \text{TIBC}) + 21$

Interpretation:

>170 mg/dL, acceptable
<170 mg/dL, deficient

(Transferrin values determined from TIBC are lower than those obtained by radioimmunodiffusion, but may be elevated in patients with severe iron deficiency anemia).

Total iron-binding capacity (TIBC)

Interpretation:

250–350 μg/dL normal
>400 μg/dL indicative of iron deficiency

Total lymphocyte count (TLC), (cells/mm^3)

$$\text{TLC} = \frac{\text{\% lymphocytes} \times \text{white blood cells (mm}^3)}{100}$$

Interpretation:

>2000/mm^3, acceptable
1200–2000/mm^3, mild depletion
800–1199/mm^c, moderate depletion
<800/mm^3, severe depletion

Creatinine/height index (CHI)

$$\text{CHI} = \frac{\text{24-hour urinary creatinine (mg)}}{\text{expected 24-hour urinary creatinine (mg)}} \times 100\%$$

Interpretation:

>90%, acceptable
80%–90%, mild depletion
60%–79%, moderate depletion
<60%, severe depletion

(continued)

Appendix 25 *(continued)*

Nitrogen balance $= \frac{\text{Protein intake, g}}{6.25} - (\text{Urinary urea nitrogen} + 4^{*})$

*Value represents an estimate of the unmeasured nitrogen lost in sweat, hair, skin, and stool

Protein loss, grams = (Urinary urea nitrogen + 4) × 6.25

Prognostic Nutritional Index (PNI)

PNI = 158 − 16.6 (Alb) − 0.78 (TSF) − 0.2 (TFN) − 5.8 (DH)

Where

Alb = serum albumin (g/dL); TSF = triceps skinfold (mm)

TFN = transferrin level (mg/dL); DH = delayed hypersensitivity skin test

Interpretation:

<30, low risk of mortality and morbidity

30–50, intermediate risk

>50, high risk

Fat-free mass (FFM), prediction from 24-hour urinary creatinine excretion

FFM (kg) = 29.08 creatinine (g/dL) + 7.38

Triceps skinfold thickness (TSF), mm

Interpretation:

>50th percentile, acceptable

40–50th percentile, mild fat depletion

25–39th percentile, moderate fat depletion

<25th percentile, severe fat depletion

Midarm circumference (MAC), cm

Interpretation:

>50th percentile, acceptable

40–50th percentile, mild fat depletion

25–39th percentile, moderate fat depletion

<25th percentile, severe fat depletion

Midarm muscle circumference (MAMC), cm

MAMC, cm = MAC, cm − (TSF, mm × 0.314)

Interpretation:

>85%, acceptable

76%–85%, mild depletion

65%–75%, moderate depletion

<65%, severe depletion

Weight/height ratio or Body Mass Index (BMI)

$$\text{BMI} = \frac{\text{Weight (kg)}}{\text{Height}^2\text{ (m)}}$$

19–34 years old: 19–25, acceptable

>25, unacceptable

35 years and over: 21–27, acceptable

>27, unacceptable

Relative body weight (RBW), as % of desirable weight

$$\text{RBW} = \frac{\text{Actual weight}}{\text{Desirable body weight}} \times 100$$

Interpretation:

<70, severe underweight
70–79, moderate underweight
80%–89%, mild underweight
90%–110%, acceptable
111%–120%, overweight
>120% obese

Weight change, as % of usual weight

$$\% \text{ weight change} = \frac{\text{Present weight}}{\text{Usual weight}} \times 100$$

Interpretation:

>120%, significant weight gain
110%–120%, moderate weight gain
90%–109%, acceptable
85%–89%, mild weight loss
80%–84%, moderate weight loss
70%–79%, severe weight loss
60%–69%, life threatening
<60%, lethal
10% unplanned or recent weight loss is a risk factor for malnutrition
20% unplanned or recent weight loss is a high risk for surgical patients

Weight change, as % of usual weight over time

$$\% \text{ weight loss} = \frac{\text{Usual weight} - \text{present weight}}{\text{Usual weight}} \times 100$$

Interpretation:

	Significant loss	Severe loss
1 week	1%–2%	>2%
1 month	5%	>5%
3 months	7.5%	>7.5%
6 months	10%	>10%

Appendix 26
Physical Assessment of Nutritional Status

Body Area	Normal Appearance	Signs Associated with Malnutrition
Hair	Shiny; firm; not easily plucked	Lack of natural shine; hair dull and dry; thin and sparse; hair fine, silky, and straight; color changes; can be easily plucked
Face	Skin color uniform; smooth, pink, and healthy appearance; not swollen	Skin color loss (depigmentation); skin dark over cheeks and under eyes; lumpiness or flakiness of skin of nose and mouth; swollen face; enlarged parotid glands; scaling of skin around nostrils (nasolabial seborrhea)
Eyes	Bright, clear, shiny; no sores at corners of eyelids; membranes pink and moist. No prominent blood vessels, sclera, or mound of tissue	Eye membranes are pale (pale conjunctivae); redness of membranes; Bitot's spots; redness and fissuring of eyelid corners; dryness of eye membranes (conjunctival xerosis); cornea has dull appearance (corneal xerosis); cornea is soft (keratomalacia); scar on cornea; ring of fine blood vessels around cornea (circumcorneal injection)
Lips	Smooth, not chapped or swollen	Redness and swelling of mouth or lips (cheilosis), especially at corners of mouth (angular fissures and scars)
Tongue	Deep red in appearance; not swollen or smooth	Swelling; scarlet and raw tongue; magenta (purplish color) tongue; swollen sores; hyperemic and hypertrophic papillae; and atrophic papillae
Teeth	No cavities; no pain; bright	May be missing or erupting abnormally; gray or black spots (fluorosis); cavities (caries)
Gums	Healthy; red; do not bleed; not swollen	"Spongy" and bleed easily; recession of gums
Glands	Face not swollen	Thyroid enlargement (front of neck); parotid enlargement (cheeks become swollen)
Nails	Firm, pink	Nails are spoon-shape (koilonychia); brittle, ridged nails

Appendix 26 *(continued)*

Body Area	Normal Appearance	Signs Associated with Malnutrition
Skin	No signs of rashes, swellings, dark or light spots	Dryness of skin (xerosis); sandpaper feel of skin (follicular hyperkeratosis); flakiness of skin; skin swollen and dark; red swollen pigmentation of exposed areas (pellagrous dermatosis); excessive lightness or darkness of skin (dyspigmentation); black and blue marks due to skin bleeding (petechiae); lack of fat under skin
Muscular and skeletal systems	Good muscle tone; some fat under skin; can walk or run without pain	Muscles have "wasted" appearance; baby's skull bones are thin and soft (craniotabes); round swelling of front and side of head (frontal and parietal); swelling of ends of bones (epiphyseal enlargement); small bumps on both sides of chest wall (on ribs)—beading of ribs; baby's soft spot on head does not harden at proper time (persistently open anterior fontanelle); knock-knees or bow-legs; bleeding into muscle (musculoskeletal hemorrhages); cannot get up or walk properly
Internal systems		
Cardiovascular	Normal heart rate and rhythm; no murmurs; normal blood pressure for age	Rapid heart rate (above 100 tachycardia); enlarged heart; abnormal rhythm; elevated blood pressure
Gastrointestinal	No palpable organs or masses (in children liver edge may be palpable)	Liver enlargement; enlargement of spleen (usually indicates other associated diseases)
Nervous	Psychological stability; normal reflexes	Mental irritability and confusion; burning and tingling of hands and feet (paresthesia); loss of position and vibratory sense; weakness and tenderness of muscles (may result in inability to walk); decrease and loss of ankle and knee reflexes

From Christakis, G.: Nutritional Assessment in Health Programs. *Am J Public Health,* 63 (Suppl.1):19, 1973, with permission.

Appendix 27
Biochemical Assessment of Nutritional Status

Nutrient (unit)	Age (years)	Criteria of Status		
		Deficient	Marginal	Acceptable
Hemoglobin[a] (g/100 mL)	6–23 mos.	Up to 9.0	9.0–9.9	10.0+
	2–5	Up to 10.0	10.0–10.9	11.0+
	6–12	Up to 10.0	10.0–11.4	11.5+
	13–16 M	Up to 12.0	12.0–12.9	13.0+
	13–16 F	Up to 10.0	10.0–11.4	11.5+
	16+ M	Up to 12.0	12.0–13.9	14.0+
	16+ F	Up to 10.0	10.0–11.9	12.0+
	Pregnant (6+ mos).	Up to 9.5	9.5–10.9	11.0+
Hematocrit[a] (% packed cell volume)	Up to 2	Up to 28	28–30	31+
	2–5	Up to 30	30–33	34+
	6–12	Up to 30	30–35	36+
	13–16 M	Up to 37	37–39	40+
	13–16 F	Up to 31	31–35	36+
	16+ M	Up to 37	37–43	44+
	16+ F	Up to 31	31–37	33+
	Pregnant	Up to 30	31–32	33+
Serum albumin[a] (g/100 mL)	Up to 1	—	Up to 2.5	2.5+
	1–5	—	Up to 3.0	3.0+
	6–16	—	Up to 3.5	3.5+
	16+	Up to 2.8	2.8–3.4	3.5+
	Pregnant	Up to 3.0	3.0–3.4	3.5+
Serum protein[a] (g/100 mL)	Up to 1	—	Up to 5.0	5.0+
	1–5	—	Up to 5.5	5.5+
	6–16	—	Up to 6.0	6.0+
	16+	Up to 6.0	6.0–6.4	6.5+
	Pregnant	Up to 5.5	5.5–5.9	6.0+
Serum ascorbic acid[a] (mg/100 mL)	All ages	Up to 0.1	0.1–0.19	0.2+
Plasma vitamin A[a] (μg/100 mL)	All ages	Up to 10	10–19	20+
Plasma carotene[a] (μg/100 mL)	All ages	Up to 20	20–39	40+
	Pregnant	—	40–79	80+
Serum iron[a] (μg/100 mL)	Up to 2	Up to 30	—	30+
	2–5	Up to 40	—	40+
	6–12	Up to 50	—	50+
	12+ M	Up to 60	—	60+
	12+ F	Up to 40	—	40+
Transferrin[a] Saturation (%)	Up to 2	Up to 15.0	—	15.0+
	2–12	Up to 20.0	—	20.0+
	12+ M	Up to 20.0	—	20.0+
	12+ F	Up to 15.0	—	15.0+
Serum folacin[b] (ng/mL)	All ages	Up to 2.0	2.1–5.9	6.0+

Appendix 27 *(continued)*

Nutrient (unit)	Age (years)	Criteria of Status		
		Deficient	Marginal	Acceptable
Serum vitamin[b] B_{12} (pg/mL)	All ages	Up to 100	—	100+
	Pregnant	Up to 80	80–200	200+
Thiamin in urine[a]	1–3	Up to 120	120-–175	175+
	4–5	Up to 85	85–120	120+
	6–9	Up to 70	70–180	180+
	10–15	Up to 55	55–150	150+
	16+	Up to 27	27–65	65+
	Pregnant	Up to 21	21–49	50+
Riboflavin in urine[9] (μg/g creatinine)	1–3	Up to 150	150–499	500+
	4–5	Up to 100	100–299	300+
	6–9	Up to 85	85–269	270+
	10–16	Up to 70	70–199	200+
	16+	Up to 27	27–79	80+
	Pregnant	Up to 30	30–89	90+
RBC transketolase TPP-effect (ratio)	All ages	25+	15–25	Up to 15
RBC glutathione reductase-FAD-effect (ratio)[b]	All ages	1.2+	—	Up to 1.2
Tryptophan load[b] (mg xanthurenic acid excreted)	Adults (Dose: 100 mg per kg body wt)	25+ (6 hours)	—	Up to 25
		75+ (24 hours)	—	Up to 75
Urinary pyridoxine[b] (μg/g creatinine)	1–3	Up to 90	—	90+
	4–6	Up to 80	—	80+
	7–9	Up to 60	—	60+
	10–12	Up to 40	—	40+
	13–15	Up to 30	—	30+
	16+	Up to 20	—	20+
Urinary *N'* methyl nicotinamide[a] (mg/g creatinine)	All ages	Up to 0.2	0.2–5.59	0.6+
	Pregnant	Up to 0.8	0.8–2.49	2.5+
Urinary pantothenic acid[b] (μg)	All ages	Up to 200	—	200+
Plasma vitamin E (mg/100 mL)	All ages	Up to 0.2	0.2–0.6	0.6+
Transaminase index (ratio)[b]				
EGOT	Adult	2.0+	—	Up to 2.0
EGPT	Adult	1.25+	—	Up to 1.25

EGOT = erythrocyte glutamic oxalacetic transaminase; EGPT = erythrocyte glutamic pyruvic transaminase.

[a]Adapted from the Ten State Nutrition Survey.

[b]Criteria may vary with different methodology.

From Christakis, G.: Nutritional Assessment in Health Programs. *Am J Public Health*,63 (Suppl 1):34, 1973, with permission.

Appendix 28
Reference Values for Blood Lipids

Age (Years)	Percentile Values in Serum, mg/dL: Total Cholesterol 50	Total Cholesterol 75	Total Cholesterol 90	LDL-Cholesterol 50	LDL-Cholesterol 75	LDL-Cholesterol 90	HDL-Cholesterol 50	HDL-Cholesterol 75	HDL-Cholesterol 90
Men									
5–9	164	180	197	93	106	121	56	65	72
10–14	160	178	196	97	112	126	57	63	73
15–19	150	170	188	96	112	127	47	54	61
20–24	170	189	210	104	122	142	46	53	59
25–29	183	208	224	119	142	162	45	52	60
30–34	196	219	246	128	148	171	46	54	61
35–39	203	230	257	135	159	181	44	50	60
40–44	209	235	258	139	162	178	44	53	62
45–49	216	241	266	145	168	192	46	54	62
50–54	216	242	269	147	167	191	45	53	60
55–59	218	242	270	149	173	197	47	57	66
60–64	216	242	267	147	170	194	50	63	71
65–69	216	240	266	150	175	205	50	64	76
70+	211	236	260	146	169	187	49	58	72
Women									
5–9	168	184	201	101	118	129	56	65	72
10–14	163	179	196	97	113	130	57	63	73
15–19	160	177	197	96	114	133	47	54	61
20–24	175	196	220	105	122	145	53	64	74
25–29	178	199	223	111	130	152	57	65	76
30–34	181	202	229	112	132	151	57	66	75
35–39	189	211	235	119	143	166	55	66	76
40–44	198	222	244	126	150	170	58	67	81
45–49	207	233	259	131	155	178	60	70	84
50–54	222	247	274	138	165	192	64	73	87
55–59	230	255	284	149	173	205	62	75	88
60–64	223	260	286	153	173	197	63	77	90
65–69	233	260	285	156	190	211	64	75	88
70+	231	259	284	151	175	195	62	73	84

Adapted from The Lipid Research Clinic population studies data book. Vol. I: The prevalence study. NIH Publication No. 80-1527, U.S. Department of Health and Human Services, Public Health Service, July 1980.

Appendix 29
Expected 24-Hour Urinary Creatinine Excretion[a]

Men[b]			Women[c]		
Height		Creatinine	Height		Creatinine
ft	in	mg	ft	in	mg
5	2	1290	4	10	782
5	3	1320	4	11	802
5	4	1360	5	0	826
5	5	1390	5	1	848
5	6	1430	5	2	872
5	7	1470	5	3	894
5	8	1510	5	4	923
5	9	1550	5	5	950
5	10	1600	5	6	983
5	11	1640	5	7	1010
6	0	1690	5	8	1040
6	1	1740	5	9	1080
6	2	1780	5	10	1110
6	3	1830	5	11	1140
6	4	1890	6	0	1170

[a]For adults less than or equal to age 54. Decrease value by 10% per decade for older persons.
[b]Creatinine coefficient (men) = 23 mg/kg ideal body weight/24 hours
[c]Creatinine coefficient (women) = 18 mg/kg ideal body weight/24 hours

$$\text{Creatinine–height index} = \frac{\text{24-hour urinary creatinine excretion}}{\text{expected urinary creatinine excretion (for same height)}}$$

Interpretation:

>90% = Adequate muscle mass
80%–90% = Mild depletion
60%–79% = Moderate depletion
<60% = Severe depletion

From Bistrian, B.R.: Nutritional assessment and therapy of protein-calorie malnutrition in the hospital. *J Am Diet Assoc*, 71:395, 1977. Adapted with permission.

Appendix 30
Reference Values for Normal Blood Constituents

	Conventional Unit	SI Unit
Physical measurements		
Specific gravity	1.025–1.030	1.025–1.030
Reaction (pH)	7.35–7.45	7.35–7.45
Bleeding time	1–5 min	1–5 min
Coagulation time, venous blood	4–12 min	4–12 min
Partial prothrombin time	20–35 sec	20–35 sec
Prothrombin time	11–15 sec	11–15 sec
Erythrocyte sedimentation rate		
Men	1–13 mm/h	1–13 mm/h
Women	1–20 mm/h	1–20 mm/h
Osmolality, serum	275–295 mOsm/kg H_2O	275–295 mOsm/kg H_2O
Viscosity, serum	1.4–1.8 × water	1.4–1.8 × water
Hematology		
Cells, differential count		
Basophils	15–50/mm^3	15–50 × 10^6/L
Eosinophils	50–250/mm^3	50–250 × 10^6/L
Lymphocytes	1500–3000/mm^3	1500–3000 × 10^6/L
Monocytes	300–500/mm^3	300–500 × 10^6/L
Neutrophils, band	150–400/mm^3	150–400 × 10^6/L
Erythrocytes (RBCs)	4.2–5.9 million/mm^3	4.2–5.9 × 10^{12}/L
Leukocytes (WBCs)	4300–10,800/mm^3	4.3–10.8 × 10^9/L
Thrombocytes (platelets)	150,000–350,000/mm^3	150–350 × 10^9/L
Reticulocytes	25,000–75,000/mm^3	25–75 × 10^9/L
Hematocrit (vol% red cells)		
Men	40%–54%	0.40–0.54
Women	37%–47%	0.37–0.47
Newborn	49%–54%	0.49–0.54
Children (varies with age)	35%–49%	0.35–0.49
Hemoglobin		
Men	13.5–17.5 g/dL	8.1–11.2 mmol/L
Women	12–16 g/dL	7.4–9.9 mmol/L
Newborn	16.5–19.5 mg/dL	10.2–12.1 mmol/L
Children (varies with age)	11.2–16.5 g/dL	7.0–10.2 mmol/L
Hemoglobin, glycosylated	4.0%–7.0%	0.04–0.07 of total
Mean corpuscular volume (MCV)	82–98 μm^3	82–98 fL
Mean corpuscular hemoglobin (MCH)	27–32 pg/cell	2.7–3.2 fmol
Mean corpuscular hemoglobin concentration (MCHC)	32–36 g/dL	19–22.8 mmol/L
Total protein, serum	6–8.4 g/dL	60–84 g/L
Albumin, serum	3.5–5.0 g/dL	35–50 g/L
Globulin, serum	2.3–3.5 g/dL	23–35 g/L
$Alpha_1$	0.2–0.4 g/dL	2–4 g/L
$Alpha_2$	0.5–0.9 g/dL	5–9 g/L
Beta	0.6–1.1 g/dL	6–11 g/L
Gamma	0.7–1.7 g/dL	7–17 g/L

Appendix 30 *(continued)*

	Conventional Unit	SI Unit
Albumin/globulin ratio	1.8–2.5	1.8–2.5
Bile acids, serum	0.3–3.0 mg/dL	0.8–7.6 μmol/L
Bilirubin, serum	3.9 mg/dL	39 mg/L
Ceruloplasmin, plasma	27–37 mg/dL	270–370 mg/L
Ferritin, serum		
Men	20–200 ng/mL	20–200 μg/L
Women	20–150 ng/mL	20–150 μg/L
Iron deficiency	<10 ng/mL	<10 μg/L
Borderline	13–20 ng/mL	13–20 μg/L
Iron excess	>400 ng/dL	>400 μg/L
Fibrinogen, plasma	200–400 mg/dL	2.0–4.0 g/L
Prealbumin, serum	16.5–40.2 mg/dL	165–402 mg/L
Transferrin, serum	220–400 mg/dL	2.2–4.0 g/L
Nitrogen constituents		
Amino acid nitrogen, blood	4–8 mg/dL	40–80 mg/L
Ammonia, blood	80–110 μg/dL	47–65 μmol/L
Creatine, serum	0.2–0.8 mg/dL	15–61 μmol/L
Creatinine, serum	0.6–1.5 mg/dL	60–130 μmol/L
Nonprotein nitrogen, blood	20–40 mg/dL	10.7–25.0 mmol/L
Urea nitrogen, blood	10–20 mg/dL	3.6–7.1 mmol/L
Urea, serum	24–49 mg/dL	4.0–8.2 mmol/L
Uric acid, blood	2.5–5 mg/dL	150–270 μmol/L
Amino acids, plasma		
Alanine	2.3–5.1 mg/dL	23–51 mg/L
Alpha-aminobutyric acid	0.2–0.4 mg/dL	2–4 mg/L
Arginine	0.4–1.3 mg/dL	4–13 mg/L
Asparagine	0.2–0.8 mg/dL	2–8 mg/L
Aspartic acid	0.01–0.07 mg/dL	0.1–0.7 mg/L
Citrulline	0.2–0.6 mg/dL	2–6 mg/L
Cysteine and cystine	1.1–1.3 mg/dL	11–13 mg/L
Glutamic acid	0.2–1.2 mg/dL	2–12 mg/L
Glutamine	4–7 mg/dL	40–70 mg/L
Glycine	1.7–3.3 mg/dL	17–33 mg/L
Histidine	0.3–1.2 mg/dL	3–12 mg/L
Isoleucine	0.4–1.1 mg/dL	4–11 mg/L
Leucine	0.7–1.7 mg/dL	7–17 mg/L
Lysine	1.5–2.2 mg/dL	15–22 mg/L
Methionine	0.2–0.3 mg/dL	2–3 mg/L
Ornithine	0.3–0.8 mg/dL	3–8 mg/L
Phenylalanine	0.8–1.8 mg/dL	8–18 mg/L
Phosphoethanolamine	0–0.6 mg/dL	0–6 mg/L
Phosphoserine	0–0.1 mg/dL	0–1 mg/L
Proline	1.1–3.6 mg/dL	11–36 mg/L

(continued)

Appendix 30 *(continued)*

	Conventional Unit	SI Unit
Serine	0.6–1.4 mg/dL	6–14 mg/L
Taurine	0.4–1.3 mg/dL	4–13 mg/L
Threonine	0.9–2.4 mg/dL	9–24 mg/L
Tryptophan	1.0–1.2 mg/dL	10–12 mg/L
Tyrosine	0.04–0.07 mg/dL	0.4–0.7 mg/L
Valine	1.5–3.1 mg/dL	15–31 mg/L
Carbohydrates		
Glucose, blood		
Fasting, adult	70–110 mg/dL	3.9–6.0 mmol/L
>60 yr	80–115 mg/dL	4.4–6.4 mmol/L
2 hr postprandial	<120 mg/dL	<6.7 mmol/L
Fructose	6–8 mg/dL	0.3–0.4 mmol/L
Glycogen	5–6 mg/dL	0.28–0.44 mmol/L
Glucuronic acid	0.4–1.4 mg/dL	0.02–0.08 mmol/L
Hexoses	70–105 mg/dL	3.9–5.9 mmol/L
Pentose (total)	2–4 mg/dL	0.02–0.08 mmol/L
Sucrose	0.06 mg/dL	1.75 μmol/L
Lipids		
Cholesterol, serum		
Total	150–240 mg/dL	3.9–6.7 mmol/L
Esters	100–180 mg/dL	2.6–4.6 mmol/L
Free	50–60 mg/dL	2.6–3.0 mmol/L
LDL cholesterol	60–180 mg/dL	3.0–4.6 mmol/L
HDL cholesterol	30–80 mg/dL	1.5–3.8 mmol/L
Fats, neutral	150–300 mg/dL	3.9–7.8 mmol/L
Fatty acids, serum		
Total	190–420 mg/dL	1.9–4.2 g/L
Free	8–30 mg/dL	0.8–3.0 g/L
Glycerol, free	0.29–1.72 mg/dL	2.9–17.2 mg/L
Lecithin	100–200 mg/dL	10.0–20.0 g/L
Lipids, serum (total)	450–850 mg/dL	4.5–8.5 g/L
Phospholipids, serum (total)	230–300 mg/dL	2.3–3.0 g/L
Plasmalogen	7–8 mg/dL	0.7–0.8 g/L
Sphingomyelin	10–50 mg/dL	1.0–5.0 g/L
Triglycerides	40–150 mg/dL	4–15 g/L
Minerals		
Base, serum (total)	145–155 mEq/L	145–155 mmol/L
Aluminum, plasma	13–17 μg/dL	2.1–2.7 μmol/L
Arsenic, blood	0.2–6.2 μg/dL	0.03–0.82 μmol/L
Bromide, serum	0.8–1.5 mg/dL	10–19 μmol/L
Cadmium	0.1–0.7 μg/dL	0.02–0.11 μmol/L
Calcium, serum	9.0–11.0 mg/dL (varies with protein concentration)	2.25–2.75 mmol/L

Appendix 30 *(continued)*

	Conventional Unit	SI Unit
Calcium, ionized, serum	4.25–5.25 mg/dL	1.06–1.31 mmol/L
Chloride, serum	340–376 mg/dL (90–106 mEq/L)	90–106 mmol/L
Chromium, plasma	0.15 μg/dL	0.02 μmol/L
Cobalt, plasma	0.7–6 μg/dL	0.11–0.94 μmol/L
Copper	100–200 μg/dL	15.7–31.4 μmol/L
Fluoride	20–100 μg/dL	3.1–15.7 μmol/L
Iodine, serum	3.5–8.0 μg/dL	0.55–12.6 μmol/L
Iron, serum		
Men	70–170 μg/dL	12.5–30.4 μmol/L
Women	40–150 μg/dL	6.8–26.5 μmol/L
Iron binding capacity, serum		
Total	250–410 μg/dL	45–73 μmol/L
Saturation	20%–55%	0.20–0.55
Lead, blood	<30 μg/dL	<1.4 μmol/L
Lithium, serum	0.8–1.2 mEq/L	0.8–1.2 mmol/L
Magnesium, serum	1.8–3.0 mg/dL	0.75–1.25 mmol/L
Manganese, serum	0.05–0.07 μg/dL	9.1–12.7 μmol/L
Mercury, whole blood	<5.0 μg/dL	<0.25 μmol/L
Molybdenum, plasma	1.3 μg/dL	0.4 nmol/L
Nickel, plasma	2–4 μg/dL	0.1–0.2 nmol/L
Phosphorus, inorganic, serum	3.0–4.5 mg/dL	0.9–1.45 mmol/L
Potassium, serum	3.8–5.0 mEq/dL	3.8–5.0 mmol/L
Selenium	7–30 μg/dL	0.4–0.9 nmol/L
Silicon, plasma	500 μg/dL	4.0 nmol/L
Sodium, serum	135–148 mEq/L	135–148 mmol/L
Sulfates, inorganic, serum	2.5–5.0 mg/dL	1.9–3.8 mmol/L
Vanadium, plasma	0.5–2.3 μg/dL	0.3–0.7 nmol/L
Zinc, plasma	100–124 μg/dL	15.3–18.98 nmol/L
Vitamins		
Ascorbic acid, serum	0.4–1.5 mg/dL	23–85 μmol/L
white blood cells	25–40 mg/dL	250–400 mg/L
Beta-Carotene, serum	40–200 μg/dL	0.74–3.72 μmol/L
Biotin, plasma	300–800 pg/mL	0.66–1.76 pmol/L
Folate, serum	1.8–9.0 ng/mL	4.1–20.4 nmol/L
erythrocytes	150–450 ng/mL	340–1020 nmol/L
Riboflavin	3.6–18 ng/mL	14.6–72.8 nmol/L
Thiamin, serum	0–2 g/dL	0.75–4 nmol/L
Vitamin A, serum	20–80 μg/dL	0.7–2.8 μmol/L
Vitamin B_6, plasma	20–90 ng/mL	120–540 μmol/L
Vitamin B^{12}, serum	180–760 pg/mL	140–560 pmol/L
Vitamin D	24–40 μg/mL	60–105 μmol/L
Vitamin D, 1,25-dihydroxy	25–45 ng/mL	11.6–46.4 μmol/L
25-hydroxy	15–80 ng/mL	37–200 nmol/L
Vitamin E, serum	5.0–50 μg/mL	17.6–115 μmol/L

(continued)

Appendix 30 *(continued)*

	Conventional Unit	SI Unit
Vitamin K, prothrombin time	10–15 seconds	10–15 seconds
Organic acids		
Acetoacetic acid	0.8–2.8 mg/dL	8–28 mg/L
Alpha-ketoglutaric acid	0.2–1.0 mg/dL	2–10 mg/L
Citric acid	1.4–3.0 mg/dL	14–30 mg/L
Lactic acid	8–17 mg/dL	80–170 mg/L
Malic acid	0.1–0.9 mg/dL	1–9 mg/L
Pyruvic acid	0.4–2 mg/dL	4–20 mg/L
Succinic acid	0.1–0.6 mg/dL	1–6 mg/L
Ketone bodies	0.3–2 mg/dL	3–20 mg/L
Enzymes		
Aldolase	1.3–8.2 mU/mL	1.3–8.2 U/mL
Amylase, serum	25–125 mU/mL	25–125 U/L
Creatine kinase (CK), serum		
Men	15–160 U/L	0.25–2.67 10^6 katal/L
Women	15–130 U/L	0.25–2.17 10^6 katal/L
Lactate dehydrogenase (LD)	100–225 U/L (37°C)	1.7–3.3 10^6 katal/L
Lipase, serum	450–850 mg/dL	4.5–8.5 g/L
Phosphatase, serum		
Acid	0.11–0.60 mU/mL (37°C)	0.11–0.60 U/L
Alkaline	20–90 mU/mL (30°C)	20–90 U/L
Transaminase, serum		
SGOT or AST	7–40 mU/mL (37°C)	7–40 mU/L
SGPT or ALT	5–35 mU/mL (37°C)	5–35 U/L
Hormones		
Adrenocorticotropic hormone (ACTH)	25–100 pg/mL	25–100 ng/L
Aldosterone, supine	3–10 ng/dL	80–275 pmol/L
standing	5–30 ng/dL	140–830 pmol/L
Calcitonin		
Men	0–14 pg/mL	0–14 ng/L
Women	0–28 pg/mL	0–28 ng/L
Corticosterone, plasma or serum	0.13–2.3 μg/dL	3.75–66 nmol/L
Corticotropin, plasma	10–80 pg/mL	2–18 pmol/L
Cortisol, plasma or serum	8–20 μg/dL	221–552 nmol/L
Estradiol, plasma or serum		
Men	8–36 pg/mL	29–132 pmol/L
Women, follicular	10–90 pg/mL	37–330 pmol/L
midcycle	100–500 pg/mL	370–1835 pmol/L
luteal	50–240 pg/mL	184–880 pmol/L
postmenopausal	10–30 pg/mL	37–110 pmol/L
Estrogens, serum		
Men	40–115 pg/mL	150–420 pmol/L
Women, cycle 1–10 days	60–400 pg/mL	220–1470 pmol/L
11–20 days	120–440 pg/mL	440–1620 pmol/L

Appendix 30 *(continued)*

	Conventional Unit	SI Unit
21–30 days	156–350 pg/mL	572–1283 pmol/L
Estrone, serum		
Men	30–170 pg/mL	111–630 pmol/L
Women, follicular	20–150 pg/mL	74–155 pmol/L
Follicle-stimulating hormone (FSH)		
Men	3–18 mU/mL	3–18 U/L
Women, premenopausal	4–30 mU/mL	4–30 U/L
midcycle peak	10–90 mU/mL	10–90 U/L
pregnancy	low to detectable	
postmenopausal	40–250 mU/mL	40–250 U/L
Growth hormone, serum or plasma		
Children	>10 ng/mL	>10 μg/L
Men	<2 ng/mL	<2 μg/L
Women	<10 ng/mL	<10 μg/L
Insulin, plasma (fasting)	6–26 μU/mL	43–187 pmol/L
Luteinizing hormone (LH)		
Men	6–23 mU/mL	6–23 U/L
Women, follicular	5–30 mU/mL	5–30 U/L
midcycle	75–150 mU/mL	75–150 U/L
postmenopausal	30–200 mU/mL	30–200 U/L
Parathyroid hormone	230–630 pg/mL	230–630 ng/L
Progesterone, serum		
Men	0.12–0.3 ng/mL	0.38–1.0 nmol/L
Women, follicular phase	0.2–0.9 ng/mL	1–3 nmol/L
luteal phase	6.0–30.0 ng/mL	19–95 nmol/L
Prolactin, serum		
Men	1–20 ng/mL	1.0–20.0 μg/L
Women	1–25 ng/mL	1.0–25.0 μg/L
Somatomedin C		
Men	0.34–1.9 U/mL	340–1900 U/L
Women	0.45–2.2 U/mL	450–2200 U/L
Testosterone, plasma		
Men	437–705 ng/dL	15.2–24.5 nmol/L
Women (midcycle)	27–47 ng/dL	0.9–1.6 nmol/L
Pregnant (36th week)	76–152 ng/dL	2.6–5.3 nmol/L
Thyroxine (T_4), serum	4.5–13 μg/dL	58–167 nmol/L
>60 yr Men	5.0–10.0 μg/dL	64–129 nmol/L
Women	5.5–10.5 μg/dL	71–135 nmol/L
Thyroxine binding globulin (TBG)	15–34 μg/mL	200–260 nmol/L
Transcortin, serum		
Men	1.5–2.0 mg/dL	15–20 mg/L
Women, follicular	1.7–2.0 mg/dL	17–20 mg/L
luteal	1.6–2.1 mg/dL	16–21 mg/L
postmenopausal	1.7–2.5 mg/dL	17–25 mg/L
Triiodothyronine (T_3), total	85–205 ng/dL	1.3–3.1 nmol/L
Triiodothyronine (T_3), uptake	25%–38%	0.25–0.38

(continued)

Appendix 30 *(continued)*

	Conventional Unit	SI Unit
Blood gases		
CO_2-combining power	21–28 mEq/L	21–28 mmol/L
CO_2 content		
Serum	21–30 mEq/L	21–30 mmol/L
Whole blood	18–27 mEq/L	18–27 mmol/L
CO_2 tension	18–40 mm Hg	18–40 mm Hg
O_2 capacity, whole blood	16–27 vol%	85–105 mmol/L
O_2 content		
Arterial blood	15–23 vol%	80–89 mmol/L
Venous blood	10–16 vol%	53–62 mmol/L
O_2 saturation		
Arterial blood	94%–96%	0.94–0.96
Venous blood	60%–85%	0.60–0.85

Compiled from Bishop, M. L. et al. *Clinical Chemistry,* 2nd ed. Philadelphia, J. B. Lippincott, 1992; Leavelle, D. E.: *Interpretative Data for Diagnostic Laboratory Tests,* Reference Laboratory of the Mayo Clinic, Rochester, 1986; Linder, M. C.: *Nutritional Biochemistry and Metabolism with Clinical Applications*. New York: Elsevier, 1985, pp. 152–153; Normal laboratory values, *N Engl J Med* 314:39–49, 1986; Rakel, R. E., *Conn's Current Therapy,* Philadelphia, W. B. Saunders, 1994, pp. 1206–1211; Weinsier, R. L., Heimburger, D. C., and Butterworth, C. C.: *Handbook of Clinical Nutrition,* 2nd ed. St. Louis, C.V. Mosby, 1989, pp. 371–373; Wyngaarden, J. B. and H. Smith, L. H.: *Cecil Textbook of Medicine,* 18 ed. Philadelphia, W.B. Saunders, 1988, pp. 2395–2401.

Appendix 31
Normal Reference Values for Urine*

	Conventional Units	SI Units
Physical measurements		
Specific gravity	1.003–1.030	1.003–1.030
Reaction (pH)	5.5–8.0 (varies with diet)	5.5–8.0
Volume	800–1500 mL/24 hours	800–1500 mL/day
Total solids	55–70 g/liter	
Osmolality	300–800 mOsmol/kg H_2O	300–800 mOsmol/kg
Overhydration	<100 mOsmol/kg	<100 mOsmol/kg
Dehydration	>800 mOsmol/kg	>800 mOsmol/kg
Creatinine clearance	70–135 mL/min/1.73 m^2 at age 20 (decreased by 6mL/min/decade)	
Glomerular filtration rate	90–130 mL/min/1.73 m^2 at age 20 (decreased by 4 mL/min/decade)	
Organic constituents		
Acetone (ketone) bodies	None	None
Albumin	10–100 mg/24 hours <150 mg/24 hours ambulatory	0.15–1.5 μmol/day
Ammonia nitrogen	30–50 mEq/24 hours	30–50 mmol/day
Bile	None	None
Bilirubin, total	None	None
Catecholamines, free	4–126 μg/24 hours	24–745 nmol/day
Coproporphyrin	50–250 μg/24 hours	77–380 nmol/day
Cortisol	10–100 μg/24 hours	27.6–276 nmol/day
Creatine		
Men	0–40 mg/24 hours	0–0.30 mmol/day
Women	0–80 mg/24 hours	0–0.60 mmol/day
Creatinine		
Men	15–25 mg/kg/24 hours	0.13–0.22 mmol/kg/day
Women	11–20 mg/kg/24 hours	0.09–0.18 mmol/kg/day
Crystine or cysteine	None	None
Epinephrine	<10 μg/24 hours	55 nmol/day
Estrogens, total		
Men	4–25 μg/24 hours	14–90 nmol/day
Women	5–100 μg/24 hours	18–360 nmol/day
Glucose	<250 mg/24 hours	<250 mg/day
Hemoglobin	None	None
Homogentisic acid	None	None
17-Hydroxycorticosteroids		
Men	3–9 mg/24 hours	8.3–25 μmol/day
Women	2–8 mg/24 hours	5.5–22 μmol/day
5-Hydroxyindoleacetic acid	2–6 mg/24 hours (women lower than men)	10–31 μmol/day

(continued)

Appendix 31 *(continued)*

	Conventional Units	SI Units
17-Ketosteroids		
Men	8–22 mg/24 hours	28–76 μmol/day
Women	6–15 mg/24 hours	21–52 μmol/day
Metanephrine	<1.3 mg/24 hours	<0.6 mmol/day
Norepinephrine	<80 μg/24 hours	<590 nmol/day
Porphobilinogen	0–2.0 mg/24 hours	<9 μmol/day
Pregnanetriol	<2.5 mg/24 hours	<7.4 μmol/day
Protein, total	10–150 mg/24 hours	10–150 mg/day
Sugar	None (in some persons, 2–3 mg per 24 hours after a heavy meal)	None
Sulfate, organic	60–200 mg/24 hours	
Urea nitrogen	6–17 g/24 hours	0.21–0.60 mol/day
Uric acid	250–750 mg/24 hours (diet dependent)	1.48–4.43 mmol/day
Urobilinogen	0.5–4 mg/24 hours	0.6–6.8 μmol/day
Uroporphyrin	10–30 μg/24 hours	12–36 nmol/day
Vanillylmandelic acid (VMA)	1–8 mg/24 hours	5–40 μmol/day
Inorganic constituents		
Arsenic (24-hour)	<50 μg/liter	<0.65 mol/liter
Borate (24-hour)	<2 mg/liter	<32 μmol/liter
Calcium		
Low Ca diet	<150 mg/24 hours	<3.8 mmol/day
Usual diet	<250 mg/24 hours	6.3 mmol/day
Chloride (as NaCl)	110–250 mEq/24 hours (varies with intake)	110–250 mmol/day
Copper	0–50 μg/24 hours	0–0.80 μmol/day
Lead	<120 μg/24 hours	<0.39 μmol/day
Magnesium	6.0–10.0 mEq/24 hours	3–5 mmol/day
Phosphate	0.9–1.3 g/24 hours (varies with intake)	29–42 mmol/day
Potassium	25–100 mEq/24 hours (Varies with intake)	25–100 mmol/day
Sodium	130–260 mEq/24 hours (Varies with intake)	130–260 mmol/day

Compiled from Borer, W. Z.: Reference Values for the Interpretation of Laboratory Tests. In: Rakel, R. E. (Ed.) *Conn's Current Therapy*. Philadelphia, W.B. Saunders, 1994; Leavelle, D. E.: *Interpretative Data for Diagnostic Laboratory Tests*. Reference Laboratory of the Mayo Clinic, Rochester, 1986; Normal Laboratory Values, *N Engl J Med* 314:39–49, 1986; Wyngaarden, J. B and Smith, L. H. *Cecil Textbook of Medicine*, 18 ed. Philadelphia, W.B. Saunders, 1988, pp. 2395–2401.

Appendix 32
Dietary Fiber in Selected Foods

Food	Serving Size	Fiber (g)	Food	Serving Size	Fiber (g)
Breakfast cereals			Plum, damson	5	0.9
All-Bran	⅓ c	8.5	Prune	3	3.0
Bran Buds	⅓ c	7.9	Raisins	¼ c	3.1
Bran Chex	⅔ c	4.6	Raspberries	½ c	3.1
Cheerios-type	1¼ c	1.1	Strawberries	1 c	3.0
Corn Bran	⅔ c	5.4	Watermelon	1 c	0.4
Cornflakes	1¼ c	0.3			
Cracklin' Bran	⅓ c	4.3	**Vegetables (cooked)**		
Frosted Mini Wheats	4 pcs	2.1	Asparagus	½ c	1.0
Graham Crackers	¾ c	1.7	Beans, string	½ c	1.6
Grape Nuts	¼ c	1.4	Broccoli	½ c	2.2
Honey Bran	⅞ c	3.1	Brussels sprouts	½ c	2.3
Nutri-Grain	¾ c	1.8	Cabbage	½ c	1.4
Oatmeal	¾ c	1.6	Carrots	½ c	2.3
100% Bran	½ c	8.4	Cauliflower	½ c	1.1
Raisin Bran-type	¾ c	4.0	Corn, canned	½ c	2.9
Rice Krispies	1 c	0.1	Kale leaves	½ c	1.4
Shredded Wheat	⅔ c	2.6	Parsnip	½ c	2.7
Special K	1⅓ c	0.2	Peas	½ c	3.6
Sugar Smacks	¾ c	0.4	Potato, w/o skin	1 med	1.4
Tasteeos	1¼ c	1.0	Potato, w/ skin	1 med	2.5
Total	1 c	2.0	Spinach	½ c	2.1
Wheat Chex	⅔ c	2.1	Squash, summer	½ c	1.4
Wheat germ	¼ c	3.4	Turnip	½ c	1.6
Wheaties	1 c	2.0	Zucchini	½ c	1.8
Fruits			**Vegetables (raw)**		
Apple, w/o skin	1 med	2.7	Bean sprout	½ c	1.5
Apple, w/ skin	1 med	3.5	Celery, diced	½ c	1.1
Apricot, fresh	3 med	1.8	Cucumber	½ c	0.4
Apricot, dried	5 halves	1.4	Lettuce, sliced	1 c	0.9
Banana	1 med	2.4	Mushrooms, sliced	½ c	0.9
Blueberries	½ c	2.0	Onions, sliced	½ c	0.8
Cantaloupe	¼ melon	1.0	Pepper, green	½ c	0.5
Cherries, sweet	10	1.2	Tomato	1 med	1.5
Dates	3	1.9	Spinach	1 c	1.2
Grapefruit	½	1.6			
Grapes	20	0.6	**Legumes (cooked)**		
Orange	1	2.6	Baked beans	½ c	8.8
Peach, w/ skin	1	1.9	Dried peas	½ c	4.7
Peach, w/o skin	1	1.2	Kidney beans	½ c	7.3
Pear, w/ skin	½ lg	3.1	Lentils	½ c	3.7
Pear, w/o skin	½ lg	2.5	Lima beans	½ c	4.5
Pineapple	½ c	1.1	Navy beans	½ c	6.0

Source: Lanza, E. and Butrum, B.: A critical review of food fiber analysis and data, *J Am Diet Assoc* 86:732, 1986. Adapted with permission.

Appendix 33
Alcohol and Caloric Content of Alcoholic Beverages

Beverage—Serving	Weight g	Alcohol[a] g	CHO g	Calories
Ale, mild—8 fl oz	230	8.9	8.0	98
Ale, mild—12 fl oz	345	13.4	12.0	147
Anisette—1 cordial glass	20	7.0	7.0	74
Beer—8 fl oz	240	8.7	9.0	99
Beer—12 fl oz	360	13.1	13.2	148
Benedictine—1 cordial glass	20	6.6	6.6	69
Brandy—1 brandy glass	30	10.5	—	73
Champagne, dry—1 champagne glass	135	13.0	3.0	105
Champagne, sweet—1 champagne glass	135	13.0	17.0	160
Cider, fermented—6 fl oz	180	9.4	1.8	71
Creme de menthe—1 cordial glass	20	7.0	6.0	67
Curacao—1 cordial glass	20	6.0	6.0	54
Daiquiri—1 cocktail	100	15.1	5.2	122
Eggnog—4 fl oz punch cup	123	15.0	18.0	335
Gin, 80% proof—1 jigger	42	14.0	trace	97
Gin rickey—4 fl oz	120	21.0	1.3	150
Highball—8 fl oz	240	24.0	—	166
Madeira wine—1 wine glass	100	15.0	1.0	105
Manhattan—1 cocktail	100	19.2	7.9	164
Martini—1 cocktail	100	18.5	0.3	140
Mint julep—10 fl oz	300	29.2	2.7	212
Muscatel wine—1 wine glass	100	15.0	14.0	158
Old-fashioned—4 fl oz	100	24.0	3.5	179
Planter's punch—4 fl oz	100	21.5	7.9	175
Port wine—1 sherry glass	30	4.0	5.0	50
Rum, 80% proof—1 jigger	42	14.0	trace	97
Rum sour—4 fl oz	100	21.5	—	165
Sauterne, California—1 wine glass	100	10.5	4.0	84
Sherry, dry—1 wine glass	60	9.0	4.8	84
Tom collins—10 fl oz	300	21.5	9.0	180
Vermouth, dry—1 wine glass	100	15.0	1.0	105
Vermouth, sweet—1 wine glass	100	18.0	12.0	167
Vodka, 80% proof—1 jigger	42	14.0	trace	97
90% proof—1 jigger	42	15.9	trace	110
Whiskey, 80% proof—1 jigger	42	14.0	trace	97
100% proof—1 jigger	42	17.9	trace	124

[a]One gram (cc) alcohol = 7 calories. Alcohol is metabolized by the body as fat. It has negligible or a slightly lowering effect on blood sugar. In dietary calculations, alcohol is considered as fat.

Source: Pennington, J. A. and Church, H. N. *Bowe's and Church's Food Values of Portions Commonly Used,* 14 ed., p. 196. Philadelphia, J. B. Lippincott, 1985. Adapted with permission.

Appendix 34
Cholesterol and Fatty Acid Content of Selected Foods (per 100 g of edible portion)

	Per 100 g Edible Portion			
Food Item	CHOL (mg)	SFA (g)	MFA (g)	PFA (g)
Cereal grains				
Barley, bran	0	1.0	0.6	2.7
Corn, germ	0	3.9	7.6	18.0
Oats, germ	0	5.6	11.1	12.4
Rice, bran	0	3.6	7.3	6.6
Wheat bran	0	0.7	0.7	2.4
Wheat germ	0	1.9	1.6	6.6
Dairy and eggs				
Cheese, cheddar	105	21.1	9.0	0.9
Cheese, Roquefort	90	19.3	8.5	1.3
Cream, half and half	37	7.2	3.3	0.4
Cream, heavy whipping	137	23.0	10.7	1.4
Creamer, imitation	0	1.5	4.3	3.8
Egg, whole	548	3.4	4.5	1.4
Eggyolk	1602	9.9	13.2	4.3
Eggwhite	0	0.0	0.0	0.0
Ice cream, medium-rich	59	10.0	4.6	0.6
Ice milk	8	1.6	0.8	0.1
Milk, whole	14	2.1	1.0	0.1
Milk, 2%	8	1.2	0.6	0.1
Milk, 1%	4	0.7	0.3	—
Sherbert	7	1.2	0.6	0.1
Yogurt, low-fat	6	1.0	0.4	0.3
Fats and oils				
Butter	219	50.5	23.4	3.0
Coconut oil	0	86.5	5.8	1.8
Cod liver oil	570	17.6	51.2	25.8
Corn oil	0	12.7	24.2	58.7
Cottonseed oil	0	25.9	17.8	51.9
Herring oil	766	19.2	60.3	16.1
Lard, rendered	95	39.6	45.1	11.8
Margarine, corn, 80% fat	0	12.6	29.6	33.8
Mayonnaise, commercial	57	11.8	22.7	41.3
MCT oil	0	94.5	0.0	0.0
Menhaden oil	521	33.6	32.5	29.5
Olive oil	0	13.5	73.7	8.4
Palm oil	0	49.3	37.0	36.6
Peanut oil	0	16.9	46.2	32.0
Rapeseed oil (Canola)	0	6.8	55.5	33.3
Rice bran oil	0	19.7	39.3	35.0

(continued)

Appendix 34 *(continued)*

Food Item	Per 100 g Edible Portion CHOL (mg)	SFA (g)	MFA (g)	PFA (g)
Safflower oil	0	9.1	12.1	74.5
Salmon oil	485	23.8	39.7	29.9
Sesame oil	0	14.2	39.7	41.7
Shortening, vegetable	0	24.7	44.5	26.1
Soybean, partly hydrogenated	0	14.9	43.0	37.6
Sunflower oil	0	10.3	19.5	65.7
Walnut oil	0	9.1	22.8	63.3
Wheat germ oil	0	18.8	15.1	61.7
Fish and seafoods				
Clam, hardshell	31	0.6	trace	trace
Crab, blue	78	0.2	0.2	0.5
Haddock	63	0.1	0.1	0.2
Halibut, Pacific	32	0.3	0.8	0.7
Lobster, northern	95	0.2	0.2	0.2
Mackerel, Atlantic	80	3.6	5.4	3.7
Mussel, blue	38	0.4	0.5	0.6
Oyster, eastern	47	0.6	0.2	0.7
Perch, white	80	0.6	0.9	0.7
Pike, northern	39	0.1	0.2	0.2
Pike, walleye	86	0.2	0.3	0.4
Salmon, chum	74	1.5	2.9	1.5
Scallop, Atlantic	37	0.1	0.1	0.3
Shrimp, unspecified	147	0.2	0.1	0.4
Swordfish	39	0.6	0.8	0.2
Trout, rainbow	57	0.6	1.0	1.2
Tuna, bluefin	38	1.7	2.2	2.0
Fruits/Vegetables				
Avocado	0	2.4	9.6	2.0
Coconut, fresh	0	29.7	1.4	0.4
Peas, blackeye, chick, cooked	0	0.2	0.5	1.1
Peas, split or lentils	0	0.2	0.2	0.5
Soybeans, dry, cooked	0	0.9	1.3	3.3
Meats (cooked)				
Bacon, regular, cooked	85	17.4	23.7	5.8
Beef, approx. 6% fat	66	2.8	2.8	0.4
Beef, approx. 30% fat	94	13.3	15.7	1.3
Bologna, beef	56	11.7	13.3	1.1
Chicken, Cornish hen, turkey, light meat, no skin	89	1.2	1.1	0.9
Duck, goose, no skin	92	4.4	4.0	1.5
Frankfurter, all beef	48	12.0	14.4	1.2
Frankfurter, beef, pork	50	10.7	13.7	2.7
Lamb, approx. 8% fat	84	2.7	3.4	0.6

Appendix 34 *(continued)*

Food Item	Per 100 g Edible Portion			
	CHOL (mg)	SFA (g)	MFA (g)	PFA (g)
Lamb, approx. 36% fat	98	16.8	14.7	2.1
Pork, approx. 24% fat	82	9.1	11.5	2.8
Salami, dry	77	11.9	16.0	3.7
Veal, approx. 6% fat	99	2.3	2.9	0.5
Veal, approx. 25% fat	101	9.2	0.2	1.3

SFA = saturated fatty acid; MFA = monounsaturated fatty acid; PFA = polyunsaturated fatty acid.

From USDA Provisional Table on the Content of Cholesterol, Fatty Acids, and other Fat Components in Selected foods. Nutrient Data Research Branch, Human Nutrition Information Service, HNIS/PT-103, May 1986.

Appendix 35
Average Caffeine Content of Selected Foods (mg)

Item	mg
Coffee beverage, per 5 fl oz	
Drip, automatic	137
Drip, nonautomatic	124
Instant	60
Instant decaffeinated	3
Percolated, automatic	117
Percolated, nonautomatic	108
Coffee flavored (from instant mixes), per 6 fl oz	
Cafe amaretto	60
Cafe francais	52
Cafe vienna	57
Irish mocha mint	27
Orange cappuccino	74
Sunrise	37
Suisse mocha	40
Coffee instant dry powder	
Decaffeinated, 1 tsp	3
Regular/freeze-dried, 1 tsp	60
Tea beverage, per 5 fl oz	
Black, 1 min brew	28
Black, 3 min brew	42
Black, 5 min brew	46
Decaffeinated	1
Green, 1 min brew	14
Green, 3 min brew	27
Green, 5 min brew	31
Instant	33
Mint flavor, 5 min brew	50
Orange and spice, 5 min brew	45
Oolong, 1 min brew	13
Oolong, 3 min brew	30
Oolong, 5 min brew	40
Tea instant dry powder	
Lemon flavored, 1 tsp	38
Regular, 1 tsp	32
Soft drinks, per 12 fl oz	
Aspen	36
Big Red	38
Big Red, diet	38
Canada Dry	30
Coca Cola	45
Dr. Pepper	40
Kick	31
Mellow Yello	53
Mountain Dew	54
Mr. Pibb	41
Pepsi Cola	38
Pepsi Light	36
Royal Crown Cola	36
Royal Crown with a Twist	21
Shasta Cola	44
Tab	45
Foods containing chocolate	
Baking choc, 1 oz	35
Choc brownie, 1¼ oz	8
Choc cake, 1 med slice	14
Chocolate candy	
Choc, german sweet, 1 oz	8
Choc kisses, 6 pieces	5
Crunch Bar (Nestle), 1 oz	7
Golden Almond, 1 oz	5
Kit Kat, 1.5 oz	5
Krackel Bar, 1.2 oz	5
Milk chocolate	6
Mr. Goodbar, 1.65 oz	6
Semi-sweet, Bakers 1 oz	12
Special dark, 1 oz	23
Sweet dark choc, 1 oz	20
Choc flavored chips, ¼ c	12
Choc ice cream, ⅔ c	5
Choc milk, 8 fl oz	5
Choc pudding, ½ c	6
Choc syrup, 2 Tbsp	4
Cocoa beverage, 6 fl oz	5
Cocoa, dry powder, 1 Tbsp	11
Cocoa, dry, Hershey, 1 oz	70
Cocoa mix, Hershey, 1 pkt	5
Cocoa mix, Nestle, 1 oz	4
Cocoa mix w/marshmallows, Nestle, 1 oz	4

Source: Pennington, J. A. and Church, H. N.: *Bowe's and Church's Food Values of Portions Commonly Used,* 14th ed., pp. 223–224. Philadelphia: J. B. Lippincott, 1985. Adapted with permission.

Appendix 36
Salt Substitutes and Seasonings: Sodium and Potassium Contents

Product	Na (mg)	K (mg)
A. High in Sodium (sodium content 600 and above mg per tsp)		
Iodized salt (reference salt for comparison)	2300	trace
Monosodium glutamate	600	—
Salt'n Spice	938	15
Season All	980	17
Season All-Garlic	652	36
Seasoned Salt	1300	11
B. Sodium-Containing (between 200 mg and 500 mg Na per tsp)		
Lemon Pepper	340	11
Lite-lite Salt	346	0.02
Pinch of Herbs	259	9
Salad Supreme	280	—
Season All Light	380	—
Soy sauce (1 tbsp)	412	12
C. Low Sodium or Sodium-free (below 10 mg Na per tsp)		
All Purpose Parsley Patch[c]	4.0	36
Basil	0.4	40
Bay Leaves	0.3	8
Black Pepper	0.2	28
Celery Seed	4.1	34
Chili Pepper	0.2	34
Garlic Powder	0.1	16
Mustard Powder	0.1	11
No Salt (salt substitute)[d,e]	0	2500[d]
Nu-Salt (salt substitute)[d,f]	0	3180
Oregano	27	trace
Salt Free 17[a]	2	35
Salt Substitute[b]	<1.0	2800[d]
Saltless[c]	<1.0	2540[d]

[a]Brand name by Lawry's
[b]Morton's
[c]Schilling's
[d]Salt substitutes that are very high in potassium are not suitable for potassium-restricted diets.
[e]RCN Products, Inc.
[f]Cumberland Corp.

Appendix 37
Composition of Milk and Selected Formulas for Infant Feeding

	kcal per oz	Composition per Liter									Renal Solute Load[c]
		Protein (g)	CHO (g)	Fat (g)	Fe (mg)	Ca (mg)	P (mg)	Na (mg)	K (mg)	Cl (mg)	
Human milk	22	11	68	38	1.5	333	133	161	507	390	97
Cow's milk	20	33	48	37	1	1200	946	506	1365	994	310
Advance®	16	20	55	27	12	510	390	230	900	520	128
Enfamil®[a]	20	15	70	37	12	458	312	182	718	417	98
Enfamil® Premature[a]	20	20	74	34	12	1104	552	260	687	562	125
Enfamil® Premature[a]	24	24	88	41	12	1323	667	312	823	677	150
Isomil®	20	18	68	37	12	700	500	320	950	430	122
Isomil® SF	20	20	68	36	12	700	500	320	770	590	131
Kindercal	30	34	135	44	10.6	850	850	370	1310	740	99
Lofenalac®	20	22	86	26	12.5	625	469	312	677	469	132
MSUD Diet Powder[b]	20	14	90	28	12	687	375	260	688	520	100
Nursoy®	20	22	66	37	13	630	420	198	693	370	136
Nutramigen®	20	19	90	26	12.5	625	417	312	729	573	124
Phenyl-free®[b]	25	42	140	14	25	1042	1042	833	2812	1917	330
Portagen®	20	23	77	31	12.5	625	469	365	833	573	145
Preemie SMA®	20	20	86	44	NA	750	400	322	741	NA	175
Pregestimil®	20	19	69	38	12.5	625	488	260	729	573	122
Product 3200A[b]	20	22	86	26	12.5	625	469	312	677	469	132
Product 3200K[b]	20	20	67	35	12.5	625	495	240	812	552	127
Product 3232A[b]	20	19	90	28	12.5	625	417	286	729	573	123
Product 80056[b]	20	0	83	26	12.5	625	344	188	708	417	NA
Prosobee®	20	20	67	35	12.5	625	495	240	812	552	127
RCF®	—	20	0	36	1.5	700	500	320	770	590	131
Similac® 13[a]	13	12	46	23	7.8	410	310	190	620	400	83
Similac® 20[a]	20	15	72	36	12	510	390	230	800	500	105
Similac® 24[a]	24	22	85	43	15	730	560	350	1100	740	152
Similac® 27	27	25	96	48	2	810	620	380	1200	820	170
Similac® 24 LBW	24	22	85	45	3	730	560	360	1220	900	161
Similac® PM 60/40	20	16	69	38	1.5	400	200	160	580	400	96
Similac® 20 SP	20	18	72	37	2.5	1200	600	310	940	640	128
Similac® 24 SP	24	22	86	44	3.0	1440	720	380	1120	710	154
SMA®	20	15	72	36	12.5	440	320	150	562	372	92
SMA® 24	24	18	86	43	14.4	510	336	179	663	451	110
SMA® 27	27	20	97	49	16.2	564	378	202	753	508	122
Soyalac®	20	21	68	37	12.8	640	186	299	780	461	134

Manufacturers: Loma Linda (Soyalac); Mead Johnson Nutritionals (Enfamil, Kindercal, Lofenalac, MSUD, Nutramigen, Phenyl-free, Portagen, Pregestimil, Product 3200AB, Product 3200K, Product 3232A, Product 80056, Prosobee); Ross Laboratories (Advance, Isomil, RCF, Similac); Wyeth-Ayerst (Nursoy, SMA).

[c]Renal solute load = (Protein, g × 4) + Na, mEq + K, mEq + Cl, mEq.[a] Formula with iron.

[b]These formulas are for specific metabolic disorders; certain nutrients may be low or absent from the formula. Information on exact composition is available from the manufacturer.

Appendix 38
Composition of Oral and Intravenous Electrolyte Solutions

A. Oral Solutions
Glucose (g) and electrolytes (mEq) per Liter

	Glucose g	Na	K	Cl	Citrate	Bicarbonate	mOsmol/ kg H_2O
EquaLYTE™	25	78	22	68	30	—	305
Gastrolyte®	20	90	20	80	30	—	NA
Hydralyte™	12	84	10	59	—	10	NA
Infalyte®	20	50	20	40	—	30	NA
Pedialyte®	25	45	20	35	30	—	250
Rehydralyte®	25	75	20	65	30	—	305
Resol®[a]	20	50	20	50	34	—	265
Ricelyte®	30[b]	50	25	45	34	—	200
WHO solution	18	90	20	80	—	30	333

[a]Also contains magnesium (4 mEq/liter) and phosphate (4 mEq/liter).
[b]As rice syrup solids.

B. Intravenous Solutions
(Electrolyte content in mEq/liter)

	Na	K	Ca	Mg	Cl	Lactate	Acetate	Gluconate	mOsmol/ kg water
Isolyte® S	140	5	—	3	98	—	29	23	295
Normosol®-R	140	5	—	3	98	—	27	23	295
Plasma-Lyte® R	140	10	5	3	103	8	47	—	312
Ringer's®	147	4	4	—	156	—	—	—	310
Ringer's®, lactated	130	4	3	—	109	28	—	—	272
Concentrates:[c]									
Hyperlyte®	25	40.5	5	8	33.5	—	40.6	5	6015
Hyperlyte® CR	25	20	5	5	30	—	30	—	5500
Lypholyte® II	35	20	4.5	5	35	—	29.5	—	6200
Multilyte®-20	25	20	5	5	30	—	25	—	4205
TPN Electrolytes	35	20	4.5	5	35	—	29.5	—	6200
TPN Electrolytes II	15	18	4.5	5	35	—	7.5	—	3400

[c]Electrolyte concentrates are for compounding of IV admixtures and are not used for direct infusion. Osmolarity is based on the concentrate; electrolyte concentrations are when diluted in 1 liter.

Manufacturers: Abbott Laboratories (Normosol, Ringer's, TPN electrolytes); Baxter (Plasma-Lyte R, Ringer's); Jayco (Hydralyte); Kendall McGaw (Hyperlyte, Hyperlyte CR, Isolyte-S, Ringer's,); LyphoMed (Lypholyte, Multilyte-20); Mead Johnson (Ricelyte); Pennwalt (Infalyte); Ross (Equalyte, Pedialyte, Rehydralyte); USV Labs (Gastrolyte); Wyeth-Ayerst (Resol)

Appendix 39
Proprietary Formulas for Enteral Nutrition

Product name	kcal per mL	Composition per Liter: Protein (g)	CHO (g)	Fat (g)	Na (mg)	K (mg)	mOsmol per kg	Vol (mL) to meet RDA
	Monomeric (elemental) Products							
Accupep HPF®	1.0	40	189	10	680	1150	490	1600
AlitraQ ™	1.0	52	165	16	1000	1200	575	1500
Criticare HN®	1.06	38	220	5	630	1320	650	1890
Crucial ™	1.5	94	135	68	1168	1872	490	1000
Peptamen®	1.0	40	127	39	500	1252	270	1500
Peptamen VHP™	1.0	62	104	39	560	1500	300	1500
Reabilan®	1.0	31	131	39	702	1252	350	2000
Reabilan® HN	1.33	58	158	52	1000	1661	490	2222
Tolerex®	1.0	21	230	>2	470	1200	550	3160
Travasorb® Hepatic	1.1	29	215	15	235	882	600	2260
Travasorb® HN	1.0	45	175	14	921	1170	560	2000
Travasorb® Renal	1.35	23	271	18	tr	tr	590	2070
Travasorb® STD	1.0	30	190	14	921	1170	560	2000
Vital® High Nitrogen	1.0	42	185	11	566	1400	500	1500
Vivonex® Plus	1.0	45	190	7	611	1055	650	1800
Vivonex® T.E.N.	1.0	38	206	3	460	782	630	2000
	Polymeric Products							
Milk-based, lactose-containing								
Carnation® Instant Breakfast	1.1	63	150	35	1062	3083	694	1373
Compleat® Regular	1.07	43	130	43	1300	1400	450	1500
Great Shake®	1.47	51	243	34	960	2685	800	885
Meritene® liquid	1.0	58	110	32	880	2600	500	1250
Meritene® powder	1.06	69	120	34	1100	2800	690	1040
Sustagen®	1.86	98	270	14	901	2883	1130	1030
Fiber-containing, lactose-free								
Advera™	1.28	60	216	23	1056	2827	680	1184
Compleat® Modified	1.07	43	140	37	1000	1400	300	1500
Ensure® with fiber	1.1	40	162	37	846	1693	480	1391
Fiberlan™	1.2	50	160	40	1012	1716	310	1250
Fibersource™	1.2	43	170	41	1100	1800	390	1500
Fibersource™ HN	1.2	53	160	41	1100	1800	390	1500
Glucerna®	1.0	42	94	56	930	1560	375	1422
Glytrol®	1.0	45	100	48	740	1400	380	1400
Impact® with fiber	1.0	56	140	28	1100	1300	375	1500
IsoSource® VHN	1.0	62	130	29	1300	1600	300	1250
Jevity®	1.06	44	152	36	930	1570	300	1321
Nutren® 1.0 with fiber	1.0	40	127	38	500	1252	303	1500
PediaSure® with fiber	1.0	30	114	50	380	1310	345	1000
Profiber®	1.0	40	147	35	800	1500	300	1250
Replete® with fiber	1.0	62	113	34	500	1560	300	1000
Sustacal® 8.8	1.06	37	148	35	930	1564	500	1321
Sustacal® w/ Fiber	1.06	46	140	35	720	1390	480	1420

Appendix 39 *(continued)*

Product name	kcal per mL	Composition per Liter					mOsmol per kg	Vol (mL) to meet RDA
		Protein (g)	CHO (g)	Fat (g)	Na (mg)	K (mg)		
Ultracal®	1.06	44	123	45	930	1610	310	1180
Vitaneed®	1.0	40	128	40	680	1250	300	1500
Lactose-free, low calorie (<1 kcal/mL)								
CitriSource®	0.76	37	150	0	150	42	700	NA
Citrotein®	0.67	41	120	<2	830	550	490	1100
Entrition™ 0.5	0.5	18	68	18	350	600	120	4000
Introlan™	0.53	22	70	18	460	780	150	2000
Introlite™	0.53	22	70	18	930	1570	220	1321
Pre-Attain®	0.5	20	60	20	680	1150	150	1600
Lactose-free, normocaloric (1–1.2 kcal/mL)								
Attain®	1.0	40	135	35	805	1600	300	1250
Boost®	1.01	43	147	28	550	1690	590	1180
Compleat® modified	1.07	43	140	37	1000	1400	300	1500
Enercal	1.0	38	144	34	500	1250	370	NA
Ensure®	1.06	37	145	37	846	1564	470	1887
Ensure® HN	1.06	44	141	36	802	1564	470	1321
Entrition™ HN	1.0	44	114	41	845	1579	300	1300
Immun-Aid®	1.0	37	120	22	575	1055	460	2000
Isocal®	1.06	34	135	44	530	1320	270	1890
Isocal® HN	1.06	44	123	45	930	1610	270	1180
Isolan™	1.06	40	144	36	897	1326	300	1250
IsoSource®	1.2	43	170	41	1200	1700	360	1500
Nutren® 1.0	1.0	40	127	38	500	1252	300	1500
Nutrilan™	1.06	38	143	37	690	1326	520	1585
Osmolite®	1.06	37	145	38	640	1020	300	1887
Osmolite® HN	1.06	44	141	36	930	1570	300	1321
PediaSure®	1.0	30	110	50	380	1310	310	1000
Precision LR®	1.1	26	250	2	700	880	510	1710
Resource®	1.06	37	140	37	890	1600	430	1890
Lactose-free, normocaloric, high protein								
Ensure® High Protein	1.0	61	140	23	900	2100	475	1000
IsoSource® HN	1.2	53	160	41	1100	1700	330	1500
IsoSource® VHN	1.0	62	130	29	1300	1600	300	1250
Isotein HN®	1.2	68	160	34	620	1100	300	1770
Meritene®	1.06	69	120	34	1100	2800	690	1040
Promote®	1.0	62	130	26	930	1980	330	1250
Protain XL™	1.0	55	138	30	860	1500	340	1250
Replete®	1.0	62	113	34	500	1560	350	1000
Sustacal®	1.01	61	140	23	930	2100	650	1069
Lactose-free, hypercaloric, high protein								
Advera™	1.28	60	216	23	1056	2827	680	1184

(continued)

Appendix 39 *(continued)*

Product name	kcal per mL	Composition per Liter: Protein (g)	CHO (g)	Fat (g)	Na (mg)	K (mg)	mOsmol per kg	Vol (mL) to meet RDA
Comply®	1.5	60	180	60	1100	1850	410	1000
Deliver® 2.0	2.0	75	200	102	800	1700	640	1000
Ensure Plus®	1.5	55	200	53	1050	1940	690	1420
Ensure Plus® HN	1.5	63	200	50	1180	1820	650	947
Isocal® HCN	2.0	75	200	102	800	1690	640	1000
Magnacal®	2.0	70	250	80	1000	1250	590	1000
Nitrolan™	1.24	60	160	40	874	1482	310	1250
Nutren® 1.5	1.5	60	170	68	752	1872	410	1000
Nutren® 2.0	2.0	80	196	106	1000	2500	710	750
Perative®	1.3	67	177	37	1040	1730	385	1155
Resource Plus®	1.5	55	200	53	1300	2100	600	1400
Sustacal® HC	1.52	61	190	58	850	1480	670	1184
TwoCal® HN	2.0	84	217	91	1310	2456	690	947
Ultralan™	1.5	60	202	50	1173	1911	540	1000
Lactose-free, low fat								
CitriSource®	0.76	37	151	0	230	63	700	NA
Citrotein®	0.67	41	120	<2	830	550	490	1100
Ross SLD™	0.7	38	137	>1	835	835	545	1200
Sustacal® Powder	1.01	64	187	>2	960	1920	700	800
Tolerex®	1.0	21	230	>2	470	1200	550	1800
Specialized Formulas								
Diabetes mellitus								
Glucerna	1.0	42	94	56	930	1560	375	1422
Glytrol®	1.0	45	100	48	740	1400	380	1400
Respiratory failure								
NutriVent™	1.5	68	101	95	752	2240	450	1000
Pulmocare®	1.5	63	106	92	1310	1730	465	947
Respalor®	1.5	76	148	71	1270	1480	580	1440
Liver disease, with BCAA								
Hepatic Aid® II	1.2	44	168	36	338	230	560	NA
NutriHep™	1.5	40	290	21	320	1320	690	1000
Travasorb® Hepatic	1.1	29	215	15	235	882	600	2260
Renal disease								
Amin-Aid®	2.0	19	366	46	339	230	700	NA
Nepro®	2.0	70	215	96	829	1057	635	947
Re/Neph	2.0	65	260	82	1240	42	700	NA
Suplena®	2.0	30	255	96	783	1116	600	947
Travasorb® Renal	1.35	23	271	18	trace	trace	590	2070
Fat Malabsorption, with high MCT								
Lipisorb® Liquid	1.35	70	161	57	1350	1690	630	1186
Portagen®	1.0	35	114	48	468	1248	320	NA
Travasorb® MCT	1.0	50	123	33	350	1000	250	2000

Appendix 39 *(continued)*

Product name	kcal per mL	Composition per Liter Protein (g)	CHO (g)	Fat (g)	Na (mg)	K (mg)	mOsmol per kg	Vol (mL) to meet RDA
Cardiac, low sodium								
Lonalac®	1.01	54	75	56	40	2000	360	NA
Stress or trauma								
Alitraq®	1.0	52	165	16	1000	1200	575	1500
Crucial™	1.5	94	135	68	1168	1872	490	1000
Immun-Aid™	1.0	37	120	22	575	1055	460	2000
Impact®	1.0	56	130	28	1100	1300	375	1500
Impact® with fiber	1.0	56	140	28	1100	1300	375	1500
Perative®	1.3	67	177	37	1040	1730	385	1155
Protain XL™	1.0	55	138	30	860	1500	340	1250
Replete®	1.0	62	113	34	500	1560	350	1000
Replete® with fiber	1.0	62	113	34	500	1560	300	1000
TraumaCal®	1.5	83	145	68	1200	1400	490	2000
Traum-Aid® HBC	1.0	56	166	12	530	1173	640	3000
Geriatric care								
ProBalance™	1.2	54	156	41	763	1560	NA	1000

Modular Supplements

Product Name	Per 100 g Powder or 100 mL Liquid: Protein	CHO	Fat	Na	K	kcal
Modular protein						
Casec®	88	0	2	120	10	370
Elementra™	75	2	5	39	1515	380
ProMod®	75	13	12	300	1300	560
Propac™	75	15	20	225	500	395
Modular carbohydrate						
Moducal®	0	95	0	70	10	380
Polycose® Liquid	0	50	0	70	6	200
Polycose® Powder	0	94	0	110	10	380
Sumacal®	0	95	0	100	0	380
Modular fat						
MCT Oil	0	0	93			770
Microlipid®	0	0	50			450

Manufacturers: Clintec Nutrition Company (Carnation Instant Breakfast, Crucial, Elementra, Entrition 0.5, Entrition HN, Glytrol, Nutren 1.0, Nutren 1.0 with Fiber, Nutren 1.5, Nutren 2.0, NutriHep, NutriVent, Peptamen, Peptamen VHP, ProBalance, Replete, Replete with Fiber, Travasorb HN, Travasorb MCT, Travasorb Hepatic, Travasorb Renal, Travasorb STD); Elan Pharma (Fiberlan, Introlan, Isolan, Nitrolan, Nutrilan, Reabilan, Reabilan HN, Ultralan); Kendall McGaw Laboratories (Amin-Aid, Hepatic Aid II, Immun-Aid, Traum-Aid HBC); Mead Johnson (Boost, Casec, Criticare HN, Deliver, Isocal, Isocal HCN, Isocal HN, Lipisorb, Lonolac, MCT Oil, Moducal, Portagen,

Appendix 39 *(continued)*

Sustacal, Sustacal HC, Sustacal with Fiber, Sustagen, Traumacal, Ultracal); Nutra Balance Products (Re/Neph); Ross Laboratories (Advera, Alitraq, Alterna, Ensure, Ensure HN, Ensure Plus, Ensure Plus HN, Ensure with Fiber, Glucerna, Introlite, Jevity, Nepro, Osmolite, Osmolite HN, Perative, Polycose, ProMod, Promote, Pulmocare, Replena, Ross SLD, Suplena, TwoCal HN, Vital HN); Sandoz Nutrition (CitriSource, Citrotein, Compleat Modified, Compleat Regular, Fibersource, Fibersource HN, Impact, Impact with Fiber, IsoSource, IsoSource HN, IsoSource VHN, Isotein HN, Meritene, Precision LR, Resource Plus, Tolerex, Vivonex Plus, Vivonex T.E.N.); Sherwood Medical (Accupep HPF, Attain, Comply, Magnacal, Microlipid, Pre-Attain, Profiber, Propac, Protain XL, Sumacal, Vitaneed); Wyeth (Enercal).

Appendix 40
Selected Amino Acid Solutions for Parenteral Nutrition

General Amino Acid Solutions	Procal-amine® 3%	Amino-syn® 5%	Troph-Amine® 6.0%	Fre-Amine® 8.5%	Trava-sol® 8.5%	Amino-syn® II 10%	Nov-amine® 15%
Amino acid concentration	3.0%	5.0%	6.0%	8.5%	8.5%	10%	15%
Total nitrogen (g/100 mL)	0.46	0.79	0.93	1.43	1.43	1.53	2.37
Essential amino acids (mg/100 mL)							
Histidine	85	150	290	240	372	300	894
Isoleucine	210	360	490	590	406	660	749
Leucine	270	470	840	770	526	1000	1040
Lysine	220	360	490	620	492	1050	1180
Methionine	160	200	200	450	492	172	749
Phenylalanine	170	220	290	480	526	298	1040
Threonine	120	260	250	340	356	400	749
Tryptophan	46	80	120	130	152	200	250
Valine	200	400	470	560	390	500	960
Nonessential amino acids (mg/100 mL)							
Alanine	210	640	320	600	1760	993	2170
Arginine	290	490	730	810	880	1018	1470
Proline	340	430	410	950	356	722	894
Serine	180	210	230	500		530	592
Tyrosine		44	140		34	270	39
Glycine	420	640	220	1190	1760	500	1040
Glutamic acid			300			738	749
Aspartic acid			190			700	434
Cysteine	<20		<20	<20			
Electrolytes (mEq/L)							
Sodium	35			10	70	45.3	
Potassium	24	5.4			60		
Magnesium	5				10		
Chloride	41		<3	<3	70		
Acetate	47	86	56	73	141	71.8	151
Phosphate (mM/Liter)	3.5			10	30		
Osmolarity (mOsm/L)	735	500	525	810	1160	873	1388

(continued)

Appendix 40 *(continued)*

Disease-Specific Amino Acid Solutions	Renal Failure				Stress			
	Amin-ess®	Amino-syn® RF	Nephr-Amine®	Ren-Amin®	Branch-Amin®	Fre-Amine® HBC	Amino-syn® HBC	Hepat-Amine®
Amino acid concentration	5.2%	5.2%	5.4%	6.5%	4.0%	6.9%	7.0%	8.0%
Nitrogen (g/100 mL)	0.66	0.79	0.65	1.0	0.44	0.97	1.12	1.2
Essential amino acids (mg/100 mL)								
Histidine	412	429	250	420	—	160	154	240
Isoleucine	525	462	560	500	1380	760	789	900
Leucine	825	726	880	600	1380	1370	1576	1100
Lysine	600	535	640	450	—	410	265	610
Methionine	825	726	880	500	—	250	206	100
Phenylalanine	825	726	880	490	—	320	228	100
Threonine	375	330	400	380	—	200	272	450
Tryptophan	188	165	200	160	—	90	88	66
Valine	600	528	640	820	1240	880	789	840
Nonessential amino acids (mg/100 mL)								
Alanine	—	—	—	560	—	400	660	770
Arginine	—	600	—	630	—	580	507	600
Cysteine	—	—	<20	—	—	<20	—	<20
Glycine	—	—	—	300	—	330	660	900
Proline	—	—	—	350	—	630	448	800
Serine	—	—	—	300	—	330	221	500
Tyrosine	—	—	—	40	—	—	33	—
Electrolytes (mEq/L)								
Acetate	50	105	44	60	—	57	72	62
Chloride	—	—	<3	31	—	<3	≤40	<3
Phosphate (mM/L)	—	—	—	—	—	—	—	10
Potassium	—	5.4	—	—	—	—	—	—
Sodium	—	—	5	—	—	10	7	10
Osmolarity (mOsm/L)	416	475	435	600	316	620	665	785

Manufacturers: Abbott (Aminosyn, Aminosyn II Aminosyn RF, Aminosyn HBC); Clintec Nutrition (Aminess, BranchAmin, RenAmin, Travasol); KabiVitrum (Novamine); Kendall McGaw (FreAmine, FreAmine HBC, HepatAmine, NephrAmine, Procalamine, TrophAmine).

Appendix 41
Intravenous Fat Emulsions

	Intralipid 10%	Liposyn® II 10%	Liposyn® III 10%	Nutrilipid 10%
Fat source (%)				
Soybean oil	10	5	10	10
Safflower oil		5		
Fatty acid content (%)				
Linoleic	50	66	55	49–60
Oleic	26	18	22	21–26
Palmitic	10	9	11	9–13
Linolenic	9	4	8	6–9
Stearic	3.5	3	4	3–5
Egg yolk phospholipid (%)	1.2	1.2	1.2	1.2
Glycerin (%)	2.25	2.5	2.5	2.21
Calories/mL	1.1	1.1	1.1	1.1
Osmolarity (mOsmol liter)	260	276	292	280

Manufacturers: Abbott Laboratories (Liposyn); Clintec Nutritional (Intralipid); Kendall McGaw (Nutrilipid).

Appendix 42
Suggested Intravenous Multivitamin Formulation

	Adults and Older Children[a]	Infants and Children under 11 years[b]
Vitamin A, IU	3300	2300
Vitamin D, IU	200	400
Vitamin E, IU	10	7
Vitamin K, mg	—	0.2
Vitamin C, mg	100	80
Vitamin B_1, mg	3	1.2
Vitamin B_2, mg	3.6	1.4
Vitamin B_6, mg	4	1
Vitamin B_{12}, μg	5	1
Folic acid, μg	400	140
Niacin, mg	40	17
Biotin, μg	60	20
Pantothenic acid, mg	15	5

[a]M.V.I.-12, M.V.I. Plus, M.V.C. 9 + 3, Berocca PN;
[b]M.V.I. Pediatric
Source: American Medical Association, Nutrition Advisory Group.

Appendix 43
Caloric Values and Osmolarities of Intravenous Dextrose Solutions

Percent Dextrose	Calories per Liter	Calculated Osmolarity
2.5	85	126
5	170	250
7.7	260	390
10	340	505
11.5	390	580
20	680	1010
25	850	1330
30	1020	1515
38	1290	1920
38.5	1310	1945
40	1360	2020
50	1700	2525
60	2040	3030
70	2380	3530

Reprinted with permission from McEvoy, G. K.: AHFS (American Hospital Formulary Service) drug information 90. American Society of Hospital Pharmacists, Inc., Bethesda, MD, 1990:1442.

Appendix 44
Nutrition Therapy in Inborn Errors of Metabolism

Disorder or Enzyme Deficiency	Nutritional Intervention
Disorders in amino acid metabolism	
Acetylglutamate synthetase deficiency	Protein restriction; arginine supplementation
Alkaptonuria	Phenylalanine and tyrosine restriction; ascorbic acid supplementation
Arginase deficiency	Protein-free diet; supplementation with amino acids, except arginine
Argininemia	Protein restriction; essential amino acids and ornithine supplementation
Argininosuccinic aciduria	Protein restriction; arginine, sodium benzoate, and essential amino acid supplementation
Beta-methylcrotonylglycinuria	Leucine restriction
Branched chain alpha-ketociduria	BCAA (branched-chain amino acid) restriction and thiamin supplementation
Branched-chain transaminase deficiency	Protein restriction
Carbamylphosphate synthetase deficiency	Protein restriction; sodium benzoate and essential amino acid supplementation
Chediak–Higashi syndrome	Ascorbic acid supplementation
Citrullinemia	Protein restriction; essential amino acids, arginine, and benzoic acid supplementation
Cystathionine synthetase deficiency	Methionine restriction; pyridoxine, folic acid, and cysteine supplementation
Cystathioninuria	Pyridoxine supplementation
Glutaric acidemia I	Protein restriction; carnitine and riboflavin supplementation
Glutaric acidemia II	Protein and fat restriction; riboflavin and carnitine supplementation
Glutathionine synthetase deficiency	Vitamin E
Histidase deficiency	Histidine restriction
Homocystinuria	Methionine restriction; cysteine, betaine, folate, and pyridoxine supplementation betaine, and pyridoxine
Hydroxy-methyl glutaryl CoA deficiency	Protein restriction
Hyperornithinemia	Protein and arginine restriction; lysine and pyridoxine supplementation
Hyperphenylalaninemia	Phenylalanine restriction
Hypervalinemia	Valine restriction
Isovaleric acidemia	Protein and/or leucine restriction; carnitine and glycine supplementation
β-ketothiolase deficiency	Moderate protein and isoleucine restriction
Leucine-induced hypoglycemia	Moderate protein (leucine) restriction; small, frequent feedings
Lysine intolerance	Protein restriction
Maple syrup urine disease	Leucine, valine, and isoleucine restriction; thiamin supplementation
Methylcrotonyl CoA carboxylase deficiency	Leucine restriction

Appendix 44 *(continued)*

Disorder or Enzyme Deficiency	Nutritional Intervention
Methylmalonic acidemia	Protein restriction (methionine, threonine, valine, and isoleucine free; vitamin B_{12} supplementation
Multiple carboxylase deficiency	Protein restriction; biotin supplementation
Ornithine transcarbamylase deficiency	Protein restriction; arginine, essential amino acids, and benzoic acid supplementation
Phenylketonuria	Phenylalanine restriction; tyrosine supplementation
Propionic acidemia	Isoleucine, methionine, threonine, and valine restriction; biotin supplementation
Pyroglutaric acidemia	Protein restriction; alkali
Tyrosinemia	Phenylalanine and tyrosine restriction; high-calorie diet; vitamin D supplementation with rickets
Disorders in carbohydrate metabolism	
Fructose-1,6-diphosphatase deficiency	Fructose restriction; frequent glucose feeding; folate supplementation
Fructose-1-phosphate aldolase deficiency	Exclusion of fructose, sucrose, and sorbitol
Galactosemia	Exclusion of galactose and lactose
Glucose 6-phosphate dehydrogenase deficiency	Avoidance of fava beans
Glycerate dehydrogenase deficiency	High-phosphate diet; pyridoxine supplementation
Glycogen storage disease	
Type I (glucose-6-phosphatase deficiency)	Frequent feeding; high carbohydrate intake
Type III (amylo-1,6-glucosidase deficiency)	Frequent glucose feeding; high protein intake
Type V (muscle phosphorylase deficiency)	Intravenous glucose; high carbohydrate intake
Type VI (liver phosphorylase deficiency)	Frequent glucose feedings
Type VIII (phosphorylase kinase deficiency)	Avoidance of fasting; high protein intake
Pyruvate carboxylase deficiency	Frequent feeding; biotin and thiamin supplements
Pyruvate dehydrogenase deficiency	Ketogenic diet; thiamin supplementation
Disorders in lipid metabolism	
Abetalipoproteinemia	Low-fat with medium chain triglycerides (MCT); vitamin A, E, and K supplementation
Acyl-CoA dehydrogenase deficiency	Fat restriction; riboflavin and carnitine suppl.
Apolipoprotein C-II deficiency	Moderate fat restriction
Familial cholesterol ester deficiency	Restricted fat diet
Familial hyperlipoproteinemias	Restriction of saturated fatty acids and cholesterol; nicotinic acid
Type I (hyperchylomicronemia)	Low-fat, high-protein diet; use of medium-chain triglycerides (MCT); no alcohol; weight control

(continued)

Appendix 44 *(continued)*

Disorder or Enzyme Deficiency	Nutritional Intervention
Type II (hypercholesterolemia)	Saturated fat and cholesterol restriction; use of unsaturated fats (MUFAs and PUFAs); nicotinic acid and high fiber
Type III (broad beta disease)	Cholesterol restriction; no concentrated sweets; PUFAs preferred; weight control/maintenance
Type IV (hyperprebetalipoproteinemia)	Weight control; moderate cholesterol and saturated fat restriction; controlled carbohydrate
Type V (mixed hyperlipidemia)	Calorie, carbohydrate, saturated fat, and cholesterol restriction; nicotinic acid; no alcohol
Glucocerebrosidase deficiency	Iron and vitamin supplementation
Hypobetalipoproteinemia	Long-chain triglyceride restriction; vitamins A, E, and K supplementation; medium-chain triglycerides
Lipoprotein lipase deficiency	Low-fat diet; medium chain triglycerides
Refsum's disease	Phytanic acid restriction
Beta-sitosterolemia	Plant sterol restriction
Tangier disease	Fat restriction
Disorders in nucleic acid metabolism	
Adenine phosphoribosyltransferase deficiency	Dietary purine restriction; high fluid intake
Gout	Weight reduction; low purine; high fluids; alcohol restriction; alkalinized urine
Myoadenylate deaminase deficiency	High-ribose diet
Orotic aciduria	Large doses of uridine
Xanthinuria (xanthine oxidase deficiency)	Purine restriction; high fluid intake; alkali
Disorders in transport	
Abetalipoproteinemia	Long-chain triglyceride restriction; MCT and vitamins A, D, E, and K supplementation
Acrodermatitis enteropathica	Zinc supplementation
Cystinosis	Methionine restriction; cysteamine, vitamin C, vitamin D, and phosphate supplementation
Cystinuria	High fluid; bicarbonate to alkalinize urine
Dibasic aminoaciduria	Protein restriction; arginine supplementation
Fanconi syndrome	Vitamin D and phosphate supplementation; control of acidosis with bicarbonate
Folic acid transport defect	Parenteral folate supplementation
Glucose–galactose malabsorption	Glucose and galactose restriction; fructose substituted for glucose
Glutamate–aspartate transport defect	Glutamine supplementation
Hartnup disease	Nicotinamide supplementation; high protein intake
Hypobetalipoproteinemia	Long-chain triglyceride (LCT) restriction; use of MCT; vitamins A, D, E, and K supplementation
Hypomagnesemia, idiopathic	Magnesium supplementation
Hypophostatemic rickets	Phosphorus and vitamin D supplementation

Appendix 44 *(continued)*

Disorder or Enzyme Deficiency	Nutritional Intervention
Lactose intolerance (primary alactasia)	Lactose restriction
Menke's kinky hair syndrome	Parenteral copper administration
Methionine malabsorption	Methionine restriction; cysteine supplementation
Pseudohypoaldosteronism, Type I	Sodium supplementation
Pseudohypoaldosteronism, Type II	Chloride restriction
Renal tubular acidosis, Type I	Potassium and bicarbonate supplementation
Renal tubular acidosis, Type II	Potassium restriction
Sucrose-isomaltose intolerance	Sucrose restriction
Transcobalamin II deficiency	Vitamin B_{12} parenterally
Vitamin D-dependent rickets, Type I	Vitamin D (1,25-DHCC)
Other disorders	
Congenital erythropoietic porphyria	Beta-carotene supplementation
Ehlers–Danlos syndrome	Ascorbic acid supplementation
Ferrochelatase deficiency	Caloric restriction; beta-carotene supplementation
Hereditary coproporphyria	High-carbohydrate diet
Hydroxykynureninuria	Nicotinic acid supplementation
Methemoglobinemia	Diet free of nitrate and nitrite; ascorbic acid and riboflavin supplementation
Porphyria (acute, intermittent)	High carbohydrate with glucose supplementation
Porphyria cutanea tarda	Low alcohol; pyridoxine and vitamin E supplementation
Pyridoxine-dependent seizures	Pyridoxine parenterally
Wilson's disease	Copper restriction

Appendix 45
Common Prefixes, Suffixes, and Symbols

Prefixes

a, an	without; as avitaminosis
ab	away from; as abnormal
ad	near, toward; as adrenal
ana	upward; as anabolism
anti	against; as antibiotic
auto	self; as autodigestion
bio	life; as biology
calor	heat; as calorimeter
cata	downward; as catabolism
chole	bile, gall; as cholagogue
chroma	color; as chromatosis
co	together; as coenzyme
di	two, double; as diplopia
dis	ill, negative; as disease
dys	difficult; as dyspepsia
ec	outside; as ectopic
encephal	brain; as encephalogram
endo	inside; as endogenous
exo	outside; as exogenous
hemo	blood; as hemopoiesis
hyper	excessive; as hyperacid
hypo	little; as hypofunction
im	not; as immature
in	not; as incurable
inter	between; as interstitial
intra	within; as intravascular
meta	change; as metaplasia
necro	dead; as necrosis
para	beside; as paravertebral
peri	around; as pericardium
post	after; as postmortem
pre	before; as prenatal
syn	union; as synthesis

Suffixes

algia	pain; as neuralgia
ase	enzyme; as amylase
cide	kill; as bactericide
clysis	drenching; as venoclysis
cule	small; as molecule
cyte	cell; as erythrocyte
ectomy	cut off; as appendectomy
emesis	vomiting; as hematemesis
emia	blood; as anemia
esthesia	sensation; as anesthesia
ism	condition; as alcoholism
itis	inflammation; as appendicitis
lysis	destruction; as hemolysis
malacia	softening; as osteomalacia
oma	tumor, swelling; as adenoma
opsy	to view; as biopsy
osis	condition; as tuberculosis
pathy	disease of; as neuropathy
penia	poverty; as leucopenia
phagia	to eat; as polyphagia
phil	to love; as basophil
phobia	fear of; as photophobia
pnea	breath; as hyperpnea
poiesis	to produce; as hemopoiesis
ptysis	to spit; as hemoptysis
rrhea	to discharge; as diarrhea
tomy	to cut; as vagotomy
trophy	growth; as hypertrophy
uria	urine; as glucosuria

Symbols

(+)	significant; uncommon
+	plus; positive; present
++	trace or notable reaction
+++	moderate amount or reaction
++++	large amount or reaction
(−)	insignificant
−−	minus; negative; absent
±	more or less; with or without
=	equal to
≠	not equal; unequal
≡	identical
≢	not identical
↑ or ∧	elevated; above; enlarged
↓ or ∨	decreased; below; depressed
↗	increasing
↘	decreasing
⇒	implies; implication
>	greater than; leads to
<	less than; caused by
≯	not greater than
≮	not less than
≥	greater than or equal
≤	less than or equal
~	about; approximately
≈	approximately equal
::	proportionate to
Δ	change
∴	therefore

Appendix 46
Common Abbreviations in Nutrition and the Medical Records

a	ante; before	BCAA	branched-chain amino acids
aa	each; of each	BCG	bacillus Calmette-Guerin
AAA	aromatic amino acid	BE	barium enema
abd	abdominal; abdomen	BEE	basal energy expenditure
ABG	arterial blood gases	b.i.d.	vis in die; twice daily
ac	ante cibum; before meals	bilat	bilateral
ACF	acute care facility	BM	bowel movement; bone marrow
ACTH	adrenocorticotropic hormone	BKA	below knee amputation
ADH	antidiuretic hormone	BMI	body mass index
ADI	averge or acceptable daily intake	BMR	basal metabolic rate
		BMT	bone marrow transplantation
ADL	activities of daily living	bp	boiling point
ad lib	ad libitum; as desired	BP	blood pressure
Adm	admission	BPH	benign prostatic hypertrophy
ADMR	average daily metabolic rate	BR	bed rest
ADR	adverse drug reaction	BRP	bathroom privileges
AF	atrial fibrillation	BS	blood sugar
A/G	albumin/globulin ratio	BSA	body surface area
AHD	arteriosclerotic heart disease	BSL	blood sugar level
AID	acute infectious disease	BSP	bromsulphalein
AIDS	acquired immune deficiency syndrome	BTL	bilateral tubal ligation
		BUN	blood urea nitrogen
AKA	above-knee amputation	BV	blood volume; biological value
alb	albumin	BW	body weight
alk	alkaline	Bx	biopsy
ALS	amyotrophic lateral sclerosis	c̄	cum; with
AMA	arm muscle area	C	centigrade
amp	ampule	C_{cr}	creatinine clearance
amt	amount	Ca	calcium; cancer; carcinoma
AODM	adult-onset diabetes mellitus	CA	chronological age
AP	anterior–posterior; angina pectoris	CAD	coronary artery disease
		CAH	chronic active hepatitis
APC	arterial premature contraction	Cal	large calorie
approx	approximately	CALD	chronic active liver disease
Aq	aqua; water	cap	capsule
ARC	AIDS-related complex	CAPD	continuous ambulatory peritoneal dialysis
ARDS	adult respiratory distress syndrome		
		CAT	computerized axial tomography
ARF	acute renal failure; acute rheumatic fever	cath	catheterize
		CBC	complete blood count
A.R.T.	accredited record technician	CBD	common bile duct
ASHD	arteriosclerotic heart disease	cc	cubic centimeter
as tol	as tolerated	CC	chief complaint
A & W	alive and well	CCF	cephalin–cholesterol flocculation
B	born; basophils		
Ba	barium	CCK	cholecystokinin

(continued)

Appendix 46 *(continued)*

CCU	coronary care unit or critical care unit	DF	dietary fiber
		DI	diabetes insipidus
c.d.	cane die; daily	Diag, Dx	diagnosis
CHD	coronary heart disease	Diff	differential
CHF	congestive heart failure	dil	dilute
CHI	creatinine height index	disc	discontinue
CHO	carbohydrate	Disch	discharge
Chol	cholesterol	DJD	degenerative joint disease
chr	chronic	DKA	diabetic ketoacidosis
ck	check	dl	deciliter
CK	creatine kinase	DL	danger list
Cl	chloride	DM	diabetes mellitus
cm	centimeter	DNR	do not resuscitate
CNS	central nervous system	D_2O	deuterium or heavy water
c/o	complains of	DOA	dead on arrival
CO_2	carbon dioxide	DOE	dyspnea on exertion
COLD	chronic obstructive lung disease	DOS	day of surgery
comp	compound	DPM	discontinue previous medication
conc	concentration	DPT	diphtheria, pertussis, tetanus
con	continued	dr	dram; drachm; 3.8 g
COPD	chronic obstructive pulmonary disease	DR	delivery room
		DRG	diagnostic related groups
C.O.T.A.	certified occupational therapy assistant	D/S	dextrose and saline
		DSD	dry sterile dressing
CNSD	certified nutrition support dietitian	DT	delirium tremens
		d/t	due to
Cpd	compound	D.T.R.	dietetic technician registered
CPK	creatine phosphokinase	DU	duodenal ulcer
CPN	central parenteral nutrition	DVT	deep vein thrombosis
CPR	cardiopulmonary resuscitation	DW	distilled water
Creat	creatinine	Dx	diagnosis
CRF	chronic renal failure	e	et; and
CRI	chronic renal insufficiency	ea	each
crit	hematocrit	EAA	essential amino acid
C & S	culture and sensitivity	E, EOS	eosinophils
C/S	cesarean section	ECG, EKG	electrocardiogram
CSF	cerebrospinal fluid		
CT	computerized tomography	ED	emergency department
C.T.R.S.	certified therapeutic recreation specialist	EDC	expected date of confinement
		EEG	electroencephalogram
CV	cardiovascular	EFA	essential fatty acid
CVA	cerebrovascular accident	Elix	elixir
CVP	central venous pressure	ENT	ear, nose, and throat
CVS	cardiovascular system	ER	emergency room
Cw	crutch walking	ESR	erythrocyte sedimentation rate
d	daily	ESRD	end-stage renal disease
db	diabetic	ETOH	ethyl alcohol
DBW	desirable body weight	exp	expired
D & C	dilatation and curettage	Expl Lap	exploratory laparotomy
D/C	discontinue; discharge	ext	external

Appendix 46 *(continued)*

extr	extract	HH	hiatal hernia
F	father; female; Fahrenheit	HIV	human immunodeficiency virus
FA	fatty acid	H & N	head and neck
FBS	fasting blood sugar	HO	house officer
fdg	feeding	H/O	history of
Fe	iron	HOB	head of bed
FFA	free fatty acid	HNV	has not voided
FH	family history; fetal heart	H & P	history and physical
Fib	fibrillation	HPI	history of present illness
fld	fluid	HPN	home parenteral nutrition
fl oz	fluid ounce	HPT	hyperparathyroidism
FMH	family medical history	HR	heart rate
FSH	follicle-stimulating hormone	h.s.	hora somni; at bedtime
ft	foot or feet	Ht	height
FTT	failure to thrive	HTN	hypertension
F/U	follow-up	ht, hgt	height
FUO	fever of unknown origin	hx	hospitalization; history
Fx	fracture	I	iodine
GA	gastric analysis	IBD	inflammatory bowel disease
gal	gallon	ibid	same as before
GB	gallbladder	IBW	ideal body weight
GBD	gallbladder disease	Ict	icterus index
GE	gastroenteritis; gastroenterology	ICU	intensive care unit
GBS	gallbladder series	I & D	incision and drainage
GC	gonococcal count	IDDM	insulin-dependent diabetes mellitus
GE	gastroenteritis		
GFR	glomerular filtration rate	IDL	intermediate-density lipoprotein
GI	gastrointestinal	IF	intrinsic factor
GIT	gastrointestinal tract	IHD	ischemic heart disease
g	gram; 15.43 grains	IM	intramuscular
GN	glomerulonephritis	Imp	impression
gr	grain	IMV	intermittent mechanical ventilation
GTF	glucose tolerance factor		
gtt(s)	gutta; drop(s)	incl	include
GTT	glucose tolerance test	int	internal
GU	genitourinary	I & O	intake and output
Gyn	gynecology	IPPB	intermittent positive pressure breathing
h	hour		
H & H	hemoglobin and hematocrit	irrig	irrigation
H & P	history and physical	IU	international unit
HA	headache; hyperalimentation	IV	intravenous
Hb, Hgb	hemoglobin	IVH	intravenous hyperalimentation
HBP	high blood pressure	IVP	intravenous pyelogram
Hct	hematocrit	J	joule
HCVD	hypertensive cardiovascular disease	K	potassium
		kcal	kilocalories
HDL	high-density lipoprotein	kg	kilogram
HEN	home enteral nutrition	kJ	kilojoule

(continued)

Appendix 46 *(continued)*

KUB	kidney, ureter, and bladder	MI	mitral insufficiency; myocardial infarction
L	liter	mL	milliliter
LAP	laparotomy	mM	millimole
lat	lateral	Mn	manganese
lb	pound	MO	mineral oil; month
LBBB	left bundle branch block	M.O.	medical officer
LBP	low back pain	MOM	milk of magnesia
LBM	lean body mass	mOsmol	milliosmole
LBW	low birth weight; lean body weight	M & R	measure and record
LCT	long-chain triglyceride	M.R.A.	medical record administrator
LDH	lactate dehydrogenase	MRI	magnetic resonance imaging
LDL	low-density lipoprotein	MS	mitral stenosis; multiple sclerosis
LES	lower esophageal sphincter	M.T.	medical technologist
LFT	liver function test	MVA	motor vehicle accident
LH	luteinizing hormone	MVI	multiple vitamin infusion
liq	liquid	MVR	mitral valve replacement
LLE	left lower extremity	n	nocte; night; normal
LLL	left lower lobe	N	nitrogen
LLQ	left lower quadrant	Na	sodium
LMP	last menstrual period	NAD	no apparent distress
LOC	loss of consciousness	N.B.	newborn
LOS	length of stay	NC	noncontributory
LP	lumbar puncture	NDF	no diagnostic findings
L.P.N.	licensed practical nurse	NE	niacin equivalent
LR	labor room	NEFA	nonesterified fatty acid
LTT	lactose tolerance test	neg	negative
LUE	left upper extremity	ng	nanogram
LUL	left upper lobe	N/G, NG	nasogastric
LUQ	left upper quadrant	NIDDM	non-insulin-dependent diabetes mellitus
lym	lymphocyte	nil	nothing
lytes	electrolytes	NKA	no known allergy
m	minim; meter; male	NP	neuropsychiatry
M	mother; monocyte; male	N.P.	nurse practitioner
MAC	midarm circumference	NPH	neutral protamine Hagedorn
MAMC	midarm muscle circumference	NPN	nonprotein nitrogen
MAO	monoamine oxidase	NPO	nil per os; nothing by mouth
M & N	mone et nocte; day and night	NS, NSS	normal saline solution
MCFA	medium-chain fatty acid	NSR	normal sinus rhythm
mcg	microgram	N & T	nose and throat
MCH	mean corpuscular hemoglobin	NTG	nitroglycerin
MCHC	mean corpuscular hemoglobin concentration	NTS	nontropical sprue
MCT	medium-chain triglyceride	N & V	nausea and vomiting
MCV	mean corpuscular volume	NWB	no weight bearing
MDR	minimum daily requirement	O_2	oxygen
mEq	milliequivalent	OB	obstetrics; occult blood
Mg	magnesium	OBS	organic brain syndrome
mg	milligram		

Appendix 46 *(continued)*

o.d.	daily	PID	pelvic inflammatory disease
O.D.	oculus dexter; right eye	PKU	phenylketonuria
OGTT	oral glucose tolerance test	PM	postmortem
oint	ointment	PMD	private medical doctor
OM	omne mone; every day	PMH	past medical history
ON	omne nocte; every night	PMI	post myocardial infarction
OOB	out of bed	PMP	previous menstrual period
OOR	out of room	PMS	premenstrual syndrome
O & P	ova and parasites	PND	paroxysmal nocturnal dyspnea
OPD	outpatient department	PNI	prognostic nutritional index
ophth	ophthalmology	po	per os; by mouth; orally
OR	operating room	POMR	problem-oriented medical record
ORIF	open reduction internal fixation	pos	positive
Orth(o)	orthopedics	Post	posterior
O.S.	oculus sinister; left eye	pp	postprandial; postpartum
O.T.	occupational therapy	PPBS	postprandial blood sugar
OTC	over the counter	PPD	purified protein derivative
O.T.R.	occupational therapist registered	PPN	peripheral parenteral nutrition
O.U.	both eyes; each eye	pr	per rectum
oz	ounce	prn	prore nata; whenever necessary
p	post; after	Pro	protein
P	pulse; phosphorus	prog	prognosis
PA	posterior–anterior (x-ray film); pernicious anemia	PS	pulmonary stenosis
		PSE	portal systemic encephalopathy
P.A.	physician assistant	PSP	phenolsulfonphthalein
PAC	premature atrial contraction	P/S ratio	polyunsaturated/saturated fatty acid ratio
PAF	paroxysmal atrial fibrillation		
PAME	preanesthesia medical exam	pt	patient
Pap smear	Papanicolaou smear	P.T.	physical therapy; prothrombin time
PAT	pregnancy at term; paroxysmal atrial tachycardia		
		PTA	prior to admission
path	pathology	PTB	pulmonary tuberculosis
PBI	protein-bound iodine	PTH	parathyroid hormone
pc	post cibum; after meals	PTT	partial thromboplastin time
PCM	protein calorie malnutrition	PU	peptic ulcer
PCV	packed cell volume	PUD	peptic ulcer disease
PE	physical examination	PUFA	polyunsaturated fatty acid
Ped	pediatrics	PVA	peripheral venous alimentation
PEG	percutaneous endoscopic gastrostomy	PVC	premature ventricular contractions
PEM	protein-energy malnutrition	PVD	peripheral vascular disease
PER	protein efficiency ratio	PWB	partial weight bearing
PERRLA	pupils are equal, round, regular, react to light and accommodation	PZI	protamine zinc insulin
		q	quaque; every
		q.d.	quaque die; every day
pg	picogram	q.h.	quaque hora; every hour
pH	hydrogen ion concentration	q.i.d.	quater in die; four times a day
PH	past history	ql	quantum libit; as much as desired
PI	present illness		

(continued)

Appendix 46 *(continued)*

qn	quaque nocte; every night	SH	social history
qns	quantity not sufficient	sibs	brothers and sisters
qod	every other day	SIDS	sudden infant death syndrome
qoh	every other hour	sig	sign; write or label
qon	every other night	SL	sublingual
qs	quantum sufficiat; quantity sufficient	SLE	systemic lupus erythematosis
qt	quart	SOAP	subjective, objective, assessment, plan
R	right; rectal	SOB	shortness of breath
RA	rheumatoid arthritis	Sol	solution
RAI	radioactive isotope	S.O.S.	if it is necessary
RBBB	right bundle branch block	S/P	status postop
RBC, rbc	red blood cell	spec	specimen
RBP	retinol binding protein	sp gr	specific gravity
R.D.	registered dietitian	ss	one half
RDA	recommended dietary allowance	S & S	signs and symptoms
RE	retinol equivalent	SSE	soapsuds enema
REE	resting energy expenditure	SSS	subscapular skinfold
RF	renal failure; rheumatic fever	Staph	staphylococcus
RHD	rheumatic heart disease	stat	immediately
RLE	right lower extremity	STD	sexually transmitted disease
RLL	right lower lobe	Strep	streptococcus
RLQ	right lower quadrant	supp	suppository
R.N.	registered nurse	S.W.	social worker
R/O	rule out	Sx	symptoms
ROM	range of motion	Sy	syphilis
ROS	review of systems	T3	triiodothyronine
R.P.T.	registered physical therapist	T4	thyroxine
RQ	respiratory quotient	T & A	tonsils and adenoids
RR	recovery room; respiratory rate	tab	tablet
RT	radiation therapy	TB, TBC	tuberculosis
R.T.	respiratory therapist	TBW	total body weight
RTA	renal tubular acidosis	TCR	turn, cough, and rebreathe
RTC	return to clinic	TF	tube feeding
RUE	right upper extremity	TG	triglyceride
RUL	right upper lobe	THA	total hip arthroplasty
RUQ	right upper quadrant	TIA	transient ischemic attack
Rx	treatment; take	TIBC	total iron-binding capacity
$\bar{s}$	without	t.i.d.	three times a day
SAH	subarachnoid hemorrhage	tinct	tincture
SBE	subacute bacterial endocarditis	TKA	total knee arthroplasty/ amputation
SBO	small bowel obstruction		
sc	subcutaneous	TLC	total lymphocyte count; total lung capacity
SCI	spinal cord injury		
SDA	specific dynamic action	TNA	total nutrient admixture
sed rate	sedimentation rate	T.O.	telephone order
SGOT	serum glutamic oxaloacetic transaminase	TP	total protein
		TPN	total parenteral nutrition
SGPT	serum glutamic pyruvic transaminase	TPR	temperature, pulse, respiration
		Trach	tracheostomy

Appendix 46 *(continued)*

TSF	triceps skinfold
TSH	thyroid-stimulating hormone
TTT	thymol turbidity test
TUR	transurethral resection
TURP	transurethal prostatectomy
TWE	tap water enema
Tx	treat
U/A	urinary analysis
UGI	upper gastrointestinal
UIBC	unsaturated iron-binding capacity
U.R.	utilization review
URI	upper respiratory infection
Urol	urology
USP	*United States Pharmacopoeia*
UTI	urinary tract infection
UUN	urinary urea nitrogen
VC	vital capacity
VD	venereal disease
VF	ventricular fibrillation
VH	vaginal hysterectomy
vit	vitamin
VLDL	very-low-density lipoprotein
VMA	vanillylmandelic acid
VO	verbal order
VNS	visiting nurse service
VP	venous pressure
VPC	ventricular premature contraction
VS	vital signs
VT	ventricular tachycardia
WB	weight-bearing
WBC	white blood count; white blood cell
wdwn	well developed, well nourished
WF	white female
WM	white male
WNL	within normal limits
wt, wgt	weight
WU	work-up
YO	year old
Zn	zinc

Appendix 47A
Cultural Food Practices[a]

Ethnic Group	Characteristic Foods and Cooking Methods
African-American ("soul food")	Corn, cornmeal; rice; hominy grits; beans; meats and fish; salt pork, sausage; lard and shortening; potatoes and yams; melons, berries; peaches; pecans; green leafy vegetables. Popular cooking methods are: boiling, braising, frying, use of gravies.
Chinese-American	Rice and rice gruel; wheat mungbean, rice noodles; soybean products such as tofu, soybean milk, soy sauce; peanuts, almonds, chestnuts; bamboo shoots, sweet potatoes; radishes; yellow and green onions; peapods; mushrooms; ginger; pickled vegetables; sea foods; tropical fruits (lychees, longans, mangoes, papayas, bananas); tiny portions of meat, fish with bones, tea as beverage. Popular cooking methods are: steaming, boiling, stir frying, frying in sesame or peanut oil.
Italian	Ricotta, mozzarella, parmesan, gorgonzola cheeses; figs; grapes; tangerines; tomatoes; olives and olive oil; eggplants; artichoke; zucchini; pastas; rice; polenta; Italian bread; pizza; breadsticks; ground beef, veal, chicken, sausages, seafood. Popular cooking methods are: baking, braising with tomato sauce, sauteing in olive oil.
Japanese-American	Rice, noodles, barley, oat; seaweeds; miso, tofu, bean paste, mungbean, and soy sauce; tropical fruits; fish with bones; seafood; sugar as seasoning; ginseng; fresh vegetables and salads. Popular cooking methods are: steaming, grilling, boiling.
Mexican-American	Goat cheese, evaporated milk, flans (custards), eggs; prickly pears, bananas, peppers, especially chili, corn, melons; oranges; tomatoes and salsas, avocados; tortillas; tacos; rice; macaroni; sweet rolls; cornmeal; hominy; various beans; meat and sausages; fish; poultry; lard; sweet chocolate and coffee drinks; assorted Mexican baked goods. Popular cooking methods are: frying, grilling, baking, steaming.
Middle-Eastern	Cow and goat milk; feta cheese; custards; yogurt; cucumber; eggplant; okra; tomatoes; grape leaves; zucchini; lemons; apricots; dates; figs; rice; cracked wheat; millet; pita bread; lamb; poultry; saltwater fish; fish roe; squid; octopus; eggs; chickpeas and beans. Popular cooking methods are: baking, steaming, stewing.
Puerto Rican	Steamed white rice; various beans; wheat breads; starchy vegetables such as yuccas, yams, breadfruit, green bananas; green peppers; tomatoes; garlic; dried, salted fish; chicken; pork; lard; olive oil; jams and jellies; sweet pastries; sugared fruit juices; coffee. Popular cooking methods are: boiling, baking, frying and brasing with "sofrito" (an orange-colored sauce high in fat).
Seminole Indians	Corn; cornmeal; coontie (flour from a palmlike plant); squashes; papayas; beef may be substituted for exotic traditional meats, such as alligator, duck, snake, and wild hog. Popular cooking methods are: roasting, boiling, baking.

Vietnamese-American	Rice and noodles; fresh breads with butter; hot peppers; curry dishes; potatoes; tropical fruits and vegetables; small portions of poultry; eggs, fish pates; nuoc nam (a strong, fermented fish sauce); sweets; sweetened drinks; coffee; tea. Popular cooking methods are: similar to Chinese and Japanese.

[a]Selected ethnic groups. This is only a partial list.

Appendix 47B
Religious Food Practices[a]

Religious Group	Characteristic Foods and Cooking Methods
Greek Orthodox	No animal products for 40 days before Christmas and Easter. Some fast only on Wednesday/Friday 40 days before Christmas and Easter.
Jewish ("Kosher Foods")	No pork products, including animal shortening and ordinary gelatin. No fish without scales or fins, such as eels, shellfish, etc. No food containing blood. No meat or poultry in combination with dairy products. No dairy containing beverages until 6 hours after eating meat or poultry. Orthodox Jewish people eat only foods having kosher markings, which indicate inspection by a rabbi to ensure proper processing and packaging laws are followed. A kosher kitchen has separate sets of dishes, pots, and so forth, for meat or poultry and for milk meals. Special cleaning and special dishes are required for certain holidays. To ensure no prior use of milk or meat containing meals that are eaten away from home, disposable containers are used. Some foods (eggs, oils, grains, fish, fruits, vegetables) are considered PARVE (neutral) and may be eaten with either dairy or meat meals. There is a 24-hour fast on Yom Kippur, except for those whose health is compromised. During Passover matzo (unleavened bread) is eaten.
Mormons	No alcoholic beverage. No stimulant beverages, such as caffeine-containing drinks.
Muslims	No pork products. No meat slaughtered by someone other than a Jew, Muslim, or Christian. (Exceptions can be made only if matter of life/death.) No alcoholic beverages or other intoxicants, except for medical reasons. Use of stimulants such as caffeine is discouraged. Fasting during the 9th month of the Islamic lunar calendar (Ramadan) from dawn to dusk.
Roman Catholic	Traditional Catholics still do not eat meat on Fridays, Ash Wednesday, or Good Friday during the Lent season (40 days before Easter). Exceptions are young children below 7 years of age, pregnant and lactating women, or any person whose health could be jeopardized from the fast.
Russian Orthodox	No animal products for 40 days before Christmas and Easter. Some abstain from meat and poultry.
Seventh Day Adventist	No pork products. No meat broth or blood. No shellfish, alcoholic beverages, or stimulants such as coffee. Substitute coffee/tea with hot-based cereal beverages. Some do not use dairy products or eggs, unless they are lacto-ovovegetarian. Use of spices and snacking is discouraged. Light evening meal is preferred.

[a]Selected religious groups.

Appendix 48
Public Health Nutrition Programs and Surveys (USA)[a]

Name of Program or Survey (Administering Agency)
Description or Main Features

Brown Bag Program (USDA)
Surplus food products given free to low-income seniors.

Cancer Prevention Awareness Program (NIH/NCI)
Cancer prevention education for U.S. adults ongoing since 1984.

Child care food program (USDA)
Offers nutritious meals to children enrolled in child- or day-care centers.

Commodity Supplemental Food Program or CSFP (USDA)
Provides commodities to low-income children, mothers, and senior citizens.

Congregate Meal Program (DHHS)
Provides nutritious meals to senior citizens in churches, civic centers, and schools with educational and social activities.

Expanded Food and Nutrition Education Program or EFNEP (USDA)
Provide education for low-income families at nutritional risk; under the auspices of local Co-op extension programs and land grant universities.

Food Basket/Food Vouchers (USDA)
Awarded to eligible indigent families; vouchers are used to buy food at a grocery store.

Food Cooperatives/Food Co-ops or SHARE (USDA)
Food worth $13 to $15 given to individuals who served 2 hours for the Co-op.

Food Labeling and Safety (FDA, FTC, USDA)
See Appendices 5 and 6.

Food Stamps (USDA)
Free to U.S. citizens and some legal aliens who qualify under specified criteria to purchase food in participating stores.

Head Start (DHHS)
Preschool children are provided with comprehensive health, educational, nutrition, social, and other services. Parents attend educational sessions.

Healthy People, Promoting Health and Preventing Disease Program (ODPHP)
Coordinates national programs on health promotion and disease prevention. Implements Health Objectives for the Year 2000. See Appendix 7B.

Home-Delivered Meals or Meals-on-Wheels (USDA)
Meals are delivered to persons who are homebound (ill or disabled).

Household Food Consumption Surveys (USDA)
See National Food Consumption Surveys (NFCS).

Indian Health Service (DHHS)
Food distribution program on Indian reservations; operates like the Food Stamp Program and is funded by USDA.

Meals-on-Wheels. See Home-Delivered Meals (DHHS)

Medicaid (DHHS)
Funded under Title XIX of the Social Security Act to provide home health services for the indigent and disabled.

Medicare (DHHS)
Funded under Title XVIII of the Social Security Act to provide insurance coverage for health services for older people.

(continued)

Appendix 48 *(continued)*

Name of Program or Survey (Administering Agency)
Description or Main Features

Multiple Risk Factor Intervention Trial or MRFIT (NHLBI)
A longitudinal, epidemiologic study dealing with the effect of diet and other risk factors related to cardiovascular diseases.

National Cholesterol Education Program or NCEP (NHLBI)
Educational campaign to modify risk factors for elevated cholesterol.

National 5 A Day Program (NCI, PHHS)
Educational campaign urging people to eat 5 servings or more of fruits and vegetables daily.

National Food Consumption Surveys or NCFS (USDA)
A survey consisting of a 24-hour recall and a 2-day food record from 48 states. The intake of 14 nutrients was related to income, employment status, household size, race, geographic region, and educational levels.

National Health and Nutrition Examination Survey I or NHANES I (DHHS)
The first nutritional survey from 1971 to 1974. Clinical examinations, biochemical tests, anthropometric measurements, and dietary intake were data used to evaluate their nutritional status.

National Health and Nutrition Examination Survey II or NHANES II (USDA)
The second nutritional survey from 1976 to 1980 consisted of subjects aged 6 months to 74 years old. Methodology as for NHANES I.

National Health and Nutrition Examination Survey III or NHANES III (DHHS)
Survey covers the years 1988 to 1994 and surveys people at regular intervals throughout their lives.

National Nutrition Surveillance System (CDC)
Started in 1973; continuous monitoring of the nutritional status of specific high-risk population groups for at least 32 states.

National School Lunch Program (USDA)
Schools provide low-cost or free nutritious lunches to school children; provides one-third of RDA.

Nutrition Assistance Program for Puerto Rico or NAP (USDA)
Operates under the same eligibility guidelines as the Food Stamp program. Sponsored by the Puerto Rico Commonwealth Government.

Nutrition Education Training Program or NET (USDA)
Trains personnel, teachers, and students in local school cafeterias about food service and planning nutritionally balanced meals.

Omnibus Budget Reconciliation Act or OBRA
Federal legislation to assure residents in nursing homes quality care.

Preschool Nutrition Survey or PNS (USDA)
Determines the nutritional status of a cross-sectional sample of preschool children at all income levels in the United States.

School Breakfast Program (USDA)
Provides nutritious low-cost breakfast to children in participating schools or institutions.

Special Milk Program (USDA)
Provides milk to children of school age in child-care centers, in schools, as well as in institutions where lunch is not served.

Special Supplemental Food Program for Women, Infants and Children or WIC (USDA)
Provides supplemental foods and nutrition education to pregnant, postpartum, and breast-feeding women and to children 5 years and under at nutritional risk.

Appendix 48 *(continued)*

Name of Program or Survey (Administering Agency)
Description or Main Features

Summer Food Service for Children (USDA)
Provides one nutritious meal to school children during summer as a substitute for National School Lunch and School Breakfast programs.

Temporary Emergency Food Assistance Program or TEFAP (USDA)
Provides commodity foods to low-income households through public or private nonprofit agencies who are approved emergency providers.

Ten-State Nutrition Surveys (DHHS)
Collected data between 1968 and 1970 on demographic, dietary, anthropometric, biochemical, and clinical assessments of 24,000 families.

Key to Abbreviations: AOA = Administration on Aging; ACS = American Cancer Society; CDC = Centers for Disease Control; CES = Cooperative Extensive Services; DHHS = Department of Health and Human Services; FDA = Food and Drug Administration; NCI = National Cancer Institute; NHLBI = National Heart, Lung, and Blood Institute; NIA = National Institute on Aging; NIH = National Institutes of Health; ODPHP = Office of Disease Prevention and Health Promotion; USDA = United States Department of Agriculture.

Appendix 49
Agencies and Organizations with Nutrition-Related Activities

AIDS Hotline. (800/342-2437) (Check local support groups.)

Al-Anon Family Group Headquarters, 1372 Broadway, New York, NY 10018. (800/344-2666)

Alcohol and Drug Abuse Hotline. (800/252-6465)

Alcoholics Anonymous Inc., 468 Park Avenue South, New York, NY 10016. (Contact Local AA office) or (800/821-4357)

Alzheimer's Hotline. (800/621-0379)

American Academy of Cerebral Palsy and Developmental Medicine, P.O. Box 11086, Richmond, VA 23230. (804/282-0036)

American Academy of Pediatrics, 141 Northwest Point Boulevard, Elk Grove Village, IL 60007. (312/228-5005)

American Anorexia/Bulimia Association Inc., 133 Cedar Lane, Teaneck, NJ 07666. (201/836-1800)

American Association for Maternal and Child Health, 233 Prospect, P-204, La Jolla, CA 92037. (619/459-9308)

American Association of Diabetes Educators, 500 North Michigan Avenue, Suite 1400, Chicago, IL 60611. (312/661-1700)

American Association of Retired Persons (AARP), Box 22796, Long Beach, CA 90801-5796. (Check local office.)

American Board of Nutrition, 9650 Rockville Pike, Bethesda, MD 20814. (301/530-7110)

American Cancer Society, 777 Third Avenue, New York, NY 10017. (800/525-3777)

American Celiac Society, 45 Gifford Avenue, Jersey City, NJ 07304. (210/432-2986)

American College of Nutrition, 345 Central Park Avenue, #207, Scarsdale, NY 10583. (914/948-4848)

American College of Preventive Medicine, 1015 15th Street NW, Suite 903, Washington, DC 20005. (202/789-0003)

American College of Sports Medicine, P.O. Box 1440, Indianapolis, IN 46024. (317/637-9200)

American Council on Science and Health, 1995 Broadway, 18th Floor, New York, NY 10023. (212/362-7044)

American Council on Transplantation, 700 N. Fairfax Street, Suite 505, Alexandria, VA 22314. (703/836-4301)

American Dental Association, 211 East Chicago Avenue, Chicago IL 60611. (202/898-2400)

American Diabetes Association, National Service Center, 1660 Duke Street, Alexandria, VA 22313. (800/232-3472)

American Dietetic Association, 216 West Jackson, Chicago, IL 60604. (800/877-1600)

American Digestive Disease Society, 7720 Wisconsin Avenue, Bethesda, MD 20814. (301/652-9293)

American Epilepsy Society, 179 Allyn Street, Suite 304, Hartford, CT 06103. (203/246-6566)

American Food Service Association, 1600 Duke Street, Alexandria, VA 22314. (800/877-8822)

American Foundation for AIDS Research, 733 Third Avenue, 12 floor, New York, NY 10017. (212/682-7440)

American Geriatrics Society, 770 Lexington Avenue, Suite 400, New York, NY 10021. (212/308-1414)

American Health Care Association, 1200 15th Street NW, Washington, DC 20005. (202/628-4410)

American Heart Association, 7320 Greenville Avenue, Dallas, TX 75231. (800/241-6993) or (800/AHA-USA-1)

American Home Economics Association, 2110 Massachusetts Avenue NW, Washington, DC 20036. (703/706-4600) or (800/424-8080)

American Hospital Association, 840 North Lake Shore Drive, Chicago, IL 60611. (312/280-6000)

American Industrial Health Council, 1330 Connecticut Avenue, Cedar Grove, NJ 07009. (201/857-2626)

American Institute for Cancer Research, 1759 R Street, NW, Washington, DC 20009. (202/328-7244)

American Institute of Nutrition, 9650 Rockville Pike, Bethesda, MD 20814. (301/530-7050)

American Liver Foundation, Cedar Grove, NJ 07009. (800/223-0179)

American Lung Association, 1740 Broadway, New York, NY 10019. (212/315-8700) or (800/LUNG-USA)

American Medical Association, Nutrition Information Service, 535 North Dearborn Street, Chicago, IL 60610. (312/645-4470)

American National Red Cross, Food and Nutrition Consultant, National Headquarters, Washington, DC 20006. (202/737-8300)

American Parkinson Disease Association, 116 John Street, Suite 417, New York, NY 10038. (212/732-9550)

American Pediatric Society, 450 Clarkson Avenue, Brooklyn, NY 11207. (718/270-1692)

American Psychological Association, 1200 17th Street, NW, Washington, DC 20036. (202/955-7618)

American Public Health Association, 1015 15th Street, NW, Washington, DC 20005. (202/789-5600)

American Red Cross AIDS Education Office, 1730 D Street, NW, Washington, DC 20006. (202/737-8300)

American Society for Allied Health Professionals, 1101 Connecticut Avenue, NW, Washington, DC 20036. (202/857-1100)

American Society for Clinical Nutrition, 9650 Rockville Pike, Bethesda, MD 20814. (301/530-7110)

American Society for Hospital Food Service Administrators, 840 North Lakeshore Drive, Chicago, IL 60611. (312/280-6416)

American Society for Parenteral and Enteral Nutrition, 8630 Fenton Street, Suite 412, Silver Spring, MD 20910. (301/587-6315)

American Society of Internal Medicine, 1101 Vermont Avenue, NW, #500, Washington, DC 20005. (202/289-1700)

Anorexia Nervosa and Related Eating Disorders, P.O. Box 5120, Eugene, OR 97405. (503/344-1144)

Arthritis Foundation, 1314 Spring Street, NW, Atlanta GA 30309. (800/523-1429)

Association of Maternal and Child Health Programs, 2001 L Street, NW, Washington, DC 20036. (202/466-8960)

Association of Schools of Public Health, 1015 15th Street, NW, Washington, DC 20005. (202/842-4668)

Association of State and Territorial Public Health Nutrition Directors, 6728 Old McLean Village Drive, McLean, VA 2201. (703/556-9222)

Asthma and Allergy Foundation of America, 1717 Massachusetts Avenue, Washington, DC 20036. (202/265-0265) or (800/728-7462)

Bulimia/Anorexia Self-help Hotline. (800/762-3334)

CDC National AIDS Clearinghouse, P.O. Box 6003, Rockville, MD 20849-6003. (800/458-5231)

Center for Ulcer Research and Education, 11661 San Vincente Boulevard, Suite 304, Los Angeles, CA 90049. (213/825-5091)

Centers for Disease Control (CDC) Information Hotline. (404/332-4555) or (800/227-8922)

Child Nutrition Forum, 1319 F Street, NW, Suite 500, Washington, DC 20004. (202/393-5060)

Community Nutrition Institute, 2001 S Street, NW, Washington, DC 20009. (202/462-4700)

Crohn's and Colitis Foundation of America. (800/343-3637)

Cystic Fibrosis Foundation, 6931 Arlington Road, #200, Bethesda, MD 20814. (800/344-4823)

Department of Health and Human Services, Office of Child Health, 1848 Gwynn Oak Avenue, Baltimore, Maryland 21207.

Epilepsy Foundation of America, 4351 Garden City Drive, Landover, MD 20785. (800/332-1000)

European Federation of the Associations of Dietitians, Tak Van Poortvlietstraat 3, NL-5344, GZ Oss, Netherlands.

Food and Agricultural Organization, 345 Park

(continued)

Avenue South, New York, NY 10016. (202/653-2400)

Food and Drug Administration, Parklane Building, 5600 Fishers Lane, Rockville, MD 20857. (800/FDA-4010)

Food and Drug Law Institute, 1000 Vermont Avenue, Washington, DC 20005. (202/371-1420)

Food and Nutrition Board, National Research Council, 2101 Constitution Avenue, Washington, DC 20418. (202/334-2238)

Food Research and Action Center, 1319 F Street NW, Suite 500, Washington, DC 20004. (202/393-5060)

Gluten Intolerance Group, P.O. Box 23053, Seattle, WA 98102-0353. (201/325-6980)

Institute of Food Technologists, 221 North La Salle Street, Chicago, IL 60601. (312/782-8424)

Institute of Nutrition of Central America and Panama (INCAP), Apartado Postal 1188, Guatemala, Guatemala

International Association of Milk, Food and Environmental Sanitarians, P.O. Box 701, Ames, IA 50010. (515/232-6699)

International Life Sciences Institute, Nutrition Foundation, 1126 16th Street NW, #300, Washington, DC 20036. (202/659-0074)

Iron Overload Disease Association, 224 Dalura Street, Suite 912, West Palm Beach, FL 33401. (305/659-5616)

Joint Commission on Accreditation of Healthcare Organizations, 875 North Michigan Avenue, Chicago, IL 60611. (301/642-6061)

Juvenile Diabetes Foundation International, 432 Park Avenue South, 16th Floor New York, NY 10016. (212/889-7575)

Kellogg Foundation, 400 North Avenue, Battle Creek, MI 49016. (800/624-6668)

La Leche League International, Inc., 9615 Minneapolis Avenue, Franklin Park, IL 60131. (800/LA LECHE)

Leukemia Society of America, Inc., 733 Third Avenue, New York, NY 10017. (212/573-8484)

Lupus Foundation of America, 1717 Massachusetts Avenue, SW, Suite 203, Washington, DC 20036. (202/328-4550)

March of Dimes Birth Defects Foundation, 1275 Mamaroneck Avenue, White Plains, NY 10602. (800/336-4797)

Maternity Center Association, 48 East 92nd Street, New York, NY 10128. (212/369-7300)

Multiple Sclerosis Association Hotline. (800/833-4672)

Muscular Dystrophy Association Inc., 810 Seventh Avenue, New York, NY 10019. (212/586-0808)

National Academy of Sciences/National Research Council (NAS/NRC), 2101 Constitution Avenue NW, Washington, DC 20418

National Association for Down's Syndrome, P.O. Box 4542, Oak Brook, IL 60522. (312/325-9112)

National Association for Home Care, 519 C Street, NE, Stanton Park, Washington, DC 20002. (202/547-7424)

National Association for Public Health Policy, 208 Meadowood Drive, South Burlington, VT 05403. (802/658-0136)

National Association of Anorexia Nervosa and Associated Disorders, Inc., Box 271, Highland Park, IL 60035. (708/831-3438)

National Association of Community Health Centers, 1330 New Hampshire Avenue, NW, #122, Washington, DC 20036. (202/659-8008)

National Association of Developmental Disabilities Councils, 1234 Massachusetts Avenue, NW, #103, Washington, DC 20005. (202/347-1234)

National Association of Pediatric Nurse Associates and Practitioners, 1101 Kings Highway, #206, Cherry Hill, NJ 08034. (609/667-1773)

National Association of WIC Directors, 1516 W. Mount Royal Avenue, Baltimore, MD 21217. (301/383-2766)

National Cancer Institute, Cancer Information Service, Building 31, Bethesda, MD 20892. (800/4-CANCER)

National Center for Health Education, 30 East 29th Street, New York, NY 10016. (212/689-1886)

National Center for Health Statistics (NCHS),

U.S. Department of Health and Human Services, Public Health Service, 3700 East West Highway, Hyattsville, MD 20782. (301/436-8500)

National Center for Nutrition and Dietetics, 216 West Jackson Boulevard, Chicago, IL 60606-6995. (312/899-0040) or (800/366-1655)

National Child Nutrition Project, 1501 Cherry Street, Philadelphia, PA 19102. (215/662-1024)

National Cholesterol Education Program, 4733 Bethesda Avenue, Ste. 530, Bethesda, MD 20814. (301/951-3260)

National Council Against Health Fraud, P.O. Box 1276, Loma Linda, CA 92340. (714/824-4690)

National Council on Patient Information and Education, 666 11th Street, NW, #810, Washington, DC 20001. (202/347-6711)

National Council on the Aging, 600 Maryland Avenue, SW, W. Wing 100, Washington, DC 20024. (202/479-1200)

National Council on Alcoholism, 12 W. 21st Street, New York, NY 10010. (800/NCA-CALL)

National Council for International Health, 1701 K Street, NW, Suite 600, Washington, DC 20006. (202/833-5900)

National Dairy Council, 6300 N. River Road, Rosemont, IL 60018. (312/696-1020)

National Down's Syndrome Society, 141 Fifth Avenue, New York, NY 10010. (800/221-4602)

National Foundation for Ileitis and Colitis Inc., 444 Park Avenue S, New York, NY 10016. (212/685-3440)

National Foundation for Long Term Health Care, 1200 15th Street, NW, Washington, DC 20005. (202/659-3148)

National Headstart Association, 1220 King Street, #200, Alexandria, VA 22314. (703/739-0875)

National Health Council, 622 Third Avenue, 34th Floor, New York, NY 10017. (800/622-9010)

National Heart, Lung, and Blood Institute, National High Blood Pressure Education Program, Information Center, P.O. Box 30105, Bethesda, MD 20824-0105. (301/251-1222)

National High Blood Pressure Education Program, 120/80 National Institutes of Health, Bethesda, MD 20892. (301/951-3260)

National Institutes of Diabetes, Digestive and Kidney Diseases, National Institutes of Health, 900 Rockville Pike, Bethesda, MD 20892. (301/496-8500)

National Kidney Foundation, 30 East 33rd Street, New York, NY 10016. (800/622-9010)

National Livestock and Meat Board, 444 N. Michigan Avenue, Chicago, IL 60611. (312/467-5520)

National Meals on Wheels Foundation, 2675 44th Street, NW #305, Grand Rapids, MI 49509. (800/999-6262)

National Medical Association, 1012 10th Street, NW, Washington, DC 20001. (202/347-1895)

National Mental Health Association Hotline. (800/909-6642)

National Multiple Sclerosis Society, 205 East 42nd Street, New York, NY 10017. (212/986-3240)

National Osteoporosis Foundation, 1625 - I Street, NW, Suite 1011, Washington, DC 20006. (202/223-2226)

National Wellness Association, University of Wisconsin-Stevens Point, South Hall, Stevens Point, WI 54481. (715/346-2172)

Nutrition Education Association, P.O. Box 20301, 3647 Glen Haven, Houston, TX 77225. (713/665-2946)

Nutrition Foundation, Inc., 1126 Sixteenth Street NW, Suite 111, Washington, DC 20036. (202/659-0074)

Nutrition Institute of America, 200 W 86th Street, Suite 17A, New York, NY 10024. (212/799-2234)

Nutrition for Optimal Health Association, P.O. Box 380 Winnetka, IL 60093. (312/835-5030)

Nutrition Screening Initiative, 2626 Pennsylvania Avenue, NW, Washington, DC 20037. (202/625-1662)

Nutrition Today Society, 428 East Preston

(continued)

Appendix 49 *(continued)*

Street, Baltimore, MD 21202. (410/358-5891)

Pan American Health Organization, 525 23rd Street, NW, Washington, DC 20037. (202/861-3200)

Parkinson's Disease Foundation Inc., 640 West 168th Street, New York, NY 10032. (212/923-4700) or (800/344-7872)

Prader–Willi Syndrome Association, 6490 Excelsion Boulevard, E-102, St. Louis Park, MN 55426. (612/926-1947)

President's Council on Physical Fitness and Sports, 450 5th Street, NW, Suite 7103, Washington, DC 2001. (202/272-3421)

Price-Pottenger Nutrition Foundation, P.O. Box 2614, La Mesa, CA 92041. (619/582-4168)

Public Health Foundation, 1220 L Street, NW, #350, Washington, DC 20005. (202/898-5600)

Sister Kenny Institute, 800 E. 28th Street at Chicago Avenue, Minneapolis, MN 55407. (612/863-4457)

Society for Nutrition Education, 1700 Broadway, Suite 300, Oakland, CA 94612. (415/444-7133) or (800/235-6690)

Society for Public Health Education, Inc., 2001 Addison Street, #220, Berkeley, CA 94704. (415/644-9242)

Spina Bifida Association of America, 343 South Dearborn Street, Suite 310, Chicago, IL 60604. (800/621-0379)

T.O.P.S. (Take Off Pounds Sensibly), P.O. Box 07360, Milwaukee, WI 53207. (Check local chapters.)

United Cerebral Palsy Association Inc., 66 East 34th Street, New York, NY 10016. (212/481-6344)

United Ostomy Association (UOA), 2001 W. Beverly Boulevard, Los Angeles, CA 90057. (213/413-5510)

United Parkinson Foundation, 360 W Superior Street, Chicago, IL 60610. (312/664-2344)

United Scleroderma Foundation Inc., P.O. Box 350, Watsonville, CA 95077. (408/728-2202)

USDA Cooperative Extension Service, Washington, DC 20250. (202/720-2791)

USDA Food and Nutrition Service (FNS), Child Nutrition Division, 3101 Park Center Drive, Alexandria, VA 22302. (800/237-6401)

USDA Food Safety and Inspection Service (FSIS), 14th and Independence Avenue, SW, Washington, DC 20250. (800/535-4555)

USDA Human Nutrition Information Service (HNIS), 6505 Belcrest Road, Hyattsville, MD 20782. (800/535-4555)

USDA Meat and Poultry Hotline. (800/535-4555)

USDA Nutrition Program, Consumer and Food Economics Division, Agricultural Research Service, Hyattsville, MD 20782. (800/282-5852)

USDA National School Lunch Program, Food and Nutrition Service, 3101 Park Center Drive, Alexandria, VA 22302. (Check local schools.)

USDA Special Food Supplemental Program for Women, Infants, and Children (WIC), Food and Nutrition Service, 3101 Park Center Drive, Room 1017, Alexandria, VA 22302. (Check local WIC.)

U.S. Department of Agriculture (USDA), 14th Street SW and Independence Avenue, Washington, DC 20250. (202/720-2791)

U.S. Department of Education (DOE), Accreditation Agency Evaluation Branch, 7th and D Street, SW, Building 3, Room 336, Washington, DC 20202. (202/708-7417)

U.S. Environmental Protection Agency (EPA), 401M Street, NW, Washington, DC 20460. (202/382-3535)

U.S. EPA Safe Drinking Water Hotline. (800/426-4791)

U.S. Government Printing Office, The Superintendent of Documents, Washington, DC 20402. (800-347-4265)

World Health Organization (WHO), 525 23rd Street, NW, Washington, DC 20037. (202/861-3200)

World Hunger, 505 8th Avenue, New York, New York, 10018-6582. (212/629-8850)

Appendix 49 *(continued)*

Trade Organizations

Abbott Laboratories, 14th Street, Sheridan Road, North Chicago, IL 60064. (708/938-5700) or (800/688-9118)

Alberto-Culver, Melrose, IL 60160. (708/450-3000)

American Egg Board, 1460 Renaissance Street, Park Ridge, IL 60068. (212/759-1811)

American Institute of Baking, 400 East Ontario Street, Chicago IL 60611.

American McGaw, Division of Travenol, 1425 Lake Cook Road, Deerfield, IL 60015. (714/261-6360)

American Meat Institute, 1700 N. Moore Street, Arlington, VA 22209. (703/841-2400)

Amgen, P.O. Box 7710, Washington, DC 20044. (800/272-9376)

Baxter Healthcare Corporation, 1 Baxter Parkway, Deerfield, IL 60015

Beech-Nut, Checkerboard Square, 1B, St. Louis, MO 63164. (800/523-6633)

Best Foods, Division of CPC International, Englewood Cliffs, NJ 07623. (201/894-2364)

Biosearch Medical Products Inc., 35 Industrial Parkway, P.O. Box 1700, Somerville, NJ 08876. (201/722-5000)

Bristol-Myers Squibb. Hot Line (800/426-7644)

Campbell Soup Company, Food Service Products Division, Campbell Place, Camden, NJ 08103-1799. (609/342-4800)

Carnation Company, Glendale, CA 91203. (818/549-6000)

Cheeseborough-Pond's Inc., 33 Benedict Place, Greenwich, CT 06830. (800/722-9294)

Clintec Nutrition Company, 3 Parkway North, Suite 500, Deerfield, IL 60015. (708/317-2800)

ConAgra Corporation, One ConAgra Drive, Omaha, NE 68102. (402/595-4000)

Delmark Food Service Company (Division of Sandoz) 5320 West 23rd, Minneapolis, MN 55416. (800/999-9978)

Del Monte Corporation, Consumer and Education Services, Box 3757, San Francisco, CA 94119. (800/221-7318)

Dietary Specialties Inc., P.O. Box 227, Rochester, NY 14601. (716/263-2787) or (800/544-0099)

Doyle Pharmaceutical Company (Division of Sandoz) 5320 West 23rd, Minneapolis, MN 55416. (800/658-5491)

Elan Pharma, 2 Thurber Boulevard, Smithfield, RI 02917. (800/733-4424)

Eli Lilly and Company, Medical Department, 307 East McCarty, Indianapolis, IN 46225

Ener-G Foods Inc., 6901 Fox Avenue South, Seattle, WA 98124. (800/331-5222)

Fleischmann's Margarine, Standard Brands Inc., 625 Madison Avenue, New York, NY 10022

Florida Citrus Commission, Box 148, Lakeland, FL 33802. (813/499-2500)

Food and Drug Law Institute, 1000 Vermont Avenue, NW, Washington, DC 20005. (202/371-1420)

Food Marketing Institute, 1750 K Street, NW, Washington, DC 20006. (202/452-8444)

Food Sciences Corporation, Consumer Center, 250 North Street, White Plains, NY 10625. (800/352-BEST)

Frito-Lay, Inc., P.O. Box 660634, Dallas, TX 75266-0634. (214/334-4000)

General Foods USA Foodservice, 250 North Street, White Plains, NY 10625. (914/335-3231) or (800/352-BEST)

General Foods Hotline 800/432-6333

General Mills Inc., 3716 Eight Avenue, Minneapolis, MN 55426. (612/729-1157). Hotline 800/328-6787

Gerber Products Company, 445 State Street, Fremont, MI 49412. (616/928-2257). Hotline 800/443-7237

Grocery Manufacturers of America, 1010 Wisconsin Avenue, Washington, DC 20007. (202/337-9400)

H. J. Heinz, Consumer Relations, P.O. Box 57, Pittsburgh, PA 15230. (412/237-5757)

Hunt-Wesson Foods, Educational Services, 1645 West Valencia Drive, Fullerton, CA 92634

Hoffman-LaRoche Hotline. (800/285-4484)

(continued)

Appendix 49 *(continued)*

Keebler Co., 140 Industrial Drive, Elmhurst, IL 60126. (708/782-2680)

Kellogg Company, Department of Home Economics Services, Battle Creek, MI 49016. (616/961-9170)

Kendall McGaw Laboratories Inc., P.O. Box 25080, 2525 McGaw Avenue, Irvine, CA 92714. (800/854-6851)

Kraft Foods Hotline (800/323-1768)

Lederle Laboratories, 1 Cyanamid Plaza, Wayne, NJ 07470. (201/831-2000)

Mead Johnson Nutritionals, 2400 West Lloyd Expressway, Evansville, IN 47721-0001. (800/429-5000)

Med-Diet Laboratories Inc., 3050 Ranchview N, Plymouth, MN 55427. (800/633-3438)

Menu Magic, 1717 West 10th Street, Indianapolis, IN 46222. (317/269-3500)

Milani Foods, Inc., "Thick-It", 2525 Armitage Avenue, Melrose, IL 60160. (800/333-0003)

Nabisco Consumer Affairs, 100 DeForest Avenue, East Hanover, NJ 07936. (800/NABISCO)

National Dairy Promotion and Research Board, 211 Wilson Boulevard, Suite 600, Arlington, VA 22201. (703/528-4800)

National Milk Producers Federation, 1840 Wilson Boulevard, Arlington, VA 22209. (703/243-6111)

National Restaurant Association, 1200 17th Street, NW, Washington, DC 20001. (202/331-5900)

Nestle's Consumer Center Hotline. (800/637-8537

Norwich Eaton Pharmaceuticals, P.O. Box 191, 17 Easton Avenue, Norwich, NY 13815-0191. (800/446-6654)

O'Brien/KMI, Department 9567, 320 Charles Street, Cambridge, MA 02141. (800/345-6039)

Pfizer Hotline. (800/869-9979)

Pillsbury Company, 200 S. 6th Street, Minneapolis, MN 55402. (612/330-4966)

Proctor and Gamble Educational Services, P.O. Box 14009, Cincinnati, OH 45214. (513/983-1100)

Ralston-Purina Co., Checkerboard Square, St. Louis, MO 63188. (314/982-1000)

Roche Laboratoriers, Division of Hoffman-LaRoche Inc., 340 Kingsland Street, Nutley, NJ 07110. (800/526-6367)

Ross Laboratories, 625 Cleveland Avenue, Columbus, OH 43216. (800/624-7677)

Sandoz Nutrition Corporation, 5320 West 23rd Street, P.O. Box 370, Minneapolis, MN 55440. (800/999-9978)

Sherwood Medical Company, 1951 Olive Street, St. Louis, MO 63103. (800/428-4400)

Sysco Corporation, 1390 Enclave Parkway, Houston, Texas 77077. (713/584-1390)

Sunkist Growers, P.O. Box 2706, Los Angeles, CA 90054. (818/379-7181)

Travenol Laboratories, 1425 Lake Cook Road, Deerfield, IL 60015. (312/940-6524)

United Fresh Fruit and Vegetable Association, 727 N. Washington Street, Alexandria, VA 22314. (703/836-3410)

UpJohn Company, 7000 Portage Road, Kalamazoo, MI 49001. (800/253-8600)

Wyeth-Ayerst Laboratories, P.O. Box 8299, Philadelphia, PA 19101. (800/666-7248)

Appendix 50
Sources of Nutrition Information

Books

Adrogue, Horacio J. and Donald Wesson. 1992. *Potassium*. Houston, TX: Libra Germini Publications, Inc.

Agostini, Rosemary. 1994. *Medical Issues of Active and Athletic Women*. St. Louis, MO: The C. V. Mosby Co.

Aitio, Antero, Antti Oro, Jorma Jarvisalo, and Harri Vainio. 1991. *Trace Elements in Health and Disease*. Cambridge, England: The Royal Society of Chemistry.

Albert, Michael B. and C. Wayne Callaway. 1992. *Clinical Nutrition for the House Officer*. Baltimore, MD: Williams & Wilkins.

Allen, Ann Moore. 1992. Food–Medication Interactions, 7th ed. Pottstown, PA: Food-Medication Interactions.

Altschul, A. M. 1993. *Low-Calorie Foods Handbook*. New York, NY: Marcel Dekker.

American Dietetic Association. 1992. *Handbook of Clinical Dietetics*. New Haven, CT: Yale University Press.

American Dietetic Association. 1992. *Manual of Clinical Dietetics, 4th ed.* Chicago, IL: The American Dietetic Association.

Aspen Reference Group. 1994. *Dietitian's Patient Educational Manual*. Rockville, MD: Aspen Publishers, Inc.

Baker, Susan B., Robert D. Baker, Jr., and Anne Davis. 1994. *Pediatric Enteral Nutrition*. New York, NY: Chapman and Hall.

Barness, Lewis A. 1993. *Pediatric Nutrition Handbook,* 3rd ed. Elk Grove Village, IL: American Academy of Pediatrics.

Behrens, Rosemary I. and Anne K. Blocker. 1994. *Continuous Quality Improvement and Nutritional Care Planning: A Manual for Long-Term Care Facilities*. Rockville, MD: Aspen Publishers, Inc.

Bender, David A. 1993. *Introduction to Nutrition and Metabolism*. London, England: VCL Press Ltd.

Berg, Frances. 1993. *Health Risks of Obesity: 1993 Special Report*. Hettinger, ND: Obesity & Health.

Berning, Jacqueline R. and Suzanne Nelson Steen. 1994. *Sports Nutrition for the 90s: The Health Professional's Handbook*. Rockville, MD: Aspen Publishers, Inc.

Blackburn, George L. and Beatrice Kanders. 1994. *Obesity: Pathophysiology, Psychology and Treatment*. New York, NY: Chapman and Hall.

Billon, Wayne E. 1991. *Clinical Nutrition Case Studies*. St. Paul, MN: West Publishing Co.

Biswas, M. R. Mamdoub Gabr. 1994. *Nutrition in the Nineties: Policy Issues*. New York, NY: Oxford University Press.

Blackman, James A. 1994. *Feeding Infants and Young Children with Special Problems*. Rockville, MD: Aspen Publishers, Inc.

Bloch, Abby S. 1994. *Nutrition Management of the Cancer Patient: A Practical Guide for Professionals*. Rockville, MD: Aspen Publishers, Inc.

Blomhoff, R. (Editor). 1994. *Vitamin A in Health and Disease*. New York, NY: Marcel Dekker.

Bounous, Gustavo. 1992. *Uses of Elemental Diets in Clinical Situations*. Boca Raton, FL: CRC Press.

Bowlby, Carol. 1993. *Therapeutic Activities with Persons Disabled by Alzheimer's Disease and Related Disorders*. Rockville, MD: Aspen Publishers, Inc.

Boyle, Marie and Diane Morris. 1994. *Community Nutrition in Action: An Entrepreneurial Approach*. St. Paul, MN: West Publishing Co.

Boyle, Maria A. and Gail Zyla. 1992. *Personal Nutrition,* 2nd ed. St. Paul, MN: West Publishing Co.

Bucci, Luke. 1993. *Nutrients as Ergogenic Aids for Sports and Exercise*. Boca Raton, FL: CRC Press.

Cataldo, Corinne Balog, Linda K. DeBruyne, and Eleanor Whitney. 1992. *Nutrition and Diet Therapy: Principles and Practice,* 3rd ed. St. Paul, MN: West Publishing Co.

Cataldo, Corinne Balog, Sharon Rady Rolfes, and Eleanor Whitney. 1993. *Understanding Nutrition,* 6th ed. St. Paul, MN: West Publishing Co.

Chernecky, Cynthia, Ruth L. Krech, and Barbara J. Berger. 1993. *Laboratory Tests and Diagnostic Procedure*. Philadelphia, PA: W. B. Saunders Co.

Chernoff, Ronni. 1991. *Geriatric Nutrition: The Professional's Handbook*. Frederick, MD: Aspen Publishers, Inc.

Cohen, I. K., R. F. Diegelmann, and W. J. Lindblad. 1992. *Wound Healing—Biochemical and Clinical Aspects*. Philadelphia, PA: W. B. Saunders Co.

Cohen, P. T. et al. 1994. *The AIDS Knowledge Base,* 2nd ed. Boston, MA: Little, Brown and Co.

Creighton, Thomas E. 1993. *Proteins—Structures and Molecular Properties,* 2nd ed. New York, NY: W. H. Freeman and Co.

Cunningham-Rundles, S. 1992. *Nutrient Modulation of the Immune Response*. New York, NY: Marcel Dekker.

David, T. J. 1993. *Food and Food Additive Intolerance in Childhood*. Cambridge, MA: Blackwell Scientific Publications.

Davidson, Mayer B. 1991. *Diabetes Mellitus*. New York, NY: Churchill Livingston, Inc.

Davis, Julie Ratliff and Kim Sherer. 1994. *Applied Nutrition and Diet Therapy for Nurses*. Philadelphia, PA: W. B. Saunders Co.

Deutsch, Ronald M. and Judi S. Morrill. 1993. *Realities of Nutrition*. Palo Alto, CA: Bull Publishing Co.

DSM-IV®. 1994. *Diagnostic and Statistical Manual of Mental Disorders*. 4th ed. Washington, D.C.: American Psychiatric Association.

Drummond, Karen Eich. 1994. *Nutrition for the Food Service Professional,* 2nd ed. New York, NY: Van Nostrand Reinhold.

Ekvall, Shirley Walberg (Editor). 1993. *Pediatric Nutrition in Chronic Diseases and Developmental Disorders: Prevention, Assessment, and Treatment*. New York, NY: Oxford University Press.

Engel, Joyce. 1993. *Pocket Guide to Pediatric Assessment,* 2nd ed. St. Louis, MO: Mosby-Year Book, Inc.

Ensminger, Audrey H., M. E. Ensminger, James E. Konlande, and John R. K. Robson. 1994. *Foods & Nutrition Encyclopedia, 2nd ed*. Boca Raton, FL: CRC Press, Inc.

Escott-Stump, Sylvia. 1992. *Nutrition and Diagnosis Related Care, 3rd ed*. Baltimore, MD: Lea & Febiger.

Fidanza, F. 1991. *Nutrition Status Assessment: A Manual for Population Studies*. New York, NY: Chapman and Hall.

Filer, Lloyd J., Ronald M. Lauer, and Russell V. Luepker. 1994. *Prevention of Atherosclerosis and Hypertension Beginning in Youth*. Baltimore, MD: Lea & Febiger.

Fishback, Frances. 1992. *A Manual of Laboratory & Diagnostic Ttests, 4th ed*. Philadelphia, PA: J. B. Lippincott Co.

Fomon, Samuel J. 1993. *Nutrition of Normal Infants*. St. Louis, MO: Mosby-Year Book, Inc.

Fomon, Samuel J. and Stanley Zlotkin. 1992. *Nutritional Anemias*. New York, NY: Raven Press Ltd.

Forse, R. Armour, George L. Blackburn, Stacy Bell, and Lynda G. Kabbash (Editors). 1994. *Diet, Nutrition, and Immunity*. Boca Raton, FL: CRC Press, Inc.

Frankle, Reva T. and Anita L. Owen. 1993. *Nutrition in the Community: The Art of Delivering Services*. St. Louis, MO: The C. V. Mosby Co.

Frisancho, A. R. 1990. *Anthropometric Standards for the Assessment of Growth and Nutritional Status*. MI: The University of Michigan Press.

Gallagher-Allred, Charlotte. 1993. *Implementing Nutrition Screening and Intervention Strategies*. Washington, D.C.: The Nutrition Screening Initiative.

Garrow, J. S. and W. T. James. 1994. *Human Nutrition and Dietetics*. 9th ed. London, England: Churchill Livingston.

Gerber, James M. 1992. *Handbook of Preventive and Therapeutic Nutrition*. Rockville, MD: Aspen Publishers, Inc.

Gibson, R. S. 1993. *Nutritional Assessment: A Laboratory Manual*. New York, NY: Oxford University Press.

Gibson, R. S. 1990. *Principles of Nutritional*

Assessment. New York, NY: Oxford University Press.

Gines, Deon J. 1990. *Nutrition Management in Rehabilitation*. Rockville, MD: Aspen Publishers, Inc.

Gottschlich, M. M., L. E. Matarese, and E. P. Shronts. 1993. *Nutrition Support Dietetics, 2nd ed*. Rockville, MD: Aspen Publishers, Inc.

Grant, Anne and Susan DeHoog. 1991. *Nutritional Assessment and Support, 4th ed*. Seattle, WA: Anne Grant/Susan DeHoog Publishers.

Grant, John P. 1992. *Handbook of Total Parenteral Nutrition, 2nd ed*. Philadelphia, PA: W. B. Saunders Co.

Grimes, Deanna E. and Richard M. Grimes. 1994. *AIDS and HIV Infection*. St. Louis, MO: Mosby-Year Book, Inc.

Hamilton, Eva May. 1991. *Nutrition Concepts and Controversies, 5th ed*. St. Paul, MN: West Publishing Co.

Hay, Jr., William W. 1991. *Neonatal Nutrition and Metabolism*. St. Louis, MO: The C. V. Mosby Co.

Hendricks, Kristy M. and W. Allan Walker, 1990. *Manual of Pediatric Nutrition*. St. Louis, MO: The C. V. Mosby Co.

Hermann-Zaidins, Minda and Riva Touger-Decker. 1989. *Nutrition Support in Home Health*. Rockville, MD: Aspen Publishers, Inc.

Hess, Mary Abbott and Anne Elise Hunt. 1993. *Review of Dietetics: Registered Dietitian's Examination Study Manual*. Rockville, MD: Aspen Publishers, Inc.

Hickey, Michael S. 1992. *Handbook of Enteral, Parenteral, and ARC/AIDS Nutritional Therapy*. St. Louis, MO: The C. V. Mosby Co.

Hickson-Wolinsk. 1995. *Nutrition and the HIV Positive Persons*. Boca Raton, Fl: CRC Press, Inc.

Himes, John H. 1991. *Anthropometric Assessment of Nutritional Status*. New York, NY: Wiley-Liss.

Homer, Elizabeth. 1993. *Nutrition for Women: The Complete Guide*. New York: NY: H. Holt and Co.

Hunt, Sara M. and James L. Groff. 1990. *Advanced Nutrition and Human Metabolism*. St. Paul, MN: West Publishing Co.

Huyck, Norma I. and Margaret Maher Rowe. 1990. *Managing Clinical Nutrition Services*. Frederick, MD: Aspen Publishers, Inc.

Iowa Dietetic Association. 1994. *Simplified Diet Manual, 6th ed*. Ames, IA: Iowa State University Press.

Kanarek, Robin B. and Robin Marks-Kaufman. 1991. *Nutrition and Behavior. New Perspectives*. New York, NY: Van Nostrand Reinhold.

Karp, R. J. (Editor). 1993. *Malnourished Children in the United States*. New York, NY: Springer Publishing Co.

Katch, Frank I. and William D. McArdle. 1993. *Introduction to Nutrition, Exercise and Health, 4th ed*. Baltimore, MD: Lea & Febiger.

Kaufman, Mildred. 1990. *Nutrition in Public Health: A Handbook for Developing Programs and Services*. Rockville, MD: Aspen Publishers, Inc.

King, F. S. and A. Burgess. 1993. *Nutrition for Developing Countries, 2nd ed*. Oxford, England: Oxford University Press.

Kinney, John M. and Hugh Tucker. 1992. *Energy Metabolism—Tissue Determinants and Cellular Corollaries*. New York, NY: Raven Press Ltd.

Kirby, Donald F. and Stanley J. Dudrick. 1994. *Practical Handbook of Nutrition in Clinical Practice*. Boca Raton, FL: CRC Press, Inc.

Klimis-Tavantzis, Dorothy J. 1994. *Manganese in Health and Disease*. Boca Raton, Fl: CRC Press, Inc.

Klurfeld, David M. 1993. *Nutrition and Immunology*. New York, NY: Plenum Press.

Langford, Herbert and Barbara Levine. 1990. *Contemporary Issues in Clinical Nutrition. Vol. 12, Nutrition Factors in Hypertension*. New York, NY: Alan R. Liss, Inc.

Lee, Robert D. and David C. Nieman. 1993. *Nutritional Assessment*. Dubuque, IA: Wm. C. Brown Publishers.

(continued)

Appendix 50 *(continued)*

Lessof, M. H. 1992. *Food Intolerance*. New York, NY: Chapman and Hall.

Linder, Maria A. 1991. *Nutritional Biochemistry and Metabolism. With Clinical Applications, 2nd ed.* New York, NY: Elsevier Science Publishing Co., Inc.

Lutz, Carroll A. and Karen Rutherford Przytulski. 1994. *Nutrition and Diet Therapy*. Philadelphia, PA: F. A. Davis Co.

Machlin, Lawrence J. 1991. *Handbook of Vitamins,* 2nd ed. New York, NY: Marcel Dekker, Inc.

Marriott, Bernadette. 1993. *Nutritional Needs in Hot Environments*. Washington, D.C.: National Academy Press.

McDonald Hugh J. and Frances M. Sapone. 1993. *Nutrition for the Prime of Life: The Adult's Guide to Healthier Living*. New York, NY: Plenum Press.

McLaren, Donald S. 1993. *Color Atlas and Text of Diet-Related Disorders*. St. Louis, MO: The C. V. Mosby Co.

McLaren, Donald S., David Burnan, Neville R. Berton, and Anthony F. Williams. 1991. *Textbook of Pediatric Nutrition,* 3rd ed. New York, NY: Churchill Livingstone.

Mitch, William E. and Saulo Klahr. 1993. *Nutrition and the Kidney*. Boston, MA: Little, Brown and Co.

Moore, Mary Courtney. 1993. *Pocket Guide to Nutrition and Diet Therapy, 2nd ed.* St. Louis, MO: Mosby-Year Book, Inc.

Morley, John E., Zvi Glick, and Laurence Z. Rubenstein. 1990. *Geriatric Nutrition. A Comprehensive Review, 4th ed.* New York, NY: Raven Press, Ltd.

Mosby Staff. 1994. *Mosby's Medical Nursing and Allied Health Dictionary*. St. Louis, MO: The C. V. Mosby Co.

Munro, Hamish and Gunter Schlierf. 1992. *Nutrition of the Elderly*. New York, NY: Raven Press Ltd.

National Research Council. 1990. *Diet and Health: Implications Reducing Chronic Disease Risk*. Washington, D.C.: National Academic Press.

National Research Council. 1989. Recommended Dietary Allowances. 10th ed. Washington, D.C.: National Academy Press.

Nelson, Jennifer K., Karen Moxness, Clifford F. Gastineau, and Michael D. Jenson. 1994. *Mayo Clinic Diet Manual: A Handbook of Dietary Practices, 7th ed.* St. Louis, MO: The C. V. Mosby Co.

Newman, Cade and Bill Horn. 1993. *Nutrition and AIDS*. Rockville, Md.: Aspen Publishers, Inc.

Nieman, D. C. 1990. *Fitness and Sports Medicine: An Introduction*. Palo Alto, CA: Bull Publishing Co.

Packer, L. and Fuchs, J. 1993. *Vitamin E in Health and Disease*. New York, NY: Marcel Dekker.

Pastors, Joyce Green and Harold Holler (Editors). 1994. *Meal Planning Approaches for Diabetes Management*. Chicago, IL: The American Dietetic Association/DCE.

Pennington, Jean A. T. 1994. *Bowes & Church's Food Values of Portions Commonly Used, 16th ed.* Philadelphia, PA: J. B. Lippincott Co.

Peragallo-Dittko, Kathryn Godley, and Julie Meyer. 1993. *A Core Curriculum for Diabetes Education, 2nd ed.* Chicago, IL: American Association of Diabetes Education.

Pipes, Peggy L. and Christine Marie Trahms. 1993. *Nutrition in Infancy and Childhood, 5th ed.* St. Louis, MO: The C. V. Mosby Co.

Prasad, Ananda S. 1993. *Essential and Toxic Trace Elements in Human Health and Disease: An Update*. New York, NY: Wiley-Liss.

Queen, Patricia M. and Carol E. Lang. 1993. *Handbook of Pediatric Nutrition*. Rockville, MD: Aspen Publishers, Inc.

Rakel, Robert E. 1992. *Conn's Current Therapy*. Philadelphia, PA: W. B. Saunders Co.

Reichert, Kathryn. 1993. *Nutrition for Recovery: Eating Disorders*. Boca Raton, FL: CRC Press, Inc.

Reiff, Dan W. and Kathleen Kim Reiff. 1992. *Eating Disorders: Nutrition Therapy in the Recovery Process*. Rockville, MD: Aspen Publishers, Inc.

Roe, Daphne. 1993. *Nutrition and Chronic*

Disease. New York, NY: Van Nostrand Reinhold.

Rokusek, Cecilia and Heinrichs Eberhard. 1992. *Nutrition and Feeding for Persons with Special Needs: A Practical Guide and Resource Manual*. Pierre, SD: Child and Adult Nutrition Services, Dept of Education and Cultural Affairs.

Rombeau, John L. and Michael Caldwell. 1993. *Clinical Nutrition: Pareneteral Nutrition*. Philadelphia, PA: W. B. Saunders Co.

Schiller, M. Rosita, Judith A. Gilbride, and Julie O'Sullivan Maillet. 1991. *Handbook for Clinical Nutrition Services Management*. Rockville, MD: Aspen Publishers, Inc.

Schiller, M. Rosita, Karen Miller-Kovach, and Mary Angela Miller. 1994. *Total Quality Management for Hospital Nutrition Services*. Rockville, MD: Aspen Publishers, Inc.

Schlenker, Eleanor D. 1993. *Nutrition in Aging, 2nd ed.* St. Louis, MO: Mosby-Year Book, Inc.

Schreiner, J. E. 1990. *Nutrition Handbook for AIDS, 2nd ed.* Aurora, CO: Carrot Top Nutrition Resources.

Schroeder, S. A., M. A. Krupp, L. M. Tierney, and S. J. McPhee. 1990. *Current Medical Diagnosis and Treatment*. Norwalk, CT: Appleton and Lange.

Shearer, Martin J. and M. J. Seghatchian. 1993. *Vitamin K and Vitamin K-Dependent Proteins: Analytical, Physiological, and Clinical Aspects*. Boca Raton, FL: CRC Press, Inc.

Shils, Maurice E., James A. Olson, and Moshe Shike. 1994. *Modern Nutrition in Health and Disease, 8th ed., Vol. 1 & Vol. 2*. Philadelphia, PA: Lea & Febiger.

Simko, Margaret D., Catherine Cowell, and Judith A. Gilbride. 1994. *Nutrition Assesesment: A Comprehensive Guide for Planning Intervention, 2nd ed.* Rockville, MD: Aspen Publishers, Inc.

Simopoulos, Artemis P., Victor Herbert, and Beverly Jacobson. 1993. *Genetic Nutrition: Designing a Diet Based on Your Family Medical History*. New York, NY: Macmillan Publishing Co.

Smith, Nathan J. and Bonnie Worthington-Roberts. 1992. *Food for Sport*. Palo Alto, CA: Bull Publishing Co.

Somogyi, J. C. and E. S. Koskinen. 1990. *Nutrition Adaptation to New Life Styles*. Farmington, CT: S. Karger Co.

Stine, Gerald. 1993. *Acquired Immune Deficiency Syndrome*. Englewood, NJ: Prentice Hall.

Surgeon General's Report on Nutrition and Health. 1988. DDHS Publication no. (PHS) 88-50211. Washington, DC: DDHS.

Suskind, Robert M. and Leslile Lewinter-Suskind. 1993. *Textbook of Pediatric Nutrition,* 2nd ed. New York, NY: Raven Press Ltd.

Thomas, Paul R. 1991. *Improving America's Diet and Health: From Recommendations to Action*. Washington, DC: National Academy Press.

Tsang, Reginald C., Alan Lucas, Ricardo Uauy, and Stanley Zlotkin. 1993. *Nutritional Needs of the Preterm Infant: Scientific, Basic, and Practical Guidelines*. Baltimore, MD: Williams & Wilkins.

Tuchman, David N. and Rhonda S. Walter (Editors). 1994. *Disorders of Feeding and Swallowing in Infants and Children: Pathophysiology, Diagnosis, and Treatment*. San Diego, CA: Singular Publishing.

Vellas, B. and J. L. Albarede (Editors). 1992. *Facts and Research in Gerontology 1992*. New York, NY: Springer Publishing.

Wardlaw, Gordon M. and Paul Insel. 1993. *Perspectives in Nutrition*. St. Louis, MO: Mosby-Year Book, Inc.

Wardlaw, Gordon M., Paul M. Insel, and Marcia Seyler. 1993. *Contemporary Nutrition: Issues and Insights, 2nd ed.* St. Louis, MO: The C. V. Mosby Co.

Watson, Ronald Ross (Editor). 1994. *Handbook of Nutrition in the Aged, 2nd ed.* Boca Raton, FL: CRC Press, Inc.

Weinsier, Roland L., Sarah L. Morgan, and Virginia Gilbert Perrin. 1993. *Fundamentals of Clinical Nutrition*. St. Louis, MO: Mosby-Year Book, Inc.

(continued)

Appendix 50 *(continued)*

Westerterp-Platenga, M. S., E. W. H. M. Fredrix, and A. B. Steffens. 1994. *Food Intake and Energy Expenditure*. Boca Raton, FL: CRC Press, Inc.

Westhof, Eric. 1993. *Water and Biological Macromolecules*. Boca Raton, FL: CRC Press Inc.

Whitney, Eleanor and Sharon Rady Rolfes. 1993. *Understanding Nutrition, 6th ed.* St. Paul, MN: West Publishing Co.

Whitney, Eleanor Noss, Corinne Balog Cataldo, and Sharon Rady Rolfes. 1994. *Understanding Normal and Clinical Nutrition,* 4th ed. St. Paul, MN: West Publishing Co.

Williams, C. and J. Devlin. 1993. *Foods, Nutrition and Sports Performance*. New York, NY: Chapman and Hall.

Willett, Walter. 1990. *Nutritional Epidemiology, Monographs, in Epidemiology and Biostatics. Vol. 15.* New York, NY: Oxford University Press.

Williams, Sue Rodwell, Bonnie Worthington-Roberts, Kathleen Mahan, Peggy Pipes, Jane Rees, and Eleanor Schlenker. 1992. *Nutrition Throughout the Life Cycle, 2nd ed.* St. Louis, MO: The C. V. Mosby Co.

Williams, Sue Rodwell. 1993. *Essentials of Nutrition and Diet Therapy, 6th ed.* Davis, CA: SRW Productions, Inc.

Williams, Sue Rodwell. 1993. *Nutrition and Diet Therapy,* 7th ed. St. Louis, MO: The C. V. Mosby Co.

Wolinsky, Ira and James F. Hickson. 1994. *Nutrition in Exercise and Sport, 2nd ed.* Boca Raton, FL: CRC Press, Inc.

Worthington-Roberts, Bonnie S. and Sue Rodwell Williams. 1993. *Nutrition in Pregnancy and Lactation, 5th ed.* St. Louis, MO: Mosby-Year Book Inc.

Zaloga, Gary P. 1994. *Nutrition in Critical Care*. St. Louis, MO: Mosby-Year Book Inc.

Zeman, J. S. 1991. *Clinical Nutrition and Dietetics*. New York, NY: MacMillan Publishing.

Selected sources for software and audiovisual aids[a]

Bioelectrical Sciences, Inc. (Body Composition Measurement System)

CAMDE Corporation (Nutri-Calc RD and Nutri-Calc HD-Plus)

Career Aids (Software, Multimedia and Video Programs for Health and Nutrition)

Case Software (Nutrition Through the Life Cycle)

Compu-Cal (Compu-Cal PRO and Compu-Cal SE)

Comp-U-Menu Company Computerized Meal Planning and Recipes)

Computrition, Inc. (Nutritional Software Library II)

Contemporary Services (Computerized Health Diets)

CyberSoft, Inc. (Nutribase:Personal Nutrition Manager)

DDA Software (RD-Designed Nutrient Analysis System 2)

Diet Research, Inc. (The Good Diet and Health Program—IBM Dietary Software, Inc. (SDI-TF 2000 and TF 3000)

Eli Lilly (Humabase)

ESHA Research, Inc. (The Food Processor II)

Expert Software, Inc. (Diet Expert)

Food–Medication Interactions (Computerized Food-Medication Interactions)

Futrex Inc. (Futrex 5000, Fitness and Body Fat Analyzer)

GeriMenu (GM/Consultant)

Hopkins Technology (Food/Analyst CD and Sante)

Kissware, Inc. (Meal Planning—IBM)

Massachusetts General Hospital (MGH Computer—The Pocket Productivity Tool)

Mistebar Computers (Diabetic, Renal and Enteral)

N-Squared Computing Company (Nutritionist IV 3.0)

Nutri-Comp Systems (The Menu Manager)

NutriData Software Corp (Diet Balancer)

Nutritional Data Resources Company (NutriPak MenuMaker and Student DietWise, EnergyWise)

Nutrition and Diet Services (NDS Computerized Menus & Diets)

Appendix 50 *(continued)*

Nutrition Assessment Software (Energy Assessment)
Nutrition Coordinating Center (Counseling NDs)
Nutrition Scientific (Quick Check for Fat and Coronary Risk)
Poly-Bytes, Inc. (Pocket R.D. II)
Positive Solutions (Diet Max for Windows and Three Squares Food Service and Nutrition Management)
Relational Technologies, Inc. (The Nutrition Manager)
Simple Solutions (Computerized Tray Cards)
Softech Computing Company (IBM-PC Dietary Assessment System)
Software Integration (Menu Manager V1.0)
The Nutrition Coordinating Center (NDS 32 U of Minnesota)
University of Texas-Houston, School of Public Health Human Nutrition Center (Food Frequency Data Entry and Analysis Programs)
Wellsource, Inc. (The Professional Dietitian)

[a]The *Annual Journal of Dietetic Software* gives current listings for software programs in dietetics. Also, various professional organizations and trade companies prepare software and other visual aids in the field of nutrition. See Appendix 49.

Journals

Advances in Food Research
Advances in Nutritional Research
Aging
AIDS
American Academy of Applied Nutrition Journal
American Family Physician
American Journal of Cardiology
American Journal of Clinical Nutrition
American Journal of Digestive Diseases
American Journal of Diseases of Children
American Journal of Epidemiology
American Journal of Gastroenterology
American Journal of Health Promotion
American Journal of Medicine
American Journal of Nursing
American Journal of Preventive Medicine
American Journal of Public Health
Annals of Human Biology
Annals of Internal Medicine
Annals of Nutrition and Metabolism
Annual Reviews of Medicine
Annual Reviews of Nutrition
Archives of Diseases of Children
Archives of Internal Medicine
Archives of Pediatrics and Adolescent Medicine
Archives of Surgery
Australian Journal of Nutrition and Dietetics
Borden's Review of Nutrition Research
British Journal of Nutrition
British Medical Journal
Canadian Journal of Public Health
Canadian Medical Association Journal
Cancer
Cancer Bulletin
Cancer Research
Canadian Medical Association Journal
Carbohydrate Research
Circulation
Clinical Gastroenterology
Clinical Nutrition
Critical Medicine
Current Concepts in Nutrition
Current Topics in Nutrition and Disease
Currents in Infant Care, Nutrition and Medicine
Diabetes
Daibetes Care
Diabetes Educator
Diabetes Forecast
Digestive Diseases and Sciences
Drug–Nutrient Interactions
Ecology of Food and Nutrition
Environmental Nutrition
Epidemiology
European Journal of Clinical Nutrition
European Journal of Public Health
Federation Proceedings
Food and Drug Administration Consumer
Food and Nutrition
Food and Nutrition Quarterly Index
Food Technology
Gastroenterology
Geriatrics

(continued)

Appendix 50 *(continued)*

Health Care Management Review
Human Nutrition: Applied Nutrition
Human Nutrition: Clinical Nutrition
Human Nutrition: Food Sciences and Nutrition
International Child Health with UNICEF and WHO
International Journal of Obesity and Related Metabolism
International Journal of Peptide and Protein Research
International Journal of Vitamin Research
Journal of Agricultural & Food Information
Journal of Applied Nutrition
Journal of Applied Physiology
Journal of Clinical Biochemistry and Nutrition
Journal of Clinical Endocrinology and Metabolism
Journal of Clinical Investigation
Journal of College & University Foodservice
Journal of Consumer Research
Journal of Food Composition and Analysis
Journal of Food Protection
Journal of Food Science
Journal of Food Technology
Journal of General Internal Medicine
Journal of Gerontology
Journal of Home Economics
Journal of Human Nutrition
Journal of Nutrition
Journal of Nutrition Education
Journal of Nutrition for the Elderly
Journal of Nutritional Immunology
Journal of Nutrition in Recipe & Menu Development
Journal of Nutrition Research
Journal of Parenteral and Enteral Nutrition
Journal of Pediatric Gastroenterology
Journal of Pediatrics
Journal of Renal Nutrition
Journal of Respiratory Diseases
Journal of Restaurant & Foodservice Marketing
Journal of Surgical Research
Journal of the American Dental Association
Journal of the American Dietetic Association
Journal of the American Geriatric Society
Journal of the American College of Nutrition
Journal of the American Medical Association
Journal of the Canadian Dietetic Association
Journal of the National Cancer Institute
Journal of the New Zealand Dietetic Association
Lancet
Mayo Clinic Proceedings
Medical Clinics of North America
Metabolism
Modern Health Care
New England Journal of Medicine
Nutrition Abstracts and Reviews
Nutrition and Metabolism
Nutrition Forum
Nutrition in Clinical Practice
Nutrition Reviews
Nutrition Today
Pediatrics
Perspectives in Applied Nutrition
Postgraduate Medicine
Preventive Medicine
Proceedings for the Society for Experimental Biology and Medicine
Proceedings of the Nutrition Society
Public Health Reports
Science
Scientific American
Seminars in Oncology
Surgical Clinics of North America
Surgical Forum
The Journal of Nutritional Biochemistry
Topics in Clinical Nutrition
Vitamins and Hormones
Western Journal of Medicine
World Review of Nutrition and Dietetics

Miscellaneous: Newsletters, Leaflets, Booklets, Video Cassettes and Tapes

ADA Courier
ADA Practice Group Newsletters
Clinical Management
CNI Weekly Report
Contemporary Nutrition
Currents in Food, Nutrition, and Health
Dairy Council Digests
Diabetes Interview
Diet and Nutrition Newsletter
Dietetic Currents
Drug-Nutrient Interactions
FDA Consumer

Appendix 50 *(continued)*

- Food Action
- Food and Nutrition News
- Harvard Medical School Health Letter
- Health and Nutrition Newsletter
- Healthline
- Hospital Food and Nutrition Focus
- National Center for Nutrition and Dietetics
- National Dairy Council Digests
- NCAF Newsletter
- Nutrition and the MD
- Nutrition Counselor
- Nutrition Forum
- Nutrition News
- Nutrition Perspectives
- Seminars in Nutrition
- Tufts University Diet and Nutrition Letter
- U.C. Berkeley Wellness Letter
- Various publications of the ADA—Dietetic Practice Groups (DPGs)
- Various publications of major drug/food corporations
- Various video cassettes and tapes